MW01282596

STILL IN SEARCH OF PREHISTORIC SURVIVORS

STILL IN SEARCH OF PREHISTORIC SURVIVORS

The Creatures That Time Forgot?

DR KARL P.N. SHUKER

COACHWHIP PUBLICATIONS
Greenville, Ohio

Still in Search of Prehistoric Survivors

ISBN 1-61646-428-3
ISBN-13 978-1-61646-428-8

Cover Design © William M. Rebsamen

CoachwhipBooks.com

CONTENTS

DEDICATION

In loving, happy memory of my family—all gone now from this life, but never forgotten and forever loved. May we all be together again one day, ever more.

And also in commemoration of my good friend Scott T. Norman (1964-2008). Your contributions to cryptozoology, Scott, will never be forgotten; I am proud to have known you, and to have been one of your friends.

A happy Sunday afternoon with my family at Bourton-on-the-Water, England, back in the late 1960s when I was a child and when life was good—with Mom (Mary Shuker), Nan (Gertrude Timmins), and Dad (Jack Shuker), and with Grandad (Ernest Timmins) taking the photograph (© Dr Karl Shuker)

A celebratory painting of Scott T. Norman (© William M. Rebsamen)

ACKNOWLEDGEMENTS

I wish to thank most sincerely the following persons, institutions, organisations and publications for their kind interest and assistance during the preparation of this book.

Academy of Applied Science (Concord, New Hampshire); Dr Victor Albert; The Anomalist @anomalistnews; Neil Arnold; Michel Ballot; Prof. Henry H. Bauer; Dr Hilary Belcher; the late W. Ritchie Benedict; Matthew Bille/*Exotic Zoology* newsletter; Birmingham Public Libraries; Hans E.A. Boos; Dr Ed Bousfield; Alan Brignall; British Library; G.G. Bryan; Markus Bühler; John and Lesley Burke; Owen Burnham; the late Mark Chorvinsky/*Strange Magazine*; Paul Clacher/*The Fossickers Network*; Karl L. Claridge; the late Prof. John L. Cloudsley-Thompson; Loren Coleman; Jay Cooney/*Bizarre Zoology* blog; Paul Cropper; *Cryptomundo*; Dami Editore s.r.l.; Adam Davies; the late Tim Dinsdale; Jonathan Downes/Centre for Fortean Zoology; Drayton Manor Park and Zoo; Dudley Public Libraries; Hugh Evans; *Express and Star* (Wolverhampton); *Falmouth Packet*; Thomas Finley; Miroslav Fišmeister; Steve Fletcher; Angel Morant Fores; *Fortean Times*; Goro Furuta; Tonio Galea; Ken Gerhard; Bill Gibbons; Michael Goss; the late J. Richard Greenwell; Craig Harris/*Crypto Chronicle*; Bob Hay; Tony Healy; David Hearder; Markus Hemmler; Brother Richard Hendrick; David Heppell; Rip Hepple/*Nessletter*; the late Dr Bernard Heuvelmans; Caroline Forcier Holloway; International Society of Cryptozoology (ISC); Brian Irwin; Orosz István; Prof. Christine Janis; Dr Andrew Jeram; Dr Andrew Johns; Jeff Johnson; Todd Jurasek; the late Clinton Keeling; Carl Kelsall; Ralf Kiesel; Dr E. Klengel; the late Prof. Grover S. Krantz; Connor Lachmanec; Brad LaGrange; Rebecca Lang; Prof. Paul LeBlond; Gerard van Leusden; Library of Congress (Washington D.C.); Dr Adrian Lister; Donald W. McNamee; the late Prof. Roy P. Mackal; the late Ivan Mackerle; Ulrich Magin; Scott Mardis; Prof. Terry Matheson; Jason McAllister; James I. Menzies; Tötös Miklós; Loes Modderman; Dr Ralph Molnar; Roderick Moore; Tim Morris; Wm Michael Mott; Richard Muirhead; Robert Mullin; Dr Darren Naish/*Tetrapod Zoology* blog; National Archives of Canada; Natural History Museum London; Natural History Museum of Los Angeles County; News Team International; Michael Newton; the late Scott T. Norman; North Queensland Naturalists Club; Hodari Nundu; Terry O'Neill; Theo Paijmans; Stuart Pike; 'Pat the Plant'; Michael

Playfair; John Potter; Judy Preece; 'Prunella'; Range Pictures/Bettmann Archives/United Press International; Dr Clayton E. Ray; Michel Raynal; Prof. Brian Regal; Mike E. Richburg; Bob Rickard; the late Dr Robert H. Rines; Dr W.D. Ian Rolfe; Gustavo Sanchez Romero; Karl G. Rose/Enigmatic Static; Lorenzo Rossi; Mark Rothermel; Samwell Rowan; Sandwell Public Libraries; Dr François de Sarre; Ron Scarlett; Dr Ben Scripture; Adam Selzer; the late Mary D. Shuker; Paul Sieveking; Malcolm Smith/ *Malcolm's Musings* blog; Michael J. Smith; G.R. Cunningham van Someren; Anette Stichnoth; the late Dora Stokes; Nick Sucik; Richard Svensson; Lars Thomas; the late Ernest and Gertrude Timmins; United Press International; Dr Leigh Van Valen; Frances Vargo; Arnošt Vašícek; Vorderasiatisches Museum; Dr David Waldron; Anthony Wallis; Jonathan Walls; Walsall Public Libraries; Sebastian Wang; John Warms; Sally Watts; Richard Wells; Travis Westley; the late Jan Williams; Paul Willison; Wolverhampton Public Libraries; the late Gerald L. Wood; Dr Michael Woodley; Prof. Bernd Wursig; Rebecca Tosh Xayasith; Zoological Society of London; Prof. George Zug.

My especial thanks and gratitude go to Chad Arment at Coachwhip Publications for his enthusiasm in seeing into print this long-awaited, extensively-updated new edition of *Prehistoric Survivors*; to Michael Newton for preparing his delightful, inspiring foreword to this new edition; and to William M. Rebsamen for very kindly permitting me to include so many of his wonderful, highly-renowned cryptozoological illustrations in it, as well as for designing its beautiful covers.

DISCLAIMER

CONVERSION TABLE

1 inch = 2.54 cm
1 foot = 0.31 m
1 yard = 0.91 m
1 mile = 1.61 km
1 acre = 0.41 hectares
1 ounce = 28.35 g
1 pound = 0.45 kg
1 ton (imperial) = 1016 kg
°F = 9 x °C + 32

FOREWORD TO THE FIRST EDITION

Prof. Roy P. Mackal PhD DSc
(1925-2013)

There are a number of definitions of cryptozoology. As a practising cryptozoologist for over 30 years, I prefer the following: *cryptozoology is the study and investigation of evidence for animals unexpected in time or place or in size or shape.* Dr Karl Shuker has addressed the first category of animals in this definition ('animals unexpected in time'), referring to possible select species—that is, surviving animal species known only from the fossil record or which were known in the past but are now thought to be extinct.

There have been many studies and books written about so-called 'living fossils', a somewhat contradictory misnomer. None in my opinion even come close to this superb work by Dr Shuker entitled *In Search of Prehistoric Survivors*. Let me explain why this book is so outstanding. First, the exceptional scholarship, which is not only comprehensive, but also scrupulously accurate, sets this book apart. Second, the analyses of data are always well balanced, compellingly reasonable and objective, without becoming boring or tedious. Last, the presentation and literary style provide a clear, understandable text even to non-professionals but is never 'written down', so that it is acceptable to scientists as well.

I always imagined that I was aware of most, if not absolutely all, available cryptozoological information, yet to my surprise and delight I found a great wealth of new data about cryptids (possible unknown animals) totally new to me as well as additional information about cryptids already familiar to me. Not since *On the Track of Unknown Animals* by Dr Bernard Heuvelmans has anyone compiled so much excellent material and integrated it so well into the corpus of cryptozoological knowledge.

The foregoing would be more than enough to establish Shuker's book as the best in recent times, yet there is more to excite the reader. Shuker has done real cryptozoology. The core cryptozoological research consists of comparing available data about a particular cryptid to all known animal groups, both living and extinct, in order to establish tentative relationships possibly leading to identification of the animal in question or at least establishing affinities. Further, an assessment must be made as to whether or not information actually is related to real unknown animals, rather than a hoax, a mistaken identification, or simply wishful thinking or fabrication.

Shuker has demonstrated encyclopedic zoological knowledge, the application of which has provided a well-reasoned, compelling argument for the identity of many of the cryptids in his book. Throughout, Shuker's approach is strictly scientific without ever relying on questionable explanations. This book will undoubtedly be a great inspiration and reference for its readers.

San Diego, California, 1995.

FOREWORD TO THE SECOND EDITION

Michael Newton
Author of *Encyclopedia of Cryptozoology: A Global Guide to Hidden Animals and Their Pursuers* (2005)

How does one improve upon a masterpiece? Only with tender loving care and by meticulous attention to detail. Dr Karl Shuker's *Still In Search Of Prehistoric Survivors* succeeds on both counts.

When I read the first edition of this work in 1995 I was enthralled—and surprised. Although I knew Dr Shuker's fine work from the pages of *Fortean Times* and *Strange Magazine*, I learned that *In Search of Prehistoric Survivors* was not his first book, but in fact his fifth published within six years. I immediately set out to redress my deficiency, beginning with his justly famous *Mystery Cats of the World* (1989), and acquiring each new work in turn, through his then-latest *A Manifestation of Monsters* (2015). As *Still In Search* appears, I feel it's safe to say that Dr Shuker has revealed, analyzed, and occasionally debunked more cryptids—"hidden animals," the subjects of cryptozoology—than any other author, living or deceased.

And I am pleased to say he isn't finished yet.

In 2005 I dedicated my own *Encyclopedia of Cryptozoology* to Dr Shuker as the pre-eminent cryptozoologist of a new millennium, and nothing since that date has changed my mind. While the late Dr Bernard Heuvelmans (1916-2001) is rightly honored as the "Father of Cryptozoology" and amasser of vast archives, he sadly published only eight books in his busy lifetime of research and exploration, four of which remain unavailable in English. Dr Shuker, by contrast, has twenty-five books to his credit thus far, and he thankfully shows no sign of slowing his pace.

Nor is it quantity alone that sets apart his work, but also *quality*. While some tomes on cryptozoology present wild tales of speculation as fact, and others offer up analyses as dry as dust, Dr Shuker strikes the perfect balance between wide-eyed wonder and true scepticism—i.e. demanding solid evidence to prove a "new" creature's existence, without dogmatically rejecting stories out of hand. And best of all, I've found that each chapter of any given book by Dr Shuker teaches me something I never previously knew. At my stage of life, that is no small achievement.

Although *Still In Search* stands as a second edition, I would argue that it might be considered a wholly new work in itself. While including nearly all of the first edition's text, it vastly expands on its predecessor's scope, presenting readers with over 600 pages to the first book's 187, plus more than 300 illustrations versus the previous volume's 150. The new material picks up from where the "Stop Press" sections of each former chapter ended with the latest-breaking news of 1995. And high time, too, since the world—to paraphrase Stephen King's *The Dark Tower*—has moved on, and dramatically so.

In the twenty-odd years since *In Search of Prehistoric Survivors* was published, science has uncovered many new fossils of prehistoric life forms, while at the same time discovering hundreds of new animal species and *re*discovering another multitude presumed extinct worldwide or extirpated from their normal ranges when the first edition appeared. And it is fair to say that literally no one knows more about rediscovery of "lost" species than Dr Shuker, as revealed in his books *The Lost Ark* (1993), *The New Zoo* (2002), and *The Encyclopaedia of New and Rediscovered Animals* (2012). Unfortunately, while he makes the information readily accessible to all, some self-styled sceptics—in truth, professional "skofftics"—still maintain that no new life forms of appreciable size remain to be revealed on Earth. Almost incredibly, some mainstream scientists take a similar view, despite each year's bountiful crop of new species. Future discoveries, they say, are almost certain to be tiny creatures—like the new species of loriciferan named for Dr Shuker in 2005—or larger animals dredged up from the bathypelagic zone of Earth's oceans. Thanks to Dr Shuker's tireless work, we know today that nothing could be farther from the truth.

Whether or not you've read the first edition of the epic work in hand, you will find much to marvel at within the pages of this new edition, sequel, call it what you will. Despite the many problems of our planet—overcrowding, climate change, ever-increasing pollution, and endless wars in territories still poorly explored—we occupy a world of wonders waiting yet to be exposed.

And we may rest assured that Dr Shuker seeks them still, to share with one and all. For that alone, and the engaging manner of his revelations, he deserves our commendation and respect.

Nashville, Indiana, 2016

AUTHOR'S PREFACE TO THE SECOND EDITION

In Search of Prehistoric Survivors—How It Came to Be, and Is Now Again (Critics Notwithstanding!)

What has been will be again.

The Holy Bible—Ecclesiastes (1:9)
New International Version (NIV)

Of all of my 25 books, none has attracted such acclaim but also such contention as *In Search of Prehistoric Survivors*. Last year marked the 20th anniversary of its original publication in 1995, and having received countless requests from readers over the years for its republication (after having been out of print for more than a decade), I am delighted that thanks to Coachwhip Publications, it is now finally back in print in this new, extensively updated, expanded, and retitled edition.

Meanwhile, and after having given the matter much thought, I feel it necessary to reveal precisely how this book came to be, because ever since it first appeared back in 1995 there has been a degree of confusion and controversy in some quarters as to where I stand in relation to its theme and contents. Consequently, I hope that all of this will be elucidated satisfactorily by the following explanation—one that I have already outlined privately to various colleagues down through the years, and also more recently via a *ShukerNature* blog post, but which I have not disclosed publicly in hard-copy print before.

In many ways, this book is the most unusual of any of mine, inasmuch as its final, published form was not how I had originally conceived it at all. Let me explain.

Following the publication in 1991 of my second book, *Extraordinary Animals Worldwide*, I was planning a major book on herpetological cryptids—everything from alleged living dinosaurs, pterosaurs, and plesiosaurs, to mystery lizards of many kinds, giant snakes and crowing serpents, chelonian cryptids of all shapes and sizes, anomalous amphibians, and even a major section devoted to the possible origin of and inspiration for the world's plethora of legendary dragons. A synopsis of this proposed book did the rounds of publishers, and Blandford Press was particularly interested in it.

Mindful, however, of the enormous worldwide popularity of Steven Spielberg's blockbuster movie *Jurassic Park* at that time (the first film had been released in 1993), Blandford suggested a fundamental change to the contents and slant of my book. Instead of confining it to herpetological mystery beasts, they proposed that I expand its range of subjects to cryptids across the entire zoological spectrum, but concentrate exclusively upon those that have been suggested at one time or another by cryptozoologists to constitute prehistoric survivors.

It was certainly a most intriguing brief, and one that I therefore decided to accept, even though—and I must emphasise this unequivocally here—I did not personally consider it likely that all of those cryptids truly were prehistoric survivors. But my personal opinion was irrelevant as far as the book's remit was concerned. What was required was for me to present a dossier of reports and native traditions for each cryptid, and then discuss it in the context of whichever prehistoric creature(s) it had been likened to in the cryptozoological literature (with theories not appertaining to prehistoric survival receiving only minimal treatment, as they were not the focus of this study). So that is precisely what I did. Consequently, out went most of the mystery lizards and amphibians, as well as the snakes and also the dragons section, and in came putative mammalian methuselahs like chalicotheres, thylacoleonids, amphicyonids, ground sloths, basilosaurines, and sabretooths, alleged lingering avians like teratorns and *Sylviornis*, the giant carnivorous megalodon shark, and even some reputed eurypterid survivors.

In Search of Prehistoric Survivors, published in 1995, the original edition of this present book (© Dr Karl Shuker)

During the years that have followed, the concept of prehistoric survivorship—or what British palaeontologist Dr Darren Naish refers to as the Prehistoric Survivor Paradigm (PSP)—has received some harsh criticism from cryptozoological sceptics. And indeed, I am the first to concede that such survival becomes increasingly untenable the further back in time from the present day any given example's most recent known fossil antecedents are (i.e. the longer its so-called ghost lineage is). However, as my books on new and rediscovered animals have disclosed time and again, some truly extraordinary, spectacular, and entirely unpredictable zoological discoveries have been made in modern times. So I remain reluctant to dismiss PSP out of hand.

Having said that, although I have often been accused of "believing" that a given cryptid is a particular type of prehistoric survivor, this is untrue, for the simple reason that it is impossible to state definitely (although certain cryptozoologists habitually attempt to do so) what a given cryptid must be. Without tangible evidence to examine (and I am referring here to physical remains, not photographic evidence, which can be convincingly faked with alarming ease nowadays), all that can be done is to pass a personal opinion as to how likely or unlikely a given identity appears to be. However, opinions are not facts, and should never be put forward as such, or be mistaken for such. In short, therefore, I do not "believe" that any cryptid is any specific identity—I merely indicate what I personally consider to be likely (or unlikely) identities for it, nothing more.

Republishing this book has posed something of a dilemma for me, because doing so meant that its original remit (and also therefore my own misgivings regarding the plausibility of prehistoric survival for certain of its cryptids) would remain fundamental to its raison d'être. The only alternative would be for me to rewrite it completely, with an entirely altered slant. However, the result of that would be not only a totally different book but also a much more extensive one—so extensive, in fact, that I sincerely doubt whether it would be financially viable for any publisher to take on. Yet whatever one's own personal opinion may be concerning prehistoric survival in any capacity, the wealth of historical reports and cryptozoological coverage presented in this book's pages is such that it would be a tragedy for it to remain out of print, especially when, as noted earlier, there is a such a very considerable demand among readers for it to reappear.

Consequently, now that I have outlined here how it came to be and why it is what it is, so that there can no longer be any confusion or contention regarding it, I am very happy to see what many people consider to be my finest cryptozoological volume back in existence, containing numerous significant updates and new illustrations but with its basic context and content otherwise unchanged. (Much as I would have liked to have updated each section of the book comprehensively, to do so would have resulted in sections so lengthy that each could have stood alone as an entire book in its own right—now, there's a thought!)

Last of all—but definitely not least of all—I wish to thank most sincerely all of this book's numerous supporters for their kind words through all of the intervening years, urging me to resurrect it—just like a veritable prehistoric survivor itself, in fact!

The Comoros coelacanth (© William M. Rebsamen)

INTRODUCTION

Lost Worlds and Living Fossils—Fiction or Fact?

Let me review the scene,
And summon from the shadowy Past
The forms that once have been.

HENRY WADSWORTH LONGFELLOW
A GLEAM OF SUNSHINE

Prior to the biological revolution that accompanied the arrival of Darwin and his theory of evolution onto the scientific stage during the mid-19th Century, zoologists and even palaeontologists firmly believed that animal species were immutable and immortal—i.e. that all were exactly the same today as they had always been, and that extinction did not occur.

The numerous fossils that had already been unearthed ought to have constituted sufficient proof that animals evolved from primitive forms into more advanced ones, and that this evolution was necessarily accompanied by extinction (eliminating species unable to survive in the face of competition from more successful forms). However, scientists of the day explained away these inconvenient remains as the bones of antediluvian monsters—ancient creatures that had perished in the Great Flood survived by Noah and were thus lost forever.

In stark contrast, the reality of evolution and attendant extinction is largely accepted by science today—thereby making the extraordinary tenacity of a certain overtly anachronistic notion all the more difficult to comprehend. The notion in question is that prehistoric creatures could not have survived undetected into the present day. Remarkably, any attempt to challenge this statement constitutes an inviolate taboo for an appreciable proportion of current zoologists—thus harking back to the discredited belief in diluvian destruction.

'Living Fossils' and the 'Lost World Syndrome'

If there were no precedents for prehistoric persistence into the modern world, the reluctance of its critics to countenance its likelihood could be readily understood. Yet the annals of contemporary zoology contain numerous instances in which creatures belonging to ancient groups hitherto believed extinct since prehistoric times, or individual species assumed to have died out long ago, have been sensationally uncovered alive and well.

Examples of such 'living fossils' that have long been known to science include: the tuataras *Sphenodon punctatus* and *S. guntheri*, New Zealand's lizard-like rhynchocephalians, the only surviving members of this once-diverse reptilian lineage that was contemporary with the dinosaurs; the horseshoe crab *Limulus*, an antiquated aquatic arthropod genus practically unchanged from fossils 300 million years old, and most closely related to the long-demised eurypterids (sea scorpions); the deceptively plant-like sea lilies (crinoids), actually constituting a taxonomic class of starfish-allied animals, which were known only from fossils, some dating back more than 400 million years, until a living species (the first of many) was discovered during 1755 in deep water off the Caribbean island of Martinique; and the chambered nautiluses, a handful of cephalopod mollusc species exhibiting little evolutionary development during the past 500 million years, and more closely related to the first cephalopods dating back to that very early time in mollusc history than to the early modern cephalopods that first appeared about 100 million years later (the ammonoids and coleoids).

Additionally, there are a number of celebrated examples that remained unknown to science until as recently as the 20th Century, as documented in three of my previous books—*The Lost Ark: New and Rediscovered Animals of the 20th Century* (1993), and its two updated editions, *The New Zoo* (2002) and *The Encyclopaedia of New and Rediscovered Animals* (2012). The first books devoted exclusively to the major zoological arrivals and revivals of the past hundred or so years, they present the histories of many lately-revealed 'living fossils'.

The most famous, and spectacular, of these is unquestionably the coelacanth. On 22 December 1938, Marjorie Courtenay-Latimer, the curator of South Africa's East London Museum,

The Encyclopaedia of New and Rediscovered Animals, with front-cover artwork by William M. Rebsamen featuring the okapi and the Vu Quang ox (© Dr Karl Shuker/William M. Rebsamen/Coachwhip Publications)

was inspecting some freshly-caught fishes at the local docks when she noticed an unusual fin sticking out from the bottom of a pile of sharks. When she succeeded in extricating the fin's recently-deceased owner, she found herself face to face with the most extraordinary creature that she had ever seen.

Measuring 5 ft long, it was a robust slaty-blue fish whose thick plates gave it an armour-plated appearance—but far stranger than its scales were its fins. The rays of its first dorsal fin were arranged like a fan, as in the fins of most other fishes. The rays in all of its other body fins, however, were borne upon fleshy lobes, so that these fins resembled short, stumpy legs. Its tail fin was equally bizarre, terminating not in two lobes but in three—unlike any other modern fish then known to science.

Unable to identify this peculiar fish, Courtenay-Latimer arranged for it to be preserved, and sent an annotated sketch of it to Prof. J.L.B. Smith—South Africa's most eminent ichthyologist. When he saw the sketch, he recognised the fish straight away, but was scarcely able to believe what he saw—because its strange features identified it unhesitatingly as a coelacanth, a member of an archaic lineage of fishes called the crossopterygians, which were previously believed to have died out even before the dinosaurs, i.e. over 66 million years ago!

When Smith examined the fish in the flesh, however, there was no room for doubt—it really was a hitherto-unknown, modern-day species of coelacanth, and in honour of its discoverer he duly dubbed it *Latimeria chalumnae*. Since then, many other specimens of *Latimeria* have been caught, most of them in the waters surrounding the Comoro Islands, near Madagascar. Perhaps the most ironic aspect of this case also emerged from the Comoros—the natives there were already so familiar with the coelacanth, which they called the kombessa, that they habitually utilised its scales as sandpaper, for mending punctures in bicycle tyres!

In 1997, moreover, a second species of living coelacanth was discovered in waters off Sulawesi, Indonesia, and has been formally dubbed *L. menadoensis*, the Indonesian coelacanth.

Other significant living fossils discovered since 1900 include the following examples:

A series of superficially limpet-like species of the genus *Neopilina*—they first came to light during the 1950s and were found to be monoplacophorans, primitive molluscs that supposedly died out 350 million years ago.

Neoglyphea inopinata—a nondescript crab-like animal scooped up from the South China Sea in 1908 but remaining un-named and unstudied until 1975, when it was sensationally identified as a glyphid, a primitive crustacean hitherto believed extinct for 50 million years. A second living glyphid, *Laurentaeglyphea neocaledonica*, was discovered in 2005, off New Caledonia.

The famous okapi *Okapia johnstoni*—discovered within what is now the Democratic Republic of the Congo's Ituri Forest in 1901 and belonging to a group of short-necked forest giraffes dating back 30 million years.

The mountain pygmy possum *Burramys parvus*—a primitive Australian marsupial known only from fossils at least 10,000 years old until a living specimen was unexpectedly captured in 1966.

The Chacoan peccary *Catagonus wagneri*—a pig-like species again known only from fossils and believed to have died out during the Ice Ages until found alive in Argentina in 1974 (Chapter 4).

Cephalodiscus graptolitoides—dredged up from the sea around the New Caledonian islet of Lifou in 1989, formally described in 1993, and widely deemed to constitute a living graptolite, thereby resurrecting those mysterious colonial invertebrates from 300 million years of official extinction.

The kha-nyou *Laonastes aenigmamus*—a fairly large, superficially rat-like rodent that first came to Western scientific attention in 1996 when a dead specimen was seen on a market stall in Laos by World Conservation Society zoologist Dr Robert Timmins, and which, after further specimens had been collected and its species formally described in 2005, was ultimately revealed to be not just a species new to science but also the only recorded living member of the taxonomic family Diatomyidae, previously known only from fossil species and hitherto believed to have died out 11 million years ago.

The kha-nyou (© Markus Bühler)

Many other modern-day methuselahs are also on record; yet for the most part the zoological world stubbornly continues to close its mind to the exciting possibility that similar such discoveries are still waiting to be made—preferring instead to discount those believing in such a scenario as suffering from acute 'Lost World Syndrome'! Originally published in 1912 but still one of science fiction's most popular novels, *The Lost World* by Sir Arthur Conan Doyle was written in the style of an authentic report (recalling such notable factual accounts of zoological discovery as Darwin's own *Voyage of the Beagle* and Henry W. Bates's *The Naturalist on the River Amazons*), and dealt with a scientific team's finding of an autonomous prehistoric world in majestic isolation on top of a lofty South America mesa, where many of the most impressive creatures of the far-distant past had somehow survived into the present day.

Needless to say, I am not for one moment suggesting that the unveiling of an entire prehistoric world is feasible in reality. However, amid the vast scientific and popular non-fiction literature are countless clues and indications—ranging from reliable eyewitness accounts to even an alleged specimen or two—that provide a thought-provoking corpus of evidence pointing to the possible but still-unconfirmed survival of various prehistoric animal types in disparate corners of the world.

The putative existence of a number of spectacular yet still-undiscovered species of animals is recognised nowadays by a steadily-increasing association of open-minded scientists and lay investigators referred to as cryptozoologists (seekers of 'hidden animals'), leading to the foundation in 1982 of the International Society of Cryptozoology (ISC)—the world's first scientific body devoted to the investigation of such creatures worldwide. Sadly, the ISC is no more (though in early 2016, veteran American cryptozoologist Loren Coleman announced the official launch of a new, comparable global organisation, the International Cryptozoology Society), but the goals of its erstwhile members (including myself) and other cryptozoological investigators worldwide still exist. Namely, to dredge records of mystery beasts out of the morass of often obscure, little-known publications in which they have lain buried for decades or longer, thus bringing them to public attention for the first time in many cases, and to seek these animals directly in the field whenever the opportunity arises.

In 2012, resurrecting the ISC's most valuable contribution to cryptozoology, the British-based Centre for Fortean Zoology (CFZ) launched a new, scientific, peer-reviewed journal devoted to cryptozoology, with myself as its editor. Four volumes of the *Journal of Cryptozoology* containing scholarly papers across this discipline's wide spectrum of subjects have been published so far, and it has become an established, recognised periodical.

The inaugural volume of the *Journal of Cryptozoology*, launched in 2012 (© Dr Karl Shuker/CFZ)

But if prehistoric mystery creatures really do exist, why have we not already found them? The answer lies primarily in the nature of their habitats' general terrain, which invariably embodies a unique combination of ecological security for its elusive denizens, and virtual inaccessibility for Western investigators wishing to confirm their existence. (This inaccessibility, moreover, sometimes owes its success as much to bureaucratic barriers as to environmental ones!)

This can be best illustrated by considering the habitats of those living fossils already known to science. Many first came to attention as fortuitous finds present in non-specific hauls of zoological specimens dredged up from the depths of the vast oceans—which indisputably harbour numerous other surprises still awaiting disclosure. As for the remaining prehistoric survivors presently on record, these usually inhabit isolated areas little-changed for thousands if not millions of years (often due to prominent topographical features that have effectively delineated such areas from their immediate surroundings).

Ensconced within these pocket havens of suspended time, primitive creatures have often been sheltered from competition with more advanced species that have evolved and established themselves elsewhere, or they have retreated here in the wake of such competition experienced elsewhere. Areas of this type include: treacherous mountain ranges; inhospitable rainforests encircled by rivers, sheer-sided valleys, or other natural barriers; remote inland stretches of water long since separated from any contact with the sea and still far removed (as yet) from encroaching human civilisation; as well as small uninhabited islands fairly/totally undisturbed by Westerners due to their distant location or lack of commercially viable prospects.

It should be no surprise to learn, therefore, that these are the self-same areas in which, on those infrequent occasions when intrepid Western explorers and scientists *have* penetrated their forbidding seclusion, tantalising glimpses have sometimes been made of what could be valid examples of prehistoric survivors still eluding official discovery and documentation.

Continuing Evolution, Ghost Lineages, and Lazarus Taxa

Consequently, it is nothing if not ironic that much of traditional zoology's reluctance to give serious consideration to the prospect of prehistoric survival stems not from its incomplete knowledge of the present day's fauna but rather from its infinitely less complete knowledge of the past's. In particular, this has resulted, bizarrely, in an absolutely crucial component of the prehistoric survivors scenario continuing to be largely or entirely ignored, or at the very least unconsidered, by many mainstream researchers. Namely, acknowledging and taking into full account the extremely significant, potentially far-reaching influence of continuing evolution upon a given fossil species' postulated future morphology, behaviour, and ecology.

I am therefore duly expanding this introduction that originally appeared in the first edition of the present book by including below a concise review of continuing evolution and its relevance to prehistoric survivorship.

Palaeontology waits for no-one, including mystery animals. In relatively recent times, various long-treasured tenets of cryptozoology have been seriously challenged by new palaeontological discoveries and interpretations, including some with relevance to putative prehistoric survivors.

Traditionally, palaeontologists assumed that those exceedingly lengthy archaeocete whales known as basilosaurines—currently believed from fossil evidence to have died out around 36 million years ago—moved through the water via vertical undulations of their elongate bodies. Accordingly, generations of cryptozoologists have cited a surviving species of basilosaurine as a much-favoured identity for serpentiform lake monsters and sea serpents—famed for their ability to flex their bodies into vertical hoops when swimming.

But wait. . . More recent palaeontological reconstructions of basilosaurines offer a very different image—postulating that the basilosaurine backbone was notably *in*flexible dorsoventrally, and hence would be incapable of executing the marked vertical undulations characterising the locomotion of serpentiform water monsters.

Does this therefore mean the end of basilosaurine-based cryptofauna as we know it? Panic not. Bearing in mind that these latest basilosaurine reconstructions are drawn from fossil evidence that is at least 36.5 million years old, if a living basilosaurine exists today it will be the product of 36 million years of further, continuing evolution—i.e. evolution that has occurred *since* those most recent basilosaurine fossils were formed.

During such a considerable span of time, continuing evolution can engender all manner of dramatic morphological and physiological modifications. Consequently, the possible existence, for instance, of an *evolved* contemporary basilosaurine with a more flexible vertebral column than any of its currently-known fossil antecedents cannot be entirely ruled out. Equally, it is not impossible that an analogous form could have evolved among one or more of the modern cetacean taxa whose backbones are already flexible (the ziphiids or beaked whales being particularly plausible candidates). In short, serpentiform water monsters and vertically-undulating undiscovered cetaceans are not irrevocably incompatible after all.

This issue is also relevant to the plesiosaur identity for long-necked water monsters. Critics of this identity enjoy attempting to expose morphological, behavioural, and physiological discrepancies between these water monsters and plesiosaurs. And indeed, some such differences do exist—but that is exactly what we should expect. The only plesiosaur material currently available for science to examine is fossilised material that is at least 66 million years old. Add to that a full 66 million years of continuing evolution, and who can possibly

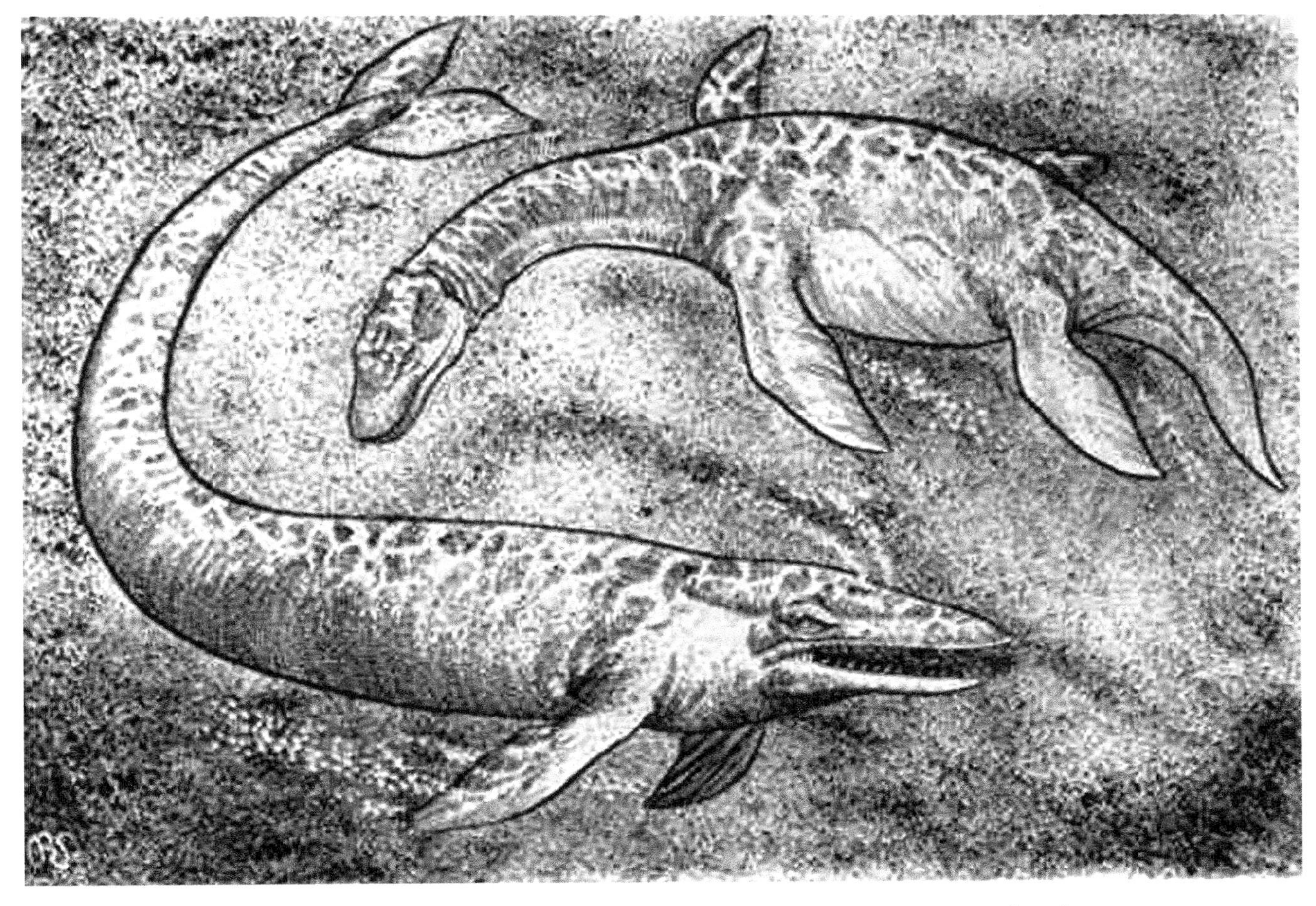

Restoration of a basilosaurine (bottom) and a plesiosaur (top) swimming underwater (© Richard Svensson)

predict accurately how the end-product would look, behave, or function?

For although it is certainly true that the evolution of living organisms is constrained to quite an extent by fundamental factors such as functional morphology, genetics, behaviour, and ecological influences, and therefore should not be looked upon as capable of engineering unlimited, unconstrained transformations, it is equally true nonetheless that, even working within such constraints, evolution has wrought all manner of exceedingly dramatic, wide-ranging morphological manipulations, especially over very long periods of time. Dare we state with imperious authority, therefore, that no modern-day plesiosaur could possibly possess dorsal air-sacs, hair-like respiratory outgrowths, or be gigantothermic when we only have 66-million-year-plus fossils for guidance?

After all, could anyone correctly envisage, for instance, the vast and precise diversity of mammalian forms alive today—from bats, kangaroos, giraffes, tigers, porcupines, and rhinoceroses to whales, pangolins, platypuses, giant pandas, humans, and tree sloths—if presented only with the nondescript shrew-like fossils of mammals that existed 66 million or more years ago? And can anyone hope to prophecy unerringly what our descendants will look like in 66 million years time, courtesy of continuing evolution—always assuming, of course, that our lineage (and planet!) survives that long?

Indeed, if long-necked plesiosaur-lookalike water monsters really are contemporary plesiosaurs, we should not be surprised by the existence of differences between them and their fossil antecedents, but rather be incited to discover why plesiosaur evolution has actually been so very conservative during the past 66 million years, resulting in modern-day plesiosaurs that still look so similar to their far-distant Mesozoic antecedents.

Incidentally, I prefer not to say that plesiosaurs became extinct (or went extinct—grammatically excruciating!) 66 million years ago, simply because we cannot be absolutely, categorically certain that they did. (After all, at least as far as I am aware, no-one has checked every single cubic inch of sea and freshwater to confirm that there is not even one plesiosaur lurking reticently anywhere.) Instead, all that we can say with certainty—at present—is that the known, examined fossil record does not contain any plesiosaur remains formally verified to be less than 66 million years old. And this is not the same thing at all—as exemplified by the coelacanth.

It was back in December 1938 when the scientifically unexpected discovery of a living species, *Latimeria chalumnae*, sensationally resurrected the hitherto 'extinct' coelacanth lineage from post-Mesozoic obscurity. Ever since then, this lobe-finned megastar has been hailed as a classic example of how a supposedly exclusively prehistoric taxon can have a modern-day representative existing undetected by science until relatively frequently, despite possessing no intervening, post-Mesozoic fossil record. (Such a taxon, incidentally, that disappears from the known fossil record for an extensive period of time before reappearing, is referred to scientifically as a Lazarus taxon, after the biblical Lazarus, who was raised from the dead by Jesus.) This in turn has often been cited in cryptozoological publications as a significant precedent for believing that other major prehistoric animal taxa with no post-Mesozoic fossil record, like plesiosaurs, non-avian dinosaurs, and pterosaurs, could also have undiscovered modern-day representatives.

However, this cryptozoological crutch seemed to have been kicked away when palaeontologist Dr Darren Naish revealed several years ago to me that some post-Mesozoic coelacanth fossils had lately been unearthed. Suddenly, portions of *Latimeria*'s long-lost Cenozoic fossil record (known technically as a ghost lineage) were lost no longer. Does that mean, therefore, that the existence of *Latimeria* cannot argue any more in support of modern-day survival of plesiosaurs, non-avian dinosaurs, etc?

On the contrary: I consider that this highly significant palaeontological discovery should not be cursed but praised by cryptozoologists everywhere. For the very fact that it has taken science such an inordinately long time to discover *any* post-Mesozoic coelacanth fossils (even though we have known definitely since December 1938 that in theory there ought to be at least a few preserved somewhere) should incite hope that there are post-Mesozoic plesiosaur, non-avian dinosaur, etc, etc, fossils still awaiting discovery too (especially as these have never been specifically looked for)—and, in turn, that there may well be living descendants still eluding detection.

Certainly, prior to the discovery of a living coelacanth courtesy of *Latimeria* in 1938, no-one had actively searched museum collections or in the wild for evidence of such a creature's existence—and why should they have done? Who could have envisaged at that time that such a famously long-vanished, exclusively prehistoric lineage still survived today when no intervening fossils had come to light during normal palaeontological fieldwork? But once *Latimeria* turned up, palaeontologists had the necessary incentive to pursue its ghost lineage—and behold, post-Mesozoic coelacanth fossils were indeed discovered (though even today, these remain exceedingly scant). Presently, I think it highly unlikely that any palaeontologist would be brave enough to apply for a grant to fund searches in museum collections for hitherto-overlooked post-Mesozoic remains of plesiosaurs, pterosaurs, or non-avian dinosaurs—and who could blame them? But what would happen if a small yet very much alive, evolved pterodactyl were captured in best Conan Doyle *Lost World* tradition in some remote territory? I predict a sudden and very emphatic surge of interest in seeking post-Mesozoic pterosaur fossils amid the world's palaeontological collections!

Having said that, it hardly needs pointing out that the fossil record is in any case very imperfect, very incomplete. Indeed, recognising that its contents are by no means comprehensive, due to the many factors working against the fossilisation of a dead creature's mortal remains, goes back at least as far as Charles Darwin, who devoted a chapter to this subject in his seminal book *On the Origin of Species* (1859). In it, he compared the entire history of our planet to the pages of a stone book in which:

> . . . of this volume, here and there a short chapter has been preserved; and on each page, only here and there a few lines. Each word of the slowly-changing language, in which the history is supposed to be written, being more or less different in the interrupted succession of chapters, may represent the apparently abruptly changed forms of life, entombed in our consecutive, but widely separated, formations.

The author alongside a bust of Charles Darwin (© Dr Karl Shuker)

And as quoted by Matthew Brace in a *Geographical* article of July 2003 on Australian fossil mammals unearthed at Riversleigh, over a century after Darwin penned those latter words, British TV naturalist Sir David Attenborough made a similar observation:

> Only in one or two places on the surface of our planet, in the course of the last three thousand million years, have conditions been just right to preserve anything like a representative sample of the species living at any particular time. Those places are the rare treasure houses of palaeontology.

But what reasons are there for the fossil record's failings? To begin with, there is no guarantee that the carcase of a dead animal, whether intact or disarticulated by scavengers, will even remain in existence for long enough to become fossilised, due to the array of organisms out there that can collectively consume not only its soft tissue but also its bones and other hard tissues. Moreover, there are considerable expanses of terrain that are either very poor at preserving fossils anyway, such as hardwood forests and jungle terrain, or are inaccessible to fossil seekers, like the ocean bed and remote mountain peaks.

For instance, although the Morrison Formation, a sequence of Upper Jurassic sedimentary rock, is renowned in palaeontological circles as the most fertile source of dinosaur fossils in the whole of North America, only a small proportion of it is actually exposed and accessible to geologists and paleontologists. Over 75 per cent has never been examined, because it remains buried under the prairie to the east, and because much of its western paleogeographic extent was eroded (thus destroying any fossil content) during exhumation of the Rocky Mountains. So how much richer and more diverse might North America's known fossil dinosaur fauna be if that 75 per cent of inaccessible and eroded formation had been available for scientific examination?

Greenwich University geologist Prof. Andrew S. Gale and two researchers from London's Natural History Museum, Drs Andrew B. Smith and Neale E.A. Monks, co-authored a paper relevant to this subject that was published by the American journal *Paleobiology* in 2001. In a media interview concerning their paper, Gale was quoted as follows concerning how some apparent extinctions may actually be artefacts—i.e. pseudo-extinctions—resulting from absence of fossils:

> Large gaps in the fossil record are often cited as evidence of mass extinctions. But there are other explanations for this lack of fossil evidence which do not point to a catastrophic annihilation of large numbers of species. During the Cretaceous period, there were periods of intense global warming which saw dramatic rises in sea levels so severe that the oceans flooded Europe, turning it into an archipelago of little islands. This forced shallow marine species and land animals to migrate from their usual habitats. Once the sea level dropped again these species migrated back with it, and the fossil record laid down in sedimentary rock during those periods of high sea level was largely destroyed over time by wind, rain and glacial erosion. The interruption in the fossil record during these periods was caused by species migration and the loss of the fossil record of that migration, and not by a mass extinction.

In addition, palaeontological excavations are necessarily methodical rather than random, due to practical as well as financial limitations. To maximise their chances of success, therefore, they naturally tend to concentrate their efforts in areas where fossils have already been found or where there are already signs of fossils becoming exposed (even if these signs are just isolated bones or bone fragments)—as opposed to simply going off somewhere that has no visible signs of fossiliferous sediments and then digging some random holes just on the off-chance that there may nonetheless be some fossils awaiting excavation there, or employing expensive ground-penetrating radar in the same haphazard manner rather than utilising it to its best advantage in more promising terrain. This focusing upon fossiliferous hotspots thereby adds further bias to the finding of fossils.

And in any case, only a small percentage of fossils exist close enough to the surface for palaeontologists to detect and disinter them, even when using ground-penetrating radar.

In short, for all of these reasons, fossils of extinct taxa that would obliterate ghost lineages may never have formed, or may have formed but not survived into the present day, or may have formed and still exist today but only in localities where they can never or will never be accessed. Consequently, all of these factors need to be firmly borne in mind when attempting to define the boundaries of any given fossil lineage. (Incidentally, still-undiscovered fossils that would close up portions or all of an existing ghost lineage if eventually found are colloquially referred to as Jimmy Hoffa taxa, after the eponymous President of the International Brotherhood of Teamsters (IBT) union who famously vanished on 30 July 1975, and has never been found since.)

So let's hear three cheers for Old Fourlegs—far from being a cryptozoological red herring (metaphorically, as opposed to ichthyologically!), long may the coelacanth remain a symbol of cryptozoological credibility, respectability, and hope.

The 'coelacanth corner' in the author's library (© Dr Karl Shuker)

SETTING THE SCOPE OF THIS BOOK

As the first work devoted exclusively to the fascinating subject of prehistoric survivors, this book (in its two editions) constitutes a unique introduction to and investigation of putative undiscovered living fossils concealed amid the many remote realms scattered across the globe—surveying a vast spectrum of anachronistic anomalies that range from non-avian dinosaurs, pterosaurs, and plesiosaurs, to giant birds, sabre-toothed tigers, mammoths, and much more too.

Having said that, and after giving the matter much thought, I have decided not to include an in-depth coverage regarding prehistoric survivors of the man-beast variety in the present book. Human evolution is, in itself, so notoriously complex and controversial a subject that if its cryptozoological aspects (constituting an additional dimension of intricacy and intrigue) were given the same depth of coverage accorded here to other animal forms, its chapter would be disproportionately long. Yet a more superficial treatment would severely truncate many anthropological and evolutionary issues that require detailed explanation for their significance to the cryptozoological topics under consideration to be satisfactorily conveyed.

Clearly, then, this is a subject requiring a comprehensive book-length treatment to itself (planned as a future companion volume to this one). In the meantime, therefore, the man-beasts are restricted in this present book to a deliberately brief resumé, which merely outlines current theories linking them to prehistoric survival; it is not intended to be equivalent to the depth of coverage given to other cryptozoological subjects here. As a result, this has freed up much-needed space in which to assess many notable examples of prehistoric survivors that have rarely if ever before received extensive documentation in a cryptozoological work.

In a book of this scope, and which traverses so frequently the far reaches and backwaters of zoology, an extensive bibliography listing the major (and those all-too-many maddeningly obscure) sources utilised in its preparation is naturally essential—so that readers interested in pursuing various aspects in greater detail can do so, without being confronted by the unenviable task of uncovering for themselves the identity of those sources. Hence a select chapter-by-chapter bibliography will be found at the end of this book's main text.

AND FINALLY . . .

In summary: those zoological old-timers dubbed 'living fossils' are instrumental in helping to reshape science's ideas regarding not only our planet's contemporary life forms but also its earlier ones and, as a natural consequence (we can but hope!), its attitude towards prehistoric survival—a trend that will gather even more momentum if any of the mystery beasts documented in the following chapters are successfully unveiled in the future.

After all, as Gerald White Johnson once wrote:

> Nothing changes more constantly than the past; for the past that influences our lives does not consist of what actually happened, but of what men believe happened.

This is just as true for cryptozoology as for any other field of human interest.

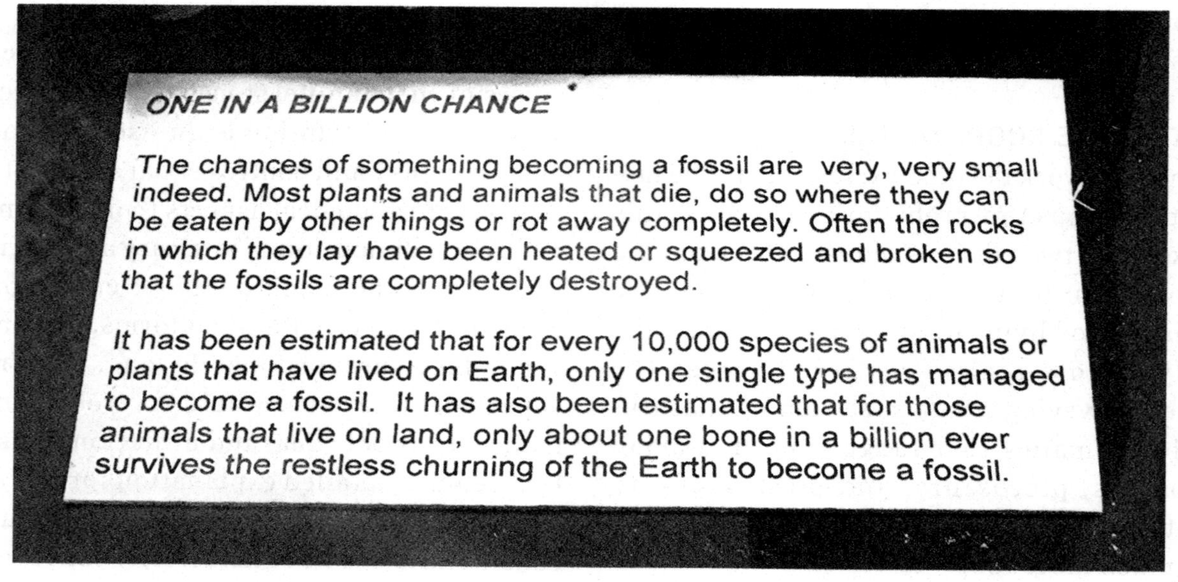

Information placard at Dudley Museum & Art Gallery, England (© Dr Karl Shuker)

ERA	PERIOD	EPOCH	MILLIONS OF YEARS AGO
CENOZOIC	QUATERNARY	HOLOCENE	0.0117
		PLEISTOCENE	2.58
	TERTIARY	PLIOCENE	5.3
		MIOCENE	23
		OLIGOCENE	33.9
		EOCENE	56
		PALAEOCENE	66
MESOZOIC	CRETACEOUS		145
	JURASSIC		201.3
	TRIASSIC		252.2
PALAEOZOIC	PERMIAN		298.9
	CARBONIFEROUS		358.9
	DEVONIAN		419.2
	SILURIAN		443.8
	ORDOVICIAN		485.4
	CAMBRIAN		541
PROTEROZOIC EON	PRE-CAMBRIAN		2500
ARCHAEAN EON			4600

GEOLOGICAL TIME CHART

CHAPTER 1

How Dead are the Dinosaurs? A Reptilian Resurrection

I can appreciate the opinion of the great majority of zoologists and paleontologists who believe that the survival of any dinosaur into the present time is improbable. I cannot agree that the idea is impossible, which is what makes the whole question so interesting.

Prof. Roy P. Mackal—*A Living Dinosaur? In Search of Mokele-Mbembe*

There's places in Africa where you get visions of primeval force. . . in Africa the Past has hardly stopped breathing.

'Trader Horn' [Alfred Aloysius Smith]—*Trader Horn*

As they are widely considered nowadays to constitute a specialised theropod subgroup, birds are therefore deemed to be living dinosaurs—the *only* living dinosaurs, in fact. For according to mainstream zoological dictum, all non-avian dinosaurs were extinct by the end of the Cretaceous Period, 66 million years ago—weren't they?

The reason(s) for the all-encompassing, all-inclusive disappearance of the dinosaurs (excluding birds!) is one of zoology's greatest unsolved mysteries. Out of the many thousands of dinosaur species that evolved during their 165-million-year reign on Earth, not a single non-avian representative survived beyond the close of the Cretaceous. Why? Since the true nature of dinosaur fossils was first recognised back in the early 1800s, almost as many theories concerning their extinction as there were once dinosaurs themselves have been proposed, discussed, and ultimately rejected—because none has so far provided a convincing explanation for the extraordinary selectivity that effected these exceedingly diverse creatures' demise while permitting various other creatures of reptilian descent to persist right up to the present day.

Why should every single species of non-avian dinosaur—large, small, semi-aquatic, terrestrial, herbivorous, carnivorous, active, sluggish, warm-blooded, cold-blooded, solitary, social, scaly, hairy, feathered, bipedal, quadrupedal—die out, whereas various other contemporary vertebrates such as the crocodilians, lizards, snakes, chelonians, birds, and mammals survive? (As will be seen in later chapters of this book, the same enigma of selectivity faces anyone seeking reasons for the wholesale disappearance by the end of the Cretaceous of the plesiosaurs, mosasaurs, and pterosaurs too.)

Perhaps there is no answer, perhaps the riddle is insoluble—for the simple reason that the scenario portrayed by it never took place. "When you have eliminated the impossible, whatever remains, *however improbable*, must be the truth" is an oft-quoted maxim of Sherlock Holmes, Sir Arthur Conan Doyle's famous detective. Holmes, of course, was fictional, but his principle of rational analysis is perfectly sound, and when applied to the case of the missing non-avian dinosaurs it provides a highly thought-provoking outcome. If, as would appear to be true, the concept of the non-avian dinosaurs' bewilderingly exclusive disappearance is so bizarre that it is simply not tenable, then it is time to give serious consideration to its outwardly improbable yet nonetheless valid alternative.

Namely, that not all of the non-avian dinosaurs did die out, that in reality a few species persisted beyond the Cretaceous, and are represented today by living, modern-day descendants concealed within certain remote regions of the world—extraordinary species familiar to the native peoples sharing their reclusive domains, but currently undiscovered by science.

What is so fascinating about this dramatic prospect is that many reports of unidentified creatures remarkably dinosaurian in alleged appearance (and therefore often dubbed 'neo-dinosaurs' by cryptozoologists) have indeed emerged from such areas, particularly in Africa, and have originated not only from so-called 'primitive' tribes (who nonetheless generally know their wildlife far more intimately than do many scientists) but also from Western explorers, settlers, and even—on a surprising number of occasions—from scientists themselves. As will be revealed in this chapter, it is genuinely possible that reports of the dinosaurs' death—just like that of several still-living human celebrities over the years by obituary-obsessed media—will prove to have been greatly exaggerated! (NB—for brevity's sake, hereafter in this chapter and book the term 'dinosaur' should be taken to mean 'non-avian dinosaur'.)

Also under consideration here are a number of non-dinosaurian members of prehistory's terrestrial and swamp-dwelling herpetological fauna that may have lingered long beyond their extinction dates—ranging from monstrous monitors to some very sizeable salamanders.

THE CONGOLESE MOKELE-MBEMBE—ILLUSIVE DRAGON, OR ELUSIVE DINOSAUR?

At around 7 o'clock one clear, sunny morning in the 1960s, a 17-year-old African hunter called Nicolas Mondongo was standing on the bank of the Likouala-aux-Herbes River upstream of Mokengi, a village in the Republic of the Congo (formerly the French Congo). Without warning, the waters parted, and a huge animal surfaced—beginning with an extremely long, slender neck and well-defined head, followed by a very bulky, elephantine body rising up on four massive legs, and finally revealing a lengthy, tapering tail.

For a full three minutes, this extraordinary animal remained above the surface of the water, which had cascaded down from its shoulders like a miniature cataract during its unannounced emergence. Standing less than 15 yards away, terror-stricken yet mesmerised by the mighty water beast's awesome presence, Mondongo estimated its total length to be around 30 ft—of which about 6 ft were accounted for by its head and neck, and a similar length for its tail. Then suddenly, just as

Artistic representation of the mokele-mbembe as a sauropod-like aquatic cryptid (© William M. Rebsamen)

abruptly as it had appeared, the creature submerged again, and within a few moments the last ripples on the water marking its disappearance were gone, leaving Mondongo with only the memory of his astonishing encounter.

A scene, perhaps, from Steven Spielberg's sci-fi blockbuster, *Jurassic Park*, or any of its sequels? Not according to the native Africans, the pygmies, and the French missionaries inhabiting the Congo's humid, inhospitable Likouala swamplands—for whom this astonishing animal is a vehemently-acknowledged reality. They call it the mokele-mbembe ('blocker of rivers'), and even though its very existence has yet to be officially verified by science it has incited an ever-increasing degree of Western excitement in recent decades—because if local eyewitness descriptions of it are accurate, the reclusive mokele-mbembe may conceivably be a living dinosaur.

The possibility of Central Africa one day hosting a sensational reptilian resurrection first received detailed consideration from veteran French cryptozoologist Dr Bernard Heuvelmans. His analyses of this region's varied catalogue of dinosaurian mystery beasts were published in *On the Track of Unknown Animals* (1958) and *Les Derniers Dragons d'Afrique* (1978), yielding inspiration to and information on all manner of modern-day dinosaur quests. But if the mokele-mbembe really is a living dinosaur, what type could it be?

An elephant-sized sauropod, akin to prehistory's *Diplodocus* and *Apatosaurus* (formerly *Brontosaurus*)? That was the opinion of the late Prof. Roy P. Mackal—an eminent biologist from the University of Chicago, who until his death in 2013 was the world's leading expert on this elusive entity from Central West Africa, and until its demise during the 1990s was also Vice-President of the International Society of Cryptozoology (ISC). After conducting extensive bibliographical research that uncovered many hitherto-unpublicised sightings (including the discovery by some French missionaries of gigantic clawed tracks, each about 3 ft in circumference and 7-8 ft apart, in a Congolese forest southwest of the Likouala sometime prior to 1776), Mackal led two expeditions to the Congo during the early 1980s, seeking evidence to confirm the mokele-mbembe's existence.

The Mackal Expeditions

On his first expedition, a month-long foray during February/March 1980, Mackal was accompanied by James H. Powell, a herpetologist from Texas. This trip introduced them to northern Congo's Likouala region, a vast and virtually uncharted, unexplored wilderness of environmentally-hostile swamps—55,000 square miles of virgin territory where an entire herd of dinosaurs would be no more conspicuous than a swarm of gnats.

Mackal and Powell believed that if dinosaurs could exist unknown to science anywhere in the world, the Likouala is where they would be, and the team's enquiries, taking them south from Impfondo to the village of Epéna, fuelled their hopes by yielding first-hand accounts of mokele-mbembe sightings gathered from more than 30 reliable eyewitnesses among the native Congolese people and pygmies. In particular, they were informed by one of the Kabonga pygmies from a village called Minganga, northwest of Epéna, that these animals had certainly been seen in a river that they called the Tebeke—which appeared to be their local name for the northern stretch of the Bai River, to the west of Epéna.

Although Mackal and Powell did not obtain any personal sightings of their quarry, Mackal was greatly heartened by the native accounts that they had collected. Accordingly, in late October 1981 he returned to the Congo to lead a second, more detailed exploration of the Likouala. His team included J. Richard Greenwell (then Secretary of the ISC), Congolese biologist Dr Marcellin Agnagna, American Baptist missionary Pastor Eugene Thomas (who had lived in the Likouala region with his wife, Sandy, for 26 years), some Congolese army personnel, and a party of pygmies as helpers. This time, Mackal decided to focus his attention upon the northern Bai River and a large shallow lake nearby called Tele, where mokele-mbembes had again been seen on many occasions according to local testimony.

After flying north of Epéna to Impfondo, the site of Pastor Thomas's mission, the team set out in native dugouts southwards along the Ubangi River, from which they crossed via a weed-choked canal into the westerly-sited Tanga River, taking them in a southwesterly direction past Epéna on the Likouala-aux-Herbes. After reaching the confluence of this river with the southernmost portion of the Bai River, they established a base at Kinami on the Bai, and travelled to several nearby villages, questioning the natives regarding mokele-mbembes.

Although the team acquired many additional eyewitness accounts, they learnt to their great disappointment that the Tebeke River as

Representation of the mokele-mbembe, based upon eyewitness descriptions (© Drawn by David Miller, under the direction of Prof. Roy P. Mackal)

referred to by the Kabonga pygmies during Mackal's first expedition was not the northern Bai after all. Instead, it was a lake, to the west of the Bai, which could only be reached from their present position via a near-impossible overland trek that would require several days of sustained travel through very inhospitable swampy terrain. Moreover, an equally forbidding overland trek to the east faced them if they decided to journey from the Bai to Lake Tele. Hence the team reluctantly abandoned their hopes of reaching either of these destinations, and after completing their studies around Kinami they retraced their journey back to Impfondo, then flew home to the U.S.A. in December 1981.

Sadly, Mackal was once again denied the excitement of observing a mokele-mbembe with his own eyes—but on one occasion he came within a hair's breadth of experiencing a very close encounter. As revealed in his fascinating book *A Living Dinosaur? In Search of Mokele-Mbembe* (1987), it took place while the expedition was travelling along the Likouala-aux-Herbes in a couple of native dugouts. Paddling around a sharp curve in the river, upstream of Epéna, their eyes scanned the river bank, whose shadow lay over the water inshore. "At that very moment," recalled Mackal, "a great 'plop' sound and wave cresting at 25 cm [10 in] washed over the dugout, directly from the bank of the shadowed area. The pygmies screamed hysterically, 'Mokele-mbembe, Mokele-mbembe'". Mackal at once urged them to turn back in pursuit of whatever unseen creature had been responsible for this enormous disturbance, but his African assistants were most reluctant to do so.

Notwithstanding their protestations, however, Mackal and party did change course, but by the time that the dugouts had reached the spot where the creature had submerged, all signs of its presence had vanished. Crocodiles do not create waves of this nature when they dive, elephants cannot submerge entirely, and hippos (which can) do not occur in the Likouala swamps—all of which lends considerable credibility to the pygmies' ready identification of the missing monster as a bona fide mokele-mbembe.

Compensating for his disappointment regarding that incident, however, during his searches in the Likouala swamps Mackal once again succeeded in gathering an impressive dossier of independent eyewitness reports (including Nicolas Mondongo's) and background details regarding their mysterious occupant. These generally corresponded closely with one another, as well as with those from his previous expedition, and can be summarised as follows.

According to local testimony and traditional beliefs, the mokele-mbembe is a massive and elusive but unequivocally real water beast, reddish-brown in colour, that inhabits deep pools in the banks of the Likouala's forested waterways. It is rarely observed, spending much of its time submerged, with only its head and neck spasmodically breaking the surface in order to feed upon bankside vegetation; but when it is seen, its appearance corresponds with Mondongo's sighting as described previously. Most sightings occur during early morning or late afternoon (and generally in the dry season), but a few have occurred at night.

Despite its size, the mokele-mbembe is extremely quiet, not giving voice to any sound, and is exclusively herbivorous, its principal food constituting an indigenous fruit known locally as the malombo, which Mackal and Powell revealed to be a species of *Landolphia* gourd (confirming a prediction made earlier by Heuvelmans). Yet despite its vegetarian lifestyle and generally placid temperament, the mokele-mbembe can become very aggressive if provoked, capsizing dugouts by rising up underneath them, and killing (but not devouring) their hapless occupants with powerful thrashes of its long sturdy tail.

Of particular significance (as will be seen later) is the claim by native eyewitnesses that on those rare occasions when the mokele-mbembe

The author with a cast of a three-toed fossil dinosaur footprint (© Dr Karl Shuker)

ventures out of the swamps onto land, it leaves behind very large and distinctive three-clawed footprints—footprints that are unlike those of any other modern-day animal.

During his two expeditions, Mackal and his fellow team members were very careful not to influence eyewitnesses' descriptions when questioning them concerning the mokele-mbembe. Their descriptions were verbal or via drawings traced in the sand—and only *after* these had been recorded were picture books of animals shown to them. Of the numerous detailed descriptions of mokele-mbembes by first-hand observers that were obtained, all readily recalled a small (i.e. elephant-sized) sauropod dinosaur, and when these eyewitnesses were shown pictures of living and extinct animals the pictures of sauropods were the ones selected by them as being most similar to the mokele-mbembe. Moreover, the eyewitnesses were from several different ethnic cultures, religions, and geographical locations.

Early Mokele-Mbembe Reports

Mackal's expeditions certainly initiated modern-day public interest in the mokele-mbembe (duly capitalised upon by Hollywood on many occasions since then, but beginning in 1985 with the release of the successful Touchstone film *Baby—Secret of the Lost Legend*, concerning the discovery of a baby mokele-mbembe by a palaeontologist popularly believed to be loosely inspired by Mackal himself). However, there are reports by Westerners specifically mentioning this creature that date back at least as far as 1913.

This was the year that saw the commencement of the Likuala-Kongo Expedition, a German exploration of northern Congo (then part of Cameroon) led by Captain Freiherr von Stein zu Lausnitz, and scheduled to last for two years. Due to World War I, however, it was never completed, and the official findings were not published, but certain sections were obtained by cryptozoological writer Willy Ley, who discovered that during their travels the team had collected native reports describing a very large water beast known to the locals in the area of the lower Ubangi, Sanga, and Ikelemba Rivers as the mokele-mbembe. Their description of its appearance and lifestyle was entirely consistent with the version that Mackal would hear again and again during his own expeditions almost 70 years later—except for one feature.

According to Stein zu Lausnitz: "It is said to have a long and very flexible neck and only one tooth but a very long one; some say that it is a horn". When Mackal queried the natives regarding a horn, they were insistent that the mokele-mbembe did not have one. Instead, this was possessed by an equally mysterious but entirely separate animal, the emela-ntouka (documented later in this present chapter). Mackal believed that Stein zu Lausnitz's error stemmed from the natives' confusing tendency to use 'mokele-mbembe' both as a specific name for the sauropod-like mystery beast and as a more general, collective term encompassing both this animal and the genuinely-horned emela-ntouka. A later mokele-mbembe seeker, Bill Gibbons, has confirmed this (see later), but there is also evidence to suggest that certain long-necked mokele-mbembe-type beasts do indeed possess a horn (again, see later).

Not merely this creature itself but even a close version of its name appears to penetrate beyond the confines of the Congo northward as far as southern Cameroon, because in 1938 a German magistrate called Dr Leo von Boxberger, who had spent many years in this region, noted that the natives here firmly believe in the existence of a gigantic long-necked water reptile that they call the mbokale-muembe.

There are also on file many Western accounts of unidentified beasts elsewhere in Africa that have totally different names, yet which are all seemingly akin to the Congo's mokele-mbembe—as shown by the following selection.

Mbilintu, Isiququmadevu, Irizima, Badigui, Amali, N'yamala, Etcetera

Someone who spent much of his life in the Intelligence and Administrative Services of East Africa and often functioned as a local magistrate is hardly the type of person prone to wild imagination or unreliable testimony—which is why we should give serious consideration to the series of articles by Captain William Hichens published during the 1920s and 1930s that discussed the possible existence in Africa of several spectacular species of animal still undiscovered by science. Among those was a water beast called the mbilintu, which has been widely reported—from Zambia's Lake Bangweulu, and Lake Mweru (bordering Zambia and the Democratic Republic of the Congo, or DR Congo for short), to the swamplands of DR Congo itself (formerly called Zaire and, before then, the Belgian Congo), and Lake Tanganyika (bordering DR Congo and Tanzania).

According to Hichens, descriptions of this creature vary considerably, but can be split into two morphologically distinct types. These evidently represent two wholly separate species that inhabit the same areas and which have become confused with one another in native traditions and recollection. Paralleling the Likouala situation noted earlier, one is a large rhino-like beast (verbally distinguished in certain localities, such as Zambia's Lake Bangweulu, as the chipekwe—see later), whereas the other is clearly allied to the mokele-mbembe, and is likened to a gigantic lizard with a neck like a giraffe, elephant-like legs, a small snake-like head, and a 30-ft-long tail (no doubt a native exaggeration).

One of the mbilintu's earliest Western investigators was the renowned animal collector Carl Hagenbeck. In his book *Beasts and Men* (1909), he recalled that two wholly independent sources (an English hunter, and Hans Schomburgk—one of Hagenbeck's most proficient animal collectors) had supplied him with details of a gigantic reptilian beast, half-dragon and half-elephant, inhabiting the interior of Rhodesia (later split into Zambia and Zimbabwe). "From what I have heard of the animal", concluded Hagenbeck, "it seems to me that it can only be some kind of dinosaur, seemingly akin to the brontosaurus". So convinced was he concerning its existence that he dispatched a formal expedition in search of hard evidence to confirm this—but repeated onslaughts by savage tribes, virulent debilitating fever, and the unimaginably vast expanses of near-impenetrable marsh and jungle requiring exploration effectively conspired to frustrate all attempts to espy their cryptic quarry.

Slightly more successful was Lewanika, the king of Barotseland (a northwestern district of what is now Zambia), who had always displayed a great interest in his kingdom's fauna. One day prior to the 1920s, three of his subjects arrived at his court house in great excitement, to inform him that they had just seen one of these great creatures, which they referred to as the isiququmadevu, lying on the edge of a marsh. Their description of it perfectly matched that of the mokele-mbembe—taller than a man, with a huge body, a long neck with a snake-like head, and sturdy legs. Once the creature had caught sight of the men, however, it had rapidly retreated, submerging into deep water.

Undaunted, Lewanika lost no time in travelling to the spot where this encounter had taken place, and although the creature did not

Two views of Lake Bangweulu in Zambia—home of the mbilintu (© Alan Brignall)

reappear for him, he could plainly see where it had been—revealed by the presence of a large area of flattened reeds, and a broad path with water flowing into it that led to the water's edge. The Barotse king was so impressed that he sent an official report of the incident and his observations to Colonel Hardinge, the British Resident, in which he recorded that the channel made by the animal's body was "as large as a full-sized wagon [roughly 4.5 ft wide] from which the wheels had been removed".

More successful still was Alan Brignall, to whom I am greatly indebted for kindly permitting me to document here his first-hand encounter with what would appear to be a mbilintu of the long-necked persuasion. During the early 1950s, Brignall was working in Kiture, Zambia (then Northern Rhodesia), and his sighting occurred one afternoon in May 1954 while taking a few days' holiday in the bush with some colleagues. He had been fishing from the shore of Lake Bangweulu, and could see two tiny islands in the lake. Covered with small trees and shrubs, they were situated about 25 yards from the shore where he was standing, and were surrounded by reeds.

Suddenly, looking up from his fishing and across at the islands, Brignall received a considerable shock, because something that seemed at first to be a huge snake had risen up out of the water on the shore side of the islands. However, it did not resemble any type of snake that he had seen before. A small head was raised roughly 4.5 ft above the water surface on a long vertical neck, approximately 1 ft wide, seemingly scaleless, and uniformly grey in colour. The neck remained stationary, but the head, which possessed a distinct brow, blunt snout, and visible jaw line, swivelled from side to side as if seeking something. The mouth remained closed, and Brignall was unable to discern any nostrils or ears. Just behind the neck was a small hump, the same colour as the neck, and which Brignall believed to be a portion of the creature's body rather than merely a fin.

Brignall recalled that after observing these details for a few seconds, he either attempted to signal to a friend nearby or put down his fishing rod to reach for his camera, Whichever action it was that he took, however, it immediately attracted the creature's attention, because it instantly sank down, submerging vertically until it was totally hidden beneath the water. Brignall did not see it again, and remains convinced that the creature was not a lizard, crocodile, or turtle, but rather something thought to be extinct.

According to a letter published in the *Neue Mannheimer Zeitung* (6 January 1934), a dinosaur-like beast resembling an island floating on Lake Tanganyika and leaving three-clawed prints on its shore had been sighted from a distance by several ships around 1928, but as soon as any attempted to draw nearer, it immediately sank beneath the water. The prints that had been observed were larger than those of an elephant, and revealed the presence of a thick tail too. Its most notable investigator was the brother of Sir Edward Grey, but despite spending many years researching the creature he never solved its mystery.

DR Congo's version of this beast is the irizima, and in a by-now-familiar tradition the natives apply its name to two different animals—the mysterious horned creature known elsewhere as the emela-ntouka or chipekwe, and the sauropod-like mokele-mbembe beast. According to a description of the latter creature recorded by Captain Hichens, it is ". . . a marsh monster with a hippo's legs, an elephant's trunk, a lizard's head, and an aardvark's tail", and inhabits Lake Edward. In 1927, a big-game hunting expedition led by Lieut.-Col. H.F. Fenn hoped to uncover the identity of this aquatic anomaly, but did not succeed in doing so.

Although Mackal's enquiries convincingly ascertained that the horn sometimes ascribed to the mokele-mbembe was actually that of the totally distinct (albeit equally undiscovered) swamp-dweller known as the emela-ntouka, there is at least one detailed account on record of a genuinely horned mokele-mbembe-type beast. In November 1979, mystery beast investigator Philip Averbuck was following up reports of creatures resembling the mokele-mbembe in Cameroon, and learnt that two such beasts had been sighted by air security officer A.S. Arrey at Kumba during 1948-9, when he was a child.

Arrey had been swimming with some friends in Lake Barombi Mbo, in which some British soldiers were also swimming, when the water at the lake's centre began to boil. Everyone immediately swam to shore, and after clambering out they looked back and saw to their astonishment that a long neck had emerged above the water's surface, bearing a small slender head at its tip. A few minutes later, a second, larger neck and head surfaced nearby, about 200 yards from their human observers. At this point, the soldiers turned and fled, but the others remained, fascinated by the extraordinary spectacle.

The neck of the larger creature, which Arrey assumed to be the male, was slightly curved,

The mokele-mbembe as a spiny sauropod (© Tim Morris)

extended 12-15 ft above the water surface, and was covered in smooth snake-like scales. Its head was about 2 ft long, and bore at the back a spiny horn or cap-like structure roughly 8 in long, which projected downwards. Interestingly, the smaller, 'female' creature was hornless. The two remained visible for about an hour, after which the female sank back down beneath the water, followed a few minutes later by the male. According to local tradition, this animal never leaves the water, and only surfaces very infrequently. If it is true that only one sex possesses a horn (thereby constituting a very notable example of sexual dimorphism), this would provide an additional explanation for why some mokele-mbembe reports gathered by Mackal referred to horned beasts and others to hornless ones—some had spied specimens of one sex, and some had spied specimens of the other sex.

Palaeontological research has confirmed that a horned or spiny sauropod would not be unprecedented. In the journal *Geology* (December 1992), dinosaur researcher Dr Stephen Czerkas from Monticello, Utah, revealed that the skin of a then-unnamed species of *Diplodocus*-related sauropod from Wyoming was patterned with spiny osteoderms, each just over 1 in across, borne upon its tail and probably extending along its backbone too. Most of the spines were small and conical, but the larger ones were flattened and up to 9 in long. Other armoured, osteoderm-bearing sauropods included *Ampelosaurus* and *Saltasaurus*.

North of the two Congo republics and east of Cameroon lies the Central African Republic—

another country reputedly harbouring mokele-mbembe counterparts. Here, such beasts are known variously as the badigui, diba, guaneru, ngakula-ngu, and songo. They inhabit rivers and lakes where they remain with their bodies completely submerged, only very occasionally coming out onto land. Instead, they raise their long slender necks far up above the water and browse upon the branches of lakeside trees like reptilian giraffes. Their heads are flattened and somewhat larger than those of pythons, their necks are smooth and patterned with snake-like markings, paler upon their under surface than upon their upper surface.

Evidence for putative surviving sauropods can come to light in the most unexpected ways, and places. In the 1960s, a leading jewel designer called Emanuel Staub was commissioned by the University of Pennsylvania's University Museum to produce replicas of a series of small gold weights obtained in Ghana, West Africa. Most of these were in the form of animals, every one different, and served as specific logos for the Ashanti gold dealers there—each dealer thus having his own, unique trademark. So well-crafted were they that the animals that they depicted could be instantly identified by zoologists—all but one, that is, which could not be satisfactorily reconciled with any known animal, until Staub saw it. A founder member of the American-based Society for the Investigation of The Unexplained (SITU), which was keenly interested in cryptozoological matters, Staub was astonished when he saw the

Ashanti gold weight recalling a sauropod dinosaur—inspired by the mokele-mbembe? (© Dr Bernard Heuvelmans)

mystifying artefact, because it seemed to bear a striking resemblance to a dinosaur!

In January 1970, SITU's journal, *Pursuit*, documented this enigmatic model (erroneously giving its origin as Dahomey—now Benin), and attempted to identify its dinosaurian similarities. Unfortunately, however, the model was photographed resting on its hind legs as if the beast in question were bipedal—an inappropriate representation, bearing in mind that all four legs of the model were of the same length and that the creature as depicted by it was quite clearly quadrupedal. Coupling this with its notably lengthy neck, the report's likening of the model to the ornithopod dinosaur *Iguanodon* (whose forelimbs were shorter than its hindlimbs, and whose neck was proportionately much shorter than the model's) was very unconvincing.

If, conversely, the model is stood on all four legs, a radical transformation occurs—from an inaccurate rendition of an *Iguanodon* to an extraordinarily precise rendition of a sauropod. The squat, elephantine body, the four sturdy legs of equal length, the long tail, the elongate neck curving upwards—all are faithfully reproduced in this remarkable artefact. As native gold dealers in Ghana are hardly likely to have expertise in palaeontological reconstruction of long-extinct animals, however, we must assume that this artefact, just like all of the others, was modelled upon a real, living animal, and one that may be indigenous to this region. In short, Ghana could harbour a creature similar to the mokele-mbembe.

Indeed, the only noticeable difference between Congolese descriptions of the mokele-mbembe and the creature represented by the Ashanti gold artefact is that the latter's head is somewhat larger in proportion to the rest of its body. During his various animal collecting trips to West Africa, however, American cryptozoologist Ivan T. Sanderson (who was also president of SITU) obtained reports of mokele-mbembe-type creatures with large heads—substantiating the accuracy of the artefact, and indicating that West Africa's representatives of the alleged surviving sauropod clan may constitute a separate species from their Central and East African counterparts.

This may not be the only enigmatic artefact of potential West African relevance to the mokele-mbembe either, as demonstrated by the following example, which has not previously been documented in the cryptozoological literature. On 7 April 2013, Brother Richard Hendrick, a Capuchin Franciscan Friar (Monk) of the Irish Province who has long been interested in cryptozoology, posted on my 'Journal of Cryptozoology' Facebook group's page a colour photograph that he had snapped a few days earlier, depicting a very eyecatching wooden statue, and which I am now reproducing here by kind permission of Brother Richard.

He also included the following details about it:

> Found this African carving in a collection held by the Rosminian Missionary Fathers in Glencomeragh House, Clonmel [in County Tipperary] Ireland. No provenance or date other than sometime this century probably from west Africa. Intriguing?

After a few replies from others in this FB group were posted in response to his photo and comment, including one from American crypto-author Matt Bille saying that it looked to him like a modern creation, possibly inspired by western interest in the mokele-mbembe (rather than an example of traditional native artwork), Brother Richard responded as follows:

Wooden carving of a scaly sauropod-like creature and a smaller short-necked mystery beast; probably of West African origin, currently housed at Glencomeragh House, Ireland (© Brother Richard Hendrick)

> I'm afraid I know very little more than I mentioned above. The Rosminian Fathers are a missionary order and worked for many years in Africa. Many of the Irish fathers brought back souvenirs of their time there on their return to Ireland. I was merely visiting the house on retreat myself and spotted it sitting dusty and forgotten amongst many other carvings. I would guess it has been there since at least the 1950s as no one currently living in the house seemed to know anything about it and didn't see it as anything extraordinary. It does have a "modern" feel but I felt it was worth sharing.

(Judging from his comment that he felt that this carving had probably been at Glencomeragh House since at least the 1950s, his previous comment that it probably dated from "sometime this century" was no doubt simply a slip of the pen, having forgotten momentarily that this is now the 21st Century, not the 20th.)

His final posts concerning the carving read as follows:

> I have to say I was more intrigued by the second creature? Not sure if it represents a "baby" or another species. I guess at the very least it demonstrates the creature being present in a "culture" even if it was introduced as an idea by those seeking it.
>
> Not sure if it is in a fighting / mating / feeding pose with the larger creature. They both appear to have the same tail, scales and cloacal opening though. A puzzle to add to the connundrum of the whole piece! Never seen anything like it before though!

When I contacted Brother Richard directly concerning this carving, I learnt from him that it was roughly 6 in tall, was quite heavy, seemed to have been carved from a single piece of wood, and although he had examined it rather thoroughly he did not see any inscriptions or dates on it.

There is no question that the larger of the two animals in this carving closely resembles a sauropod dinosaur in body shape. However, it also sports certain highly unusual features for any such identity. Most obvious of these is its array of extremely large, almost pangolin-like overlapping scales, which are absent only from its face, feet, and, most noticeably, the entire extent of its long tail, and which have not been reported by mokele-mbembe eyewitnesses in the Congo (though crocodile-like scales have been claimed by the Baka pygmies for their Cameroon version, the li'lela-bembe). Nor has the large, even more intriguing—and very odd—rayed fin-like structure visible on the underside of its tail's basal portion, which Brother Richard suggested may be a cloacal opening. And its long pointed teeth are not what one would expect from a herbivore, though these may simply be non-specific, stylised representations.

As for the carving's second, smaller animal, standing upright on its hind legs, possibly supported somewhat by its tail, but also resting directly upon the sauropod-like beast's rear body portion, this does not resemble any creature known to me and is something of a paradox. For whereas it sports exactly the same type of body scales as the 'sauropod', and the same kind of tail too, even bearing an identical rayed fin on its tail's basal underside, the body scaling is absent from its lower limbs, it totally lacks the long neck of the 'sauropod', and its face and jaws do not match those of the latter beast either.

Is this small animal meant to represent a young version of the 'sauropod', I wonder? The presence of the same type of scales and in particular the very distinctive underside tail fin (which is unlike anything that I know of in vertebrates present or past), would certainly support such a possibility. Yet its virtually non-existent neck argues in an equally persuasive manner against this option. Even the pose in which the 'sauropod' has been depicted in relation to the small animal is somewhat ambiguous, as recognised by Brother Richard too, and could be construed equally as one of anger toward a potential attacker or as one of response (feeding?) toward its own offspring.

British mystery beast researcher Penny Odell made a very interesting, original contribution to the series of comments posted beneath Brother Richard's photo of the carving that may explain the small creature's contradictory form:

> I think it is supposed to be the same species. However the artist only had a certain space and size of wood and ability to carve with. Had the smaller animal been given a longer neck it would have changed the carving making it harder to carve into the space[;] therefore the artist made the smaller animal with a shorter neck for convenience.

With virtually no historical information concerning this fascinating artefact available, there is little more that can be said about it. Despite being for many years a keen visitor to antique/collector/ethnic-craft fairs and shops, and an equally avid online browser relating to such areas of interest, I have never encountered anything even similar, let alone identical, to this carving before. And if it does date back to at least the 1950s, this long precedes the Mackal mokele-mbembe expeditions from the early 1980s that incited the continuing modern-day wave of interest in the latter Congolese cryptid, thereby offering support for believing that this carving's creation was not sparked by any Western interest in such a creature after all. Yet if it is a traditional native depiction of a mokele-mbembe or comparable cryptid, how can the scaling, the tail fin, and the short-necked smaller beast be explained?

The late 1960s and early 1970s were a fruitful period for controversial dinosaur depictions in Africa. Just a month before *Pursuit*'s account of the Ashanti artefact was published, newspapers worldwide were carrying reports of some bewildering cave paintings discovered in the Gorozomzi Hills, 25 miles from Harare (then Salisbury), the capital of what is now Zimbabwe. They had been painted by bushmen who had inhabited the area from 1,500 BC to about 200 years ago, and who always based their work upon real-life animals rather than imaginary ones, which is why one of the depictions was so astounding—because it portrayed a creature remarkably similar to a sauropod dinosaur. Surrounded by readily identifiable pictures of familiar species such as the African elephant, giraffe, and hippo, the anachronistic anomaly had been depicted extending its long neck up out of a swamp.

Nor is this the first time that cave paintings of creatures reminiscent of dinosaurs have been discovered. Three long-necked, long-tailed quadrupedal beasts that certainly seem to be something other than crocodiles (their 'official' identity) appear in a painting found in a cave at Nachikufu in Zambia (reproduced in Mackal's book *A Living Dinosaur?*).

In his eponymous book, African adventurer Trader Horn (real name Alfred Aloysius Smith) claimed that he had chiselled from some rocks in Cameroon a painting of a swamp-dwelling beast called the amali or jago-nini, and had given it to President Ulysses Grant of the U.S.A. Elsewhere in his book, Horn referred briefly to the footprints left behind by the amali—"about the size of a good frying-pan in circumference and three claws instead of five". This compares well with Congolese descriptions of the mokele-mbembe's spoor.

The similarity of their names and lifestyle makes it plausible that the amali of Cameroon is the same species as the n'yamala of Gabon—a small country south of Cameroon and west of the Republic of the Congo. It was here where James Powell came to study crocodiles in 1976 and first learnt of the mysterious n'yamala, and his continuing investigations of this creature in 1979 stimulated Mackal to combine forces with him in 1980 to seek its Congolese counterpart, the mokele-mbembe.

During his studies, Powell had been showing one of the witch doctors from the Fang tribe of Bantus a number of pictures of animals in various books, in order to learn which species frequented their territory, and while flicking through the pages he came to a picture of a *Diplodocus*. To his amazement, the witch doctor stated that such a creature existed in Gabon and later claimed that he had seen one himself in 1946!

Dwelling only within the deepest lakes and rarely observed but nonetheless a real animal, Powell's informant called it the n'yamala, and also identified as the n'yamala a picture of a plesiosaur—a long-necked aquatic reptile that officially died out alongside the dinosaurs, and whose body and long slender neck superficially resemble those of a sauropod. During the course of his crocodile researches, Powell took

James Powell (left), with native guide and Prof. Roy P. Mackal (© Prof. Roy P. Mackal)

the opportunity to question Gabonese natives from several different tribes and cultural backgrounds, and obtained the same result every time—pictures of sauropods and plesiosaurs were consistently identified as the n'yamala.

Convincing him of their veracity was the fact that whereas they were adept at identifying from pictures all of the creatures that Powell already knew to be native to the area, they were equally honest in admitting that they did not recognise the animals if he showed them pictures of creatures totally foreign to the region, such as camels and bears. Consequently, Powell concluded from his enquiries that the Gabonese natives genuinely knew of a creature inhabiting their country that bore a close resemblance to a living sauropod dinosaur.

Inevitably, a subject with as much potential for attracting media attention as the discovery of a modern-day dinosaur has not avoided the slings and arrows of outrageous hoaxes and sensationalised documentation. During the early 1920s, for example, newspapers all over the world vied with one another to print the most lurid accounts of a thoroughly ludicrous dinosaur hunt that had begun in 1919 when a railway construction supervisor called L. Lepage allegedly encountered in the then Belgian Congo a 24-ft-long dinosaur with tusk-like horns and a short horn above its nostrils. Inexplicably termed a brontosaurus by the press, this triceratopsian terror inspired varied searches and stories, including claims that a Belgian prospector called Gapelle had attempted to shoot a similar Congolese monster.

Even today, versions of these incidents still appear in cryptozoological publications, yet they have no foundation in reality—no more, in fact than their two protagonists. As avid wordsmiths will have already spotted, 'L. Lepage' and 'Gapelle' are simple anagrams of one another. The 'great brontosaurus hunt', as dubbed by the media, was merely a great hoax.

A more recent spoof, but this time of the deliberately humorous, harmless variety, occurred in 1989 when English newspaper *Today* published an exclusive report describing the public debut at London Zoo of Pizza, a living brontosaur from Guinea. The report featured a somewhat blurry photo—but that did not prevent readers from readily seeing through this charade. After all, they had only to look at the date of the newspaper—1 April.

Operation Congo (1-2), and Other Post-Mackal Mokele-Mbembe Expeditions

In September 1981, an expedition led by Californian consulting engineer Herman Regusters travelled to Lake Tele, and upon their return in December Regusters announced that he and his wife Kia had not only seen a mokele-mbembe but had also obtained photographs of it. When developed, however, these potentially invaluable pictures were found to be greatly under-exposed, containing little or no detail, so it was impossible to determine whether they had indeed captured the image of their dinosaurian debutante. Recordings of a loud bellowing sound believed to be that of a mokele-mbembe were also obtained (yet according to Mackal's investigations, this animal is generally silent). A second Regusters expedition was planned for April 1985, but never took place. Regusters had initially agreed to participate in Mackal's second expedition, but withdrew during the planning stages and led his own instead.

In spring 1983, the first mokele-mbembe search by Congolese scientists took place, when Lake Tele was visited by a team from the Congo's Ministry of Water and Forests, led by zoologist Dr Marcellin Agnagna from the Parc de Zoologie. Spanning April and May, it returned to Brazzaville with news that Agnagna and two local helpers from a village called Boha sited near the lake had spied one of these beasts on 1 May. According to Agnagna, he had been filming monkeys in a lakeside forest with a movie camera when one of the villagers called to him in great excitement to come to the lake. When he arrived, he looked across it to where the villager was pointing—and to obtain a better view, unimpeded by shoreline vegetation, he strode into the water. It was then that he allegedly perceived the back and neck of a huge animal, visible above the surface, and about 700-900 ft away—and which turned its head as if seeking the source of the noise made by Agnagna as he approached it.

Raising his camera straight away, Agnagna began filming—but when he was unable to see anything through the viewfinder he realised to his horror that his camera was on the 'macro' setting. And by the time that he had switched to the correct setting, his film had run out! Nevertheless, making the most of a very maddening situation, Agnagna was able to obtain a very detailed view of the creature by observing it through his camera's telephoto lens, watching it turn its head and neck as it looked slowly from side to side. Its head was very small, reddish in colour, with a pair of oval, crocodile-like eyes and a slender protracted muzzle (emphasised in a sketch drawn later by Agnagna), and was borne upon an elongate neck that was raised about 3 ft above the water and was brown at the upper end, black at the base. The 7.5-9-ft-long hump that surfaced just behind the neck was also black.

As Agnagna and two of the villagers waded towards it, the creature submerged, but soon afterwards its neck re-emerged, and stayed above the surface for 20 minutes before disappearing again, this time for good. During his observations, Agnagna attempted to take photos of the creature using a small 35-mm camera, but the resulting pictures were not distinct enough for positive identification of their subject. In his opinion, the animal was certainly reptilian, rather than mammalian or avian, but was not a crocodile, python, or soft-shelled freshwater turtle—the only large reptiles *known* to frequent this locality. Instead, he considered it to be most similar in appearance to a sauropod dinosaur.

The author wearing his Operation Congo t-shirt back in the 1990s (© Dr Karl Shuker)

Britain's contribution to this long-running pursuit of the prehistoric commenced with two expeditions featuring Scottish explorer Bill Gibbons. The first, Operation Congo, was conducted during spring 1986, and was in reality a joint Anglo-Congolese venture. Its British members were Gibbons, biologist Mark Rothermel, Territorial Army sergeant Joe Della-Porta acting as medic, and teacher Jonathan Walls as French interpreter. The Congolese contingent consisted of Agnagna and his assistant D'jose D'jonnie. Sadly, the expedition's progress was hampered from its onset by bureaucratic problems—as a result, three months were lost following the four Britons' arrival in Brazzaville on 30 December 1985 before Operation Congo reached Epéna, the starting-point of its quest for mokele-mbembes.

This began four days later, when the team travelled southwards down the Likouala-aux-Herbes in dugout canoes to the village of Dzeke, and thence to a second, much more remote village called Mbouekou. Following their return to Dzeke, the Britons were taken by a retired elephant hunter called Emmanuel Moungoumela to a wooded area near the river where he claimed to have found a mokele-mbembe trail in 1982. Its footprints, a series of round depressions each with a 12-in diameter, had been well preserved in the earth, and had also been spied by the second Mackal expedition. These were accompanied by flattening of the tall river grass (to a height of 4.5 ft) through which they passed, as if thrust aside by a heavy tail. The trail led into the river—but Moungoumela had been unable to discover any further tracks leading out of it again. Thus he considered that elephants, the only scientifically-recognised species in the area with comparably-sized tracks, had not produced these.

The team's next destination was the village of Boha, situated just to the north of Dzeke. Its inhabitants claim ownership of Lake Tele—the principal focus of Operation Congo's attention—and after obtaining the services of some native porters here, the expedition progressed through swampy terrain to the lake itself. During the three-day journey, their Boha guides permitted Rothermel and Della-Porta to spend a full day on the shores of three small sacred lakes never before visited by Westerners but where mokele-mbembes allegedly live. Although a wide variety of fauna was spied here, no evidence for the presence of mokele-mbembes was obtained. The team was informed by the Boha that it is these small bodies of water which are the homes of the mokele-mbembes, not Lake Tele itself—the creatures supposedly use this merely as a source of food and as a place of transit, through which they pass when travelling from one area of the forest to another.

The team spent five days at Lake Tele, during which time they took the opportunity to examine three lobes of water stretching into the forest from the lake and which are reputedly utilised as routes into the lake by mokele-mbembes moving from their browsing areas in the swamps. The Boha guides were very fearful of these lobes, but no evidence for the presence of mokele-mbembes here was uncovered by the team. After spending five days at the lake, followed by a two-day trek back to Boha, the team spent a further three days at Boha—where they were shown a water-hole said to have been inhabited by a mokele-mbembe in 1984—before returning northward along the Likouala-aux-Herbes to Epéna, and ultimately to Impfondo. Here the Britons recuperated from the exhausting journey from Boha, thanks to the hospitality of Pastor Eugene Thomas, before setting forth on the expedition's final stretch, to Brazzaville, where it formally terminated on 5 May 1986—a total of 55 days.

Despite increasing conflict between the British and Congolese contingents, Operation Congo was a worthy successor to the previous mokele-mbembe quests, and provided medical assistance to the local Congolese. It also deposited a collection of biological specimens with London's Natural History Museum, including a specimen of what was claimed to be a hitherto-unknown subspecies of the agile mangabey monkey *Cercocebus galeritus*.

Operation Congo 2 took place in late 1992, and included Gibbons and optometrist Liz Addy among its members. It was primarily a

The mokele-mbembe and a mangabey (© William M. Rebsamen)

reconnaissance trip to prepare for what may be a more extensive foray at some future date, but during its journey up the Bai River from Kinami it succeeded in reaching two hitherto-unexplored lakes, Tibeke and Fouloukou (a known mokele-mbembe haunt, according to the locals), situated to the west of Lake Tele. It also delivered around 270 lb of medical supplies to the mission dispensary at Impfondo, together with clothing, vehicle spare parts, and educational aids for other missions and churches.

Moreover, although Operation Congo 2 did not provide the team's members with any first-hand mokele-mbembe observations of their own, it must surely have increased very substantially the likelihood of future sightings by the native Congolese. For during the team's visit to four villages, Addy handed out 200 free pairs of spectacles!

In the first Operation Congo, the team encountered at Epéna an American journalist called Rory Nugent, making his way by plane, boat, and foot on a solo journey to Lake Tele to seek out the mokele-mbembe. After reaching Tele, Nugent spied a black periscope-like object at a distance of about 1,000 yards. It was just beyond his camera's range, but he took several photos in case he could nonetheless capture something identifiable on film. As his book *Drums Along the Congo* (1993) reveals, however, the results were far too blurred.

During the 1980s, two Japanese expeditions visited the Likouala swamps in search of the mokele-mbembe. The first took place in September 1987, and included Hideyuki Takano and Tohru Mukai from the Waseda University Expedition Club, with Agnagna and D'jose D'jonnie as Congolese advisors. The team visited Epéna, Boha, Dzeke, Minganga, Mokengui, and the Mondongouma River.

The second Japanese expedition, an 11-strong team led by Takano but also including Agnagna and Tokuharu Takabayashi, arrived at Brazzaville on 10 March 1988, leaving 12 days later for the northern Likouala region, and arriving via Epéna at Lake Tele on 30 March. Here the team spent the next 35 days, before returning to Brazzaville. Despite daylight surface surveillance with telephoto-equipped cameras, all-night surface surveillance with a Starlight scope, and sonar scans from a rubber boat, no first-hand evidence for the mokele-mbembe was obtained by the team. During this period, however, two Boha villagers claimed to have spied a large black unidentified object floating at the lake's centre, and a great deal of anecdotal evidence was collected from other villagers.

In 1989, British traveller-writer Redmond O'Hanlon led a four-month, two-man expedition to Lake Tele, and concluded that reports of mokele-mbembes were based merely upon sightings of elephants crossing rivers with their trunks curved up above the water surface.

In February 1993, a ten-man British expedition organised by children's TV presenter Tommy Boyd and led by explorer Barry Marshall conducted a five-week exploration of river caves supposedly frequented by Cameroon's version of the mokele-mbembe. Dinosaur

Painting of Bill Gibbons, watched closely by
his long-necked quarry! (© William M. Rebsamen)

pictures that the team showed to the natives elicited no response, nor did the name 'mokele-mbembe' (why didn't the team try 'amali' or 'jago-nini'?).

The Dino2000 expedition to Lake Tele in 2000 featured British cryptozooological explorer Adam Davies and Swedish cryptozoologist Jan-Ove Sundberg among others, during which plenty of video footage was obtained, but no mokele-mbembe sightings. In November 2000, Gibbons and American businessman David Woetzel journeyed to Cameroon, where they penetrated the hitherto-unexplored forest and swamps along the Boumba and Loponji Rivers, and gathered numerous accounts from the Baka pygmies concerning their mokele-mbembe version, which is known to them as the li'lela-bembe. Interestingly, the Baka claim that this cryptid has armoured skin like a crocodile (hence scaly), and that it exhibits sexual dimorphism, inasmuch as the neck of the male is shorter than that of the female but bears dermal spikes, which the female's neck lacks.

In 2001, Gibbons returned to Cameroon, as part of the joint CryptoSafari-BCSCC (British Columbia Scientific Cryptozoology Club) expedition featuring fellow crypto-researchers Scott T. Norman, John Kirk, and Robert Mullin, and added further anecdotal details to their neodinosaur dossier, but no sightings of their own. Three further mokele-mbembe expeditions by other investigators, two in 2006 (one to a Cameroon river near the Congolese border) and one in 2008 (seeking Zambia's equivalent in Lake Bangweulu), also obtained local testimony but failed to provide first-hand evidence.

A Baka pygmy's dirt drawing of the li'lela-bembe (© Robert Mullin)

In 2009, Gibbons revisited Cameroon with a team that once again included Robert Mullin, as part of the cryptozoology-based television series *Monster Quest*, but no sightings occurred, although they did obtain sonar readings of very long, unidentified serpentine shapes underwater. Mullin also photographed a Baka pygmy's drawing in the dirt of the li'lela-bembe. And in 2011, another cryptozoological television series, *Beast Hunter*, sent an expedition to the Congo Basin, but no findings were made.

In mid-January 2016, French field cryptozoologist Michel Ballot visited Cameroon, accompanied by fellow explorer Serge Martin, seeking its mokele-mbembe counterpart. They searched for it in the Nki Falls area and the Lobéké National Park, and spent a fortnight there after having first met up on site in extreme southern Cameroon with the local team of helpers and guides that they were using, but no first-hand sightings were made.

When Could a Dinosaur not be a Dinosaur?

Understandably, many scientists are decidedly uncomfortable with the theory that the Congo's mokele-mbembe and its counterparts elsewhere in Africa are living dinosaurs, insisting instead that there must surely be other, less radical explanations for sightings of these creatures—but what other explanations are on offer?

Whatever their specific identity may be, they are clearly either mammals or reptiles. The only known mammals deserving of even the slightest consideration as identities for the mokele-mbembe clan are the common hippopotamus, African manatee, and African elephant (nowadays split into two closely-related species).

The hippo, however, can be readily discounted—it is found neither in the Likouala nor in several other localities allegedly frequented by these putative dinosaurs, and in any case it bears no resemblance to the long-necked mystery beasts under investigation here.

Although the African manatee *Trichechus senegalensis* does occur in parts of the Congo and elsewhere within these latter cryptids' apparent distribution range, with a total length not exceeding 12 ft it is far too small. Moreover, sporting an almost neck-less, seal-like body equipped with a front pair of flippers but lacking hind limbs, and possessing a spatulate horizontal tail fin rather than a true tail, this herbivorous inspiration for certain mermaid legends looks even less like the creatures documented here than does the hippopotamus.

As for the African elephant—although its size is closer to the mokele-mbembe's than is that of any other African mammal officially recognised by science, and its distribution range is compatible, its morphology is very different. And the prospect that mokele-mbembe reports could be based upon sightings of elephants wading through swamps with only their trunks and the tops of their heads projecting above the water surface is one that surely beggars belief. Such encounters are hardly likely to be very frequent, and in my opinion it is both insulting and credulous to assume that experienced native hunters (including eyewitness Emmanuel Moungoumela, someone who has spent his life hunting elephants in the Likouala swamplands) are incapable of recognising a partially-submerged pachyderm.

The only other mammalian contender worthy of contemplation is some wholly novel species that happens to look very like a sauropod dinosaur—but as no mammal of this nature is known from either the present or the past, the chances of such a species lurking in the Likouala yet still undiscovered by science must assuredly be far less likely than the unverified survival of a sauropod, whose onetime reality is at least well-documented in the fossil record. Worth noting here is that cryptozoologist Dr Bernard Heuvelmans hypothesised that sightings in temperate zones of long-necked Nessie-type lake monsters, as well as of morphologically similar sea serpents, involve an undiscovered species of giant, long-necked seal (see Chapter 3). Could such a beast explain the mokele-mbembe sightings too? There is no palaeontological evidence, however, that such a seal has ever evolved, and the reports of three-clawed footprints left behind by mokele-mbembes, plus their possession of sturdy legs (rather than flippers) and a long discrete tail, all argue against such an identity for them.

The outlook for finding a satisfactory explanation for the mokele-mbembe beasts is not greatly increased by turning to the range of reptilian candidates either. Of the three principal contenders, two—the Nile crocodile *Crocodylus niloticus* and the African rock (water) python *Python sebae*, both native to the Likouala and other relevant localities—can be swiftly eliminated, as their notable size cannot compensate for their lack of similarity to the beasts described by eyewitnesses as the mokele-mbembe.

Somewhat more promising is the third candidate—the African soft-shell turtle *Trionyx triunguis*. According to the textbooks, this long-necked freshwater species rarely exceeds 3 ft in length, but Agnagna claims that specimens twice as long, known locally as the ndendecki, occur in the Likouala, and some natives have reported seeing specimens with shells 12-15 ft in diameter! Although this latter claim is undoubtedly an exaggeration, it does suggest that individuals considerably larger than textbook examples may exist here. Moreover, when spied at a distance with only their long neck and their shell's uppermost portion above the surface of the water, such creatures could certainly resemble a partially-submerged sauropod.

Reconstruction of the mokele-mbembe's possible appearance if assumed to be a giant long-necked varanid (© Drawn by David Miller, under the direction of Prof. Roy P. Mackal)

Equally worthy of note is that this species possesses a noticeably elongated snout, yielding a snorkel-like tube for breathing purposes—and as mentioned earlier, Agnagna's sketch of the supposed mokele-mbembe spied by him in 1983 depicts it with a distinct, elongated snout. Could it have been a *Trionyx*?

Also, as pointed out by auditory sciences researcher Dan Gettinger (*ISC Newsletter*, winter 1984), Agnagna's description of the creature moving its head around as if to locate the source of the noise when he was approaching it at a distance of around 700 ft is one that favours a *Trionyx* identity rather than a sauropod. Fossil evidence reveals that many archaic reptiles possessed a relatively large stapes (middle ear bone), which would have thus severely restricted their auditory sensitivity—limiting it to the detection of fairly intense vibrations of the substrate in the immediate vicinity (e.g. their own footsteps on solid ground).

Perhaps, therefore, *certain* reports of mokele-mbembes have indeed been based upon poorly-spied specimens of this very large soft-shelled turtle—but not in those cases when the creature was seen at close range, and when raised mostly if not completely out of the water.

Being well aware of just how radical the concept of a living dinosaur existing undiscovered by science would seem to the world at large, in his book's analysis of the mokele-mbembe's zoological identity Mackal also examined the more conservative possibility that it is simply an extremely large varanid or monitor lizard.

As no varanid with a neck even remotely as long in proportion to its body as the mokele-mbembe's is known to science from either the present day or prehistoric times, however, it would certainly be a major new species. Yet as Mackal's morphological comparisons in his book lucidly revealed, it would still not correspond as closely with eyewitness descriptions as does a sauropod dinosaur. Even its footprints would be incorrect, because varanids have five well-developed claws on their feet, whereas the mokele-mbembe's tracks show only three—a diagnostic characteristic, conversely, of certain sauropods. Also, virtually all varanids are carnivorous, whereas the mokele-mbembe is reportedly exclusively herbivorous. Moreover, as monitors are common in the Congo and therefore very familiar creatures here, if the mokele-mbembe were simply an extra-large version it seems strange that the Congolese natives have never described it as such.

Holding a model of the mokele-mbembe's most popular, sauropod reconstruction (© Dr Karl Shuker)

When faced with such overwhelming objections to all of the more orthodox identities on offer, yet having readily to hand, in the shape of the average-sized sauropod, a creature offering such a close correspondence with the morphology of the mokele-mbembe, is it any wonder that the concept of a living dinosaur as the answer to this reclusive mystery beast of the marshlands is given such credence within the cryptozoological community?

Secrets of the Mokele-Mbembe's Success

Cryptozoological sceptics routinely disparage the mystery of the mokele-mbembe on the grounds that if this beast genuinely exists, its great size and spectacular appearance would have guaranteed its formal scientific discovery long ago. If only it were as simple as that!

As I revealed in one of my earlier books, *The Lost Ark: New and Rediscovered Animals of the 20th Century* (1993), and also in its updated editions *The New Zoo* (2002) and *The Encyclopaedia of New and Rediscovered Animals* (2012), many of the world's most spectacular animals had remained undiscovered and undescribed by science until the past hundred years or so.

Quite aside from the coelacanth, these include: the okapi (DR Congo's short-necked giraffe; discovered in 1901), giant forest hog (world's largest wild pig; 1904), Komodo dragon (world's largest lizard; 1912), megamouth shark (a huge bizarre species with a gigantic mouth; 1976), clusters of giant worms in 8-ft-tall tubes with long scarlet tentacles (thriving on the seafloor near the Galapagos Islands; 1977); the Vu Quang ox (the largest new mammal to be discovered for over 50 years; 1992); the giant muntjac *Muntiacus vuquangensis* (world's largest barking deer; 1994); the giant peccary *Pecari maximus* (world's biggest peccary, 2004); and the Kabomani tapir *Tapirus kabomani* (one of South America's largest living mammals; 2013). Hence the mokele-mbembe's ability to remain hidden from scientific detection even in modern times is not unprecedented.

The okapi (top) and Komodo dragon, both discovered after 1900 ((cc) Eric Kilby / (cc-sa) Poppet Maulding)

Equally significant is the creature's surroundings. Few people without first-hand experience can even begin to comprehend the exceptionally inhospitable nature of the Likouala swamps—one of the most forbidding, environmentally hostile areas to be found anywhere in the world today. It may be home to the mokele-mbembe, but it is certainly not designed for human comfort—boasting such dubious delights as stinking mud that continually envelops travellers unfortunate enough to be wading through it with liberal clouds of foul methane and hydrogen sulphide; sweltering temperatures exceeding 89°F (32°C) and accompanied by omnipresent humidity sometimes approaching 100 per cent; terrain infested by highly venomous snakes (including mambas) and swarms of malaria-transmitting mosquitoes; plus a constant threat of other debilitating ailments and diseases.

During Operation Congo, for instance, and despite being young and very fit at the expedition's onset, all four British participants subsequently experienced bouts of severe illness—in particular, Gibbons lost 24 lb in weight through chronic diarrhoea and various skin diseases, Rothermel contracted malaria, and Della-Porta suffered bouts of dysentery.

Clearly, therefore, any attempt to travel through the swamps is fraught with difficulties, thereby rendering active pursuit of their more elusive denizens (especially ones capable of submerging themselves beneath the water) a near impossibility. Even the native Congolese and pygmies avoid the worst areas, so the likelihood of a Western expedition successfully exploring them is remote in the extreme—the inevitable consequence of which is that any animals inhabiting such regions will obviously remain undiscovered by science. A similar scenario awaits anyone seeking marsh-dwelling cryptids elsewhere in Africa too.

Another contributing factor to the mokele-mbembe's continued scientific anonymity is the powerfully potent fusion of fear and veneration incited among the native people by this formidably large creature. During Mackal's expeditions, some natives sought to convince them that magic was required in order to catch sight of a mokele-mbembe, and the team discovered on several occasions that although there were people in the vicinity who had indeed observed this species, they were unwilling to tell them about it because they feared that they would be attacked by their village elders. Apparently, there is a longstanding tradition among the native Congolese that if anyone sees a mokele-mbembe and afterwards speaks about his sighting, death will inevitably follow. Equally, to kill a mokele-mbembe is to incur disaster, not only for its killer but also for his entire tribe—thus reducing to an absolute minimum any chance for scientists visiting native villages to find skulls or other remains of mokele-mbembes preserved there (in the way that those of many other, less-feared/venerated animals are).

Any activity that could be perceived as a sacrilegious intrusion upon the mokele-mbembe's existence is also taboo. Thus, when Rory Nugent spied what he and his native guides believed to be a mokele-mbembe on Lake Tele, he was convincingly persuaded against rowing his boat nearer to it in the hope of obtaining better photos—one of his guides pointed a loaded shotgun at his head, and another held a spear poised in readiness to impale him with it, if he should violate this "supreme jungle deity" by approaching it.

When faced with such problems as these, it is little wonder that cryptozoologists often experience difficulty in eliciting data regarding mystery beasts from the local populace!

There could be another, much more tangible reason for the natives' great fear of killing one of these beasts—the flesh of mokele-mbembes may be poisonous or may harbour virulent microbes. During his first expedition, Mackal learnt that in 1959 a tribe of pygmies living near Lake Tele had actually killed a mokele-mbembe. They had speared it to death, after it had attempted to force its way through a barricade erected to prevent these animals from entering the lake via one of its lobes and disturbing the pygmies' fishing activities. If only its remains had been salvaged! Instead, it was summarily hacked apart, and its flesh eaten—after which all of the partakers became violently ill and died! A warning to all who might be considering the novelty of dining upon a dinosaur?

Yet if the mokele-mbembe really is a living dinosaur, how did its ancestors survive the mass extinction that had annihilated all other

Artistic reconstruction of an incident featuring the killing by pygmies of a mokele-mbembe that was forcing its way through their barricade (© William M. Rebsamen)

dinosaurian lineages by the close of the Cretaceous? In fact, as this chapter of my book will reveal, cryptozoologists are far from certain that this is the *only* type of dinosaur that may be alive today—possibilities from several other localities worldwide are also on file. Relative to the mokele-mbembe, however, there are two major factors that would doubtlessly have assisted the survival of sauropods in those African localities where the long-necked mystery beasts documented here have been reported. Those factors are centres of endemism, and certain physiological adaptations exhibited by sauropods.

Louis Jacobs's book *Quest For the African Dinosaurs* (1993) recounts his discovery of a major fossil site in the hills of Malawi in Central Africa during the 1980s, and the many highly significant fossil dinosaur finds subsequently made there. However, it also includes a chapter surveying the history of and searches for supposed living sauropods like the mokele-mbembe in this same geographical area. Jacobs states that although dinosaurs *could* be living somewhere in Africa, "the odds are so great against it that the probability of finding a living sauropod has, for all intents and purposes, reached zero". His principal argument against modern-day dinosaur survival is that the swamplands of Central Africa from where so many reports of such cryptids have emanated are unlikely to have been there ever since the end of the Cretaceous, that they have most probably arisen only in much more recent times attendant with changing river-drainage systems in this region, and that like everywhere else in the world the continent of Africa must certainly have experienced considerable climatic changes during the 66 million years since then. Consequently, in Jacobs's view, sauropods couldn't have simply lingered in environmentally stable, swamp-maintaining conditions from the end of the Cretaceous onwards, as has often been claimed by cryptozoologists. (Similar comments also appear in Wikipedia's 'Living Dinosaur' page.)

Although Jacobs's view is reasonable and informed, based as it is upon known precedents recorded elsewhere due to the effects of plate tectonics, wobbles and tilts of the Earth upon its axis through time, and much else besides, it still incorporates considerable speculation and generalisations nonetheless—because there is much about the specifics of post-Mesozoic evolution and transformation of Africa's environmental conditions that remains unverified. But even if Jacobs is correct in all that he has speculated, his view is still not incompatible with modern-day sauropod survival. He notes that, based upon fossil evidence, sauropods in the Cretaceous were terrestrial forms; if so, an absence of swamplands in Central Africa at the end of the Cretaceous would therefore *not* be problematic for them. Moreover, if the cryptids reported in modern times from the Congolese Likouala swamps and other similar habitats elsewhere in Central Africa are indeed sauropods, this merely indicates that once such swamplands *had* finally come into being, sauropods existing in their vicinity gradually adapted to this new type of habitat until it became their principal domain. Animals have evolved to survive in new habitats throughout the entire history of life on Earth; hence there is nothing unreasonable in supposing that sauropods might have become secondarily aquatic if confronted with formerly drier habitats becoming wetter ones. But that is not all.

As Jonathan Kingdon expounded in his book *Island Africa* (1990), mainland Africa is far from being the zoogeographically homogeneous continent customarily portrayed in nature books. Even the standard partitioning of its wildlife into four principal ecological zones —desert, savannah, forest, and upland—obscures the extraordinary abundance of enclaves that Kingdon calls 'centres of endemism', each containing localised (often rare) species found nowhere else. Although these centres of endemism are land-locked, their wildlife's discrete, individual nature is such that, for all practical purposes, they can be looked upon as islands—effective autonomies isolated from environmental and evolutionary pressures elsewhere.

One such area that falls within this category is the crescent-shaped region stretching from the Likouala marshes to the southern portion of Chad. Flanked by the middle Congo River's west bank tributaries, it also takes in the Central African Republic, plus the bordering areas of Gabon and Cameroon. Here, amid these largely inaccessible jungle swamplands, is a world mostly unchanged for over 60 million years. With a post-Cretaceous history notably lacking earthquakes, mountain formation or subsidence, glaciations, or climatic perturbations, its wildlife has enjoyed a degree of environmental stability experienced by few other zoogeographical communities in Africa—a stability reflected by the number of ancient species that have lingered on into the present day here while dying out long ago elsewhere.

Consequently, echoing the belief of Mackal and Powell, if there is anywhere in Africa where dinosaurs might persist, little-changed morphologically or behaviourally, and undisturbed

Close encounter with a mokele-mbembe (© Richard Svensson)

by outside forces, here indeed is where it would be. And as this is also where reports of putative dinosaurs emanate from native peoples with no knowledge of evolutionary and geophysical stability, palaeontology, and so on, it is little wonder that cryptozoologists pay attention to such reports.

In addition, the possibility that it may be of sauropod descent will have greatly increased the mokele-mbembe's chances of survival here. In *Dynamics of Dinosaurs and Other Extinct Giants* (1989), Leeds University zoologist Prof. R. McNeill Alexander, a renowned expert in biophysics and biomechanics, investigated the conflicting views as to whether dinosaurs were endotherms ('warm-blooded', thus exhibiting a mammalian-type metabolism) or ectotherms ('cold-blooded', thus exhibiting a reptilian-type metabolism like crocodiles, snakes, lizards, and turtles). He did this by conducting two parallel series of calculations determining how much dinosaurs of various body sizes would have been heated by their own metabolism. One series assumed a mammalian-type metabolism for them, the other a reptilian-type.

The results indicated that unless they lost a great deal of water via evaporation, terrestrial dinosaurs of sauropod size with mammalian-type metabolism and inhabiting hot climates would be liable to overheat. If, conversely, they had typical reptilian-type metabolism, overheating would not be a problem. As the mokele-mbembe spends part of its time on land, feeding and moving from one lake to another, based upon Alexander's predictions it would therefore function most effectively with reptilian metabolism—as confirmed by the known presence in the Likouala of other sizeable reptiles with this type of metabolism, including Nile crocodiles, rock pythons, and some very large monitor lizards. If these can survive here, why not a 'cold-blooded' sauropod dinosaur?

Moreover, Alexander calculated that despite their body size, such dinosaurs would use energy at roughly the same rate as mammals with only one-fifth of their body mass—i.e. for every 5-ton elephant sustained by a given amount of vegetation, that same amount could also have sustained at least one 25-ton *Diplodocus*! Thus there would be ample food for mokele-mbembes in this vast area, and this comparison also explains how beasts as large as these could thrive on such a seemingly unsubstantial diet as *Landolphia* gourds.

Having said all of that, in more recent times the belief that the giant sauropods of prehistory like *Brachiosaurus* were ectothermic has been overturned by the discovery made from further fossil finds that large air sacs connected to the lung system were present in the neck and trunk of such creatures, invading the vertebrae and ribs, and thereby greatly reducing the overall density, as documented by American palaeontologist Dr Mathew J. Wedel in 2003. This in turn means that earlier estimates of available cooling surfaces in these dinosaurs were *under*estimated. When this was coupled with realisations that earlier estimates of body mass for these dinosaurs had been grossly *over*estimated, it is now accepted that in reality giant sauropods could be endothermic after all, because their smaller estimated bulk and greater estimated ability to cool down meant that they would have been able to obtain sufficient food to sustain an endothermic metabolism but without overheating.

As the mokele-mbembe is not described as being anywhere near as large as these prehistoric giant sauropods, however, it may be that even if this cryptid is a sauropod it could still survive perfectly adequately as an ectotherm anyway. Also, if it possesses the large air sacs of its presumed sauropod ancestors, these would help it stay submerged without its lungs collapsing from the surrounding water pressure, as long as the water was not too deep and its period of submergence not too lengthy.

Also worth noting here is that East-Central Africa has quite a rich fossil history of dinosaurs, most famously represented by Tanzania's Tendaguru Formation, dating from the late Jurassic, and including such forms as the sauropods *Giraffatitan* and *Tendaguria*, the theropod *Elaphrosaurus*, the iguanodontian *Dysalotosaurus*, and the stegosaur *Kentrosaurus*. Fossils of an early Jurassic sauropod called *Vulcanodon* have also been found in Zimbabwe, southern Africa.

All in all, the concept of a contemporary species of dinosaur inhabiting the northern Congo and its environs is actually a long way removed from the improbable, unrealistic vision so swiftly discounted by cryptozoological sceptics. However, even if the mokele-mbembe is indeed exposed one day as a surviving sauropod, this extraordinary saga will still be far from over. While questioning the Congolese natives during his expeditions, Mackal learnt to his amazement that the Likouala may also conceal two other, very different species of living dinosaur—known respectively as the emela-ntouka and the mbielu-mbielu-mbielu.

THE EMELA-NTOUKA—'KILLER OF ELEPHANTS'

Another cryptic creature from the Congo sorely in need of a satisfactory explanation, the formidable emela-ntouka was alluded to by the

Representation of the emela-ntouka, based upon eyewitness descriptions (© Drawn by David Miller, under the direction of Prof. Roy P. Mackal)

The emela-ntouka reconstructed as a ceratopsian-like cryptid (© Richard Svensson)

natives almost as frequently as the mokele-mbembe when they were questioned by Mackal. Semi-aquatic once again, it is said to be at least the size of an elephant, but can be instantly differentiated from the latter by virtue of its long, heavy tail and in particular by the very large, sharp horn borne upon its snout.

Although an apparent vegetarian, its temperament is far from placid, according to local testimony—for when submerged in a lake, this remarkable creature will not hesitate to attack any elephants or buffaloes that unwisely choose to enter or travel through it, utilising its deadly horn to stab and tear apart its hapless victims. Indeed, this belligerent behaviour is responsible for its native name—'emela-ntouka' translates as 'killer of elephants'.

It must also be the creature noted in a *Mammalia* article (December 1954) by Lucien Blancou, who had been a senior game inspector in the Likouala area of the Congo (then known as French Equatorial Africa). According to Blancou:

> Around Epéna, Impfondo, and Dongou, the presence of a beast which sometimes disembowels elephants is also known. . . A specimen was supposed to have been killed 20 years ago at Dongou.

In the past, zoologists seeking a possible identity for this animal have suggested that it might be some form of aquatic rhinoceros. However, its heavy tail opposes this possibility—as does the alleged appearance of its horn, which is said to resemble the ivory tusk of an elephant rather than the compressed mass of hair that constitutes a rhino's horn.

From his own investigations, Mackal cautiously suggested that the Congo's aquatic elephant killer might conceivably be a present-day species of ceratopsian dinosaur instead, comparable to the single-horned *Monoclonius*, which also possessed a long, heavy tail. (NB—certain nomenclatural controversies regarding this latter genus's validity and possible synonymity with *Centrosaurus* are noted, but lie outside this book's scope.)

If so, it would therefore also be related to prehistory's more famous ceratopsians, such as the three-horned *Triceratops* and the multi-frill-horned *Styracosaurus*. Having said that, however, it should be noted that no fossil

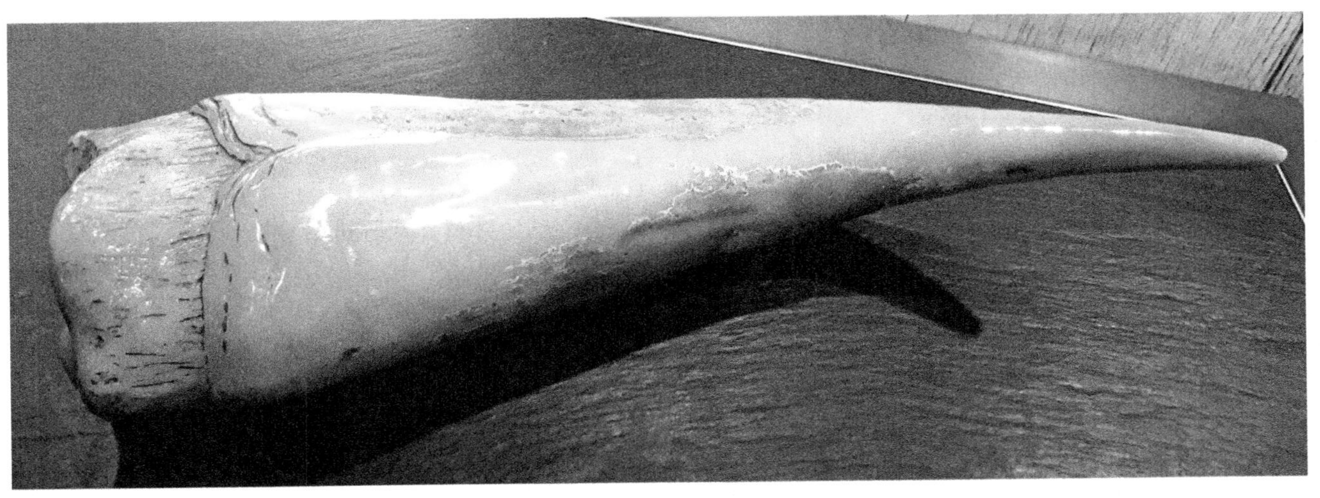

Reconstruction of a ceratopsian horn as it probably looked in life (© Dr Karl Shuker)

Representation of the chipekwe sharing its East African swampland home with hippopotamuses (© Dr Karl Shuker)

remains of ceratopsians have ever been documented from anywhere in Africa.

The nasal horns of ceratopsians were true horns, formed from the fusion and upgrowth of the nasal bones, and would thus resemble ivory at all but the closest of observation ranges—as confirmed via a reconstruction of the likely appearance in life of a ceratopsian horn as viewed by me in 2014 while visiting London's Natural History Museum (shown previous page).

The Chipekwe, an Eastern Equivalent

Just as the mokele-mbembe appears to have counterparts far beyond the Likouala swamplands, so too does the emela-ntouka. The most notable of these is the chipekwe, which has long been reported from East Africa's Lakes Bangweulu, Mweru, and Tanganyika, and the Kafue swamps. According to a *Daily Mail* letter of 26 December 1919 by C.G. James, who had lived in Africa for 18 years, this aggressive inhabitant of the deepest swamps is of enormous size, possesses a single huge ivory tusk with which it kills hippopotamuses (but does not eat them), and leaves behind footprints resembling those of its victims.

The nature of the chipekwe's 'ivory tusk' was clarified by John G. Millais, who noted in his book *Far Away Up the Nile* (1924) that according to big game hunter Denis Lyall, these three lakes harboured at least until recently a mysterious animal that he referred to as a water rhinoceros and which he described as

". . . some large pachyderm, somewhat similar in habits to the hippopotamus, but possessing a horn on the head".

Furthermore, as someone who had spent a great many years on the shores of one of this animal's principal localities, J.E. Hughes was able to substantiate these testimonies in his own book *Eighteen Years on Lake Bangweulu* (1933), reporting that the Wa-Ushi natives had once killed just such a beast as this in the Luapula River, linking Mweru with Bangweulu:

> It is described as having a smooth dark body, without bristles, and armed with a single smooth white horn fixed like the horn of a rhinoceros, but composed of smooth white ivory, very highly polished. It is a pity they did not keep it, as I would have given them anything they liked for it.

So too would any curious-minded zoologist—bearing in mind that whereas no known rhino, either from the present or from the past, has a horn composed of or even resembling ivory, this description recalls the horns of ceratopsian dinosaurs. Who knows—perhaps there is a chipekwe horn or two preserved in a local chief's dwelling somewhere in East or Central Africa? Meanwhile, French cryptozoologist Michel Ballot has lately found what may be the next best thing, as now revealed.

Artistic Representations of the Emela-Ntouka?

Since 2004, Ballot has conducted a number of excursions into Cameroon, seeking evidence for the existence of unknown aquatic beasts, including this country's version of the mokele-mbembe. In 2005, during his second expedition, travelling through a region of northeastern Cameroon bordering the Central African Republic, he visited a village where he saw (and purchased) a large, truly remarkable wooden carving. It depicted in great detail a strange beast with four sturdy legs, a long heavy tail, and a head whose nose bore a long horn.

Although this carving doesn't match any known animal alive today, as can be seen from the photograph of it reproduced here by kind permission of Michel Ballot, it is a faithful representation of the emela-ntouka:

The emela-ntouka wooden carving purchased in 2005 by Michel Ballot in Cameroon (© Michel Ballot)

Interestingly, this carving portrays the emela-ntouka with a pair of small frilly ears, almost like miniature elephant's ears, a feature not previously reported for this cryptid but which, if genuinely possessed by it, indicates a mammalian rather than a reptilian identity. Moreover, on his blog (http://mokelembembe expeditions.blogspot.co.uk/2009_10_21_archive.html), Ballot revealed that in 2005 he had actually found not one but two such carvings, in totally separate locations and created by separate artists, but identical in appearance. Judging from photos of the second one that I have seen on his website, however, it is less detailed and less well-executed than the first, more famous carving.

On 5 July 2014, in a world-exclusive report on my own *ShukerNature* blog, I revealed a third, independently-obtained but undeniably corroborative piece of iconographical evidence for the veracity of this specific morphological identikit relative to the emela-ntouka.

That remarkable piece of evidence came from the Dzanga Sangha Protected Areas (APDS), situated in the Central African Republic (CAR). To quote their website (www.dzanga-sangha.org), the Dzanga Sangha Protected Areas:

> . . . are internationally known for their beautiful rainforests, host to a remarkable diversity of wildlife, comprising western lowland gorillas, forest elephants, bongo antelopes, forest buffalos and a multitude of bird species. Furthermore, a rich local culture, comprising the Sangha Sangha fishermen as well as hunting and gathering BaAka, are present in the area. Apart from conservation and local development efforts, Dzanga Sangha operates as an eco-tourism and research centre. A variety of well developed tourism activities and a beautiful hotel complex, overlooking the Sangha River, are at your disposal.
>
> Sharing borders with Cameroun and Congo, the Dzanga Sangha Protected Areas are part of the Trinational Sangha (TNS) complex, currently in the process to become a UNESCO World Heritage Site. Roundtrips to the other National Parks (Lobéké in Cameroun and Nouabalé Ndoki in Congo) can be organized with ease.

On 8 August 2012, I receieved a fascinating email and attached set of four photographs from Anette Stichnoth of Hannover, Germany. The photographs, taken by a friend of hers from the APDS and named Cem Kok, were of four drawings on display at the first, now-dismantled Dzanga Sangha Exhibition (Kok was currently working on the second one), held in the APDS—with whom Stichnoth was working in the capacity of utilising its exhibition's artworks as designs on souvenirs, such as t-shirts and mugs, that could be sold to APDS visitors in order to raise money for the area's continuing conservation. In relation to this, she asked me if I had any idea what the entities were that the drawings depicted, because she wanted to produce an information card for each one but no-one whom she had previously contacted had been able to identify them.

Jean Claude Thibault's emela-ntouka drawing (© Jean Claude Thibault, courtesy of Anette Stichnoth)

The artist responsible for these four drawings was a Frenchman named Jean Claude Thibault, who had produced them during the early 1990s or late 1980s. He had lived in the Central African Republic (a former French colony) for many years, but could not be contacted personally concerning his drawings because he had died in Bangui, the CAR's capital, in or around 2012.

After examining Thibault's drawings, it was clear to me that they were nothing if not interesting from a cryptozoological standpoint. Three of them depicted humanoid or semi-humanoid beings, but the fourth one was very different, and is reproduced here by kind permission of Anette Stichnoth.

As can be seen, it portrays one elephant fleeing in the background, plus a second one that has been stabbed in its underside by a horned beast bearing an uncanny resemblance to Michel Ballot's Cameroon-procured emela-ntouka carvings!

In order to compare those works of art directly, I have horizontally flipped one of Ballot's photographs of the more detailed of the two Cameroon emela-ntouka carvings, thereby enabling its orientation to match that of the creature in Thibault's drawing.

And as can be readily discerned, there is no doubt whatsoever that these two artworks are indeed depicting the same species of animal—whatever that may be! Every major morphological feature—from those strange little frilly ears, and sharp vertical snout-horn, to the very long, broad, sweeping tail with its distinctive dorsal ridge or keel, and the relatively short, sturdy, bent legs with well-differentiated digits—is portrayed in an identical manner by the carving and the drawing.

Fascinated by this extraordinary correspondence between works of art created in two entirely different countries by artists very unlikely to have ever seen each other's work, I swiftly emailed Stichnoth for additional information. Unfortunately, however, she didn't have anything further of substance to offer me at that time, but promised to contact me again once she had obtained more details—and almost exactly a year later, she did.

On 18 August 2013, she emailed me a series of descriptions for the four drawings that

Michel Ballot's emela-ntouka carving (top) and Jean Claude Thibault's emela-ntouka drawing (bottom) (© Michel Ballot / Jean Claude Thibault, courtesy of Anette Stichnoth)

she had lately been given by a CAR local with knowledge of his country's legendary creatures and entities. The one depicting the emela-ntouka was labelled as 'Mokele-Mbembe', and the creature was said to inhabit the deepest stretches of the Ndoki River. Referring to it as a mokele-mbembe may on first sight seem strange. However, as I have discussed earlier in this chapter of the present book, in central Africa (i.e. not just the CAR but also the two Congos, Cameroon, etc) the long-necked lake-dwelling cryptid (mokele-mbembe) and the horned lake-dwelling cryptid (emela-ntouka)

Michel Ballot with his emela-ntouka carving from Cameroon (© Michel Ballot)

are often conflated in local reports, with features of one sometimes being wrongly attributed to the other, so this is not as surprising as it might otherwise seem.

Due to how astonishingly similar Thibault's CAR emela-ntouka drawing was to Ballot's Cameroon emela-ntouka carvings in terms of the creature's morphology, I had initially wondered whether the former had been directly copied from the latter. Perhaps Thibault had seen online images of the Cameroon carvings? However, when I learnt that Thibault had produced the drawing a decade or more before Ballot had even encountered the carvings, let alone photographed them and brought them to public attention by uploading the photos onto the internet, it is surely evident that they are of independent origin, neither one influenced by the other. After all, it is highly unlikely that the Cameroon villagers responsible for the carvings had ever seen Thibault's drawing in the Central African Republic, which had only been exhibited publicly within the CAR, never outside it.

Equally relevant is that although the carved animals and the drawn one possess identical morphologies, the specific poses respectively adopted by them are not the same at all. Both carved animals are standing stationary, in a neutral behavioural pose, with the head held at a normal height, the hind limbs close together, and the tail (curving to the right in the original, non-flipped version of Ballot's photograph of the first carving, to the left in the second carving) held laterally for much of its length and very close to the body. The drawn animal, conversely, is in ferocious attack mode, with its head lowered as it belligerently drives its horn into the body of its hapless pachyderm victim, its hind legs splayed well apart in order to brace itself as it performs the powerful thrusting stab with its horn, and its tail (curving to the left) held out further away from its body in a much wider arc and portrayed throughout its length from above. In short, the two carved animals and the drawn animal indisputably portray the same species, but the two carved animals' shared pose is very different from the drawn animal's—thereby further indicating that the drawing was not copied from or influenced by the carvings or viceversa.

On 9 July 2014, I received an extremely interesting email from French cryptozoologist Michel Raynal, which provided me not only with additional information concerning the emela-ntouka drawing by Thibault documented here, but also with sight and details of a second one prepared by him, as now revealed.

The drawing already documented here was #12 of twelve Thibault drawings of mythological/legendary entities from the CAR that in 1996 were featured in a special calendar produced by the Worldwide Fund For Nature (WWF) in the CAR to raise funds for the Doly-Lodge Project in Bayanga, the largest village within the APDS. So this was not a worldwide calendar, but one specifically for the APDS in the CAR. The emela-ntouka drawing subsequently reappeared in 2000 within an article concerning this cryptid published in issue #3 of the Bangui-based anthropological periodical *Zo*, and written by Alfred J-P. Ndanga, from Bangui University's Department of Anthropology and Palaeontology. Intriguingly, throughout the article Ndanga referred to the creature not as the emela-ntouka but rather as the mokele-mbembe—another instance of conflating these two great water cryptids of central Africa.

As for the second Thibault drawing of the emela-ntouka, brought to my attention by Raynal: this appeared in a Congolese newspaper (date and title currently unknown to me), and is now reproduced here (note that once again this cryptid is mislabelled as a mokele-mbembe). As can be seen, the depicted creature possesses the very same, specific features as displayed by the two carvings and Thibault's first emela-ntouka drawing:

Jean Claude Thibault's second emela-ntouka drawing (© Jean Claude Thibault)

Consequently, as the two carvings and the two drawings all correspond with one another so closely morphologically speaking, we must conclude that the image of the emela-ntouka provided by them is an accurate one—that it really does possess small frilly ears, a vertical snout-horn, bent legs, and a very broad, lengthy, powerful tail. The ears alone are enough to demonstrate that it is evidently mammalian in nature (as they are clearly bona fide ears and not, for instance, a pair of abbreviated ceratopsian bony frills), thereby eliminating a surviving ceratopsian dinosaur from serious consideration in the future. Yet it does not correspond with any known species of mammal. Indeed, it is not even possible to allocate this mysterious creature with ease to any existing taxonomic order of mammals. Having said that, however, the image of it yielded by the carvings and the drawing does remind me a little of *Arsinotherium zitteli.*

Dating from the late Eocene, this was a massive elephantine horn-bearing species of fossil African ungulate belonging to the extinct order Embrithopoda. Named after eminent palaeontologist Dr Karl Alfred von Zittel and Queen Arsinoe I, the wife of the Egyptian pharaoh Ptolemy II (its remains were found in present-day Egypt's Faiyum Oasis), it was believed to have been aquatic in lifestyle, spending much of its time wading and swimming in rainforest swamps rather than walking on land, as it was unable to straighten its legs (thus recalling the emphatically bent legs of the emela-ntouka depicted in the drawing and carvings). Moreover, it is particularly famous for its pair of truly enormous, laterally-sited snout horns,

Vintage illustration of *Arsinotherium* by Heinrich Harder

composed of bone but hollow and covered in keratinised skin.

According to their known fossil record, embrithopods officially died out almost 30 million years ago (*Arsinotherium* was their last known genus), but could the emela-ntouka possibly be a single-horned, scientifically-undiscovered, modern-day representative? The hefty, lengthy tail, however, poses a notable problem—why would an embrithopod evolve such a decidedly non-ungulate feature?

Going full circle, a more conservative option would be a species of semi-aquatic water rhinoceros, as first suggested by some cryptozoologists many years ago. Yet in spite of its single horn, the heavy-tailed creature portrayed in artwork bears scant resemblance to any of the diverse array of rhino forms on record from either the present day or prehistoric times. Currently, therefore, the emela-ntouka is something of an anomaly, but at least it is one that now appears to have a well-defined albeit extremely perplexing morphology.

Finally, the name 'irizima', as noted earlier (p. 40), is applied by natives living in the vicinity of DR Congo's Lake Edward not only to long-necked water beasts of the mokele-mbembe variety but also to what may be an emela-ntouka equivalent—described by Captain Hichens as ". . . like a gigantic hippopotamus with the horns of a rhinoceros on its head". His use of "horns", i.e. plural, is intriguing—just an error of reporting, or an indication that the Lake Edward creature is visibly distinct from the emela-ntouka and chipekwe unicorns?

Disentangling the Emela-Ntouka from the Mokele-Mbembe

The confusing use of the name 'irizima' by Lake Edward natives as noted in the previous paragraph is just one of several instances in which there has been terminological ambiguity between the horned, superficially rhino-like emela-ntouka (plus its non-Congolese equivalents) and the (usually) non-horned, superficially sauropod-like mokele-mbembe (plus its non-Congolese equivalents), as I have pointed out earlier. So it is clearly time to deal with this etymological issue once and for all, and by far the best, most authoritative elucidation of it that I have seen was posted online by Bill Gibbons to the cryptolist@yahoo groups.com cryptozoological chat group on 26 May 2004. It reads as follows:

> Now regarding the possible identity of Mokele-mbembes (MM's) being some sort of giant rhino, let me explain why this is simply not possible. In the last twenty years of exploring the Congo-Cameroon area (and receiving feedback from missionaries in neighboring Gabon and the Central African Republic), the rhino must be dismissed. There are at least 250 ethnic groups living throughout Congo Brazzaville [=Republic of Congo], Congo Kinshasa [=DR Congo], Cameroon and Gabon. Although 700 local languages and dialects are spoken, the population and thus the dialects spoken tends to thin out in the more remote locations, especially close to the border areas. The main languages spoken in the border areas are Kikongo, Tshiluba, Swahili, Lingala and Baka. The Bantu tribes include the Mongo, Luba, Kongo, BaKongo Bangala, Baluba, Botamingi and the Fang. The Pygmy groups include the Aka, Baka, Bangombe, Bagyeli, Bambenjele, Batwa, Binga, Bongo, Cwa, Gieli, Mbuti, and the Twa (no, not the airline).
>
> Having questioned at least 12 different ethnic and cultural groups on MM, they are all consistent with their description of an animal that is at least the size of a forest elephant, but possessing a long neck, small snake-like head and a long flexible tail. The animal is a herbivore, is seen only very occasionally out of the water (but sticks close to the river when browsing), and the young are rarely seen (I only have two reports of young MMs observed with their parents, with one report observing a calf and a presumed female on a sandbank). These reports come from people who have very little, if any contact with one another (they cannot communicate with one another anyway, unless they speak the official languages of Gabon, the two Congos, and Cameroon, which is French. However, the more remote tribal groups that we have been dealing with (especially the pygmies) do not speak French at all. Their languages or dialects, religious beliefs, cultural practises and social structure can vary widely from group to group even in a 50 mile radius. Yet their description of MM remains consistent even though the animal is known by different names such as Jago-Nini, N'Yamala, La'Kelabembe, Mokele-mbuemba, Mokele-mbembe, etc. Yet the description, colour, behaviour, preferred food supply, and habitat remain consistent.
>
> In both Congos, two slight variations of Lingala are spoken, but the general word for animal is "Yama." Only one group of pygmies that I know of, the Aka, use the name "Mokele-mbembe" as a generic term to describe any large strange animal. The rhino or horned creature the Emela-Ntouka was once referred to by the Aka as "mokele-

mbembe" when interviewed by a Frenchman who had lived in the Congo for 20 years. Other groups that I have interviewed have made a clear distinction between Mokele-mbembe proper and the elephant-sized horned animal. Roy Mackal recalled his visit to the president of Epena, who tried to tell him that "Mokele-mbembe" meant "Rainbow." In the Congo, the tribespeople will often present misleading information simply because they do not like white outsiders asking questions about animals that they fear greatly. Bantu or pygmy hunters think nothing of bringing down an enraged bull elephant with a ten foot spear thrust into its abdomen, yet they become quite fearful when pressed to discuss Mokele-mbembe.

Missionaries such as Gene Thomas, who has spent 42 years in the Congo and has travelled its most remote waterways and jungle paths, has collected an impressive wealth of knowledge on Mokele-mbembe and other strange animals. Yet it has taken him a lifetime to gain the trust of a few people.

Cameroon is quite different however. The tribespeople there have not such reservations about sharing their knowledge on Mokele-mbembe with us. They fear them greatly, but do not attribute magical powers to the animals like the tribespeople of the Congo. Hence information gathering was much easier.

Hopefully, this should finally delineate the two great Congolese aquatic cryptids once and for all—but they are not the only sizeable cryptids on record from this vast portion of Africa.

THE MBIELU-MBIELU-MBIELU AND CO—SURVIVING STEGOSAURS?

A third dinosaur-lookalike of the Likouala is the mbielu-mbielu-mbielu. According to one of its alleged eyewitnesses—a young woman called Odette Gesonget, from the village of Bounila—this triple-named cryptid is a semi-aquatic beast "with planks growing out of its back". In a bid to identify it, Mackal showed Gesonget several illustrated books depicting animals from the present and also from the distant past—and the picture that she unhesitatingly selected was of the prehistoric plate-bearing dinosaur *Stegosaurus*. Comparable descriptions were offered independently by natives encountered elsewhere during Mackal's Congolese travels too.

Yet although at least one stegosaur genus, *Kentrosaurus*, is indeed represented by fossil remains found in tropical Africa, there is no suggestion from fossil evidence that stegosaurs exhibited any amphibious inclination. But who can say whether an evolved stegosaur might have become secondarily aquatic?

It is possible that this beast, whatever its taxonomic identity may be, is related to (or is even one and the same as) another mystery animal from the Likouala swamplands—the nguma-monene, reported from the Mataba tributary of the Ubangi River. According to native descriptions, it resembles a colossal snake (at least 130 ft long!), but bears a serrated dorsal ridge along most of its body's length consisting of numerous triangular protrusions, possesses four short legs, and can walk upon land, with a low-slung body and forked tongue. Could this be a snake-like dinosaur, or perhaps a primitive reptile descended from the ancestral forms that gave rise to lizards and snakes?

A remarkable modern-day sighting of the nguma-monene by a Westerner was recorded

Representation of the mbielu-mbielu-mbielu, based upon eyewitness descriptions (© Drawn by David Miller, under the direction of Prof. Roy P. Mackal)

Representation of the nguma-monene, based upon eyewitness descriptions (© Drawn by David Miller, under the direction of Prof. Roy P. Mackal)

Artistic representation of the Bombays' encounter with a muhuru (© William M. Rebsamen)

by Gibbons in his 2010 book *Mokele-Mbembe: Mystery Beast of the Congo Basin* (which also documents the various other Congolese and Cameroon cryptids sought by him during his expeditions). The eyewitness in question was Pastor Joseph Ellis, who informed Gibbons personally of his encounter. It took place one clear sunny day in November 1971, when Ellis was journeying north in his 30-ft-long motorised canoe on the Motaba River. Suddenly he spied a huge, elongated, snake-like creature with a series of ridges like the edge of a saw running the length of its back. It was moving just across the river from the right bank, only around 100 ft away. Turning off his canoe's engine, an astonished Ellis watched as the creature, which he could see was at least as long as his canoe, swam slowly across the river to its left bank, onto which it clambered, crawling through the thick grass and then disappearing into the jungle.

Ellis never saw its head, but just the section of its body that was visible to him was at least 30 ft long, and was greyish-brown in colour. Although he had personally seen many of this region's large species of animal currently documented by science (including elephants, monitors, pythons, turtles, and crocodiles), he had never spied anything like this creature before, and, having no interest in cryptozoology, had not previously known of its existence here. What could it have been?

Mackal favoured a single, very large, and radically new species of monitor lizard as the most satisfactory explanation for both the nguma-moneme and the mbielu-mbielu-mbielu. Nonetheless, the latter's stegosaurian parallels are evidently difficult to dismiss absolutely—because as he confessed in his book: "For me, mbielu-mbielu-mbielu remains an enigma".

Nor are they the only ones. One day in summer 1961, missionary Cal Bombay and his wife were driving through the Rift Valley on their way to Nairobi, Kenya, when they had to pull up sharply in order to avoid hitting an extremely large reptile apparently sunning itself in the middle of the road. The Bombays estimated that the creature, which was dark grey in colour, measured around 10 ft long, had a snake-like head, four stubby legs, and, most remarkable of all, bore a series of diamond-shaped serrations running down its entire mid-dorsal line, from the back of its head to the tip of its tail.

After about 20 minutes, thus giving its observers plenty of time to peruse it closely, this languorous mystery beast finally stood up, and sauntered lazily into the bush. When some Kenyan natives heard of their encounter, they referred to the creature as a muhuru, but nothing like it is known to Western zoologists.

THE NGOUBOU—ONE SAVANNAH TYPE, ONE AQUATIC TYPE

When Gibbons conducted his initial expedition to Cameroon in November 2000, he discovered not only that this country apparently possessed its own version of the mokele-mbembe but also that it allegedly harboured two extraordinary horned mystery beasts akin to but even more spectacular than the emela-ntouka, and both referred to as the ngoubou (aka n'goubou) by the local Batu (BaBinga) people. They claimed

Rendition of the savannah ngoubou, with human silhouette for scale (© Connor Lachmanec)

An attack by the river-dwelling ngoubou, envisaged as a modern-day *Arsinotherium* (© William M. Rebsamen)

that one type of ngoubou is terrestrial and inhabits savannah areas to the west of the Boumba River, whereas the other is aquatic and exists in rivers there. They are also known in the Sanga region near the Central African Republic, and the two types are readily distinguished from one another morphologically.

Looking through pictures of living and fossil animals, some of the local Batu pointed to pictures of the Cretaceous's famous three-horned ceratopsian dinosaur *Triceratops*, and stated that this resembled the savannah ngoubou. They stated that it was the size of an ox, sported a large frill around its neck (which differs slightly in the female), a beaked mouth, and bore several horns, but in a different manner from those of *Triceratops*.

This baffled the team, until one of the locals drew an image of what the savannah ngoubou looked like. This revealed that in addition to a single nasal horn it bore a series of six horns around the edge of its frill—a feature which, as team member John Kirk later realised, made it look irresistibly similar to a late Cretaceous ceratopsian called *Styracosaurus*.

Moreover, according to Gibbons's coverage of this incident in his book, when the team showed them a selection of images of modern-day and prehistoric animals the locals selected an *Arsinotherium* picture as being most similar in appearance to the river-dwelling ngoubou version. In particular, they claimed that it bore not just one nasal horn but two, and that they were positioned laterally (as in *Arsinotherium*), not one behind the other (as in two-horned species of rhinoceros). This description is also reminiscent of an earlier-mentioned aquatic horned cryptid from DR Congo, the irizima. Might the aquatic ngoubou and the irizima thus be one and the same mystery beast? Curiouser and curiouser, as a cryptozoological Alice might well have said!

Ironically, during the very same month that Gibbons led his team to Cameroon, a chance to determine once and for all the taxonomic identity of at least one type of ngoubou was apparently only very narrowly missed. Two days before the arrival of missionary Pierre Sima to the Cameroon village of Ndelele, an aquatic ngoubou had been shot and killed by hunters from the village, with its carcase swiftly butchered and its prestigious horns sawn off and sold to a French employer of a logging company. If only he could be traced! Not only that, Sima allegedly ate a meal with the villagers that included meat from the slaughtered ngoubou, which he said tasted like pork. I wonder what a DNA analysis conducted upon a sample of his meal would have yielded?!

'KASAI REX' AND SANDERSON'S SAUROPOD

Two alleged living dinosaurs of the decidedly dubious kind are the following pair of African examples. On 16 February 1932, while travelling through the Kasai Valley in what is now DR Congo, Swedish plantation overseer J.C. Johanson and his native helper encountered what he subsequently described in a letter published by the *Rhodesia Herald* newspaper as "something incredible—a monster, about 16 yards [48 ft] in length, with a lizard's head and tail". This 'monster' then rapidly vanished. Later that same day, however, while crossing a big swamp on their way back home, Johanson and his helper re-encountered the giant reptile:

> There in the swamp, the huge lizard appeared once more, tearing lumps from a dead rhino. It was covered in ooze. I was only about 25 yards away. . .I thought of my camera. I could plainly hear the crunching of rhino bones in the lizard's mouth. Just as I clicked, it jumped into deep water."

J.C. Johanson's photograph of the supposed reptilian monster encountered by him, standing upon a presumed animal carcase

The photograph snapped by Johanson was published alongside his letter in the *Rhodesia Herald*, but is unquestionably a crude fake, as noted by Dr Bernard Heuvelmans in his own coverage of this incident in *On the Track of Unknown Animals* (1958). It consists of a Komodo dragon *Varanus komodoensis* that has been cut out of some other photo and pasted onto one of a jungle setting, with its feet resting almost on tip-toes upon an unidentifiable object presumed to be an animal carcase.

Having said that, it should be pointed out that nowhere in his letter did Johanson actually refer to the supposed giant reptile as a

Dramatic artistic rendition of 'Kasai rex' attacking a rhinoceros (© Hodari Nundu)

dinosaur, merely as a huge lizard. The reason why this cryptid, clearly a hoax in the light of Johanson's fake photo of it, is deemed to have been a living dinosaur—hence its subsequently-coined soubriquet of 'Kasai rex'—is a comment accompanying Johanson's letter that was written by one of the newspaper's own staff. This comment read: "Johanson stumbled upon a unique specimen of a dinosaur family that must have lived millenniums [sic] ago". Many, many millennia ago!

Tellingly, Johanson made no mention of any local name for this spectacular creature, despite being accompanied by a native helper, whose people would surely be only too well aware of such an enormous, ferocious beast (if such a beast truly existed), and would therefore have certainly given its species its own local name. Consequently, this is another clue that the entire story is just that, a work of fiction.

Online, various websites claim that Johanson later stated that the monster was reddish in colour with blackish stripes, sported a long snout, numerous teeth, thick legs that reminded him of a lion's, "built for speed"; and that he had decided that it was a *Tyrannosaurus*. However, no original sources for these additional comments claimed for Johanson are given, and they were not mentioned by Heuvelmans in his coverage.

In 2007, an additional alleged Johanson photograph of 'Kasai rex' perched upon its prey also materialised online. This was far clearer and much more professional than the original one, showing a readily identifiable theropod dinosaur on an equally recognisable prone rhinoceros. Needless to say, however, this was no more genuine than the first photo, but I will not call it a hoax. This is because it was produced by a known photo-manipulator called Finbar (using an image of an *Allosaurus* model superimposed upon a rhino photo), who (as he always does) openly admitted that it was indeed a photo-manipulated composite picture produced by him, and hence did not prepare it with any intention to deceive.

During the 1950s and 1960s, Ivan T. Sanderson, an American zoologist, animal collector, and author of several popular books documenting his journeys to exotic locations worldwide in search of animals, was also famous as America's premier cryptozoologist (and has already been mentioned here regarding the sauropod-like Ashanti gold weight). Sadly, however, he was equally (in)famous for his regrettable tendency to exaggerate quite profoundly when documenting his travels, and to entertain unsubstantiated and sometimes quite bizarre lines of zoological speculation.

He also claimed to have sighted a truly unparalleled array of cryptids. These included: a huge Cameroon bat with twice the wingspan of any known bat species; a giant three-toed penguin on a Florida beach; a large pink salamander in a pond on his New Jersey farm (see later in this present chapter); skins of an undescribed Mexican wild cat with a huge ruff of fur round its neck but which happened to be destroyed by water before anyone else was able to see or examine them; an invisible catfish and a bioluminescent lizard on Trinidad; a herd of mysterious miniature wild horses in Haiti; and, albeit only briefly, what was alleged by him to be a living dinosaur in tropical Africa.

Any one of these would be a spectacular sight to behold for any cryptozoologist, but for all of them to be seen by the same single person? That takes some believing! Indeed, commenting upon Sanderson in *A Living Dinosaur?* (1987), Mackal succinctly wrote:

> Sanderson started his career as a brilliant zoologist (in my opinion), but, over the years, became more and more sensational and exploitative in his writings . . . [though he] rarely deluded himself, regardless of how unscrupulous he was in his later writings.

Sanderson died in 1973, aged 62, and it is often claimed (although some dispute this claim) that the cause of death was a brain tumour. This allegation has in turn led to speculation that, if true, perhaps the tumour's pernicious effects may explain at least in part Sanderson's tendency towards making sensationalised claims, especially during the later part of his life.

Be that as it may (or may not), Sanderson's alleged sighting of a living sauropod supposedly occurred in 1932, during the Percy Sladen Expedition to Cameroon. Here is Mackal's account of Sanderson's reputed close encounter of the cryptid kind:

> When Sanderson, in the company of the American naturalist Gerald Russell, arrived at Mamfe Pool on the Mainyu River, they came to a place that had many caves in the cliff-like river banks, many partially or almost completely filled with river water. They reported a loud, noisy disturbance, as of fighting beasts, coming from one of the caves.

> Both saw the back of something larger than a hippopotamus break the surface, immediately submerging after only a momentary display. Farther upstream near the confluence of the Cross River, they saw "vast hippo-like tracks: although there were no hippopotami [sic] in the area". Sanderson was told there were no hippos because this creature, the "embulu-em'bembe" (Sanderson's spelling), drove them away. Sanderson stated that the tracks they found on the Mainyu River could not possibly have been made by a crocodile. He believed that what was observed rising in the cave was the head of the creature. In 1971, Sanderson, in a letter to James Powell, wrote that its head "was bigger than a whole hippo, and the tracks were sauropod".
>
> This reference by Sanderson to a large head belonging to a sauropod always bothered me a great deal. Sanderson knew very well that all sauropods had long necks with small heads. However, as we discovered during my expeditions, the term Mokele-mbembe is used for the long-necked, small-headed sauropod animal, on the one hand, and in a generic sense for other unidentified animals.

To my mind, whatever the animal was, it is far more likely to have been its back that briefly surfaced rather than its head, because any animal whose head alone was bigger than an entire hippopotamus would have been a veritable behemoth—unless, once again, it owed its gargantuan size to the imagination of Sanderson? It would be interesting to read Russell's testimony concerning this sighting, but as yet I have been unable to locate anything written by him about it.

'Kasai rex' and Sanderson's very suspect sauropod notwithstanding, there is little doubt that, whatever their precise taxonomic identities may be, there are some major zoological discoveries still waiting to be unfurled by a real-life Professor Challenger in the depths of the Dark Continent, but especially amid the Lost World of the Likouala—a vast, primeval wilderness constituting an ideal haven for secluded survivors from a past age.

DRAGONS OF BABYLON, AND DINOSAURS IN THE BIBLE

In the summer of 1983 I visited the Vorderasiatisches Museum, part of Berlin's Staatliche Museums, to gaze upon one of the most spectacular monuments from ancient history—the magnificent Ishtar Gate of Babylon. Many other visitors were also peering intently at this marvellous edifice, sumptuously decorated with life-like depictions of various animals, but to me it had an extra significance—for out of all of the people there, it is possible that I alone realised that we may be looking at the portrait of a living dinosaur!

In Search of the Sirrush or Mushussu

During his reign (605-562 BC), King Nebuchadnezzar II of Babylonia in Mesopotamia oversaw the creation of his empire's capital, the holy city of Babylon, dedicated to Babylonia's supreme deity—the god Marduk. Babylon was encircled by huge walls, wide enough for chariots to be driven along their summits, and pierced by eight huge gates. The most magnificent was the Ishtar Gate (named after the Babylonian goddess of love), through which visitors passed in order to enter the city.

Befitting such an important edifice, the Ishtar Gate was a spectacular sight, consisting of a colossal semicircular arch, flanked by enormous walls and leading to a breathtaking

The sirrush or mushussu as depicted upon the magnificent Ishtar Gate of Babylon (© Vorderasiatisches Museum, Berlin)

Processional Way, along which visitors walked to reach the city's religious centre. The gate, its walls, and the processional walls were covered by a brilliant panoply of highly-glazed enamelled bricks, yielding a backdrop of vivid blue for numerous horizontal rows of eye-catching and very realistic bas-reliefs of animals. On the gate and its flanking walls, six rows of fierce grey bulls alternated with seven rows of grim golden dragons, and along the processional walls were two rows of haughty marching lions, but the most important member of this trio of mighty beasts was the dragon—for this was the sacred beast of Marduk.

Following the eventual fall of Babylonia, its walls and gates became buried underfoot, and their glory was hidden for many centuries—until 3 June 1887 when German archaeologist Prof. Robert J. Koldewey, during a visit to the site of Babylon, found a fragment of an ancient blue-glazed brick that stimulated his curiosity and led to a full-scale excavation beginning in 1899. Three years later, the animal-adorned Ishtar Gate rose up from the dust of the past like a cobalt phoenix, revealing its bulls, its lions—and its exalted but enigmatic dragons.

Most commonly referred to as the sirrush or mushussu (two different transliterations of an Akkadian word loosely translated as 'splendour serpent'), the Ishtar dragon was a source of great bewilderment to Koldewey. For whereas archaeologists were well aware that the depicted appearance of all other seemingly fabulous, mythical animals in Babylonian tradition had changed drastically over the centuries, depictions of the sirrush (as also present on seals and paintings predating the Ishtar Gate by at least a millennium) had remained the same—just like those of real animals, such as the lion and the bull. Did this mean, therefore, that the sirrush was itself a real-life species? But if so, what could it be?

Certainly, it did not—and still does not—resemble any animal known to be alive today. After all, what modern-day species has a slender scaly body, with a small head bearing a pointed horn (or a pair—the Ishtar sirrush is only depicted in profile) on its forehead and ringlet-like flaps of skin further back, a long slender neck, a pair of forelimbs with lion-like claws, a pair of hindlimbs with eagle-like claws, and a long tail? Some authors have suggested a giant monitor lizard, but the sirrush's horn(s), ringlets, and extremely long neck contradict this identity.

Boldly, Koldewey announced in 1913 that, in his opinion, the creatures to which the sirrush most closely corresponded were the dinosaurs. Moreover, he deemed it possible that in order to explain the unchanging nature of sirrush depictions, and also various mentions of dragon-like beasts in the Bible, some such creature must have been kept within one or more of Babylon's temples by the priests of Marduk. By 1918, he had refined his belief, identifying the ornithopod dinosaur *Iguanodon* as the closest fossil relative of the sirrush. If, however, the sirrush was truly a creature of historic, rather than prehistoric, times, where had it originally come from—there is no evidence that giant reptiles occurred in Mesopotamia during the Babylonian age—and how could it have evaded scientific detection?

This mystery greatly intrigued cryptozoological chronicler Willy Ley, who suggested that the only locality from which such a creature could have been originally transported to Babylon during the reign of Nebuchadnezzar II yet remain wholly unknown to modern-day science was Central Africa, and in his book *Exotic Zoology* (1959) he recalled some of the accounts given earlier by me here concerning swamp-dwelling dinosaurian beasts reported from this portion of the Dark Continent. In addition, when Schomburgk returned to Europe from Central Africa during the early 20th Century with tales of living dinosaurs, he also brought back a glazed brick that he had found there—a brick just like those in the Ishtar Gate. Is this where the far-travelling ancient Babylonians had obtained them, along with stories—and perhaps even the successful capture from time to time—of real-life dragons?

Others have since expanded upon Ley's views, and the prospect that the sirrush was a living dinosaur has attracted cryptozoological interest. However, opinion as to the precise type of dinosaur has moved away from *Iguanodon* toward a sauropod.

On first sight, the sirrush is hardly reminiscent of such a creature. If, however, the Chaldean artists responsible for the Ishtar Gate bas-reliefs and other sirrush portrayals had not actually seen a living sauropod with their own eyes but were relying solely upon descriptions of one, then it is not too difficult to accept the resulting sirrush as nothing more dramatic than a distorted depiction of a sauropod, no doubt embellished by its creators' imagination.

Bel and the Behemoth

One biblical reference that inspired Koldewey's belief in the onetime existence of a real-life sirrush maintained by Babylon's temple priests is an episode documented in the *Apocrypha* concerning Daniel, who, after discounting an earlier supposed deity as nothing more than a

16th-Century engraving of Daniel and the venerated temple dragon

brass idol, was shown a mysterious creature housed within the temple of the Babylonian god Bel, and which was venerated by the fearful populace:

> And in that same place there was a great dragon or serpent, which they of Babylon worshipped. And the king said unto Daniel, Wilt thou also say that this is of brass? Lo, he liveth, he eateth and drinketh; thou canst not say that he is no living god: therefore worship him. Then said Daniel. I will worship the Lord my God: for he is a living God. But give me leave, O king, and I shall slay this dragon without sword or staff. The king said, I give thee leave.

True to his word, Daniel accomplished his vow—via the unusual if effective expedient of choking the creature to death by forcing lumps of bitumen, hair, and fat down its throat—a brave act if genuinely faced by a conflagrating dragon, but one that will not endear him to cryptozoologists if it is ever shown that his adversary was nothing more rapacious than a morose herbivorous mokele-mbembe a long way away from its humid tropical home amid the Congolese swamplands.

Another biblical monster that has never been satisfactorily identified with any known animal alive today is the behemoth, which is described in the *Book of Job* (40:15-24) as follows:

> Behold now Behemoth, which I made with thee; he eateth grass as an ox.
>
> Lo, now, his strength is in his loins, and his force is in the navel of his belly.
>
> He moveth his tail like a cedar: the sinews of his stones are wrapped together.
>
> His bones are as strong pieces of brass; his bones are like bars of iron.
>
> He is the chief of the ways of God; he that made him can make his sword to approach unto him.
>
> Surely the mountains bring him forth food, where all the beasts of the field play.
>
> He lieth under the shady trees, in the covert of the reed, and fens.
>
> The shady trees cover him with their shadow; the willows of the brook compass him about.
>
> Behold, he drinketh up a river, and hasteth not: he trusteth that he can draw up Jordan into his mouth.
>
> He taketh it with his eyes: his nose pierceth through snares.

Over the centuries, four principal identities have been touted by theological and zoological scholars—the ox, Nile crocodile, elephant, and hippopotamus. Least popular is the ox—apart from its herbivorous nature, it has no similarity to the behemoth. Only the New English Bible supports the crocodile's candidature—certainly, the concept of a vegetarian crocodile is an implausible one, to say the least. The elephant's supporters are also few—only Prof. George Caspard Kirschmayer in *Un-Natural History of Myths of Ancient Science* (1691) and Dr Sylvia Sikes in *The Natural History of the African Elephant* (1953) have seriously sought to link the two great beasts with one another.

The most popular and (until quite recently) most favourable pairing of the behemoth has been with the hippo—whose cavernous mouth, prodigious drinking capacity, mighty build, sturdy skeleton, swamp-dwelling lifestyle, herbivorous diet, and status as the largest animal native to the Bible lands compare satisfactorily with the behemoth—but not conclusively. How, for example, can the hippopotamus "moveth his tail like a cedar"? This description implies a very long, powerful tail—not the puny, inconspicuous appendage sported by the hippo.

And then came a late entry in the identity stakes—a living sauropod. As Mackal persuasively pointed out in *A Living Dinosaur?*, not only the description of the behemoth's tail but also all of the features hitherto likened to the hippopotamus are equally applicable to one of these giant vegetarian dinosaurs. Moreover, the great size attributed to the behemoth, while far exceeding that of the hippo, would be consistent with a sauropod of mokele-mbembe proportions. Compare the Bible's description of the behemoth (given three paragraphs earlier here) with Mackal's defence of his sauropod identity for it (given as follows), and judge for yourself.

> The behemoth's tail is compared to a cedar, which suggests a sauropod. This identification is reinforced by other factors. Not only the behemoth's physical nature, but also its habits and food

A vaguely hippo-like behemoth (top) with the leviathan (bottom) in this famous 19th-Century illustration by William Blake

preferences are compatible with the sauropod's. Both live in swampy areas with trees, reeds and fens (a jungle swamp). Indeed, the identification of the biblical behemoth as a sauropod dinosaur provides excellent correspondence between the descriptive features in the biblical text and the characteristics of these dinosaurs as inferred from the fossil record.

Equally interesting concerning this sauropod link is that the *Book of Job* was written some time between 2,000 and 700 BC, thereby considerably predating the Ishtar Gate's depictions of the sirrush. Clearly, then, the gate did not inspire the behemoth account—instead, this was based upon something very large and visually impressive that was known in the Middle East long before the birth of Nebuchadnezzar II (in c.634 BC).

Finally: while preparing this updated, much-expanded new edition of *In Search of Prehistoric Survivors*, I was idly re-reading the behemoth's description in the *Book of Job* when my attention was suddenly seized by the very last feature listed for this cryptid: "his nose pierceth through snares". How could the

Job and a sauropod Behemoth (© William M. Rebsamen)

nose of any of the five identities presently on offer—ox, Nile crocodile, elephant, hippopotamus, and sauropod—achieve this action? Conversely, a rhinoceros's nose, for instance, bearing a long sharp horn upon it, could certainly do so, but a rhino's tail is far too slight for it to be moved "like a cedar".

Then, suddenly, an image came to mind—an image of a wooden carving depicting an extremely large beast allegedly herbivorous in diet but exceedingly formidable in nature and inhabiting jungle swamps, which sports a horn-bearing nose *and* a long powerful sweeping tail. The carving is owned by Michel Ballot, and the creature that it depicts is, of course, the emela-ntouka! Is it conceivable, therefore, that the behemoth was actually the emela-ntouka, which may even have been native to the Middle East in ancient times (*Arsinotherium*, for instance, a proposed identity for this mystery beast, is known from fossils in Egypt) but became extinct here prior to the rise of Babylonia? Might it even be that the sirrush is based upon distant memories and lore concerning this horn-bearing, long-tailed amphibious cryptid? Obviously, this is all entirely speculative, but the emela-ntouka certainly compares more closely with the behemoth than does any other contender examined in the past.

Initially, the Bible must seem the *last* place where zoologists would expect to find details of living dinosaurs—but if living dinosaurs (or travellers' reports of them) were known in this region of the world at this particular time in humanity's history, the Bible is unquestionably the *first* place where zoologists should look for evidence of their existence.

Alternatively, it may be that the Babylonian dragons were nothing more than monitor lizards, albeit very sizeable ones. Pertinent to this prospect is a report from 1173 AD by Castillian traveller Benjamin of Tudela. According to him, he had spied many Mesopotamian 'dragons', which he also claimed were so infesting the ruins of Nebuchadnezzar II's palace as to render them inaccessible. These again were undoubtedly large monitors, which are indeed native to this region of the Middle East. One notable example is the desert monitor *Varanus griseus*, which can attain a total length of almost 6 ft; and the widely-distributed Bengal monitor *V. bengalensis*, occurring in south-eastern Iraq, can reach 5.75 ft. Both of these would certainly appear somewhat dragonesque in form to any non-zoological observer.

Most recently, I have learned that Koldewey himself allegedly saw a giant lizard near Baghdad, and thereafter wondered if sightings of this unidentified species (a very large monitor?)—as opposed to his earlier notions regarding modern-day dinosaurs—had inspired the Ishtar Gate's dragon depictions, and if a captured specimen had similarly inspired the Apocrypha story of a worshipped captive dragon. This information is apparently contained in an article published by the Dutch archaeological magazine *Jaarbericht Ex Oriente Lux* during the 1930s, but I have no specific reference to it.

However, this does call to mind a little-known giant mystery lizard from the Middle East that was briefly reported by explorer Sir Wilfred Thesiger in his book *The Marsh Arabs* (1964), and more recently by me in my own book *A Manifestation of Monsters* (2015).

Also known as the Madan, the Marsh Arabs inhabited the marshlands of the Tigris and Euphrates rivers in the south and east of Iraq, and along the Iranian border—formerly a vast area of wetland covering more than 5.8 square miles. According to Thesiger, who had lived among them intermittently for eight years during the 1950s prior to the Iraqi revolution of 1958, the canoe-borne Madan claimed that the marshes at the mouth of the Tigris in Iraq was home to a monstrous lizard, which they termed the afa. Little else appears to have been documented concerning it, but as various monitor species are already known to exist in this region, the afa might have been either an extra-large version of one of them or possibly even a giant still-undiscovered relative.

Sadly, however, the question of the afa's taxonomic identity may be nothing more than academic now. This is because following the Gulf War in 1991, the Iraqi government initiated a major programme to divert the flow of the Tigris and Euphrates Rivers away from the marshes in retaliation for a failed Shia uprising among the Arabs living there. This not only eliminated the Madan's food sources, forcing them to move elsewhere, but also turned the marshes themselves into a desert. Hence the afa may have been exterminated, especially if it were primarily aquatic, as I am not aware of any post-1991 reports alluding to it.

As for living dinosaurs, there is more to being a putative sauropod survivor than having a long neck and an ancient African provenance, as aptly demonstrated by the Narmer Palette palaver. I documented this enigmatic artefact in detail within my book *Dr Shuker's Casebook* (2008), just a couple of years after personally viewing it in Egypt. Here is what I wrote:

SERPOPARDS, PSEUDO-SAUROPODS, AND THE NARMER PALETTE

Surrounded by spectacular sarcophagi, mummies, and other necrological relics of every conceivable size, age, and nature, the last thing that I expected to encounter during my visit to the Egyptian Museum, Cairo, in January 2006 was an artefact of cryptozoological controversy. However, while walking around Gallery 43 on the museum's ground floor, this is precisely what happened. Suddenly, I found myself in front of a large glass case containing a greyish-green, shield-shaped exhibit, and as I looked in surprise at the pair of bizarre beasts carved upon one side of it I realised that I was looking at the extraordinary Narmer Palette—one of Egypt's oldest, and most enigmatic, historical objects.

Composed of dark schist, measuring 25 in high and 16.5 in wide, and richly adorned on both sides with elaborate, finely-wrought carvings, this remarkable artefact was discovered during 1898 by archaeologist James E. Quibell in the Upper Egyptian city of Nekhen (nowadays Hierakonpolis) while excavating the royal residences of various ancient Egyptian rulers. Despite dating back to c.3,200 BC (the Old Kingdom), the palette has survived intact, and was a votive (gift) offered up by King Narmer to the sun god Amun-Ra. What makes this artefact so significant historically is that it not only bears some of the earliest-known examples of Egyptian hieroglyphics but also commemorates a major event in ancient Egyptian history—the unification of Lower Egypt and Upper Egypt into a single land, with King Narmer as the first ruler of both lands.

On one side of the palette, King Narmer is vividly portrayed as ruler of Upper Egypt smiting his Lower Egypt enemy, and facing his own incarnation as the falcon deity Horus, god of the sky. On the other side, there are various depictions celebrating Narmer's triumph after capturing the crown of Lower Egypt, thereby unifying Upper and Lower Egypt. The largest, most striking image on this side, however, does not feature Narmer at all. Instead, it portrays a pair of inordinately long-necked creatures whose flexible necks entwine around one another, forming a border round a central circular reservoir that some researchers believe may have been used to hold perfume, or to serve as a receptacle within which such cosmetics were manufactured in situ.

Both sides of the Narmer Palette, the serpopards depicted on the side shown in the right-hand photograph

These extraordinary beasts are generally referred to as serpopards (though in at least one reference source they are termed mafedets), and for good reason. For whereas their necks are decidedly serpentine in appearance, their heads are very leopard-like. As for their bodies: I have seen them likened variously to panthers, lions, and even baboons. After having finally witnessed the palette at first-hand (prior to then, I knew of its images only from various internet pictures of varying quality), I agree that their bodies, long limbs, and lengthy tails certainly possess a degree of simian similarity, more than I had previously realised when simply viewing pictures of them. But what were these serpopards meant to be—wholly symbolic, a purely legendary beast, perhaps a very distorted portrayal of some known animal, or something more than any of these options?

The reason why I was already familiar with the Narmer Palette is that in the past it has attracted a degree of cryptozoological speculation that the serpopards may conceivably represent a stylised or alternatively a distorted depiction of some mokele-mbembe-type species of surviving long-necked dinosaur that was alive at least at the time of King Narmer. Much as I would like to admit it to this select crypto-company, after viewing the palette's serpopards up close and personal, however, I was left in no doubt whatsoever that these necking entities were unquestionably mammalian, not even remotely reptilian—in short, pseudo-sauropods.

The serpopard head, complete with ears, is indeed leopard-like, not leonine as some have suggested, but the toes of the feet, posture of the body and limbs, as well as the limbs' relative lengths, and the shape and carriage of the tail all struck me as rather more monkey-like than feline. As for the palette pair's disproportionately long, impossibly flexible necks, it seems likely that they were intertwined not only to symbolise the union of Upper and Lower Egypt (as well as the eastern and western heavens?), but also for practical purposes—to fit neatly around the palette's central reservoir. Interestingly, each of the two depicted serpopards is held on a leash by a handler, who may be a slave, or a tribute, indicating perhaps that the serpopards were a gift to King Narmer, or possibly even domesticated?

Significantly, depicted serpopards are not restricted to the Narmer Palette. Another early Hierakonpolis palette, known as the Oxford or Two Dogs Palette and retained at Oxford University's Ashmolean Museum, also bears a pair of these striking creatures on one side, plus a single one on the other side. The paired serpopards on this artefact have even longer necks than those on the Narmer Palette, but this time they are not entwined—instead, they are held in a painful-looking zig-zag pose above their bodies, one on each side of a central reservoir. On the Two Dogs Palette (named, incidentally, after the two superficially canine—but quite possibly hyaenid—beasts constituting the upper section and outer sides of the palette, though the head of one is missing), the necks of the serpopards are striped, and there are stripes on their foreparts too.

Another serpopard-depicting palette is the Four Dogs Palette held at the Louvre, Paris, and there is also a preserved cylinder seal from Susiana, the high country of the ancient Persian civilisation of Elam, that depicts a series of very long-tailed neck-entwined serpopards. Clearly, therefore, the serpopards of the Narmer Palette were clearly not just an invention of its sculptor, devised merely as a decorative motif for bordering and highlighting the palette's central reservoir and/or as a symbol of King Narmer's unified Egypt.

A more conservative identity than a mokele-mbembe but no less intriguing is that perhaps the serpopards were poor representations of a giraffe (a species that once existed in Egypt), possibly based upon indirect descriptions of what this exceptional creature looked like rather than personal observations. If this were so, however, surely the giraffe's long legs would have been mentioned and described to the sculptor, not just its long neck. Yet although the serpopards' legs are fairly long, they are far shorter than one would expect for a giraffe, whereas their necks are much too long. In any case, it just so happens that there is absolute iconographical proof readily to hand to confirm that the serpopard and giraffe are totally discrete animals.

On the reverse side of the Two Dogs Palette, a wide range of creatures are depicted, including readily-identifiable lions, antelopes, goats, a hartebeest- or gnu-like ungulate—and not only a serpopard but also a clearly-recognisable giraffe, the latter beast complete with long inflexible erect neck, small horns as well as ears on its head, long giraffe-like legs, hoofed feet, and downward-pointing tail. Just above it, offering a perfect opportunity for direct comparison, is a flexible-necked, hornless, leopard-headed, shorter-legged, toe-footed, upward-tailed serpopard—indisputably a wholly different animal.

Equally worthy of note is that alongside portrayals of real animals on the Two Dogs

Palette is not just a serpopard but a winged griffin too, depicted in traditional composite form with leonine body, eagle's head, and feathered wings. This provides immediate proof that ancient Egyptian sculptors carved real and fabled animals together, so the presence elsewhere of serpopards depicted alongside people and real animals cannot be taken as firm evidence that the serpopards themselves must also be real.

Accordingly, the most reasonable solution to the mystery of its identity is that the serpopard is nothing more than another composite (albeit exotic-looking) mythical beast, just like the griffin, as well as certain other Egyptian monsters such as the hippo-bodied crocodile-headed ammut, and the venomous winged snakes that reputedly swarmed across ancient Egypt each year like locusts (indeed, perhaps they were directly inspired by actual locust swarms). After all, not all beasts of legend are creatures of cryptozoology in disguise—from centaurs and minotaurs to yokai and sciapods, the inventive human imagination is more than sufficiently capable of summoning forth from its uncharted depths a veritable menagerie of wholly original monsters surpassing even the wildest excesses of Mother Nature.

NEODINOSAURS—OR LATTER-DAY DRAGONS—IN POLAND?

Needless to say, the prospect of living dinosaurs anywhere in Europe is not very promising. So it should come as no surprise to learn that reports of European cryptids resembling such creatures are extremely sparse—but not unprecedented.

Perhaps the most intriguing case, featuring some ostensible examples from Poland, is one that my friend Miroslav 'Mirek' Fišmeister from the Czech Republic alerted me to in May 2007. It was documented in a passage from a 1998 book entitled *Tajemná Minulost* ('Mysterious Past'), written by renowned Czech traveller Arnošt Vašícek (a mutual acquaintance of ours). Vašícek had first learnt of these baffling creatures in 1996, albeit by chance. While filming a documentary at Niedzica Castle, near to where they had reputedly existed, he was informed of them by a very elderly man who had lived at the castle for 70 years, and who also showed him local magazine items about them.

Mirek kindly translated the relevant passage in Vašícek's book from Czech into English for me:

> One of the latest reports of the presence of unknown creatures comes from Poland. In the area of Novosady Dolina between the promontory of Beskids and Gorce, a pack of strange aggressors used to trouble the local shepherds since time immemorial. They were roughly the size of a human, they looked like lizards, and they were bipedal. Usually, albeit very seldom, they were seen in the forests on the hillsides, which are full of many caves. They hunted the stray cattle or sheep on the little-guarded remote pastures. As for wolves, which were among the most feared animals of those areas, they could easily kill them. One priest from Rabki owned a fur of a giant wolf, which, together with its female mate, did not want to give up their prey—a stray lamb. The pair of wolves had their throats bitten through and it was apparent that they had faced a most powerful opponent.
>
> Only one human was attacked by these mysterious creatures. One shepherd wanted to protect his herd and was knocked down and terribly bitten by one of the creatures. The existence of these

unknown animals is confirmed by records from a local monastery, and village chronicles record the testimonies of eyewitnesses. It seems that the pack was getting smaller and smaller as time went by, until only solitary specimens remained. The last time that one was seen was in 1897.

Information on the 'dragons' of Rabki was being collected by a well-known researcher Mieczyslaw Wojcieszyn at the end of the 1920s. Some of the witnesses tried to draw the creatures. They were simple highlanders. None of them read books or even newspapers, let alone knowing anything about palaeontology. Despite that, they all produced similar pictures of creatures which looked like carnivorous dinosaurs. Their survival into modern times, if we admit that the reports are true, was enabled by the special microclimate conditions of the area. The creatures were said to have been inhabitants of caves and holes dug near the hot springs. They survived the cruel winters, probably in the same way as some other animals, by hibernating. They were only observed from spring to autumn, and their tracks were never seen in the snow.

The prospect of hibernating dinosaurs has been a controversy in itself. Back in the Cretaceous, Australia was much closer to the South Pole than it is today, with portions of it actually inside the Antarctic Circle. Consequently, for several months of the year it would have been plunged into total darkness and would have experienced very cold temperatures too. Consequently, back in 1999 Australian palaeontologists Prof. Patricia Vickers-Rich and Dr Thomas Rich speculated that dinosaurs living here may have hibernated during those extreme, sub-polar conditions. In 2011, however, these same researchers were part of a team who published a comparison of bone microstructure across 18 different species of dinosaur, some from Australia and some from non-polar regions of the world, which revealed that dinosaurs living near the South Pole were not physiologically different from those living anywhere else. This in turn means that hibernation is unlikely to have occurred after all.

So although the above-documented history of putative Polish neodinosaurs may well be nothing more dramatic than the last vestige of some obscure, localised, rural dragon legend, it remains a fascinating account nonetheless, and is certainly one that is very deserving of inclusion here, just in case it triggers any recollections of further details or even sightings among this present book's readers.

THE GOLDEN SERPENTS OF TASEK BERA

Outside Africa, one of the most promising sites for dinosaurian encounters may be an extremely large, reedbed-encircled freshwater swamp in the Malaysian state of Pahang, called Tasek Bera or Bera Lake (with a maximum length of 22 miles and a maximum width of 12 miles, it is the largest in the whole of Peninsular Malaysia). Having said that, to most persons in the Western world it is no more than a name on a map, but in 1951 it was visited by explorer Stewart Wavell, anxious to record the distinctive tribal music and customs of the little-known Semelai people who inhabited this lonely, near-impenetrable locality deep within the southeast Asian jungles. During his sojourn here, he gained the acceptance and trust of the Semelai, who disclosed many of their traditional beliefs to him—by far the most remarkable of which was their claim that the lake housed giant cobra-like beasts that occasionally reared above the lake's vegetation-fringed borders.

Very interested by such stories, Wavell pursued the matter with the Semelai, and obtained the following description of these monsters. Known as the ular tedong (a name also applied to normal, known cobra species in this region) and able to rise up out of the water to the height of a palm tree, Tasek Bera's 'giant serpents' have a very big snake-like head equipped with a pair of short, soft horns like those of snails, and borne upon a long neck that thickens out to a width of about 6 ft near the water surface. The head and neck are covered in thumbnail-sized scales, but the Semelai have never seen the body of one of these animals as they never come out onto land. However, the widening of the neck at its furthest point from the head suggests that it is attached to a much bulkier body, permanently submerged beneath the water. A long tail is sometimes spied above the surface. Of particular interest is the Semelai's claim that the ular tedong changes colour as it grows older. When young, its scales are dull slaty-grey and as fine as a snake's, but by the time that it is fully mature they have acquired a bright golden hue and appear thicker like a fish's.

Although he was unable to follow up this matter further before returning to his home in Kuala Lumpur, Wavell never forgot about it, and several years later his interest was re-awakened when he spoke to a Malayan police officer who had an extraordinary tale to tell concerning his own recent visit to Tasek Bera. One late afternoon, he had been swimming by himself in a deep, open portion of the lake near the headland of Tanjong Keruing, when he happened to look over his shoulder, and discovered that he was no longer alone. About 40 yards behind him, what appeared to be the head of a gigantic snake was rising high above a 15-ft-tall clump of Rassau palm, borne upon a massive slate-coloured neck that seemed smooth in texture. Moreover, two humps were also rising above the water surface, which the officer assumed to be portions of its body. Needless to say, he fled in horror to his canoe, but the great creature made no attempt to pursue him, content instead to remain poised above the water, watching his hasty retreat in unperturbed stillness. This passive manner substantiates the Semelai's testimony, who told Wavell that despite their formidable size the ular tedong are harmless plant-eating beasts that do not attack humans.

After being assured by no less an authority than the Chief Police Officer of Negri Sembilan that this officer was utterly reliable and was not given to exaggeration, Wavell decided to return to Tasek Bera. So he utilised a nine-day holiday to set out by canoe with some local guides along the River Pahang and begin an earnest search for this lake's secret inhabitants.

At the conflux of the Pahang with the Sungei Bera, the river that would take him south to Tasek Bera itself, he encountered four locals who provided him with some very unexpected, additional information regarding the ular tedong. Whereas the vocal repertoire of most genuine snakes is normally limited to a plethora of sibilant hisses (but see my book *Extraordinary Animals Revisited*, 2007, for some controversial exceptions), the 'golden serpents' of Tasek Bera very occasionally give vent to an extremely loud cry—a single echoing boom that is said to resemble an elephantine, trumpet-like blast.

Thankful that his equipment included not only a camera but also a tape recorder, Wavell continued on his way along the 20-mile Sungai Bera, flanked by uncharted jungles and swamps, until he at last reached the lake. After spending a little time at Tanjong Keruing, scene of the policeman's sighting, he travelled on to the village of Kampong Ba'apa, home of his Semelai friends. Here he drew a picture of a *Diplodocus* sauropod and, although the tribal chief conceded that he didn't know if the ular tedong had legs as they had never been spied out of deep water, the picture received a favourable response. After three days, he bid farewell to Kampong Ba'apa, and returned to Tanjong Keruing, having decided to bring his search to a close by spending an afternoon and evening scanning Tasek Bera from this relatively unobstructed vantage point, in a last attempt to espy its reclusive denizens.

One of the commonest regrets of persons unexpectedly encountering a mystery beast is that they did not have a camera to hand—which makes Wavell's experience all the more ironic. After sitting quietly looking out over the lake's still waters for quite a time, Wavell decided to pick up his camera, just in case—and, just as he did so, it happened. Without any warning, an extraordinary trumpeting blast suddenly shattered the silence of this tranquil scene—a single staccato cry emanating from the centre of the lake. As Wavell later recalled in his book *The Lost World of the East* (1958):

> It was a snort: more like a bellow—shrill and strident like a ship's horn, an elephant trumpet and sea lion's bark all in one. I was momentarily petrified, then frantically switched on the recorder, held up the microphone and waited for the next cry.

With cryptozoological inevitability, the next cry never came. The moment had passed, and everything was just as it had been before, but for Wavell, that moment had been sufficient. More like sirens than sauropods, the ular tedong had lured him to their watery domain via native descriptions of their unique voice—and now, at last, he had heard it for himself.

As they have never been observed on land, it is conceivable that these giant creatures are long-necked plesiosaurs (i.e. elasmosaurs) rather than sauropod dinosaurs. However, whereas the neck of sauropods does indeed widen noticeably at its point of junction with the body, in elasmosaurs this widening is far less pronounced, thus implying a sauropod identity—substantiated further by their allegedly herbivorous diet (snakes are not plant-eaters). Also, their trumpet-like bellow recalls the bellow reported by Herman Regusters for Africa's mokele-mbembe. As for the snail-like horns of the ular tedong, these could be respiratory snorkels that the animals poke just above the water surface, thereby enabling them to breathe air while remaining almost entirely submerged—as long as the water was not too deep (otherwise the pressure would collapse their lungs). (As mentioned earlier with the mokele-mbembe, even if prehistoric sauropods were indeed primarily terrestrial as is presently believed by most palaeontologists, I see no reason why 66 million years of continuing, post-Cretaceous evolution could not yield a modern-day sauropod of modest size and amphibious tendency.)

Whether plesiosaur or dinosaur, however, they do not appear to be the only examples of such creatures reported in this part of the world. Many miles northeast of Tasek Bera, and sited alongside the River Pahang, is another supposed monster lake, called Tasek Chini, and again claimed by the natives to harbour 'giant snakes' of the ular tedong variety, According to the Semelai, they are actually born at the top of Gunong Chini, a mountain northeast of Tasek Bera—to which they migrate via various interlinking mountain streams, living many

A life-sized model of a hadrosaur (© Dr Karl Shuker)

years afterwards here and maturing into fully-grown adults.

CHINESE DUCK-BEAKS AND TIBETAN YAK-ATTACKERS

Reports of putative dinosaurs alive and well in Asia are not confined to Malaysia. Lake Tianchi (aka Tian Chai) is ensconced within the crater of Jilin Province's 6,400-ft-high volcano Baekdu (aka Baitoushan), along China's border with North Korea, and on 23 August 1980 a team of Chinese meteorological researchers claimed a close encounter here with a peculiar beast—said to be bigger than a cow, with an elongate neck measuring more than 3 ft in length, a head shaped like that of a cow or dog, and, most memorable of all, a flat duck-like beak (an unusual characteristic recalling the duck-billed dinosaurs or hadrosaurs). The researchers' close encounter almost became a fatal encounter, for the creature—because one member of the team, Piao Longzhi, attempted to shoot it, but his aim was too high, and he only grazed the top of its head. Hardly surprisingly, the beast lost no time in diving back into the lake!

Many years earlier, another of these creatures had not been so fortunate—spotted by a party of six observers, it was shot in the stomach by one of them, and after emitting a deafening howl it sank at once to the bottom of the lake. In 1981, a Chinese author called Lei Jia announced that he had seen a creature here twice on consecutive days, and described it as a black reptile about 6 ft long with a long neck and an oval head. Sightings date back more than a century, and at least 500 eyewitnesses have been recorded since then, some of whom claim that this species has horns on its head, and a white belly. Several scientific expeditions have sought to obtain sonar traces of these beasts, but without success so far.

Following a flurry of new sightings in 1994, describing a golden/black creature with square head, horns, and long neck, a gregarious gathering of Chinese and Japanese scientists, tourists, and film crews, as well as a team of scientists from North Korea, has become a near-permanent fixture around the lake's shores. Yet despite their evident interest and enthusiasm, prior to autumn 2007 all that they were able to show for their efforts was a photograph of the beast that, in classic cryptozoological tradition, was so blurry that it could be just about anything.

During the morning of 6 September 2007, however, a veritable flotilla of finned mystery beasts was videoed swimming across the lake. TV reporter Zhuo Yongsheng filmed no fewer than six animate objects for 20 minutes and also took some still photographs of them before they finally disappeared from sight at around 7 am. One of the stills showed the black dot-like objects swimming parallel in three pairs, the others showed them grouped closely together and leaving deep ripples on the surface of the lake. Zhuo alleged that the objects he filmed were seal-like beasts but with huge fins that were longer than their body length, and that they could swim as fast as yachts but sometimes disappeared completely beneath the water surface. He sent photos to Jilin's provincial bureau for evaluation, but I am not aware of any statement having been made public.

What makes this case even more puzzling is that the volcano housing Lake Tianchi erupted in 1702, so all species living here today have been in residence for less than three centuries. Moreover, monster sceptics have claimed that it is too cold to sustain large animals. Nevertheless, it is worth noting that before the Communist Revolution, which led to a name-change, this lake had been known locally as Dragon Lake. Also, with at least one sighting on record featuring five of these creatures together, there could well be a thriving population here—but from where did these duck-billed denizens originate?

Equally intriguing is a report from June 1980 concerning a long-necked marauder indigenous to central Tibet's Lake Wembo (also spelled 'Wenbu', 'Wampo', 'Weng-po', or even 'Menbu' in some publications). This enormous but remote expanse of water, occupying about 310 square miles, and 300 ft deep in parts, is rich in fish life—and may also be home to some much more sizeable life forms. For according to the *Peking Evening News*, a huge creature resembling a sauropod or plesiosaur with a body the size of a house is the scourge of local yak herdsmen here.

Among its various misdeeds, it was blamed in the report for the mysterious disappearance of a yak that had been left for a short while to graze near the lake's shore by a district secretary of the Communist Party. When its owner returned a while later, to take it to market, only traces of its body could be found, but these apparently indicated that the unfortunate animal had been seized, hauled into the lake, and consumed there by some much larger creature. In another incident, the beast was also held responsible for the disappearance of a local villager while rowing on the lake; his body was never found. As sauropods are vegetarians, this precludes them from contention as the

The famous Cambodian 'stegosaur' glyph at Ta Prohm (© John and Lesley Burke)

identity of Tibet's aquatic flesh-eater; but if these report are genuine, might *some* form of giant reptile indeed be involved?

Also of note is the supposed sauropod of western Java's Lake Patenggang. Clearly more experienced than the Tibetan herdsmen in thwarting prehistoric aggressors, the local fishermen pacify this 18-ft-long monster by burning opium on the lake's shores.

THE CAMBODIAN STEGOSAUR—AN ANGKOR WAT ANACHRONISM?

Cryptozoological riddles turn up in the most unlikely places, but few can be as unexpected as Cambodia's alleged dinosaur carving. One of this southeast Asian country's most beautiful edifices is the jungle temple of Ta Prohm, created during the early 12th Century, i.e. around 900 years ago, and part of the enormous Angkor Wat temple complex—collectively the world's largest religious monument. Like other temples from this time period and location, it is intricately adorned with images from Buddhist and Hindu mythology as well as many depictions of animals. These latter include numerous circular glyphs each containing the carving of some local creature—but Ta Prohm also has one truly exceptional glyph unique to itself. Near one of this temple's entrances is a circular glyph containing the carving of a burly quadrupedal beast ostensibly bearing a row of plates along its back—an image irresistibly reminiscent of a stegosaurian dinosaur!

This anomalous carving is very popular with local guides, who delight in baffling Western tourists by asking them if they believe that

A life-sized model of North America's very familiar *Stegosaurus* from the late Jurassic Period (© Dr Karl Shuker)

dinosaurs still existed as recently as 900 years ago and then showing this glyph to them. Could it therefore be a modern fake, skilfully carved amid the genuine glyphs by a trickster hoping to fool unsuspecting tourists? Or is it a bona fide 900-year-old sculpture? Having communicated with a number of people who have visited Angkor Wat and have viewed this glyph close-up at Ta Prohm, I am assured by all of them that it looks of comparable age to the other glyphs surrounding it, with no visible indications that it has been carved any more recently than any of the others there.

So how can this very intriguing, seemingly anachronistic depiction be explained? Some cryptozoologists cite it as proof that a stegosaurian lineage must have survived into modern times somewhere in this vicinity but has remained undiscovered by science (the notion that this carving may portray a living stegosaur appears to have been first promoted during the late 1990s, in a couple of books on Angkor Wat written by Michael Freeman and Claude Jacques). Others have suggested that perhaps it was inspired by the temple's architects having seen some fossilised stegosaur remains. And there also is the option that it is a stegosaur only by accidental design, i.e. that its plates are not a physical component of the creature, but merely background decoration inside the circle containing it, and that to associate them with the animal is therefore a mistake. Let's consider each of these possibilities.

If we ignore its plates, the rest of the creature does not actually look much like a stegosaur as depicted in palaeontological restorations, certainly not as depicted in modern restorations (i.e. in contrast to those dating from several decades ago, but which are still the ones commonly brought to mind by laymen who may not be familiar with up-to-date versions in palaeontological publications). In particular, its apparent lateral cranial horns are decidedly non-stegosaurian, and the stegosaurs' distinctive, characteristic thagomizer (the arrangement of long pointed spines on these dinosaurs' tail) is conspicuous only by its absence in this glyph. Also contrasting with fossil stegosaurs are its relatively large head and short tail—the reverse condition to that most commonly exhibited by the former dinosaurs.

The stegosaur glyph (arrowed) in situ with other animal glyphs, including a water buffalo directly above it, an unidentified animal directly below it, and a mythological demon directly below that (© John and Lesley Burke)

Then again, if a stegosaurian lineage has indeed somehow survived into modern times, such differences from fossilised stegosaurs as those noted here are certainly not so radical that they could not have arisen during the very lengthy period of continuing evolution (in excess of 100 million years) that will have occurred from the early Cretaceous (the age of the most recent confirmed fossil stegosaurs) to the present day. As I've already noted elsewhere in this book, one only has to compare, for instance, the relatively unspecialised range of mammals or birds existing during the Cretaceous to the vast morphological diversity of mammalian or avian forms alive today to see just how extensively evolution can modify outward morphology during that particular period of time.

However, if anything as dramatic as a living stegosaur does indeed exist anywhere within the area of Cambodia (or has done so until very recently), one might reasonably expect rather more pictorial evidence of such existence than a single small carving tucked away amidst a myriad of other animal carvings. Yet I am not aware of any comparable design anywhere else in Asian art. To my knowledge, there is no suggestion of stegosaurian creatures in Cambodian mythology or folklore either, nor, indeed, in that of any other corpus of Asian traditions (thus contrasting very markedly, for example, with the extensive native beliefs associated with the mokele-mbembe in the Congo). And there is certainly no documented physical evidence for such a creature's reality—no preserved plates, skeletal remains, etc, described in any publication that I have ever encountered or seen any mention of during my researches.

If only the Cambodian stegosaur were indeed real . . . 'Angkor's Way' (© Michael J. Smith)

Moreover, even fossil stegosaur remains so far disinterred in Asia are restricted to China (predominantly) and India (very controversially—much of these proved upon closer inspection to be derived from plesiosaurs instead!). This in turn reduces the likelihood that the 'stegosaur' glyph was carved 900 years ago by a local sculptor who had previously seen fossil remains of such a creature, unless (and which is certainly not impossible) the sculptor had visited China and had seen such remains there?

Yet even if it does not represent a living contemporary (or a prehistoric fossil) stegosaur, might it conceivably depict some still-undiscovered modern-day animal that superficially resembles a stegosaur? If so, however, there do not appear to be any local sightings or lore on record concerning it. Another option is that as there is a varied mixture of the factual and the fictitious among the fauna depicted at Angkor Wat, perhaps it is some local mythological beast. Yet once again I am unaware of any from this region of the world that match its appearance.

To my mind, by far the simplest and most plausible explanation for this enigmatic carving is that its resemblance to a plate-backed stegosaur is an artefact—i.e. it is simply some form of local present-day known creature that has been carved with a plate-like decorative motif in the background, but which in turn has been wrongly associated directly with the creature. The reason that I favour this explanation is that such a motif can also be seen surrounding other carved animals of several different types enclosed within their respective glyph circles at Angkor Wat. These include birds, a water buffalo, deer, monkeys, and even mythological demons.

Although the plates surrounding the alleged stegosaur do seem somewhat better-defined (but might this indicate some very selective modern-day enhancement by a hoaxer seeking to enhance its superficial stegosaur appearance?), their general shape and size are much the same as those surrounding other carved animals. In addition, this same plate motif is also present encircling the outer perimeter of the glyph circles enclosing the carved animals, including that of the 'stegosaur' (as seen in its close-up illustration at the beginning of this present section).

Looking closely at the latter creature, its head in particular is shaped very like that of a rhinoceros, as has also been commented upon elsewhere by German cryptozoologist Markus Bühler and various others. Even its 'cranial horns' resemble the long pointed ears of such mammals. Conversely, its back seems more arched than is true of rhinos, but this discrepancy could merely be due to stylising, or once again may simply be a design artefact, the creature having been depicted in this unnatural, hardly life-like pose (for a rhino) simply in order for it to fit more readily inside its circular setting.

Incidentally, adapting the shape of an animal during its depiction in order to fit it more snugly within a designated space for it is an option that I have already explored in my book *A Manifestation of Monsters* (2015) with regard to a second anomalous Angkor Wat carving—the so-called Cambodian moa.

Returning to rhinos and the suspect stegosaur: on the latter creature's body are indications of the skin pleats exhibited by Asian rhinos of the genus *Rhinoceros* (i.e. the great Indian *R. unicornis* and the Javan *R. sondaicus*, the latter of which definitely still existed in Cambodia 900 years ago, with the former possibly doing so too). Even the creature's lack of a nasal horn is not an obstacle to identifying it as a rhino of this genus, because female Javan rhinos are sometimes hornless.

Another line of speculation that has been proposed by some investigators is that the creature actually represents a very stylised portrayal of some form of lizard, suggestions having included a chameleon (though there is none in southeast Asia) or one of the several species of southeast Asian agamid known as mountain horned dragons *Acanthosaura* spp. However, any similarities between the carving and such reptiles seem far less apparent to me (if indeed present at all) than those readily visible between the carving and a stylised and/or modified-to-fit rhinoceros. Equally, whereas an even better fit for the creature's 'cranial horns' than the pointed ears of a rhino would be the horns of a wild ox, the rest of the creature's depiction is a better fit for a rhino than for an ox.

Of course, we shall never know for sure the intended taxonomic identity of the supposed stegosaur in this perplexing carving. However, it does seem much more likely to be a stylised depiction of some local known species rather than anything more radical. After all, it surely couldn't have been based upon a sighting of a real-life stegosaur . . . could it?

THE BEAST FROM PARTRIDGE CREEK

The Arctic wastelands of the Yukon Territory, on the borders of Canada and Alaska, are surely the last place anyone might expect to meet a dinosaur—which is why the following case merits a prominent place among the dubitanda of cryptozoology.

The extraordinary tale of the beast from Partridge Creek was apparently first told on 15 April 1908, within the pages of a French journal called *Je Sais Tout*, by one of the creature's alleged eyewitnesses, French traveller Georges Dupuy. According to Dupuy, the adventure had

The Partridge Creek beast as depicted in the original *Je Sait Tout* article from 15 April 1908

begun one day in 1903, when a banker from San Francisco called James Lewis Buttler and a local gold prospector named Tom Leemore were hunting three large moose amid the marshy tundra near the Yukon's Clear Creek, about 100 miles east of Dawson City. Suddenly, one of these burly deer raised its head as if it had heard something unexpected in the vicinity, and just as it did so a second member of the trio emitted a loud bellow of alarm, sending all three of them fleeing southwards with all speed.

As the world's largest living species of deer, standing 7 ft at the shoulder and weighing up to half a ton, the moose *Alces alces* is not a creature to be readily intimidated by anything—which is why the two hunters were so surprised by their quarry's impromptu flight. When they reached the spot from where the moose had fled, however, the reason for these creatures' uncharacteristic behaviour became clear. There in the snow was the clear impression of an enormous body belonging to some unidentifiable beast of monstrous proportions.

The beast's belly had ploughed into the river bed's swampy mud a massive furrow 30 ft long, 12 ft wide, and 2 ft deep, which was flanked by gigantic footprints measuring 2.5 ft across and 5 ft in length, each also yielding impressions of sharp 1-ft-long claws. Completing the clues left behind by this unseen goliath was the imprint of a mighty 10-ft-long tail spanning 16 in across its middle.

Notwithstanding the nightmarish visions conjured up by such dramatic dimensions, the two men decided to follow the monster's tracks, which led after several miles to a gulch known as Partridge Creek, where the tracks simply came to an end—giving the men cause to speculate that it must have leapt directly up into the gulch's encompassing cliffs.

Following this, they made their way by canoe along the McQuesten River to the nearby outpost of Armstrong Creek—an Indian village where Buttler had earlier arranged to meet Dupuy to take him hunting, and which was home to the Reverend Father Pierre Lavagneux, a French-Canadian Jesuit priest. When Dupuy and the priest heard about the monster, they were highly sceptical at first, but eventually Buttler persuaded them to return to Partridge Creek with Leemore and himself in a second search for it, aided by some local Indians.

After a day's intensive but fruitless search of the immediate area and some distance beyond it in all directions, however, the party of weary beast hunters had resigned themselves

Artistic rendition of the Partridge Creek beast, with human silhouette for scale purposes (© Connor Lachmanec)

to failure, and decided to establish camp at the top of a rocky ravine. Thoughts of confronting monsters had been largely suppressed by the more practical goal of preparing a warm meal—until, wholly without warning, they received a terrifying reminder of the reason for their presence there. A hideous roar broke the Arctic stillness, and as the startled hunters reached for their rifles, one of them pointed with shaking hand to the opposite side of the ravine—where the crimson rays of the dying sun revealed a heart-stopping sight. A veritable monster in every sense of the word was clambering up its slope!

Black in colour, at least 50 ft long, and estimated by Buttler to weigh about 40 tons, their horrific visitor continued to climb upwards for a time, mercifully unaware of its terrified human eyewitnesses' close proximity. When it was about 200 paces away, it paused for 10 minutes or so, furnishing them with ample opportunity to observe it at close range in all its spine-chilling glory. Although profusely splattered with thick mud, its hide could be seen to bear many coarse, grey-black bristles like those of a wild boar, and a hideous mass of blood and saliva oozed from its cavernous jaws. Most remarkable of all, however, was the rhino-like horn perched prominently if somewhat incongruously near the end of its snout.

The Indians cowered in terror behind a large rock, and the others were rendered momentarily speechless by the monstrous apparition, but when it became clear that it had not detected them the priest gained sufficient control of his voice to whisper between still-chattering teeth: "A ceratosaurus . . . It's a ceratosaurus of the Arctic Circle!". *Ceratosaurus* was a large flesh-eating bipedal theropod dinosaur of the Jurassic Period, thus related to the later, larger theropod *Tyrannosaurus rex* and readily identified by the large horn near the tip of its snout.

As they continued to watch, the beast suddenly reared up onto its massive hind legs like a hellish kangaroo, gave voice to another ear-splitting roar, and then disappeared back down into the ravine via a single immense bound. Not too surprisingly, the beast hunters found that the thrill of pursuing this creature had lost its former charm, and the following morning they journeyed back to Armstrong Creek. Dupuy later attempted to interest the Canadian governor at Dawson City in supplying them with a large-scale hunting party of mules and 50 armed men, but when this failed, he returned home to France.

The matter was not quite over, however, for in January 1908 he received a letter from Father Lavagneux, dated Christmas Day 1907, in which the priest informed him that on Christmas Eve, and while in the company of ten Indians, he had seen their monstrous quarry once again—this time racing at speed over the creek's frozen river, with what seemed to be the carcase of a caribou clamped between its great jaws. Tracks identical to those found back in 1903 were clearly visible in the deep mud, and were followed by Lavagneux and company for at least two miles before being obliterated by falling snow.

That appears to have been the last time that the beast of Partridge Creek was spotted, either by Lavagneux or by anyone else, for there does not seem to be any further sightings on file alluding to this extraordinary creature. However, the story itself as given above has been reprinted many times since then in various American newspapers. (One of my Fortean correspondents, Theo Paijmans, has informed me that he has no fewer than 31 such reports on file, dating from 1908 to the 1920s.)

Intriguingly, on 9 May 1908, Washington DC's *Evening Star* newspaper published a very lengthy and extremely curious report written by someone identified only as 'An American In Paris', which claimed that England's Duke of Westminster was secretly planning a covert expedition to Partridge Creek to seek and even attempt to capture this locality's supposed *Ceratosaurus*. However, I am not aware of any follow-up reports. Certainly, even if the duke did launch such an expedition, nothing of this nature was ever procured.

In any event, as such an animal's existence requires belief in post-Cretaceous persistence of an endothermic *hairy* dinosaur in what must surely be the least compatible habitat for *any* type of large reptile living today, its lack of cryptozoological credibility shouldn't come as a great surprise!

A DINOSAUR CALLED 'PINKY'

If there are any regions in North America with prospects (however slim) for concealing living dinosaurs, they must include the humid swamplands of Florida—especially as a creature reputedly akin to such animals has been reported from there on several different occasions.

On 10 May 1975, an outboard motor boat transporting five people on a fishing trip along Florida's 300-mile-long St Johns River experienced a close encounter at a spot between Jacksonville and the Atlantic Ocean with something very peculiar—but certainly *not* pretty—in pink! At around 10 am, one of the passengers, Brenda Langley, saw the head and neck of an extraordinary creature surface a mere 20 ft from their vessel, and shortly afterwards it was also spied by the other four passengers when they turned the boat around—to escape what seemed to be an oncoming storm, heralded by the arrival of some dark clouds.

According to subsequent press reports of their encounter, quoted by mystery beast chronicler Mark A. Hall in *Wonders* (December 1992), Dorothy Abram likened it to ". . . a dinosaur with its skin pulled back so all the bones were showing . . . [and] pink. Sort of the color of boiled shrimp". She also noted that its head was at least the size of a man's, with a pair of snail-like horns bearing knob-like structures at their tips, there were flaps reminiscent of gills or fins hanging down from the sides of its head, its mouth turned downwards, its large eyes were dark and slanted, and its neck had protruded about 3 ft out of the water, revealing a seemingly serrated upper surface. Brenda Langley agreed with this description, and added that it was an ugly creature recalling pictures of dragons. Its behaviour had seemed to its eyewitnesses be inquisitive, returning their intrigued stares with an equal extent of keen observation during its 8-second appearance—after which it submerged so effortlessly that it did not even leave behind a ripple.

Soon to receive the inevitable nickname of Pinky, the St Johns River water beast attracted sufficient media interest to elicit accounts from several other people claiming sightings of just such a beast in this same area as far back as the mid-1950s. This, of course, is hardly an uncommon feature of cryptozoological cases receiving media exposure. However, Mark Hall has good reason for believing in the validity of these earlier Pinky reports—because 20 years *before* the encounter of May 1975, he had seen newspaper accounts from various Jacksonville papers that described sightings of a similar creature and which themselves dated back a number of years.

Moreover, Hall discovered from an *Argosy* article written by American cryptozoologist Ivan T. Sanderson that during the 1960s, while bow-hunting along the St Johns River, biology student Mary Lou Richardson, her father, and a friend had all seen a very peculiar animal with a great flat head, a rather small neck, and (in Sanderson's opinion) the overall appearance of a donkey-sized dinosaur. Four other groups of tourists independently saw it during that same day, and enquiries revealed that this animal was well known to local fishermen and hunters. In short, Pinky was far from being the cryptozoological newcomer that it had initially seemed.

Taking his lead from Sanderson, Hall has very tentatively sought to reconcile the perplexing Pinky with an undiscovered, living species of bipedal dinosaur called *Thescelosaurus*—an 11-ft-long ornithopod the size of a small car, with five fingers on each hand and four toes per foot, a long stiff tail, and characterised by rows of bony studs set in the skin along its back that may have given it a somewhat uneven, serrated appearance. One of the last known dinosaurs, it lived during the very late Cretaceous of western North America, and was related to the larger, more familiar *Iguanodon*. Based upon the preservation and completeness of many of the fossil specimens found, it may have preferred to live near streams.

Personally, however, I very much doubt that we need nominate anything as dramatic as a living dinosaur when seeking to unveil Pinky's identity—at least not until we have surveyed reports describing prominently pink mystery beasts of a herpetological persuasion from elsewhere in the U.S.A., because such beasts have been recorded far beyond the St Johns River of Florida.

Horns and Hellbenders

Two centuries ago, for example, strange creatures referred to loosely as giant pink lizards were frequently reported from south-central Ohio's Scippo Creek by the first white settlers here. Hall's book *Natural Mysteries* (1991) presents a detailed investigation of these animals, concluding that they could well be the larval form of some undiscovered giant amphibian.

They were said to be at least 3 ft long but generally averaged 6-7 ft, were invariably pink in colour, sported large horns ("like a moose", according to one very startled eyewitness, a young carpenter), and were always associated with water. When their habitat suffered a great drought some time before 1820 that dried out many streams and wells, and suffered additional devastation by way of a terrible fire, the outcome was their extinction. If we equate these animals' 'horns' with prominent, branching external gills, rather like those of the much smaller axolotl of Mexico, these 'pink lizards' could indeed have been larval salamanders—but their size far exceeds any known species in North America, or anywhere else.

Nature writer Herbert Sass was a rather more recent observer of a pink mystery beast in North America. While boating in or around 1928 with his wife Marion on Goose Creek lagoon, near Charleston, South Carolina, Sass saw something moving under the water, and when he succeeded in lifting part of its heavy bulk up out of the water on an oar they saw that it was bright salmon-pink and orange in colour, as thick as a man's lower thigh, with a smooth tail, and a pair of short legs like those of an alligator or salamander. Within moments, however, their mysterious captive had slipped off the oar, and back into the water.

A 2-ft-long portion of a large worm-like beast of similar colour to Sass's creature was allegedly spied briefly by Ivan T. Sanderson and his wife Sabina during the early 1970s. They saw it amid the dense water vegetation in a pond created from an artificial swamp on their farm at Warren County in New Jersey.

Speculation concerning the existence in North America of giant salamanders might seem just as risky as speculation regarding the presence here of living dinosaurs, were it not for the indisputable fact that the U.S.A. is already *known* to harbour one species of giant salamander—the euphoniously-named hellbender *Cryptobranchus alleganiensis*. Up to 29 in long, it is most closely related to a pair of even larger species, native to Asia's Far East—the Japanese giant salamander *Andrias* (=*Megalobatrachus*) *japonicus* (up to 5 ft long), and the Chinese *A. davidianus* (up to 6

Artistic rendition of a giant pink mystery salamander (© William M. Rebsamen)

ft long, the world's biggest salamander species alive today).

It is interesting to note that in his own report of his sighting, Sass described the creature that he briefly captured as a kind of giant hellbender, at least 5-6 ft long, because this grotesque species is also reminiscent of Pinky from Florida's St Johns River. Just like Pinky, the hellbender has a large flattened head, a long downward-curving mouth, loose folds of skin on its neck, and wrinkles along much of its body that could conceivably be mistaken for protruding bones during as swift a sighting as that of 10 May 1975. Moreover, the hellbender does indeed inhabit fast-flowing streams and rivers, and even in their well-aerated water it must still surface to gulp air every so often, because unlike some salamanders it does not have external gills as an adult. And with a lifespan for this species of at least 30 years, a single specimen could have been responsible not only for the Pinky sighting of May 1975 but also for those pre-dating it by more than 20 years.

Even its dissimilarities are not irreconcilable. Although the hellbender's official distribution range is from the Great Lakes through the eastern United States to Georgia and Louisiana, it would not be impossible for a population to remain undetected amid the little-traversed Florida swamplands. Interestingly, the concept of out-of-place giant salamanders has at least two notable if somewhat controversial precedents.

In c.1939, a 25-30-in-long *Andrias* salamander was captured by a commercial fisherman in California's Sacramento River. Maintained alive for a time within a wooden trough suspended in the fisherman's own bathtub, it was examined by Stanford University herpetologist Dr George S. Myers, who felt that it differed in colouration from both of the two known modern-day species of *Andrias*. Thus he speculated that although it could simply be an escapee from captivity, it may represent an unknown, native New World species—a relic from prehistoric times, when the zoogeographical range of the giant salamanders was much greater than is their fragmented modern-day equivalent.

19th-Century colour-tinted engraving of a Japanese giant salamander underwater

This latter identity, however, was challenged by Chico State College herpetologist Dr Thomas L. Rodgers, who announced that he too had inspected the Sacramento giant salamander, but had later learnt that it was in reality an absconded Chinese giant salamander called Benny—one of three *A. davidianus* specimens purchased by fish fancier Wong Hong somewhere in China. According to Charles Bjork, captain of the steamer *Isleton*,

Benny had escaped while being transported through the straits on the way to Stockton Harbour.

Even so, this does not explain the sighting of what eyewitnesses described as the "head of a gigantic lizard" emerging from the Sacramento River way back in 1891. Nor can it counter the longstanding belief that a deep lake in California's Trinity Alps is home to giant salamanders 5-9 ft long. Attorney Frank L. Griffith claimed to have spied five of them and even to have hooked one in the 1920s. Due to its great size, however, he was unable to haul it out of the water. His story attracted Dr Rodgers's interest, who visited the lake four times hoping to see these beasts. Despite denouncing Myers's idea regarding the Sacramento specimen, he speculated that they may be a relict population of *Andrias*, but all that he found were some *Dicamptodon* salamanders—none more than 1 ft long.

Other searches have taken place since then, whose participants included in 1960 the Texan millionaire Tom Slick, a keen amateur cryptozoological investigator, but as no specimen of *Andrias* or the hellbender has so far been found here, the case for their existence remains unproven. Yet in view of its secretive, principally nocturnal lifestyle, it would not be too surprising for populations of hellbenders outside their species' known distribution range to have escaped detection, especially in little-explored swamps or high mountain lakes.

In March 2015, moreover, I received a detailed account of an alleged giant salamander seen just a few years ago in California by an eyewitness whose identity I have on file but which, in accordance with her request, I shall not reveal in print. Instead, I shall refer to her here simply as Prunella (not her real name).

Prunella saw the creature while walking with a child in Redwood Park, Arcata, early one morning in 2005, just after some rain. She claimed that it was huge, measuring 4-5 ft long (hence much bigger than even the hellbender), walking slowly with its body raised well off the ground, with very smooth, shiny, slimy skin, and possessing a newt-like head rather than the typically flat head of known giant salamanders. It was orange in colour with black markings, and Prunella considered that although far greater in size and with much bigger legs, it looked in basic form and colouration somewhat (but not entirely) like a known species from this region—California's coastal giant salamander *Dicamptodon tenebrosus* (though as this species' known maximum length barely exceeds 1 ft, the adjective 'giant' in its vernacular name is something of a misnomer). Prunella also stated that a slightly smaller but otherwise similar specimen to the one that she'd encountered had previously been seen in the same location by the boyfriend of one of her work colleagues.

Could these creatures have been freakishly over-sized specimens of *D. tenebrosus*, with their huge size (if estimated accurately) having possibly been induced by hormonal hypersecretion? Or might they have been representatives of a still-undescribed fourth species of true giant salamander?

Having said that, the extra-large size of some of North America's unidentified salamander-like beasts (pink or otherwise) is not an intractable problem when seeking to identify them as hellbenders. For although the largest recorded specimens are under 3 ft, there may have been much larger ones in bygone times, whose greater size offered them up as targets for early Western settlers eager to test their shooting skills upon these inoffensive, sluggish beasts. Record-size specimens of many different species have elicited similar attention from hunters in the past, the systematic killing of such specimens bringing about an eventual decrease in their species' average size (the gradual reduction in average total length of the much-hunted European giant catfish or wels *Silurus glanis* over the past century exemplifies this trend). In remote areas little-frequented by humans, however, some reclusive giant specimens could still thrive, undisturbed.

What a pink-coloured hellbender may look like (© Dr Karl Shuker)

Even the distinctive pink colour of most of the strange creatures reported here does not oppose a hellbender identity. Albinism is not uncommon among salamanders, and albinistic specimens normally exhibit a pink sheen due to the presence of blood coursing through their blood vessels beneath the pallid outer skin layer. In the hellbender, the skin's supply of blood vessels is particularly pronounced; this extensive vascularisation enables the animal to obtain much of its oxygen requirements by direct absorption through its skin from the surrounding water of its aquatic domain. Hence an albinistic hellbender would be conspicuously pink. And because the genetics of albinism in salamanders (as in many other animals) are such that albino salamanders can only yield more albinos, all-pink populations could rapidly arise.

Moreover, while browsing on the internet I have found several photographs of salmon-pink and even bright orange specimens of the Chinese giant salamander (at: http://tinyurl.com/zncrkdz), thus providing notable support for the likely occurrence of comparably-hued hellbender specimens too. Some of those above-mentioned photos are copyrighted to the International Cooperation Network for Giant Salamander Conservation, and certain others to the Zoological Society of London (ZSL).

In contrast to a genetic origin, Hall noted that the pink colour of the mystery beasts documented here could be due to their diet, as with the pink plumage of flamingos. This latter example, however, is a very specific one, and there is no evidence to suggest that a similar phenomenon occurs in salamanders—even when pet albino axolotls, for example, are fed upon bright-red tubifex worms and various pink crustaceans, they do not acquire the pigmentation of their prey.

Overall, therefore, an albinistic hellbender identity for the pink mystery beasts is the most satisfactory explanation available. Even so, there is still the matter of the Scippo Creek beasts' bizarre moose-like horns to resolve, because hellbenders have no such structures—when adult, even their gills are internal. Consequently, if the Scippo creatures' 'horns' are actually large external gills, these animals would much more closely resemble a giant form of axolotl *Ambystoma mexicanum*—the best-known of the Mexican neotenic forms belonging to the tiger salamander complex.

Under standard conditions, most salamanders metamorphose normally from the gilled larval salamander into the true adult, reproductive form. In the case of the axolotl, however, and especially if the pools in which it thrives are low in iodine, this metamorphosis is often halted, and the animal retains its larval form throughout its life, but nonetheless acquires the capability to reproduce and is thus said to be neotenic.

A extra-large neotenic version of albinistic hellbender might thus explain the pink mystery beasts of Scippo, but no such specimens have so far been formally documented. Moreover, not only are neotenic salamanders exclusively aquatic, but if they are lifted out of the water their gills flatten against their neck, thereby losing their antler-like form.

And in any case, what of Pinky—with its very different, snail-like horns, and noticeably large eyes? Hellbenders have no such horns, and their eyes are very small. Could Pinky's horns be breathing tubes? If so, a hellbender with snorkels and protruding eyes is clearly taxonomically separate from the known species. So although a thriving *Thescelosaurus* dinosaur abroad in North America remains highly unlikely, the possibility of an unknown species of giant salamander lurking amid its vast swamplands may be worth further investigation.

FROM MINI-REX TO MOON COW— A RIVER DINOSAUR RIDDLE

In January 2000, veteran American cryptozoologist Ron Schaffner was emailed by someone using only a generic, non-specific email address and who identified himself simply as 'Derick', but whose information was anything but generic. Derick attracted Schaffner's interest straight away, thanks to his emailed report containing the following intriguing introductory statement:

> I live in Pueblo, Colorado. I moved out here when I was six and since then I've heard stories of the prairie devil, the pig man and the mini-rex; there's even old Indian legends of evil river demons. You get older and you try not to believe in monsters, however not even the high school kids will have a kegger [outdoor keg beer-drinking teenage parties] down by the river without a raging fire and a lot of people [presumably to ensure that 'monsters' keep well away]. It's not like people don't see things, people see them they just don't make a big deal of it. If you live by the river like me you just get used to it.

Following that enticing preamble, Derick went on to describe his encounter with what appears to have been one of this region's mysterious 'mini-rexes'. He claimed that while he and a friend were riding the latter's dirt-bike close to the Fountain River near Pueblo one day in July 1998, they suddenly saw a truly extraordinary bipedal creature run across the clearing in front of them:

> It was three to four feet long, greenish with black markings on its back, and a yellowish-orange under belly. It walked on its hind legs, never dragging its tail, its front limbs (I call them limbs because they were more like arms than anything) were smaller in comparison to the back ones and it had four or three claws/fingers. I'm not sure for it was seen at a great distance. It also had some kind of lump or horn over each eye. When it noticed our presence it let out a high pitched screech or some sort of bird chirping, that pierced my ears, and then took off.

Derick and his friend rode back at once to Derick's house to fetch a camera, then returned to the location of their encounter and photographed the creature's three-toed, 2-in-diameter tracks (with a Marlboro Red cigarette placed alongside them in some photos for scale purposes). Derick also stated that he subsequently heard of other sightings, and discovered that a friend had actually photographed one such creature. Following some persuasion, the latter friend allowed Derick to send scans of those photos, together with his own track photos and July 1998 sighting account, to Schaffner, who in turn showed them to fellow American cryptozoologist Chad Arment, who has a particular interest in American mystery reptiles. However, as the photos showed little detail, even when magnified, Schaffner and Arment agreed that their subjects could easily be dinosaur models.

Nevertheless, Arment remained sufficiently intrigued to email Derick and state that if such beasts were indeed real, better evidence would be needed in order to confirm this. He didn't expect to receive a reply, but in April 2000 he did, in the form of an email enclosing two scans of this truly remarkable photo:

Photograph allegedly depicting a shot mini-rex or 'river lizard' held by an unidentified figure—emailed by 'Derick' to Chad Arment in April 2000 (copyright owner unknown/identity and current location of 'Derick' unknown, so could not be contacted; thus reproduced here for educational /review purposes only, on a strictly Fair Use basis)

Referring to the creature in this photo as a 'river lizard' (but which is presumably the same type as the mini-rex that he saw), Derick stated that it had taken him some time to obtain scans of the photo, and that he didn't know when or specifically where it had been snapped, only "somewhere" in Colorado. As the creature does not resemble any species currently recognised by science, Arment later attempted to email Derick for further details, but received an automated reply that Derick's email address was no longer in operation, and has never heard from him again.

However, Arment has learnt from another source that the term 'river dinosaur' has apparently been used in relation to such creatures, and even that an individual who had previously collected Colorado reptiles for some of Arment's friends working in the pet trade had also offered to capture for them some 'river dinosaurs' (whose description compared closely to the appearance of the 'river lizard' in the photo supplied by Derick and reproduced here). Unfortunately, Arment's friends had been forced to decline this exciting offer due to lack of funds.

The 'river lizard' photograph supplied by Derick has been floating around online for many years now, and has attracted much attention and numerous comments on the many websites where it has appeared. The consensus seems to be that it is a hoax, but this has never been confirmed. In addition, I have conducted several Google-image searches, concentrating variously upon the whole image, the section of it featuring the person, the section of it featuring the creature, etc, but all to no avail. However, it certainly contains some anomalous details, especially in relation to the creature's appearance, as commented upon by Arment in an account of this and other bipedal 'dinosaur' sightings reported from the U.S.A.

(*North American BioFortean Review*, vol. 2, #2, 2000).

For example: if the 'river lizard' had only recently been shot when this photograph was snapped, and bearing in mind that recently-dead reptiles are normally very limp, and also bearing in mind that it was being held vertically, why is its tail curving inward rather than simply hanging straight down? And why is its mouth gaping open rather than being held closed, or at least nearly so, as one would expect under the conditions given here? As Arment also pointed out, very convincing life-like rubber models of dinosaurs can be readily obtained nowadays, so we cannot be sure that the 'river lizard' in the photo was ever a living entity anyway.

And even if it was once alive, thanks to the photo-manipulation computer software that was already available back in 2000 there is no certainty that this creature's appearance hasn't been profoundly modified digitally from whatever it was originally. Moreover, the entire photograph might conceivably be a cleverly-constructed montage, i.e. an original photo of a person holding a gun into which a second, digitally-manipulated photo of some animal has been deftly incorporated.

Certainly, whenever I've looked at it, I've been struck by just how very odd, how very unnatural this photo's image seems to be, and I don't just mean the bizarre appearance of the 'river lizard' itself but the whole image. Even the person's face is so obscured by the shadow of their hat that I'm not exactly certain whether they are male or female. But as far as the 'river lizard' itself is concerned: having spent a lifetime observing animals in photos and in the living state, at the risk of being accused of sounding unscientific and overly reliant upon gut instinct it just doesn't look 'right' to me.

In particular, the incongruous hind limbs sported by this creature seem entirely out of place on its body, their curiously flat, disproportionately large, and flared, oddly triangular (rather than oval) haunches looking not only unrealistic but also as if they have been crudely glued (or photo-applied) there rather than being a natural feature of its anatomy. And when viewed in magnified form, the lower portion of each of these hind limbs seems to be segmented, rather like that of an arthropod invertebrate. Very strange indeed. Also worrying is the artificial appearance of the creature's open mouth, almost as if the lower jaw has been manually added—and who knows, perhaps it has been.

Thus I consider it plausible that this 'river lizard' photo is indeed a hoax, my line of thought being that its mystifying creature constitutes some form of manufactured composite (either physical or photo-manipulated), perhaps a much-modified dead monitor lizard, sub-adult crocodile, or even dinosaur model, for instance. Moreover, if my reasoning so far is correct, then I further suggest that the creature's bizarre-looking hind limbs as seen in the photo are fake, that they have been seemingly less than skilfully attached to (or superimposed upon) its body, and that they presumably replaced whatever hind limbs it may have originally possessed, this substitution having been done to create the impression that these limbs are much bigger than the front limbs, thus enhancing even further the contrived exotic, unfamiliar appearance of the creature for this photo.

Having said all of that, however, this is nothing more than speculation on my part, and I am well aware that I could be entirely wrong, with the 'river lizard' potentially being a bona fide cryptid carcase after all. So I'd be most interested to receive any comments, views, or additional information concerning it from other investigators and readers.

Speaking of which: one American cryptozoologist who has investigated this and other such cryptids with particular zeal, and success, since 2001 is Nick Sucik, who duly published his very thought-provoking findings in a chapter devoted to this subject in a most interesting compendium edited by Chad Arment and entitled *Cryptozoology and the Investigation of Lesser-Known Mystery Animals* (2006). As discovered when reading Sucik's chapter, a fair number of roughly consistent cases have been documented from a range of river-related U.S. localities over the years, but with particular frequency across Colorado.

The size of such beasts may differ, as may eyewitness recollections of colouration, but by and large the same image of a miniature bipedal dinosaurian creature with sizeable hind legs, much smaller, delicate forelegs, a long study tail, small but very sharp teeth, and bare skin crops up time and again, with eyewitnesses comparing what they have seen to certain fossil theropods (at least in basic outline), including *T. rex* and *Compsognathus* (the latter small theropod being of very comparable size to most of the mini-rex specimens reported).

Sucik's chapter contains far too many cases and far too much information to review comprehensively here, but I found the following cases to be especially interesting. His chapter opens with a detailed account of a mini-rex sighting that occurred one warm July evening in 2000 as three women of successive generations were driving along a country road near the rural community of Yellow Jacket, 15 miles north of Cortez, Colorado. The creature entered their headlights from the side, and the driver braked, thinking that it was a fawn, but when the headlights lit it up, its two astonished eyewitnesses in the front of the vehicle realised that it was something very different indeed from any deer. As described by Sucik in his account of their description of it:

> Its body appeared smooth, devoid of fur or feathers. Its height perhaps was three feet and the small head was bent downward on a slender neck. The creature ran on two skinny legs with its tiny forelimbs held out in front of its body as it ran. Its body tapered down into a lengthy tail that, combined with the head and neck, made it about 5 feet long. The movement of the animal was noted as graceful, the head not bouncing as it ran.
>
> The animal quickly passed in front of them and disappeared into the darkness.

Its eyewitnesses were shocked by what they had seen, but after they had regained their composure one of them joked that it must have escaped from some local Jurassic Park. Although a light-hearted remark, it is nonetheless telling, as it serves well in underlining just how very dinosaurian the creature must have appeared to them—as opposed to rather more mundane alternatives, such as an exotic escapee from captivity (like a wallaby, but which is furry and jumps, not runs; or a rhea or emu, which do run, but are feathered).

Could it be, however, that eyewitnesses not well-versed in natural history are simply seeing various scientifically-recognised species that they themselves do not recognise? After all, certain lizards as known to adopt a bipedal running gait on occasion. Perhaps the most famous of these is the Australian frilled lizard *Chlamydosaurus kingii*, up to 2.75 ft long and which has often been likened to a dinosaurian raptor when sprinting bipedally. So too have North America's collared lizards (genus

Life-sized models of *Compsognathus* (© Dr Karl Shuker)

Crotaphytus), which measure around 1 ft long, and its common chuckwalla *Sauromalus ater*, which is up to 20 in long.

In South America, the basilisks (genus *Basiliscus*), up to 2.5 ft long, can do the same as well, even sprinting bipedally across stretches of water. Moreover, these are sometimes kept as pets in the U.S.A., so specimens may escape into the wild here on occasion. Indeed, the description of one bright green, crested, 2-ft-long, lengthy-tailed bipedal "baby dinosaur" allegedly caught then released by an 11-year-old boy in 1981 alongside some railway tracks in New Kensington, Pennsylvania, sounds just like one of these lizards, which by then were becoming quite popular as exotic non-native pets in the States.

However, there are some notable behavioural and morphological differences between such lizards as these and the mysterious mini-rex. For instance, none of these lizards is habitually bipedal, whereas the mini-rex apparently is. Moreover, their forelimbs are held laterally when they run bipedally, whereas those of the mini-rex are held in front of its body. Also, these lizards are smaller than all but the smallest mini-rexes on record, whereas some of the bigger examples of the latter cryptid are at least the height of an adult human. True, certain of the large monitor lizards are popular pets that often escape, can grow several feet long, and can run bipedally, but only for very short periods, not in the habitual and very rapid manner described by eyewitnesses for all mini-rexes, even the biggest ones.

One 'not-so-mini' mini-rex was reported by a lady called Myrtle Snow, who seems blessed with an extraordinary ability to encounter these mystery reptiles, judging at least from the fact that she claims to have done so on several occasions throughout her life in and around Pagosa Springs, Colorado, as documented by Sucik. Perhaps the most dramatic incident described by her allegedly took place during or around the late 1930s when, following the loss of several lambs to an unknown predator, a Pagosa Springs rancher armed a shepherd and asked him to guard the remaining flock. This he did, very successfully—by shooting dead a large, mysterious creature deemed to be the predator, whose carcase was then placed on a sled and hauled back to the ranch by one of its Apache ranch hands, using a team of mules.

Basilisk running bipedally (CC-SA Ryan Somma)

After being deposited inside a barn there, this specimen was viewed by many local farmers, including Snow's grandfather, who took Snow (then still a girl) with him, so that she could see it too. In 1982, following its publishing an article on the subject of whether dinosaurs had been cold-blooded, Snow wrote to the *Rocky Mountain Empire Magazine* (the Sunday supplement of the *Denver Post* newspaper), describing all of her alleged encounters with supposed modern-day dinosaurs, including her close-up viewing of the deceased mystery beast shot by the shepherd. She described it as:

> . . . about 7 feet tall, gray in color, had a head like a snake, short front legs with claws that resembled chicken feet, large stout back legs and a long tail.

Curiously, when interviewed more recently by Sucik, Snow also claimed that its body had been covered in fine grey hairs.

If so, then parsimoniously this tends towards a mammalian rather than a reptilian identity for the creature, though I am aware that certain fossil pterosaurs were hairy (but these 'hairs' were quite different structurally from true mammalian hairs and are called pycnofibers), and also that the plumes of certain types of fossil feathered non-avian dinosaur, e.g. *Sinosauropteryx*, were filamentous and therefore quite hair-like. Snow stated that the only observer who was apparently familiar with this extraordinary creature was the Apache ranch hand, who claimed that it was what elders on his reservation referred to as a moon cow, and which they said had been seen periodically in the past but were nowadays rare.

Worth noting here is that although the name 'moon cow' seems strange, it just so happens that a very similar name, 'moon calf', is a rural term used widely across Europe for an aborted, teratological foetus of a cow (and sometimes that of other farm animals too). The term derives from the once-popular folk belief that these malformed creatures resulted from a sinister effect of the moon. It has also been used more generally to refer to anything monstrous or grotesque in form. Could this distinctive term have reached North America with the original Western settlers and subsequently become incorporated into Amerindian parlance too, or might the notion of lunar influence upon the occurrence of malformed creatures have simply arisen independently on both sides of the Atlantic? In view of the creature's alleged hairiness, I cannot help but wonder whether this was its true identity, i.e. a malformed mammal of some kind.

Ironically, there may have been a chance of finding out for sure, because according to Snow the creature's carcase was packed in ice and sent by train to the Denver Museum for examination and identification. When this claim was pursued on Sucik's behalf by an archivist at the museum, however, no record of any such specimen ever even being received there could be found.

Conversely, what had definitely been sent to this museum and formally examined there, as Sucik discovered, was a box containing two alleged 'baby dinosaur' skeletons—one having been found inside a mineshaft, and the other inside a cave near to it, in Cortez, Colorado. Such creatures were apparently familiar to the local Navajo people, and during the early 1960s one of the skeletons was displayed at Cortez Museum. It was following the latter museum's eventual closure that the box was then sent to Denver Museum, but when the skeletons were examined there they were found to be composites, constructed from the bones of various different species of mammal. In other words, they were hoaxes.

Perhaps the most remarkable mini-rex claims of all, however, are a couple attesting to the alleged capture of living specimens, one of which was even supposedly kept as a pet for a while by its captors. This latter case featured testimony given to one of Sucik's correspondents by his aunts. They claimed that some time between the end of the Great Depression in America (late 1930s) and the early 1940s, while following the crop harvests from state to state as they travelled out west, what they called a 'baby dinosaur' and which seemed to them to resemble a tiny *T. rex* would come to their camp when their mother was cooking outside. One day, they succeeded in catching it, and afterwards kept it inside an old bird cage for a time, feeding it on leftovers, and finding that it would eat both meat and vegetables. According to Sucik's account of this fascinating little animal:

> It was described as having sharp little hooks on its hands and very sharp teeth, like that of a kitten. Its skin was like a lizard's but felt warm. It never tried to bite or scratch but it did not like being held. The animal behaved "like a tame squirrel." During the time they kept their pet, it grew from the size of a kitten to roughly the size of a cat, by which time it was far too big for the cage.

Eventually their family had to move elsewhere in order to follow the crops, so their father told them to leave their pet behind. Of especial interest here is that when the aunts first told their nephew about their most unusual former pet, back in the 1970s, they remembered that when it ran it "flattened out, stretched its head out front, tail out back and was really fast". As pointed out by Sucik, whereas this running posture is widely accepted nowadays by palaeontologists for bipedal fossil dinosaurs, it wasn't back in the 1970s. So if the aunts had been making up their story, they were remarkably prescient concerning this particular facet of dinosaur behaviour. Just a coincidence?

The other case of reputed mini-rex capture came to light when a lady wrote to Sucik to inform him that her three boys had once caught such a creature in New Mexico. The specimen in question was unusually large but slow-moving, and when they caught it they could see that it was old, with fainter body colouration than other, smaller, faster specimens that they had previously seen (but had never succeeded in capturing). Its body alone measured 20-24 in, i.e. not including its lengthy tail, but after admiring this impressive creature for a while, the boys released it. Interestingly, they referred to it as a mountain boomer, a colloquial name normally applied to collared lizards—but these latter reptiles never attain the size of the creature that the three boys had caught.

What could the mini-rex be? As with so many cryptids, it may well be a non-existent composite, i.e. it has been 'created' by the erroneous amalgamation of reports featuring various totally different species. The smaller individuals may indeed be nothing more than lizards sprinting bipedally, yet species like the collared lizards and common chuckwalla are so abundant and familiar that it seems difficult to believe that people living in locations where they occur would not recognise them for what they were. Perhaps, however, some mini-rex reports involve less familiar lizard species that are currently not known by scientists to be capable of bipedal locomotion. Yet the posture and habitual bipedalism reported for these cryptids do not accord well with lizards anyway, regardless of species.

A common collared lizard *Crotaphytus collaris* (CC-SA Ejohnsonboulder)

So is it possible that somewhere in the more rural regions of Colorado and elsewhere in the United States there really are bona fide 21st-Century bipedal dinosaurs (albeit of quite modest dimensions), sprinting along in blissful ignorance of their official demise 66 million years ago? Needless to say, it seems exceedingly unlikely. Yet there is surely little doubt that something corresponding at least ostensibly with such an identity is indeed out there.

The last word on this most tantalising of topics should go to its foremost investigator, the indefatigable Nick Sucik, whom I have known for many years and greatly admire for his diligent, meticulous, and, above all else, tenacious pursuit of answers to cryptozoological riddles from all around the globe:

> . . . such tales, whether they be true or not, add to the folklore of what our world would be like if dinosaurs lurked in secret and only stepped out into the clearing momentarily to be seen. If there were or are such things, they perhaps would possess the safety of being too unbelievable.

An excellent point—for there can be no doubt that it is this precise quality of unbelievability that serves so many cryptids so effectively in keeping them out of the clutches of scientific recognition. After all, who is going to admit to having seen creatures that sound too unlikely to be real, and who is going to seek them even if their eyewitnesses do admit to having seen them? Thank Heavens, therefore, for the Nick Suciks of this world, who choose dispassion over disbelief. May their zeal be rewarded one day with the discovery of those cryptids that they seek.

Before moving on from encounters with mystery bipedal reptiles in North America, mention should also be made of a veritable 'maxi-rex' of sorts, in the form of the Trimble County giant lizard. This is the name given to an allegedly huge but unidentified bipedal reptile reported on several occasions at Canip Creek near Milton in Trimble County,

Kentucky, during July 1977, investigated by the local *Trimble County Banner* newspaper as well as by American cryptozoologist Mark A. Hall.

The first sightings of this cryptid occurred in the grounds of the Blue Grass Body Shop, a junk and wrecking yard, by the yard's co-manager, Clarence Cable. He stated that it had "big eyes similar to a frog's. . . . Beneath its mouth was an off-white color and there were black and white stripes cross ways of its body with quarter-sized speckles over it". He also said that it was similar to, but not exactly like, a monitor lizard, was nearly 15 ft long, and ran on its hind legs. Yet if it were indeed a monitor and its size had been estimated accurately, it was far bigger than any known species (other than top-end specimens of Salvadori's monitor *Varanus salvadorii* from New Guinea), and as already noted in this chapter, monitors only run bipedally for very brief periods. Certain superficially monitor-like lizards native to South America known as tegus have escaped from captivity in North America (and even established wild populations in Florida). Moreover, they are indeed patterned with spots and crosswise black and white stripes, but are no more than 4 ft long and are not habitually bipedal.

Following an unsuccessful search of the yard and the surrounding area for it during early August 1977, no further reports of Trimble County's giant lizard emerged—and neither did the lizard. Consequently, its origin and identity remain unsolved cryptozoological riddles to this day.

THE TANTALISING TYRANNOSAUR OF HAVA SUPAI

A cave paintings of a sauropod in Darkest Africa may be one thing, but the ancient depiction of

Scale reconstruction of the Trimble County giant lizard's possible appearance (© Connor Lachmanec)

a tyrannosaur in an Arizona canyon is another matter entirely—which is why the following case has generated so much interest, and controversy, over the years.

In 1924, the Doheny Archaeological Expedition, led by Dr Samuel Hubbard from California's Oakland Museum, made a remarkable discovery in the Hava Supai canyon of northern Arizona. Incised in the solid rock at the canyon's gorge were some ancient carvings, one of which seemed to bear more than a passing resemblance to the side-on outline of a bipedal dinosaur! Most publications documenting this remarkable intaglio liken it to a tyrannosaur standing erect on its long hind legs and supported by its tail (but Brian Newton in his book *Monsters and Men* favours a dome-crested hadrosaur called *Corythosaurus*), equipped with noticeable jaws, a fairly long neck, and stout body.

The carving is not particularly large—the 'dinosaur' stands 11.2 in tall, with a 3.8-in-long hind limb, and a 9.1-in-long tail. Nevertheless, it is certainly eyecatching, because its outline is bright red, which yields a striking contrast with the black background colour. The black background is in reality a glossy varnish-like substance, produced by gas that gradually seeped through the canyon's sandstone and coated its outermost layer. Conversely, the red shade of the carving itself is iron oxide—normally concealed beneath the sandstone's surface but exposed in the grooves of the carving, because its creator chiselled through the varnish layer, thus penetrating the iron oxide underneath.

This enigmatic petroglyph has been offered up as evidence for all sorts of radical ideas—ranging from the suggestion that a race of tyrannosaurs somehow survived right up to the time of human entry into the New World, to the notion that *Homo sapiens* is far older than evolutionists have claimed and was actually alive at the time of the dinosaurs back in the Cretaceous. To my mind, however, even if the carving does depict a dinosaur (and I am definitely not convinced about this), there is a much simpler explanation than either of those noted so far here.

Early humans certainly encountered fossils, and were undoubtedly curious to know what they were. When a near-complete skeleton was visible, it would have been tempting to try to reconstruct its appearance in life, and the Hava Supai carving may have been one such attempt—undertaken by an artist after seeing the partially-exposed skeleton of a tyrannosaur or some other bipedal dinosaur, or after receiving its description from someone else.

Plausible precedents for such a situation are not unknown. In 1991, Princeton classical scholar Dr Adrienne Mayor suggested that legends concerning those fabulous lion-bodied, eagle-headed beasts known as griffins may have been influenced by early gold prospectors in central Asia's Altai Mountains encountering exposed, fossilised skeletons and eggs of a late Cretaceous ceratopsian dinosaur called *Protoceratops*—a lion-sized quadruped beast with an eagle-like beaked head. And back in 1916, J. O'Malley Irwin entertained a similar line of speculation linking the Chinese dragon's origin with discoveries of Oriental sauropod fossils.

No less contentious than the supposed tyrannosaur petroglyph of Hava Supai, Arizona, is the leaden sword found with other objects in a lime kiln near Picture Rocks, Tucson, Arizona, in 1924 by Charles E. Manier and his family—because a sauropod-like beast is inscribed across over half of its blade's length. Inscriptions on some of the other objects imply a Roman-Judaic origin—so how can their burial in an Arizona desert (let alone the sword's depicted

The Hava Supai petroglyph

Acámbaro figurine reminiscent of a stegosaurian dinosaur (CC-SA Brattarb)

dinosaur) be explained? These artefacts have incited considerable controversy, gaining supporters and sceptics, some of the latter citing a number of locals, including Manier himself, as the creators or originators of these finds. Others, conversely, dispute such claims, and believe the Tucson artefacts to be genuine.

A strange creature resembling a somewhat long-limbed sauropod is one of several images carved into a strange granitic statuette nowadays termed the Granby Stone. Measuring 14 in high, it was dug up from a depth of 6 ft in 1920 by some ranch hands working on the William M. Chalmers estate near Granby in Colorado. The statuette has since been lost, but some good-quality photographs still survive. Even more curious than its supposed sauropod image, however, is one that experts claim to be of a stylised human figure—bearing on his chest an ancient Chinese inscription, believed to date back to around 1,000 BC.

ANACHRONISMS FROM ACÁMBARO, ICA, AND ELSEWHERE

Controversial pre-Columbian portraits of dinosaur-like beasts in the New World are not confined to the U.S.A., as demonstrated by the following examples from Latin America.

Between 1945 and 1952, storekeeper Waldemar Julsrud and various locals that he hired as assistants unearthed a vast cache of exotic-looking figurines—numbering some 32,000 and variously composed of jade, obsidian, ceramic, and stone—at Acámbaro, in the Mexican state of Guanajuato. Whereas many represent typical animals of that region, others appear to be wholly unknown beasts, and some are faithful renditions of such prehistoric stalwarts as stegosaurian dinosaurs and *Camelops*—America's last true camel, generally believed extinct since the Pleistocene epoch, which ended 11,700 years ago (but see Chapter 4 for details of a single more recent skull on record).

In 1952, archaeologist Charles DiPeso paid Julsrud a short visit to examine some of the countless figurines stored at his home, and he also observed a couple of excavators digging up more figurines on site. From his brief observations, DiPeso concluded that these artefacts were fake, claiming that their surfaces displayed no sign of age, that they lacked any dirt in their crevices, and that there were no fragments missing from examples that were broken.

Conversely, when American college historian Prof. Charles Hapgood sent some of the figurines to a New Jersey laboratory in 1968 for radiocarbon dating, the results obtained suggested dates ranging from 5430 BC to 1640 BC—none of which is consistent with the general opinion among archaeological circles that the figurines are modern-day forgeries.

More recently, a number of thermoluminescence (TL) dating tests have been conducted upon some of these enigmatic artefacts, but with varying results and interpretations. One such test gave results indicating that those figurines examined by it had been fired as recently as the late 1930s. In general, however, it would appear that dates obtained for them via such techniques are unreliable and chronologically insignificant.

In March 1988, a *Nueva Acrópolis Revista* paper by Spanish archaeologist Prof. Jorge A.

Ica stone depicting alleged dinosaurs (CC-SA Brattarb)

Ica stone depicting a human stabbing a sauropod-like animal (CC-SA Brattarb)

Livraga Rizzi described his discovery in western Guatemala of several small, ceramic animal heads bearing more than a passing resemblance to those of certain famous dinosaurs, including *Tyrannosaurus*, and the hadrosaurs *Corythosaurus* and *Anatotitan* (aka *Anatosaurus* and *Trachodon*). Laboratory analyses have yielded an age of 3,000 years or so for them—if correct, they were clearly not influenced by modern-day dinosaur memorabilia or depictions.

The jungle caves of Cuenca in Ecuador have offered up hundreds of exquisite artefacts, including many gold pieces, during the past century. Securely held within the Church of Maria Auxiliadora, where they have been personally cared for by one of its priests, Father Carlo Crespi, some depict Egyptian and Assyrian motifs, and one—a baked-clay tablet—bears upon its surface an excellent portrait of a sauropod dinosaur.

The most notable, and notorious, series of anachronistic illustrations in the New World, however, is unquestionably the enormous collection of decorated stones obtained by anthropologist/surgeon Dr Javier Cabrera Darquea from Ica Province in Peru, and from the Ocucaje Indians. Composed of an extrusive igneous volcanic rock called andesite, they number in excess of 20,000, range greatly in size, and occupy an entire museum devoted to their extraordinary depictions. These aesthetically-beautiful engraved stones were first unearthed in 1955, and bear an unparalleled selection of inexplicable images. Dinosaurs of many different types can be clearly discerned—sauropods, tyrannosaurs, stegosaurs—as well as prehistoric sea lizards called tylosaurs, pterodactyl-like beasts ridden by humans, and even (in the opinion of Cabrera) advanced techniques in surgery such as heart transplants!

Hardly surprisingly, most archaeologists have expressed grave doubts as to their authenticity, and Cabrera himself was well aware that his own interest in these items stimulated a number of enterprising local artists to create their own versions in the hope that he would purchase these too. Indeed, even the farmer Basilio Uschuya from whom Cabrera purchased many of his earliest-obtained stones eventually claimed in interviews that he had carved them himself, copying the images from comic-books, magazines, and text books. Then again, he also revealed that selling genuine archaeological artefacts would result in his being jailed by the Peruvian authorities, so it would make sense for him to claim publicly that what he was selling were only modern-day forgeries.

In any case, Cabrera asserted that he could readily distinguish the ancient originals from modern-day versions. Moreover, after he submitted some of the ancient examples to the Mauricio Hochschild Mining Company in Lima for formal examination, the report that he received back in June 1976 from company geologist Dr Eric Wolf revealed that the stones and the grooves of the depictions upon them were covered in a fine but natural oxidation film—implying that stones and depictions alike were of a very great age. In 2000, Cabrera published a book documenting and illustrating his contentious collection, entitled *The Message of the Engraved Stones of Ica*.

How can all of these anachronistic iconographical enigmas be explained? The various tests performed upon samples of them seem to dismiss hoaxes as a comprehensive answer to their existence. If, then, at least some of them are as old as the tests indicate, only two answers appear to remain. Either the depictions were based upon a rich diversity of *living* dinosaurs that have somehow survived into modern-day Peru undetected by science (highly unlikely to say the least!), or they were created by artists inordinately well-versed in interpreting the likely appearance in life of animals whose fossils they occasionally encountered.

WHERE "NOTHIN' WOULD SURPRISE"—MONSTROUS REPORTS FROM SOUTH AMERICA

In Sir Arthur Conan Doyle's novel *The Lost World*, adventurer Lord John Roxton, one of the members of the formidable if fictitious Professor Challenger's expedition, acknowledged that in the vast Mato Grosso area of Brazil and elsewhere amid the scarcely penetrable rainforests and mountainlands of South America ". . . nothin' would surprise"—a sentiment expressed by many real-life scientists too. But does this optimism extend to the prospect of living dinosaurs existing there?

Strangely, despite the environmental potential for such creatures (as expressed at some length during the 1970s, for example, by Dr Silvano Lorenzoni in the magazine *Pursuit*), South America has not attracted a fraction of the cryptozoological interest that has been focused upon such issues in Africa. Consequently, reports on file tend to be fewer in number and less precise in description. Also, whereas Africa's putative dinosaurs fall into certain well-defined morphological categories, South America's are more diverse and ill-defined.

For instance, every so often a distressingly vague account describing giant amphibious reptiles emerges from this continent's immense northwestern section, and over the years all manner of identities have been put forward. Some such beasts, with relatively short necks, have been likened to various ornithopod dinosaurs, such as hadrosaurs or iguanodontians. One case in point was a supposed *Iguanodon* reported in Colombia's Magdalena River during 1921. (Incidentally, some years before he died in 1938, Mayan archaeologist Dr Thomas Gann encountered tracks near one of the ancient jungle-shrouded cities in Mayan Yucatan that he too thought were from a surviving *Iguanodon.*)

Another dinosaurian cryptid was dubbed the Mesquita camptosaur, after an old Brazilian Indian named Alvaro Mesquita. He informed Swedish naturalist Rolf Blomberg that one night on the shore of a swampy lake around the Rio Purús/Rio Juruá area in the Amazon basin he had encountered a huge green bipedal reptile "larger than a cow", which was likened to the large, heavy, plant-eating ornithischid dinosaur of the late Jurassic from which this cryptid subsequently derived its colloquial name. Mesquita also claimed that he had shot at the creature, but missed, whereupon it fled into the lake and disappeared from sight. Blomberg viewed Mesquita's testimony with considerable scepticism, but nonetheless documented it in his book *Rio Amazonas* (1966).

Artistic rendition of Mesquita camptosaur, with human silhouette for scale purposes (© Connor Lachmanec)

Even more perplexing was the beast allegedly spied in or around 1931 by Swedish explorer Harald Westin. While travelling down the Marmore River in Brazil's vast Mato Grosso, he reputedly saw a most peculiar animal walking along its shore. Greyish in colour, about 20 ft long, with a head like an enlarged alligator's, it apparently reminded him a little of a distended boa constrictor—but unlike any boa constrictor, it possessed four lizard-like feet, and its small eyes glowed scarlet as it raised its reptilian head to look at him.

Taking fright at encountering a veritable monster in such close proximity, Westin shot it—but apart from making a clucking sound as it hit its target, the bullet apparently had no effect at all. Instead, the creature simply continued upon its way, seemingly unharmed and undisturbed. Whatever it was, it certainly does not bring to mind images of ornithopods—or of anything else currently known to science, for that matter.

It's always good to learn of a novel cryptid, and Chile's Arica Beast is novel in every sense! As recounted in a number of media articles from 2004, several different motorists driving along the main road linking Iquique and Arica, through the Atacama Desert, have reported witnessing an extraordinary bipedal creature over 6 ft tall, with sharp teeth, leathery skin, and three-toed footprints, which has been variously likened to a very large velociraptoresque dinosaur or even a 'dinosaur kangaroo'! In the words of one eyewitness, Hernan Cuevas: "A weird animal looking like a dinosaur with two legs and huge thighs crossed the road in front of my car". Not surprisingly, the local authorities were, and remain, very puzzled.

Following an unsuccessful search for this cryptid that was featured in America's *Destination Truth* television series in October 2009, a large three-toed footprint was obtained, and the team consulted a palaeontologist who considered that it was probably a common (greater) rhea *Rhea americana*—South America's large three-toed ostrich-like ratite. However, the Arica Beast is allegedly bare-skinned, whereas the rhea is profusely feathered, a major morphological discrepancy that would seem, therefore, to place the latter bird out of contention after all as an identity for this bipedal mystery creature. Moreover, just to make matters even more perplexing, prehistoric velociraptors are now known to have been feathered too.

Reports of long-necked reptiles are no easier to deal with either. Unlike the distinctly sauropod-like forms on file from Africa, most of South America's versions seem more akin to plesiosaurs, in both appearance and behaviour. A selection of these can be found in Chapter 3. However, there are a few possible exceptions.

In October 1907, a German traveller called Franz Herrmann Schmidt journeyed up Colombia's Solimoes River with Captain Rudolph Pfleng, reaching a remote region of swamps and jungle valleys—and ultimately encountering a long-necked monster of uncommonly intriguing appearance. It was the twelfth day of their expedition, and after arriving in a long valley that had become a shallow lake they spied some huge tracks on its shore.

After paddling across the lake to reach them, the travellers inspected the tracks and

Does the Arica Beast truly resemble an extra-large featherless *Velociraptor*? (© Dr Karl Shuker)

Might a rhea explain reports of the Arica Beast? (Greenshed)

concluded that they were the impressions left behind by a huge animal the size of an elephant or hippo, leaving and then re-entering the lake. At one site, three tracks were found together—one large, the other two only half the size, as if an adult and two youngsters had been there together. They also noticed that great quantities of vegetation had been torn down, some from a height of 14 ft above the ground. Something appeared to have been browsing there—and the next morning the explorers and their helpers found out what it was.

That day, they had come upon a muddy track, one indicating something leaving and then re-entering the water, and so fresh that it was still wet. Suddenly, some monkeys in the forest nearby began to shriek, and the frightened men sitting in their boat were just able to discern a dark shadow shoot up among them before one of the Indians began paddling away as fast as he could. In moments, they were about 100 ft away from the shore—which was probably no bad thing, for as they looked back, they saw a barrel-sized head rise up over some 10-ft-tall bushes on a long snake-like neck.

Its eyes were small, and it possessed a tapir-like snout, but although it moved towards them the creature did not seem alarmed by their presence. As it gradually emerged from among the foliage, they could see that it was 8-9 ft tall at the shoulders, but instead of possessing fully-formed forelegs and feet it was equipped with a pair of massive flippers. Unlike those of plesiosaurs, however, these bore claws.

Although it did not appear aggressive, Pfleng decided that he was taking no chances with such a behemoth of a beast, so he began shooting at it, and Schmidt swiftly followed his lead. Yet although several bullets hit it, none seemed to have any effect—other than to induce the great creature to leave the forest and dive into the water—briefly revealing a heavy blunt tail furnished with rough horny lumps. Its head remained visible above the surface, and during its swift progression from land to lake its total length was estimated by the travellers to be 35 ft, of which at least 12 ft constituted the head and neck.

After carefully checking this astonishing account (which had appeared on 19 April 1913 as a major front-page article by an apparently uncredited Schmidt in the *Globe* newspaper of Sydney, Australia), Prof. Roy Mackal was inclined to accept it as authentic. Even so, what can we say about a creature that seems unable to decide whether it is a sauropod or a plesiosaur—tantalisingly combining the former's

Pfleng's flippered mystery beast that appeared in the *Sydney Globe* newspaper

overall form and plant-eating proclivity with the latter's flippered forelimbs?

I hesitate to suggest that it could have been anything as radical as a highly-evolved sauropod whose limbs were in the process of transforming into flippers, just as those of plesiosaurs, cetaceans, ichthyosaurs, and various other aquatic animals had done at some stage during their own respective evolution—but such an identity would fit the bill very satisfactorily. Having said that, in his *Globe* article Schmidt also claimed that the team had encountered a sleeping boa constrictor measuring an estimated 65-70 ft long. Consequently, I am not entirely sure about how seriously we should rate their testimony.

In 1975, while holidaying in the Amazon, a Geneva businessman (name unknown) met Sebastian Bastos—a 75-year-old guide who, by a lucky coincidence, had been educated in Switzerland and was therefore able to converse fluently with him. During one conversation, Bastos reluctantly revealed that not only were there areas of this vast jungleland that harboured enormous long-necked monsters, but he had once almost been killed by one!

Several years earlier, after beaching his canoe by a river deep in the rainforest during a period when the water level was exceptionally low, Bastos had been walking alongside it towards an Indian friend whom he had arranged to meet there—when suddenly he heard a tremendous noise behind him. Looking round, he saw to his horror that an enormous monster had risen from the water, and was in the process of ripping his boat apart. As the two terrified men ran away, they briefly looked back to see if the creature was following them, but saw to their great relief that it had submerged beneath the surface. According to some of Bastos's Indian acquaintances, beasts of this type frequented certain deep water holes in the jungle's heartland, and came out onto land occasionally at night. Their heads, necks, and backs were about 18 ft long, and the Indians took pains to avoid them.

There are native accounts from Africa telling of mokele-mbembes destroying boats, such creatures are of similar size to Bastos's beast, and they do appear to venture onto land at rare intervals. Even so, there is insufficient detail in Bastos's account to discount entirely a plesiosaur identity from consideration.

Even vaguer is a mention in *Exploration Fawcett* (1953) by famous lost explorer Lt-Col. Percy H. Fawcett. Alluding to unidentified beasts in the forests of Bolivia's Madidi, he noted:

Photograph of British artillery officer and explorer Lt-Col. Percy H. Fawcett, who vanished without trace in or after 1925 while leading an expedition through the uncharted jungles of Brazil seeking a long-lost city

> . . . some mysterious and enormous beast has frequently been disturbed in the swamps—possibly a primeval monster like those reported in other parts of the continent. Certainly tracks have been found belonging to no known animal—huge tracks, far greater than could have been made by any species we know.

The mention of tracks suggests footprints—which favour dinosaurs over flippered plesiosaurs—but little else can be deduced from such a brief description.

Fortunately, an earlier, more detailed Fawcett-authored account on this same subject also exists—a letter by him that was published in London's *Daily Mail* newspaper on 17 December 1919. The relevant portion is as follows:

> The Congo swamps are not the only region suspect of harbouring relics of the Miocene age.
>
> As I hinted in lectures in London some years ago, a similar beast is believed to exist in South American swamps. A friend of mine, a trader in the rivers and for whose honesty I can vouch, saw in somewhere about Lat. 12 S. and Long. 65 W. [Bolivia-Brazil borderland] the head and neck of a huge reptile of the character of the brontosaurus. It was a question of who was scared most, for it precipitately withdrew, with a plunging which suggested an enormous bulk. The savages appear to be familiar with the existence and tracks of the beast, although I have never come across any of the latter myself. . . These swamps over immense areas are virtually impenetrable.

Incidentally, when American explorer-adventurer Col. Leonard Clark documented a South American expedition of his own, through the dense, poorly-charted jungles of Peru in 1946, within his classic travelogue *The Rivers Ran East* (1954), he briefly referred to Fawcett's account of supposed swamp-dwelling dinosaurs. So too, in his introduction to Clark's book, did United States consul Lewis Gallardy, who also added that the existence of such creatures "is confirmed by many of the tribes east of the Ucayali [a Peruvian river], a region covered by Clark". Sadly, Clark did not encounter any himself, but as an avowed, fearless trailblazer who was certainly intrigued by these tantalising native tales, it is not beyond the

Charles Knight's spectacular if superseded 1897 painting of Mesozoic aquatic sauropods

realms of possibility that, had he lived, he would have launched an expedition at some stage in the future to investigate them. Tragically, however, Clark died in 1957, just three years after his book's publication, aged only 50, during an expedition through Venezuela's jungles seeking diamonds for mining.

Even so, in view of its author's eminence, Fawcett's account is a particularly welcome addition to South America's sparse file on putative living dinosaurs (his Miocene error notwithstanding!). Moreover, it acquires additional importance in the light of the following, much more recent report.

According to media accounts from early 1995, a party of geology students investigating quartz deposits in the Sincora Mountain range of eastern Brazil had lately spied two strange dinosaur-like creatures bathing in the shallows of the Paraguaçu River passing through the Plain of Orobo. According to these eyewitnesses, the animals were each around 30 ft in total length, with a huge body, fearsome head, a long neck measuring approximately 6 ft, and an 8-ft tail.

Reading these accounts reminds me irresistibly of veteran dinosaur illustrator Charles Knight's iconic illustration from 1897, depicting river- and swamp-dwelling prehistoric sauropods—long since rendered obsolete, alas, in palaeontological circles where it is now believed that such dinosaurs were more likely to have been predominantly terrestrial back then.

But what about today? If the sauropod lineage has indeed persisted and continued to evolve from the end of the Mesozoic through a further 66 million years into modern times within the depths of South America, might its evolved members have become secondarily amphibious? Perhaps Knight's painting is still relevant after all.

THE STOA AND THE SUWA—DINOSAURIAN SURVIVORS FROM THE *REAL* 'LOST WORLD'?

It is not widely known, but when writing his famous novel *The Lost World* (published in 1912), in which dinosaurs, pterosaurs, and other Mesozoic reptiles have survived into the present day amid a totally isolated realm present on the plateau at the summit of a very high tepui (a vertically-sided, flat-topped or table-topped mountain in South America), one of Sir Arthur Conan Doyle's inspirations was a real but still highly mysterious tepui known as Kurupira. Named after the curupira, a legendary Amazonian man-beast-like entity, this particular tepui stands 3,435 ft above sea level, and is situated on the Venezuelan-Brazilian border.

Conan Doyle had learnt of Kurupira from the previously-mentioned explorer Lt-Col. Percy Fawcett, who in 1908 had led an expedition to some great sandstone tepuis in Bolivia known as the Franco Ricardo hills. There are more than 100 tepuis in South America, and at 9,220 ft above sea level Mount Roraima is the highest (and also the largest) in the Pakaraima chain on the borders of Brazil, Venezuela, and Guyana. Although they did not encounter any prehistoric creatures in Bolivia, Fawcett and his team did receive various native reports of frightening monsters said to inhabit Kurupira and its environs by the local Waiká Indians who inhabit the jungle area around the vicinity of its base, and Fawcett's recollections of these reports provided Conan Doyle with further plot ideas during his novel's preparation.

In particular, he was enthralled by Fawcett's tales of an exceedingly voracious bipedal reptile known to the Waiká as the stoa, which was investigated more recently by Czech zoologist Jaroslav Mareš, who documented some of his findings in his cryptozoological encyclopaedia *Svet Tajemných Zvírat* ('*The World of Mysterious Animals*'), published in 1997. Mareš spent time residing at Kurupira's base during an expedition there in 1978 (sadly, their attempts to scale this tepui's steep sides proved unsuccessful), and learnt about the Waiká Indians' belief in the stoa and other alleged monsters here. They described the stoa as measuring up to 25 ft long and superficially resembling a giant-sized caiman (several species of these South American alligator relatives are known, but all are of far smaller size). However, they also stated that it can be readily distinguished from such reptiles by way of the following major differences.

First and foremost of these was the very notable claim that the stoa is exclusively bipedal, moving entirely upon its two gigantic hind legs, because its front limbs are so short that it cannot stand upon them. Its jaws are much shorter than a caiman's too, but its head is taller, and it bears a pair of prominent horns above its eyes, which are somewhat reminiscent of those sported by the South American horned frogs *Ceratophrys* spp. The Waiká likened its body colouration to theirs too (i.e. green or golden-brown with darker stripes), but its mouth is not as wide as that of these famously wide-mouthed frogs, and its skin is covered with hard, non-overlapping, tubercular scales. Above all, they affirmed that there is never any hope of escape if pursued by a stoa.

Moreover, Mareš revealed that this Indian account was confirmed by the missionaries from the Porto da Maloca settlement on the upper Rio Mapulau, located approximately 15 miles from Kurupira as the crow flies. However, they did not believe that the stoa is real. For them, it is just a part of Waiká mythology.

Mareš has also written three books specifically devoted to Kurupira and its mysteries—*Hledání Ztraceného Sveta* (*'In Search of The Lost World'*), which documented his 1978 expedition and was published in 1992; *Hruza Zvaná Kurupira* ('The Horror Named Kurupira'), published in 2001; and *Kurupira: Zlovestné Tajemství* ('Kurupira: Sinister Secrets'), published in 2005. In the second of these three, Mareš mentioned meeting during spring 1997 at Boa Vista (capital of Roraima, Brazil's northernmost state) a Scottish gold-prospector whose real name Mareš has not publicly disclosed, referring to him instead only by the pseudonym 'Reginald Riggs'. Mareš had previously met Riggs in 1978, during his previously-mentioned expedition to Mount Kurupira, and in his 2001 book he revealed that while Riggs was prospecting in the vicinity of Kurupira he had befriended a Waiká tribesman named Retewa, who supplied him with information concerning the stoa, another dinosaurian cryptid called the suwa, and a pterosaur-like beast termed the washoriwe (see Chapter 2).

According to Retewa (via Riggs), the stoa's most common prey are tapirs. Apparently, it conceals itself in dense forest close to a riverbank where these large horse-related ungulates bathe, then abruptly emerges to attack them when they arrive there. It will also devour capybaras, those sizeable pig-like rodents that occur here too. One account related by Retewa to Riggs concerned a reputed confrontation between some hunters from his village and a stoa that they inadvertently encountered while it was looking out for prey. They shot at it with their arrows, but they failed to penetrate its hard, scale-protected skin, and the enraged stoa killed several of them before the others fled.

In an attempt to explain both the origin of the Waiká's firm belief in the stoa and (as he also discovered during his investigations) the complete absence of any such belief among Indian tribes living further out from Kurupira, Mareš has cautiously offered the following thought-provoking theory. He suggests that if the stoa is indeed real, perhaps its species is normally confined entirely to this tepui's lofty isolated plateau, but that a single individual may very occasionally find its way into their ground-level territory via a crack or fracture leading down the tepui from its summit to its base, after which the Waiká live in great fear of it, thereby maintaining and reinforcing its presence in their minds and lore for another generation or so until the next accidental visitation occurs.

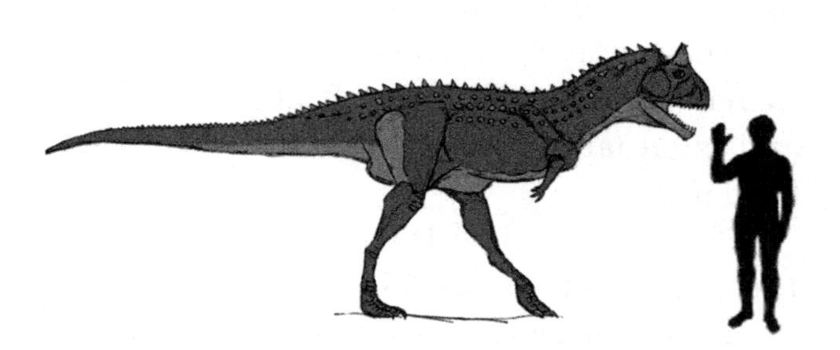

Artistic rendition of the possible appearance in life of the stoa, alongside a human for scale purposes (© Connor Lachmanec)

As for what the stoa may be, taxonomically speaking, if it does truly exist: in his cryptozoological encyclopaedia, Mareš noted that during the Cretaceous, South America was home to a taxonomic family of theropod dinosaurs known as the abelisaurids, which were bipedal, carnivorous, and, in some cases, extremely large. The most famous abelisaurid was *Carnotaurus sastrei*, which was up to 30 ft long, and as noted by Mareš it also happens to be potentially relevant to the stoa for two very different but equally intriguing morphology-based reasons. Firstly: dating from the late Cretaceous and disinterred in 1984 from the La

Restoration of the likely appearance in life of *Carnotaurus sastrei* (CC Lida Xing and Yi Liu)

Colonia Formation in Argentina's Chubut Province, its only recorded but exceptionally well-preserved fossilised skeleton shows that this particular abelisaurid species bore a pair of sharp pointed horns above its eyes, just like the stoa (*Carnotaurus* translates as 'flesh-eating bull'). Secondly: this skeleton is so well preserved that it reveals that the skin of *Carnotaurus* bore hard non-overlapping scales all over it, just like the stoa.

Coupled with the overall similarity in outward form and size between *Carnotaurus* and the stoa, these unexpected matching features led Mareš to speculate as to whether this abelisaurid's lineage may have escaped the mass extinction at the end of the Cretaceous and has possibly lingered on through the Cenozoic Era into the present-day here in this very remote South American location, isolated atop a high tepui except for rare occasions when one might find its way down into the junglelands at Kurupira's base.

The stoa was not the only putative dinosaur of Kurupira spoken about by Retewa to Riggs. He also claimed that up on this tepui's plateau lives another very strange creature, known to the Waitá as the suwa, a picture of which he drew in the sand for Riggs to see, and a copy of which Riggs in turn drew in his diary, later seen by Mareš. The picture shows a bulky, long-necked, quadrupedal creature, which Riggs likened to a sauropod dinosaur or even a plesiosaur (however, its limbs were clearly portrayed in the drawing as legs, not flippers).

Yet amidst all of these claims of Mesozoic monsters alive and well and living in splendid isolation on Kurupira's lofty plateau, there is a key question desperately needing to be asked. For even if we accept that a stoa may very occasionally find its way down from this tepui's summit to its base, the very burly, quadupedal, sauropod-like form of the suwa unequivocally debars this cryptid from following suit—so how can the Waiká be aware of its existence? Interestingly, Riggs actually asked Retewa how his people could know what exists on the plateau at the top of Kurupira, but Retewa was unable to provide an answer. So perhaps—as surmised by the missionaries—all of their claims regarding monsters are indeed based upon nothing more substantial than traditional Waiká mythology, with no basis in reality.

Alternatively, could it be that at least in earlier days, some of the Waiká's bravest warriors actually scaled Kurupira's daunting height, explored its plateau, and then returned to their tribe back on the ground with stories (exaggerated or otherwise) of what they had seen there? And, if so, perhaps what they saw there was so terrifying that they have never returned, but the original eyewitness reports have been preserved in their tribal lore down through succeeding generations. Who can say?

I wish to thank my friend Miroslav 'Mirek' Fišmeister from the Czech Republic for so kindly translating into English for me all of the relevant passages regarding Kurupira and the stoa, suwa, and washoriwe (see Chapter 2) from Mareš's books (plus the earlier-documented Polish neodinosaurs from Vašícek's). This has enabled me to present here the most extensive, accurate coverage of these cryptids ever produced in English.

Previously, the only English-language reports concerning the stoa, suwa, and washoriwe that I have been aware of, all of them online, have been sparse, confused, and sometimes entirely inaccurate. The principal reason for this inaccuracy stemmed from the fact that a prehistoric monster actually called the stoa appears in Conan Doyle's novel *The Lost World*, in which it is described as a warty-skinned, toad-like reptile, leaping on its hind legs, but larger than the largest elephant, and of frightful, horrible appearance.

This has inspired some erroneous online speculation, i.e. that there is no cryptozoological basis for the stoa, that it is entirely fictitious, a baseless invention of Conan Doyle for his novel. In reality, however, as I have now revealed here, it is the exact reverse scenario that is true. Namely, the stoa in his novel was directly inspired by reports of Kurupira's cryptozoological stoa as told to him by Fawcett.

Yet another longstanding example of online cryptozoological confusion finally elucidated and resolved.

MUNGOON-GALLI AND *MEGALANIA*

Several types of mysterious giant reptile have been reported from New Guinea. However, some of them are certainly not destined to join the ranks of prehistoric survivors.

The six-legged snakes reputedly inhabiting a mountain on the island of Waigeu (west of New Guinea), for instance, owe more to mythology than to Methuselah. A huge dinosaurian cryptid termed the row, allegedly encountered in New Guinea's Sterren Mountains by Charles Miller and his wife during their honeymoon in the 1940s, was dismissed long ago as a hoax by Heuvelmans, as discussed later in this present book. And the capture of a young specimen in 1980 during 'Operation Drake', led by Colonel John Blashford-Snell, revealed that reports of a dragonesque Papuan beast called the artrellia were based merely upon oversized specimens of Salvadori's monitor *Varanus salvadorii*.

Salvadori's monitor, the artrellia (Vassil)

Speaking of giant monitors, however: some such creatures have also been reported from many eastern portions of mainland Australia. Yet whereas those of New Guinea belong to a known, modern-day species, their Australian counterparts are believed by some to be bona fide prehistoric survivors—belonging to a giant species that at present is indeed known only from eastern Australia but is 'officially' deemed to have died out long ago.

A particularly important sighting took place in early 1979, because this was when one of these monstrous mainland lizards made itself known to a professional herpetologist. After a day of conducting field research deep amid the Wattagan mountains of New South Wales, Frank Gordon had climbed back into his Land Rover, parked next to a 6-ft-high embankment, when he saw what looked like a large log lying on top of the embankment—and then, as he started his vehicle's engine, the log moved! As Gordon stared in amazement, the 'log' revealed itself to be an immense monitor lizard, far longer than his 17-ft-long Land Rover, which scuttled away into the nearby forest on four powerful legs, leaving behind a stunned herpetologist who estimated the creature's total length at 27-30 ft.

Artistic representation of Frank Gordon's encounter with an Australian giant mystery monitor (© William M. Rebsamen)

Gordon was not the first person to record seeing such creatures here—eyewitness accounts of giant monitors in the Wattagans date back at least as far as 1830, when these mountains were reached by the first cedar timber cutters of European descent. Reports have also emerged quite regularly from New South Wales's even more remote Wallangambie Wilderness—where a lorry driver came to an abrupt halt one night in early 1980 when a 20-ft monitor stepped out in front of him as he was driving along the Bilpin Road on the wilderness's southern border. In the swampier portions of northern New South Wales, such creatures are well known to the aboriginals, who call them the mungoon-galli. It is likely that these are also responsible for the anomalous reports of 'crocodiles' periodically sighted in this state's forests—an unlikely setting for genuine crocodiles.

Nor are descriptions of giant monitors confined to New South Wales. Precisely the same type of reptile is spasmodically recorded from the terrain surrounding Lake Alexandrina in Victoria, and also from the forests of northern Queensland. This latter locality saw the arrival in June 1978 of a monitor-seeking expedition led by veteran Australian cryptozoologist Rex Gilroy, then curator of the Mount York Natural History Museum in New South Wales. He chose to explore the Atherton Range, and although he made no sightings himself, he amassed a sizeable file of accounts from the area's farmers, who claim that such goliaths are relatively common amid the mountains' more remote, inaccessible stretches, and attain lengths of 15-20 ft. Such dimensions are considerably greater than those of any species of monitor known to exist Down Under today. (Australia's largest living species of lizard is the perentie or giant monitor lizard *Varanus giganteus*, which attains a total length of up to 8 ft.)

Western Australia is also reputedly home to extra-large varanids, which are specifically referred to here by the aboriginals as the jillawarra. This term distinguishes them from the smaller yet still imposing Gould's (sand) monitor *V. gouldi*, widely distributed across Australia, which they know as the bungarra.

The latter lizard has an average length of 4.5 ft, and can weigh up to 13 lb, but the jillawarra is reputed to grow to at least 10 ft. This equals the longest scientifically-confirmed specimen on record for the Komodo dragon *V. komodoensis*, the world's biggest known species of modern-day lizard (and which just so happens to have actually existed in Australia too, as recently in fact as the mid-Pleistocene 300,000 years ago, confirmed by numerous fossil remains of this species that have been found here in recent years and documented in a 30 September 2009 *PLoS ONE* article by a research team featuring Queensland Museum geoscientist Dr Scott A. Hocknull).

So what could these various giant lizards be? A conservative identity would be Salvadori's monitor of New Guinea, the world's longest lizard with a *known* maximum length of 15 ft—but its presence in Australia has never been confirmed. However, some of the localities from which the giant mystery lizards mentioned here have been recorded are so little-explored that such a future eventuality is not impossible.

Gilroy, however, supports an even more thought-provoking identity—suggesting that Australia's giant monitors are surviving representatives of a gargantuan Antipodean species called *Megalania prisca* (=*Varanus priscus*), nowadays officially believed to have died out around 40,000 years ago. The largest meat-eater of Australia's Pleistocene epoch and the largest terrestrial lizard of all time, this mighty reptile probably preyed upon huge, herbivorous marsupials called diprotodontids.

Based upon estimates obtained from fossil reconstructions presented by Queensland Museum's now-retired fossil reptile specialist Dr Ralph Molnar in *Dragons in the Dust* (2004), *Megalania* may well have grown to a length of 23-26 ft, with an average weight of around

Megalania skeleton at Melbourne Museum, Australia (Cas Liber)

700 lb. This is nearly five times the average 150-lb weight of the Komodo dragon. Moreover, Molnar provides an estimated maximum weight for *Megalania* of 4,300 lb, which is more than eleven times the 370-lb maximum recorded weight for a wild Komodo dragon!

Traditionally, there seemed little doubt that a living *Megalania* could have satisfactorily explained the sightings of giant monitors in eastern Australia—but then came a highly unexpected revelation. In 1990, Molnar announced that some newly-documented cranial portions from a Pleistocene *Megalania* unearthed in southeastern Queensland's Darling Downs site revealed that the specimen in question had possessed a sagittal crest. That is to say, an external ridge of bone ran along the top of its skull, from front to back, and when present in other species such a ridge normally supports (in the living animal) a crest of skin—as in various present-day iguanas.

Yet there do not appear to be any reports on file that describe modern-day encounters with crested giant monitors in Australia, which would seem to argue against a contemporary contingent of *Megalania* as their identity after all—unless, that is, the crest was a feature of one sex only, and that the sex lacking the crest was the one usually seen by farmers, etc.

Another reason for not entirely discounting any reconciliation of Australia's unidentified giant monitors with a modern-day *Megalania* is Molnar's speculation that as living species of lizards with a sagittal crest are often amphibious, so too may have been *Megalania*. For in addition to accounts of terrestrial giant monitors, there are also reports of gigantic freshwater counterparts, inhabiting several tropical lakes and rivers in eastern Australia.

Crested or otherwise, therefore, *Megalania* still seems to offer a potential solution to the lizard giants Down Under. Even so, the issue cannot be resolved until a specimen is captured and identified—but how is such a daunting task likely to be accomplished? Australian cryptozoologists would greatly welcome details from anyone with a promising plan!

RAINBOW SERPENTS AND THE WONAMBI

During the 1970s, when Australian herpetologist Richard Wells was working for the Northern Territory Museum in Darwin, an amateur naturalist brought in an extremely sizeable python, roughly 13 ft long, of a previously unknown species, which was later formally described and named *Liasis* [now *Morelia*] *oenpelliensis*, the Oenpelli python. Subsequent specimens of this new snake have been recorded that vie in size with the mighty amethystine python *M. amethistina*—Australia's largest species of snake, and the fourth largest in the world, which can attain a total length of up to 23 ft.

It seems remarkable that a snake as sizeable as *M. oenpelliensis* could remain undiscovered by science until as recently as the 1970s, and even more so if local aboriginal claims are correct—namely, that specimens greatly in excess of 23 ft exist. Nor is the evidence for this dramatic prospect wholly anecdotal. According to Richard: ". . . the obvious tracks of a huge python have been found within vast underground caverns under Arnhem Land, and these were so large as to indicate that it may attain around 10 metres in length".

Moreover, Richard has received consistent reports from local tribes of a massive aquatic python-like snake, whose description leads him to consider that a bona fide prehistoric survivor may be the answer—the wonambi *Wonambi naracoortensis*. This remarkable beast was a giant species of constricting snake, but belonged to an ancient, primitive group known as the madtsoiids, and was deemed to have been at least partially aquatic. At one time, madtsoiids existed on all continents formerly part of the southern super-continent Gondwanaland, but they entirely died out around 55 million years ago—except in Australia, that is, where they continued to diversify, culminating in *Wonambi*, which became extinct only within the last 50,000 years.

In a fascinating *Nature* paper of 27 January 2000, moreover, in which a second wonambi species, *W. barriei*, was formally described and named, zoologists Drs John D. Scanlon from New South Wales University and Dr Michael Lee from Queensland University proposed that preserved memories of bygone encounters between early aboriginal settlers Down Under and the wonambi may actually have inspired their legends of the great Rainbow Serpent. This latter mythical monster is a major component of aboriginal Dreamtime lore, and for which 'wonambi' is actually a South Australian aboriginal name. Measuring 10-16 ft long, with a cross-section the size of a large dinner plate, the wonambi would have been big enough to swallow very large creatures, and thus could have readily given rise to legends of immense, all-consuming rainbow serpents.

But could the wonambi still exist today? The concept of surviving madtsoiids is not restricted to Australian cryptids. As noted in my book *The Beasts That Hide From Man* (2003), English cryptozoologist Richard Freeman has offered this identity as an explanation for reports of gigantic water snakes inhabiting the Mekong River comprising Thailand's northeastern border with Laos, and known as nagas (not to be confused with the ancient snake deities of the same name), but until now there had been no reports of comparable aquatic mystery snakes in Australia.

Conversely, when I communicated with palaeontologist Dr Ralph Molnar, formerly Curator of the Queensland Museum and an expert in Australian fossil reptiles, concerning these cryptids, he favoured the modern-day carpet (diamond) python *Morelia spilota* as a more plausible candidate taxonomically, though he is not surprised to learn that reports of extra-large constrictors have emerged here. Personally, I look towards the amethystine python, a northern species already known to attain immense lengths, which is also a good swimmer and usually occurs near water. The prospect of exceptionally-large aquatic specimens, their huge size buoyed by their liquid medium, is not untenable.

DINOSAURS OF THE DREAMTIME?

In the native Australians' traditional beliefs, Alcheringa—the Dreamtime—was the Time of Creation. As described by Mudrooroo Nyoongah in *Aboriginal Mythology* (1994), it "symbolizes that all life to the Aboriginal peoples is part of one interconnected system, one vast network of relationships which came into existence with the stirring of the great eternal archetypes, the spirit ancestors who emerged during the Dreamtime".

In the Dreamtime, all of today's Australian animals existed in human form, as kangaroo-men, emu-men, koala-men, even starfish-men, and so forth, only later transforming into animals. However, there were also many much stranger beings—some humanoid or part-humanoid, and some truly monstrous versions. These are discounted as fictitious by Westerners and are largely unknown outside Australia. However, this vast continent's native people firmly believe that they still exist even today, and can occasionally be seen—if you know where, and how, to look for them. Among the most intriguing examples from a cryptozoological standpoint are those that bear a bemusing similarity to certain dinosaurs.

Take, for instance, the tantalising similarity between *Tyrannosaurus rex* of North American prehistory and a bipedal reptilian monster known as the burrunjor, which is said by the local aboriginals to inhabit a remote expanse, also called Burrunjor, in Arnhem Land, Northern Territory. There are even depictions of what is claimed by locals to be this terrifying 30-ft-long beast in native Arnhem Land cave art. Not only that, very large three-toed prints said to be burrunjor footprints have been logged widely across the Australian Outback, in southern as well as in northern regions, earning this mystery beast the nickname 'Old Three-Toes' in this island continent's eastern extent. Photographs and casts obtained of such prints reveal a consistent bipedal track with three toes, measuring 2-3 ft in diameter, and having been discovered along river banks, watering holes, even dirt roads, but most frequently by cowmen out looking for missing livestock, who have claimed that the burrunjor will seize cows and carry them away.

Despite this scenario's ostensibly improbable likelihood, several such cases have been

Artistic representation of the burrunjor alongside a human silhouette for scale (© Connor Lachmanec)

documented by Rex Gilroy in various of his books on mystery beasts Down Under, one of which, published in 2013, is devoted specifically to this particular cryptid. Entitled *Burrunjor! The Search for Australia's Living Tyrannosaurus*, it contains details of several instances where cows (sometimes several at a single kill site) have been found with their bodies apparently bitten entirely in half, and with large limb bones broken by immensely powerful jaws. Tellingly, moreover, the area surrounding these kill sites has contained many immense three-toed footprints.

Perhaps, therefore, it is no bad thing that in recent years burrunjor reports have become far less frequent than was formerly the case. Could whatever was responsible for them, and in particular for the mystifying giant tridactyl prints, be dying out—or has it simply moved away to even more remote regions, seldom visited even by the most itinerant indigenous people? In 2007, a team from the television show *Destination Truth* conducted a field search for the burrunjor in northwestern Australia, but no sightings were made. Conversely, truckers driving at night along the Stuart Highway, which traverses long, lonely sections of Australia's barren desert heartlands, have claimed to have seen burrunjors. The principal north-south route through the continent's central interior, this famous highway, often referred to simply as 'The Track', runs from Darwin, the capital of Northern Territory, in the north, via Tennant Creek and Alice Springs, to Port Augusta, South Australia, in the south, covering a distance of 1,761 miles.

Needless to say, however, it is highly unlikely that *T. rex* has a surviving cousin Down Under. Rather more plausible is that stories of the burrunjor stem from sightings of hefty (but nonetheless size-exaggerated) goannas (Australian varanids or monitor lizards) rearing up and running bipedally on their hind legs—a common occurrence known as tripoding. Interestingly, one of the burrunjor's supposed modern-day southern strongholds is the Flinders Mountain range, where it just so happens that a particularly large, formidable species of varanid, the afore-mentioned perentie (giant monitor lizard), is known to exist. The fourth largest lizard species in the world, it can measure up to 8 ft long and sometimes weigh over 44 lb, and is known to favour more remote arid or rocky areas not regularly penetrated by humans. Who knows, therefore, whether in such rarely-explored localities, there may even be some extra-large specimens hitherto unconfirmed by science?

One Flinders Ranger interviewed in connection with the burrunjor recalled observing a massive perentie swallow a young goat whole, and then rear up onto its hind legs. A trucker or someone else perhaps not well-acquainted with herpetological taxonomy seeing such a dramatic sight while driving alone at night along some long, lonely highway in the middle of nowhere could certainly be forgiven for wondering if he had seen some dinosaurian descendant from the prehistoric age—but what about aboriginal eyewitnesses, who would surely be very familiar with perenties?

Is it really conceivable that these observers could be so readily misled, especially when the perentie features extensively as a totemic creature in several aboriginal cultures, and was even favoured as a food item among various desert-dwelling tribes, who also used its fat for medicinal and ceremonial purposes? Moreover, whereas perenties do not run or even remain on their hind legs for any length of time, there are a few genuine-sounding modern-day reports of huge, habitually bipedal reptiles on file from the coastal borderlands between Queensland and the Northern Territory.

Even so, certain other goannas are well known for being able to run bipedally. Australian cryptozoologist Debbie Hynes posted the following pertinent observation to the cz@yahoo groups.com cryptozoological chat group on 5 December 2002:

> We saw a small striped goanna, maybe 30 cm [1 ft] long take off on two legs. The front of the body rises up with the arms dangling down and the hind legs go like crazy in a circular motion. The whole body sort of wobbled a bit from side to side while it was running. They're pretty fast. Maybe it's an aerodynamic thing? ie lighter weight at the front so more lift when they get up to speed?

This elicited the following comment from Chris Kaska:

> I had a medium sized *V. dumerilii* [Dumeril's monitor] that I witnessed do it on two occasions. Both times it had been spooked.

This Asian species typically attains a total length of 4 ft, but somewhat larger specimens have occasionally been recorded.

In any case, the burrunjor is not the only neodinosaur on file from Australia. How do we explain away the kulta? According to native lore in Central Australia, long before this region became a barren desert it was carpeted in lush vegetation, and was browsed by a huge but inoffensive creature known as the kulta, which possessed a very long slender neck and tail, a huge bulky body, and four sturdy legs. This description immediately recalls the sauropods. Moreover, the fossilised remains of various sauropods have indeed been excavated in Australia, confirming the prehistoric existence of these long-necked dinosaurs here. They include *Rhoetosaurus brownei* from the mid-Jurassic, estimated to have measured more than 45 ft long; and *Austrosaurus mckillopi*, an early Cretaceous titanosaur that may have exceeded 50 ft long. But sauropods Down Under may not be limited to fossils, nor even to ancient myths. Several modern-day eyewitness accounts of cryptids seemingly bearing much more than a passing resemblance to these reptilian behemoths have also been documented, including the following two, both of which are included in Rex Gilroy's book *Out of the Dreamtime* (2006).

At around 3 pm on 30 September 1976, bushwalker Steve Taggart was trekking through bush along the northern shoreline of a certain very large, reedy swamp near Singleton, in a valley within the Wollemi National Park (an immense expanse of wilderness ensconced in the northern Blue Mountains and Lower Hunter regions of New South Wales). Suddenly he heard some very loud crashing sounds amid the scrub, and, peering in their direction, he was horrified to see a huge, 30-ft-long creature moving through the trees. It had a long neck, snake-like head, long tail, big body on four legs, was mottled-grey in colour, and reminded him of "one of those brontosaur-type dinosaurs". He watched it as it moved away through the foliage, and later informed Gilroy that it was definitely not a monitor lizard.

In January 1981, a similar beast was encountered by Peter Garland near this same valley as he was driving south of Singleton. Having stopped for a call of nature in some bushes, Garland was petrified when, hearing a rustling in some shrubbery close by, he looked around and saw an extraordinary reptilian creature, greyish, scaly, and "something like a brontosaurus", which was not only looking directly at him but also approaching him. Garland could see it sported a long thin neck, large serpent-like head, four powerful legs, and from what he could see of it he guessed that its tail was probably long, yielding a total length for the creature that he estimated at being up to 25 ft. He fled to his car, and was very relieved to see that it too was fleeing, in the opposite direction. He lost no time in driving away.

Although not dinosaurian, the yarru or yarrba should also be mentioned here for comparison purposes. In 1990, after learning from the elders of far-northern Queensland's Kuku Yalanji people about this traditional Dreamtime monster, which formerly inhabited rainforest water holes, missionary Dennis Fields asked one of the tribe's artists to paint a picture of it. This aboriginal artist had very little formal education, and no knowledge at all of palaeontology and what prehistoric animals looked like, only the descriptions of the yarru as preserved in his tribe's ancient lore—which makes it all the more remarkable that the creature painted by him bears an uncanny resemblance to a long-necked, flipper-limbed plesiosaur.

Is it conceivable that native Australians have sometimes found giant fossil bones belonging to such creatures as these, and that their myths of the kulta and yarru have arisen as a means of explaining where these bones came from? An alternative (albeit far more radical) possibility is that perhaps sauropods and plesiosaurs in Australia did not die out alongside other giant reptiles by the end of the Cretaceous Period 66 million years ago, but persisted here into much more recent times, to be sighted and marvelled at by the first humans to reach this island continent from Asia—and seemingly by some much later ones too, judging at least from the reports in Gilroy's books.

A final Antipodean anomaly worthy of brief mention here is the gauarge or gowargay. According to Gilbert P. Whitley's 1930 survey of Australian mystery beasts, this is said to resemble a featherless emu that sucks down and drowns in a whirlpool anyone daring to bathe in one of its water holes. Heuvelmans noted in *On the Track of Unknown Animals* (1958) that a featherless emu is a good description of some of the bipedal bird-mimicking dinosaurs like *Struthiomimus* ('ostrich-mimic').

However, these were terrestrial, cursorial forms, with no aquatic adaptations—and an evolved version that did have such adaptations would therefore not resemble a featherless emu. In addition, some dinosaur authorities nowadays believe that *Struthiomimus* may have been feathered, and as in terms of posture this dinosaur may have paralleled ratite birds (such as the emu), it means that *Struthiomimus* might actually have looked much more like a feathered emu than a featherless one. But would an evolved, modern-day descendant have done? As Shakespeare may well have said (if asked), that indeed is the question.

NEODINOSAURS IN NEW GUINEA?

North of Australia is the extremely large island of New Guinea, still plentifully supplied with little-explored expanses of rainforest and mountainland. It is also the setting for one of the most bizarre episodes in the history of cryptozoology.

During the late 1930s, Java-born explorer/camera-man Charles C. Miller and his newly-married wife, former American society girl Leona Jay, spent their honeymoon visiting the Sterren Mountains in what was then Dutch New Guinea (the western half of New Guinea, now known as Irian Jaya or Indonesian New Guinea). Here they allegedly encountered not only a hitherto-unknown tribe of cannibals called the Kirrirri but also what Miller believed to be a living dinosaur. Their introduction to this latter beast came about in a somewhat unusual manner—courtesy of a coconut de-husker used by one of the native women.

Leona noticed that the tool in question, roughly 18 in long and 20 lb in weight, resembled the distal portion of an elephant tusk or rhino horn, but as there are no elephants or rhinos in New Guinea she was very perplexed as to its true identity and origin. When she told her husband, he made some enquiries and was shown several of these curious objects, which were made of a horn-like substance present in cone-shaped layers—i.e. resembling a stacked pile of paper drinking cups, one cup inside another. When pressed for more details, some of the natives drew a strange lizard-like creature in the sand, whose tail terminated in one of these horns. They called this beast the row (after its loud cry), and said that it was 40 ft long.

Although Miller was initially sceptical of their claims, he could not deny the evidence of the horns and could offer no alternative explanation for their origin, and so when he learned that the hills to the northwest of the Kirrirri camp reputedly harboured these gigantic beasts, he set out with his wife and a native party in the hope of filming them. After a couple of days' journey, they reached a triangular swamp situated between two plateaux and occupying an area of roughly 40 acres. As Miller sat there, looking at a bed of tall reeds a quarter of a mile away, the reeds suddenly moved. Something was behind them. Hardly daring to breathe, Miller waited for them to move again, camera in hand—and when they did, the result was so shocking that Leona collapsed to the ground, almost fainting with fear.

A long thin neck bearing a small head fringed with a flaring bony hood had risen up through the reeds, followed by a sturdy elephantine body bearing a series of huge triangular plates running along its backbone, and a lengthy tapering tail bearing at its tip one of the mysterious horns that Miller had come to know so well. Its front limbs were shorter than its hind limbs, and while Miller was filming it, the row unexpectedly paused, raised itself up onto its hind limbs, and peered in the party's direction, almost as if it sensed the presence of these human interlopers within its private, prehistoric domain. In colour, it was precisely the same shade of light yellow-brown as the surrounding reeds, no doubt affording it excellent camouflage should it seek anonymity, but it was presently intent upon more extrovert behaviour—rearing up on two further occasions before disappearing from sight behind a clump of dwarf eucalyptus trees, just as Miller's film ran out.

In 1939, his extraordinary adventure was first published in book form—*Cannibal Caravan*. Yet despite containing many interesting pictures, there was none of his most spectacular discovery, the row. There was not even a photo of one of the horns. Similarly, although Miller claimed to have shown the film to various (unnamed) authorities, nothing more has ever emerged regarding it, or the Kirrirri either, for that matter, as this tribe has apparently never been encountered again by any other explorer. Equally odd was that in *Cannibals and Orchids* (1941), Leona Miller's own book recalling their ostensibly highly eventful New Guinea honeymoon, she relegated the row to just a single short half-paragraph, and which contained none of the descriptive details given by her husband in his book. Needless to say, this is hardly what one might expect from someone who had supposedly encountered (and been thoroughly unnerved by) a living dinosaur!

Perhaps the most paradoxical aspect of this entire episode, however, concerns the row itself. For although palaeontologists currently recognise the former existence of several hundred different species of dinosaur, collectively yielding a myriad of shapes, sizes, and forms, not one compares even superficially with the row—and for very good reason. As Dr Bernard Heuvelmans pointed out in *On the Track of Unknown Animals* (1958), the row's morphology is truly surrealistic—because it combines the characteristics of several wholly unrelated dinosaur groups.

Little wonder, then, why cryptozoologists are reluctant to countenance any likelihood of this morphologically composite creature's reality. Of course, their denunciation could be premature—but as long as Miller's film remains as elusive as the beast that it allegedly depicts, how can we blame them for remaining unconvinced?

Over the years, alleged sightings of sauropod-like mystery beasts have been reported from various tiny islets off the southwestern coast of the much larger island of New Britain (and also from New Britain itself), situated in the Bismarck Archipelago to the east of New Guinea, but little information concerning their precise appearance has been recorded. Conversely, I know of at least one reputed encounter with a very different type of supposed living dinosaur on one of these specks of land for which a detailed description is indeed on file, and which has been likened to a highly distinctive if decidedly surprising fossil form.

In January 2008, Australian cryptozoologist Brian Irwin visited the island of Ambungi (aka Umbungi), and while there he interviewed one of two eyewitnesses who claim to have seen an extraordinary animal in 2005/2006, and which has apparently been sighted here and on a neighbouring isle called Alage at least nine times since the early 1990s. Robert, whom Irwin interviewed (the absent eyewitness was named Tony Avil), stated that the creature was approximately 30-45 ft long, possessed smooth brown shiny skin, a long tail, and also a long neck, but was bipedal, and resembled a huge wallaby in overall appearance, except for its head, which was turtle-like. When walking slowly on its hind legs, the top of its

Statue representing the postulated appearance in life of *Therizinosaurus* (© Dr Karl Shuker)

head was estimated to be "as high as a house", and the vertical distance from its underbelly to the ground was estimated to be equal to the height of an adult man. It was observed from a distance of around 150 ft in the late afternoon, and for some considerable time, while it ate vegetation before eventually walking away, entering into some water, followed cautiously at a distance by its eyewitnesses.

When shown pictures of creatures, Robert selected a restoration of the possible appearance in life of the theropod dinosaur *Therizinosaurus* as most closely resembling what he and Avil had seen that day—except for the head, which was depicted as horse-like in the illustration. As Irwin has commented, however, the head's morphology in that picture was entirely speculative, because no skull identified as being from a *Therizinosaurus* has ever been documented, so the appearance of its head is currently unknown. Indeed, the only portions of this very large theropod from the late Cretaceous that *are* known from fossil evidence are its limbs and some ribs, so much of its likely appearance is merely deduced from related forms.

Ironically, however, its most famous confirmed attributes, and which must have been truly spectacular in life, are conspicuous only by their absence from Roger's description of the cryptid seen by him and Avil—because *Therizinosaurus* possessed incredibly long claws on its hands, probably up to 3 ft long (only incomplete versions are currently on record). In short, combining this startling absence from Roger's description with the relatively undetermined appearance of *Therizinosaurus* as a whole anyway, his identification of the latter dinosaur's illustration as being most similar to the cryptid that he saw clearly cannot be taken literally in any sense (although it has been on some websites), and can do no more than offer a basic idea of the latter beast's general form.

During late December 2015 through early January 2016, Irwin was in New Britain, accompanied by American cryptid investigator Todd Jurasek, to continue Irwin's earlier researches. Jurasek's summary of what they learnt while there is included exclusively here as follows, with their kind permission:

> 1) Ambungi Island—We visited Ambungi Island examining the caves reportedly used by a sauropod in recent years. Brian and I and [a] large group [of] islanders went to the caves at night. Conflicting reports from the native divers led me to suspect it wasn't really a deep one. I went back and physically examined the cave the next day in daylight. The water surrounding the entrance was maybe 15 ft at its deepest point, the cave maybe about 10 ft wide and deep. I placed a trail camera for a week above a secondary purported cave with no success. The last reported dinosaur sighting around the island was back in July of 2015 by an adult male who wished to remain anonymous. He watched a brown long necked creature with a saw like ridge on his back moving in the open ocean in the afternoon while in a canoe. Ambungi Island appears to be visited at times by these creatures but I saw no surface caves capable of hiding an animal larger than an adult human. The island is comprised of pocketed limestone that has the appearance of Swiss cheese or iron/steel slag discard from an iron or steel mill. (I'm guessing most of the islands in New Britain if not all appear this way.) Just like Swiss cheese there are no real continual holes to be found, just many odd-shaped pot-marked ones of various sizes. The only sizable holes on the island that I saw were along the shores where water erosion has occurred on [a] consistent basis creating small bluffs or overhangs.
>
> 2) Aiu Island (nearest island to Ambungi, also owned by the Ambungi people). According to [an eyewitness named] Davis who lives on the island, he and others had been chased out of the sea on multiple occasions at night by something emanating a bright white light. They were spearfishing at night when a bright white light would come out of the horizon and chase them to shore. Davis couldn't tell if the light was an animal or not. After chasing the group to shore the light would then fly away to heights of the island. The men and boys spearfish at night off canoes. My guess is whatever the light was [it] was attracted to their flashlights maybe even more so than their presence or movement. (Flashlights are used to both guide their boats and underwater for spotting fish and predators.) The light fits the descriptions of the New Guinea pterosaur-like cryptids known as ropen [see Chapter 2]. Flying brightly-lit nocturnal creatures were also reported to have been seen in Karadian in the past; one such story was told to me by [local missionary] Bryan

Girard's son, Rist. Another person in Karadian told me about an encounter with lights there at night. No planes fly in PNG [Papua New Guinea] at night so they couldn't have been aircraft. As Brian and I travelled to Karadian along the Armio road I met a young ex-school teacher from the island of Bali, PNG (not Indonesia) whose name I forget (have picture of him). He told me of similar creatures on his island. He said a bright light flew over the ocean or travelled partially submerged in the water like an octopus with its head sticking out from Bali at night to another nearby island. He was familiar with the subject of living pterosaurs and brought up Umboi Island to me as well as Roy Mackal's famous New Britain lake cryptid [already known for his mokele-mbembe expeditions, Prof. Roy Mackal also investigated the migo of New Britain's Lake Dakataua during the 1990s—see Chapter 3].

3) Akinum. Brian and I visited the [New Britain] village of Akinum where Michael Hoffman filmed the "West New Britain Carcass" video that was posted on YouTube in January of 2014. We were led to believe the rotten carcass was buried by a back hoe at some point after washing ashore; however, it appears the remains may have just washed back into the sea. Michael accompanied us to Akinum as well as to Ambungi Island. A mechanic from Karadian who viewed the decaying remains said it was built like a wallaby with a saw on its back, had small front arms with four fingers on little hands and very large back legs. The legs were so large that two men had [a] hard time lifting and moving one of them. This was reported to me by missionary Bryan Girard, the poster of the YouTube video. There were conflicting reports as to what happened to the remains. Brian Irwin and I went to the village under the impression the remains were buried on the spot due to the stench. We also heard they were picked over by curiosity seekers and that the remains had just washed out slowly back to sea. It is my opinion based from talking to the locals that this is what most likely happened. The natives that we spoke with told us they had never seen the animal before and were adamant it did not live anywhere around there.

4) Crocodile Point. Brian and I looked into a story of a man (Graham Sangeo) who reportedly had fed fish to a small bipedal dinosaur for years near Crocodile Point. The animal turned out to be a male primate of some sort that walked primarily on two legs according to our guide Leo Sangeo, Graham's father. He guided us to the cave which is currently abandoned. Leo described the creature as brown colored, about a meter to a meter and a half tall, big muscular arms and shoulders. The arms were shorter than the legs and its knees and big legs could be seen. The animal's feet were like a dog's hind feet with five toes (I asked Leo repeatedly about this feature to make sure I understood him correctly), it had very small to no tail, and canines like a monkey or ape does. The creature would come down out of the cave at night [and] scrounge around, walking on two legs at least a part of the

time. It could be seen at times seemingly staring out to sea as if it was watching the horizon. Leo and Graham and a few others would attach cooked fish to tall branches and lift them up to it. It would then eat the food out [of] its hands. Leo said the animal grew bigger over time. The creature eventually brought two babies. He said he never saw the female. I'm not sure if the others saw the female or not. Graham discovered the creature in 2011, feeding it until he left for school in 2014 or 2015; others continued afterward but eventually stopped and the creature disappeared. Based on the description I'm inclined to believe this was a small ape of some sort or possibly a small bigfoot like creature. I was told by at least one other person along the distant Andru River that wild hairy men could be found in the Whitman range.

5) Aivet Island. In 1992 John Manlel of Aivet Island had a startling encounter along the mangrove strewn shores of the island with what he described as a bright green dinosaur. He had been canoeing along the shores of the island around 4 pm when he accidentally startled the creature from about 20 or so yards away in open water. The animal attempted to submerge quickly but struggled because of how it was built. John said he watched it for about 5 minutes. He said he knew it lived on land because of the way it was built. The creature had two short hand-like front legs and much bigger back ones. The body was about 12 ft long with a very thick 5 to 6 ft tail that was about 8 in wide. The animal moved its tail back and forth as it moved through the water. John said the head looked like that of a dinosaur, the skin was rough like a crocodile and over-all build was kangaroo in shape. The creature had a small saw like structure on its back that became much bigger from back legs to the end of tail. There may have been an outlying ridge of small saw like structures along its tail like a crocodile has. If I can remember correctly the central ridge originating from the back ran between these. John said he never spoken to anyone but family about this encounter until he told Brian and I. He was frozen terrified by [it] when he sighted the creature and was adamant it was a dinosaur of some sort.

For the most part, the serrated-back water creatures sound very much like large crocodiles, but one would expect the local people to be very familiar with such beasts and not deem them to be anything other than crocodiles. Also, the bright green version reputedly spied by John Manlel, which had much shorter forelegs than hind legs, does not recall any crocodilian species known to exist today. Having viewed the 'West New Britain Carcass' video on YouTube (which can be accessed at: https://www.youtube.com/watch?v=acOF9jYOeXk), in my opinion it is the highly-decomposed remains of a large whale rather than anything reptilian, and various other zoologists and cryptozoologists who have seen it hold the same opinion, though some others favour a reptilian identity, ranging from a large crocodile or giant lizard to a bona fide dinosaur. Sadly, no physical samples from it were made available for formal scientific analysis, so the video is the only visible testament to this intriguing entity.

As for the mysterious bipedal ape-like entity fed by Graham Sangeo: no species of monkey or ape is known from anywhere in New Guinea, but there have long been reports from here of a mysterious miniature bigfoot-like creature known locally as the kayadi. So if such a creature truly exists, perhaps this is what Sangeo had been feeding. Also of note is that Sangeo claimed that the adult female was never seen, i.e. indicating that they believed the adult individual bringing the two babies to have been a male. However, it may be that the latter was actually a female but with a large clitoris that its eyewitnesses had mistaken for a penis (in some primate species, famously including the spider monkeys *Ateles* spp., the adult female's clitoris is indeed noticeably large and superficially penis-like). After all, it is far more likely to have been an adult female than an adult male that was caring for the babies.

Brian Irwin and Todd Jurasek have asked me to announce that if anyone reading this account here has information concerning any of the mystery beasts sought by them in New Britain and its outlying islets, please contact them via Todd's email address: hunterfox743@gmail.com

POST-CRETACEOUS DINOSAUR FOSSILS?

The concept of present-day non-avian dinosaur survival would become considerably more palatable even to the most hardline sceptic if there were some fossil evidence for post-Cretaceous non-avian dinosaur survival. This would, however, contradict the ostensibly inviolate maxim that all non-avian dinosaurs had perished following the alleged asteroid crash-landing on Earth at the end of the Cretaceous Period, marking the close of the Mesozoic Era—which is why, while researching the present book, I was both delighted and surprised to discover that some such evidence does exist, although it is still viewed by many palaeontologists as being highly controversial.

In 1977, Chicago University evolutionary researcher Dr Leigh Van Valen and Minnesota University geologist Dr Robert E. Sloan queried the traditional view that the mass reptilian extinctions on land and in the seas at the end of the Cretaceous were contemporaneous, by offering evidence that some southern-based terrestrial extinctions occurred later than northern-based ones. This gained momentum in 1984, when Panjab University palaeobiologist Prof. Ashok Sahni pointed out that the Takli Formation, traditionally deemed to be of Palaeocene age (66-56 million years old) contained titanosaurid fossils. Consequently, either the formation's age was incorrect (and there seemed to be no evidence to indicate this) or in India these sauropods had persisted beyond the Cretaceous into the Palaeocene epoch.

Later, in a *Science* paper of 2 May 1986, Van Valen, Sloan, and other colleagues revealed that data from a variety of different locations in addition to India, such as the Pyrenees, Peru, New Mexico, and Montana, suggested that dinosaurs survived well into the early Palaeocene within the tropics. The Montana evidence stems from the Hell Creek Formation, which contains a channel sandstone whose upper level (Ferguson Ranch) is approximately 4.5 ft above the local K/T (Cretaceous-Tertiary) boundary, and contains not only some mammalian teeth, but also Palaeocene pollen and the teeth of seven different genera of dinosaur.

The big issue here is whether the dinosaur teeth were genuine members of the Ferguson Ranch sediment (and therefore truly of post-Cretaceous age), or reworked from older sediments. Van Valen and colleagues supported the former possibility, first of all because the teeth had not been transported very far—post-mortem stream abrasion on them was minimal. Moreover, the hollow cavity at the base of these teeth is empty or filled only with channel sand—not with floodplain silts or clay, as would be expected if they had entered the Ferguson Ranch sediment due to erosion of floodplain sediments adjacent to this channel.

And as they also pointed out, it is even less feasible that they were eroded from older channel sands, because the usual occurrence in channel sands is near the sand's base—and the Ferguson Ranch channel is not cut deeply enough into lower channels to rework many teeth. Lastly, but of particular importance is the absence within the Ferguson channel of any species belonging to the Cretaceous mammalian genera *Baioconodon* and *Oxyprimus*—because these are far more numerous than the dinosaur teeth in the only channels from which the latter could have originated if reworked (and therefore would be expected to have been reworked in quantity alongside the less common dinosaur teeth if the latter had truly been reworked). As the researchers concede, the most reasonable conclusion from all of this is that these are indeed early Palaeocene dinosaurs.

Incidentally, a specimen or taxon that has been mistakenly thought to have been living long after its presumed extinction due to reworked fossils having deceived those who discovered them is known rather memorably as a zombie taxon. This term was coined by palaeontologist Dr David Archibald in his book *Dinosaur Extinction and the End of an Era* (1996).

In 1980, French palaeontologist Dr Eric Buffetaut and co-workers documented an iguanodontian tooth that had been obtained in the late Miocene deposits or 'Faluns' of Doué-la-Fontaine (Maine-et-Loire), in France's Loire Valley, and these same deposits were also the source of two small theropod teeth that in 1989 were referred to *Carcharodontosaurus saharicus*. Again, both finds were deemed to be probably reworked specimens, this time from underlying Cenomanian beds. Moreover, what is apparently the only dinosaur fossil on record from Louisiana in the U.S.A. just so happens to be a single dromaeosaur tooth that was found in a Miocene deposit in this state's Pascagoula Formation. Predictably, it is deemed to have been reworked from a Cretaceous deposit.

In more recent times, American palaeontologist Dr James E. Fassett of the U.S. Geological Survey has published several very thought-provoking scientific papers presenting what he deems to be evidence for select fossil dinosaur persistence into the Palaeocene. At the end of April 2009, for instance, Fassett published a fascinating paper in the online journal *Palaeontologia Electronica* in which he documented a collection of 34 fossil bones all belonging to a single 40-ft hadrosaur (duck-billed dinosaur) specimen that had been found 80 years earlier in rocks which have since been dated by radioactive and magnetic technology to be from the very early Palaeocene epoch, half a million years after the end of the Cretaceous, and not merely reworked Cretaceous rocks. The fossils came from the Ojo Alamo Sandstone in the San Juan Basin—an area now bordered by Colorado and New Mexico in the U.S.A., where, based upon this intriguing finding, it has been suggested that perhaps a small tribe of hadrosaurs did indeed survive into the Palaeocene.

During the past two centuries, countless reasons have been proposed for the wholesale extinction of non-avian dinosaurs by the end of the Cretaceous, but to my mind none of them can satisfactorily explain why at the very least a few straggler individuals (and possibly even a few entire species) did not manage to persist into the Palaeocene, and possibly even beyond that. True, the adaptive radiation of the mammals during the Cenozoic saw many ecological niches formerly occupied by the dinosaurs now taken over by the mammals, but the survival of several major reptilian groups (including birds) alongside them readily demonstrates that there were still sufficient niches available for other land vertebrates too. Perhaps, therefore, Fassett *et al.* are correct to keep an open mind and actively seek evidence supporting the prospect of post-Cretaceous non-avian dinosaur survival.

If nothing else, the famously imperfect, incomplete nature of the fossil record is such that there may indeed have been such survivors but whose remains simply have not been preserved, or have not survived, or have not been discovered. As an example of just how selective palaeontological finds can be, today the former existence of non-avian feathered dinosaurs is well documented—but it may come as a surprise to learn that the first fossil remains of these creatures to come to scientific attention only did so as recently as the mid-1990s. And although 42 different feathered species are currently on record, virtually all of them have been found in the very same, single fossil site in Liaoning, China. It seems highly unlikely that no feathered dinosaurs existed anywhere else in the entire world, yet no unequivocal fossils of such creatures have been found outside China (some late Cretaceous dinosaur feathers have been found encapsulated in Canadian amber). So in view of such a major precedent for missing dinosaur fossils, perhaps the same is indeed true for Palaeocene—and much later?—non-avian dinosaur fossils too. Who can say?

AND FINALLY . . .

As noted at the beginning of this chapter, the mere concept of any type of non-avian dinosaur surviving into the present day is one that has provoked cynical disbelief within the more conservative corners of the scientific establishment. If now, after having reached the end of this chapter, there are still some readers who will not concede that such a possibility (however faint) does exist, I can do nothing more than recall to them the following words of Captain William Hichens from a *Chambers's Journal* article (1 October 1927), 'On the Trail of the Brontosaurus and Co.':

> The level-headed man at home will be tempted to exclaim, "But this is rubbish, this talk of undiscovered monsters." So at first thought it may seem; but it must be borne in mind that such an obvious animal as the okapi, a beast that looks as though it had escaped from a jigsaw zoo, remained undiscovered up till the beginning of the present century, when Sir Harry Johnston startled the scientific world by sending home the skin and skeleton of one. Scientists gasped: this cannot be, they said. They said the same about the man who first reported the giraffe; in fact, they tortured him for a downright romancer of the worst description. But there were children feeding giraffes in the London Zoo this week!

Even the most die-hard sceptic would be hard-pressed to disagree with that!

CHAPTER 2

Things With Wings—From Pterosaurs to Teratorns

When in Northern Rhodesia I heard of a mythical beast . . . which intrigued me considerably. It was said to haunt formerly, and perhaps still to haunt, a dense, swampy forest region in the neighbourhood of the Angola and Congo borders. To look upon it . . . is death. But the most amazing feature of this mystery beast is its suggested identity with a creature bat- and bird-like in form on a gigantic scale strangely reminiscent of the prehistoric pterodactyl. From where does the primitive African derive such a fanciful idea?

CAPTAIN CHARLES R.S. PITMAN—*A GAME WARDEN TAKES STOCK*

He was hitting at it with his fists. If he hadn't, it never would have dropped him.

RUTH LOWE, REGARDING HER SON'S ABDUCTION BY A GIANT MYSTERY BIRD—*CHICAGO DAILY NEWS* (27 JULY 1977)

Whereas the dinosaurs reigned supreme on land during the Mesozoic Era, the kingdom of the skies was ruled by the pterosaurs—those flying reptiles of such grotesque appearance that they were described by the French palaeontologist Baron Georges Cuvier as ". . . the result of a diseased imagination rather than the normal forces of nature".

Like the dinosaurs and crocodiles, the pterosaurs were archosaurs—reptiles possessing a skull containing two large holes or temporal fossae on each side (the diapsid condition)—but were readily distinguished from all others by their many adaptations for flight. These included their membranous wings supported by an immensely elongated fourth finger, hollow limb bones to reduce body weight, and a large keeled breast-bone for attachment of the powerful wing muscles needed for true active flapping flight (as opposed to mere passive gliding).

Pterosaurs had a long slender beak, like some modern birds, but unlike theirs it was often brimming with sharp teeth (though certain genera, e.g. *Pteranodon*, *Quetzalcoatlus*, were toothless). Most fed upon fishes and other aquatic fauna as adults (some scavenged meat from dead dinosaurs), and snapped up flying insects as juveniles.

Did all pterosaurs die out when an asteroid hit the Earth at the close of the Cretaceous Period 66 million years ago? (Donald E. Davis/NASA)

Of particular interest is that except for their wings, the bodies (and heads too) of certain pterosaurs (e.g. *Sordes pilosus*) were covered in superficially hair-like structures known as pycnofibers, analogous to bats' fur—although these two groups were quite unrelated. They were almost certainly warm-blooded too. All of this is what one would expect of creatures faced with fulfilling the high energy requirements demanded by flight.

Pterosaurs have traditionally been split into two suborders (having said that, there is currently much controversy regarding pterosaur taxonomy, because the first of these two suborders is nowadays deemed to be paraphyletic and therefore not a natural grouping; however, I am retaining their usage here, but as morphology-based groups, rather than strictly taxonomic ones). Arising in the late Triassic just over 200 million years ago, the rhamphorhynchoids were mostly small, with long tails. The pterodactyloids, debuting in the late Jurassic around 162 million years ago, were larger, but had very short tails. Based upon what is known from the documented fossil record, by the mid-late Cretaceous, a little over 90 million years ago, the rhamphorhynchoids had all died out; equally, although the pterodactyloids had become very diverse during the Cretaceous, all of these had become extinct by its end, 66 million years ago. Consequently, no pterosaurs of either group persisted beyond it—or did they?

In places where fossilisation is not likely to have occurred or where fossils cannot be readily searched for today—localities such as tropical rainforests, the peaks of high, inaccessible mountain ranges, or the sea bottom—could there be much more recent pterosaur remains? Might there even be *living* pterosaurs in certain remote areas, aloof from scientific detection and avoided by fearful human neighbours? There is some compelling evidence on file to suggest that there may be much more to such a possibility than fanciful speculation.

Also investigated here is the equally startling prospect that science has overlooked some monstrous forms of giant bird, whose onetime reality is again known from fossil evidence—and in this case only dating back a few thousand years.

In addition, could there be among these prehistoric survivors of the high-flying variety a giant species of vampire bat loose in Venezuela, not to mention a couple of avian deities once dismissed as fables in feathers? For these, and other things with wings, read on.

LIZARD-BIRDS ALOFT AND WHALE-HEADS WITH WINGS—UNMASKING THE KONGAMATO?

Popularly claimed by cryptozoologists to be the likeliest land to harbour living dinosaurs, the Dark Continent is surely a promising place to begin any investigation of contemporary pterosaurs—and there are ample reports from tropical Africa to demonstrate that such an assumption is fully warranted.

The subject first attracted serious Western attention in 1923, with the publication of a book called *In Witchbound Africa*, written by traveller Frank H. Melland, who had spent some time in charge of native affairs in what was then Northern Rhodesia (now Zambia). In his book, Melland spoke of the Kaondé tribe's fervent belief in a terrible winged monster called the kongamato, allegedly inhabiting the Jiundu (aka Jiwundu) swamp in western Northern Rhodesia's Mwinilunga District, and of his own interest in learning more about this mysterious creature. Upon questioning some of the natives, a remarkable conversation ensued, with some astonishing revelations:

> "What is the 'Kongamato'?"
>
> "A bird."
>
> "What kind of bird?"
>
> "Oh well, it isn't a bird really: it is more like a lizard with membranous wings like a bat." . . .
>
> Further enquiries disclosed the "facts" that the wing-spread was from 4 to 7 feet across, that the general colour was red. It was believed to have no feathers but only skin on its body, and was believed to have teeth in its beak: these last two points no one could be sure of, as no one ever saw a *kongamato* close and lived to tell the tale. I sent for two books which I had at my house, containing pictures of pterodactyls, and every native present immediately and unhesitatingly picked it out and identified it as a *kongamato*. Among the natives who did so was a headman (Kanyinga) from the Jiundu country, where the *kongamato* is supposed to be active, and who is a rather wild and quite unsophisticated native.
>
> The natives assert that this flying reptile still exists, and whether this be so or not it seems to me that there is presumptive evidence that it has existed

An imagined kongamato confrontation (© William M. Rebsamen)

within the memory of man, within comparatively recent days.

This thought-provoking account intrigued Uppsala University palaeontologist Prof. Carl Wiman, who sought to explain it away by proposing that the kongamato tradition was a product of ethnological 'contamination'. He speculated in 1928 that it had originated from natives who had assisted German palaeontologists unearthing East African pterosaur fossils at Tendagaru in Tanganyika (now mainland Tanzania) for several years prior to World War I, and who were shown pictures of reconstructed pterosaurs—accounts of which gradually travelled westwards as far as Northern Rhodesia and were ultimately incorporated into the native folklore of this region.

Melland's information, however, was not the only source from the 1920s that touched upon the possible modern-day survival of pterosaurs in Africa. In 1925, the distinguished English newspaper correspondent G. Ward Price was accompanying the future Duke of Windsor on an official visit to the then Northern and Southern Rhodesias when he received details from a regional civil servant in Southern Rhodesia (now Zimbabwe) concerning a sinister incident that had lately occurred within one of the country's vast, impenetrable swamps. This particular swamp was greatly feared by the natives as an abode of demons, and they claimed that no-one venturing inside was ever seen again.

One exceptionally brave native, however, decided to challenge the swamp's terrors and boldly penetrated its depths. He also succeeded in coming out of it alive—but only just, because he returned home with a great wound in his chest. When asked what had happened, the injured man stated that he had been attacked in the swamp by a huge bird with a long beak—a bird of a type that he had never before seen. Curious to learn more, the civil servant showed the man a book of prehistoric animals, which the latter flicked through in a desultory manner—until he came to the page depicting a pterodactyl. He took one look at this, emitted a bloodcurdling howl of terror, and fled from the civil servant's home.

In 1928, game warden A. Blayney Percival noted that a huge creature whose tracks only revealed two feet and a heavy tail was believed by the Kitui Wakamba natives to fly down to the ground from Mount Kenya at night.

Another game warden, Captain Charles R.S. Pitman, briefly documented in 1942 the existence of a pterodactyl-like creature said to inhabit swampy forests near the border between Angola and what was then the Belgian Congo (now DR Congo)—as quoted at the opening of this chapter. It is presumably one and the same as the Jiundu kongamato documented by Melland, whose account was briefly reprised in the Royal African Society's journal by Dr Mervyn D.W. Jeffreys in 1943.

In 1956, ichthyologist Prof. J.L.B. Smith contributed a thought-provoking snippet from Tanganyika. In his book *Old Fourlegs*, he commented:

> The descendants of a missionary who had lived near Mount Kilimanjaro wrote . . . giving a good deal of information about flying-dragons they believed still to live in those parts. The family had repeatedly heard of them from the natives, and one man had actually seen such a creature in flight close by at night. I did not and do not dispute at least the possibility that some such creature may still exist.

Remembering that Prof. Smith was the codiscoverer of the first known species of living coelacanth *Latimeria chalumnae* ('Old Fourlegs'), it is neither surprising nor unwarranted that he should be more sympathetic than most scientists regarding the possible existence of anachronistic animals.

It was in 1957, however, that tales of living pterosaurs finally attracted media interest—sparked off by the publication of G. Ward Price's book, *Extra-Special Correspondent*. In it, he recounted the incident from 1925 described earlier here—which received front-page coverage on 26 March 1957 by Salisbury's premier newspaper, the *Rhodesia Herald*.

Its effect was electrifying—exciting the general public, but exasperating the Dark Continent's leading zoologists. On 29 March 1957, the *Herald* published a searing response from Dr Reay Smithers, director of Southern Rhodesia's National Museum, who averred that there was more likelihood of meeting a living specimen of our long-distant African ancestors, the australopithecines, in a city office than there was of encountering a living pterodactyl in an African swamp! This memorable denouncement, however, proved ineffective against the inevitable allure of flying reptiles winging forth from Mesozoic times into modern times—because another enigmatic encounter was swiftly committed to print by the *Herald* on 2 April.

The encounter itself had taken place in January 1956, in the wilds of Northern Rhodesia along the Luapula river, and featured engineer J.P.F. Brown. After a visit to Kasenga in the Belgian Congo, he was returning home by car to Salisbury and had reached Fort Rosebery—just to the west of Northern Rhodesia's Lake Bangweulu—when he decided to stop and fetch a thermos flask from his car's boot. It was about 6 o'clock in the evening, and he had just got out of the car when he saw two creatures flying slowly and silently directly overhead. Brown described them as birds that had an incredibly prehistoric appearance. They had a wingspan of 3-3.5 ft, a long narrow tail, and a narrow head, rather like a dog's with a long muzzle. When one of them opened its beak, Brown saw that it contained a large number of pointed teeth. He estimated the creatures' length from beak-tip to tail-tip at around 4.5 ft.

Following this, a Mr and Mrs D. Gregor announced that they had spied some 2.5-ft-long flying lizards between Bulawayo and Livingstone in Southern Rhodesia. And Dr J. Blake-Thompson recalled that the Awemba tribe, who live near the source of the great Zambezi river, at the border of the Belgian Congo and Northern Rhodesia, claim that huge creatures resembling winged rats inhabit the river gorge caves and fly out of their lairs to attack travellers.

In a *Herald* interview of 5 April, Dr Smithers fielded a sundry selection of identities in a final bid to demystify these sightings and consign this pterosaurian phalanx back into the prehistoric past.

Brown's 'things with wings' were, in Smithers's view, a pair of shoebills *Balaeniceps rex*. This peculiar species of stork-like bird, with slaty-grey plumage and a short tuft-like crest, has been allied with the storks, herons, and currently the pelicans, but is instantly distinguished from *all* birds by its grotesque, 8-in-long, hook-tipped beak, variously likened to a clog, an upturned boat, and even the head of a whale—hence *Balaeniceps* ('whale-head'). However, it does *not* contain teeth . . .

Smithers felt that the Gregors' flying lizards were probably big-beaked birds called hornbills, and he deemed the Awemba's giant flying rats to be nothing more startling than a type of gliding rodent called the scalytail—superficially similar (but only distantly related) to the familiar 'flying' squirrels of Eurasia and North America.

In June 1959, a magazine called *Le Chasseur Français* published an article by Zoé Spitz-Bombonnel concerning mystery animals, and within this she recorded an incident very pertinent to the pterosaur saga. In 1957, a hospital at Fort Rosebery allegedly received a local African who had a severe wound in his chest. When enquiries were made as to its cause, the man stated that the culprit was a great bird in the Bangweulu swamps. The doctor treating him was so curious about this that he gave the man a crayon and some paper, and asked him to sketch his attacker. To everyone's shock, the result was a silhouette that corresponded, feature by feature, with that of a pterodactyl!

Towards the close of the 1950s, *Daily Telegraph* correspondent Ian Colvin mentioned the Fort Rosebery attack in a report dealing with the flooding of the Zambezi Valley as a result of the Kariba Dam hydro-electric project. Furthermore, while in the Zambezi Valley at that time, he photographed a controversial creature that one observer believed to be a pterodactyl.

In more recent years, attention has switched from East and Central Africa to south-western Africa, specifically Namibia itself—where pterosaur-like beasts as big as Cessna aeroplanes have been reported! They intrigued American cryptozoologist Prof. Roy Mackal so

The enormous beak of a shoebill (© Dr Karl Shuker), and a shoebill painting from 1901

greatly that in summer 1988 he and a party of fellow investigators travelled to Namibia in search of further details—and possible sightings—of these winged wonders. Their focus was an isolated private property, owned by German settlers, in a desert area. After interviewing some of the area's locals, the team was able to describe the creature in question as a featherless beast with a colossal wingspan close to 30 ft, which mainly glided through the air but was capable of true flight too, and was usually spied at dusk, gliding between crevices in two kopjes (hills) about a mile apart.

Despite daily periods of observation, the team was not initially successful, and Mackal eventually returned home to the United States. No sooner had he done so, however, than one of his colleagues, James Kosi, who had stayed on in Namibia, allegedly saw the creature. From a distance of about 1,000 ft, he caught sight of something that he described as a giant glider, principally black, but with white markings.

That, then, is a concise history of Africa's putative pterosaurs—but there are still a number of important points to consider. Take, for example, Wiman's proposal that the Zambian kongamato was simply a mythicised recollection of pterosaur reconstructions shown to native helpers at the Tendagaru digs. In reality, however, native belief in this beast apparently dates back far beyond the early years of the 20th Century.

In addition, when a verbally-transmitted description of anything—let alone beasts so bizarre (at least to natives supposedly unfamiliar with them) as pterosaurs—travels distances as great as the many hundreds of miles separating Tendagaru from the Jiundu swamp, it inevitably experiences the 'Chinese whispers' effect. That is to say, it becomes greatly distorted along the route. Thus, one would have expected the Jiundu version to have been very different from its East African source. Yet descriptions of bat-like birds or flying lizards with long beaks brimming with teeth are very similar all along the route, and have also been gathered to the north and south of it. This suggests that the animal is real, not mythological, and that it occurs over a wide range. Nor does it seem to be a beast created merely to please the enquiring Westerners—a not-uncommon practice by native tribes in Africa and elsewhere. Of relevance is an excerpt from Melland's book:

Scalytails (1875 chromolithograph)

> The evidence for the pterodactyl is that the natives can describe it so accurately, unprompted, and that they all agree about it. There is negative support also in the fact that they said they could not identify any other of the prehistoric monsters which I showed them . . . The natives do not consider it to be an unnatural thing like a *mulombe* [demon], only a very awful thing, like a man-eating lion or rogue elephant, *but infinitely worse*.

Could it be, however, that the natives exaggerate its description (due to their fear of it)—so that the result is reminiscent of a pterodactyl, whereas the *real* creature is actually rather less spectacular, and may already be known to science? This was the line of thought pursued by Dr Smithers during the series of articles published by the *Rhodesia Herald*.

Of the various identities that he proposed, the least plausible is certainly the scalytail. Only a couple of feet long at most, and wholly harmless, it is very hard to believe that such a diminutive, diffident animal could have inspired the Awemba natives to dream up monstrous winged beasts that issue forth from caves to attack passing travellers. Their comparison of these anonymous but antagonistic entities to giant flying rats more readily conjures forth images of huge bats—and zoologists Gerald Russell and Ivan T. Sanderson actually claimed to have seen an enormous jet-black bat at close range, skimming across a river in Cameroon, during one of their animal collecting trips in 1932. They estimated its wingspan to be a colossal 12 ft—twice that of the world's largest known bat species.

According to the natives, who greatly feared it, this Brobdingnagian bat is called the olitiau.

Olitiau onslaught (© William M. Rebsamen)

Perhaps, therefore, some purported pterosaur accounts actually feature occasional sightings of an undiscovered species of giant black bat—whose nocturnal activity, dark colouration, and the superstitious dread that it might well evoke among local tribes would ensure that it remained undisturbed and unseen by all but the most tenacious of seekers. And if it also just so happened to be quite hideous in appearance, even better for deterring humans.

This has led to speculation among some cryptozoologists that if the olitiau really does exist, perhaps it may be an extra-large version of the hammer-headed bat *Hypsignathus monstrosus*. Named after its grotesque, monstrously large head, and widely distributed through equatorial Africa, this is Africa's largest known species of bat, boasting a wingspan that can slightly exceed 3 ft, and occurring in swamps as well as riverine forests like the one traversed by Russell and Sanderson. Despite its size, however, it is targeted by hunters as it is relatively harmless, being principally frugivorous rather than carnivorous, but perhaps a larger form would appear too daunting to hunt.

Nevertheless, however large and savage they may be, bats do not have long beaks with which they spear the chest of luckless human trespassers within their marshy realm. If the creature responsible is not a pterosaur, then we have no option but to look among the birds for a possible contender. Smithers nominated

The hammer-headed bat

two—a hornbill, and the shoebill. The former, however, is very much a non-starter—apart from their long pointed beak, hornbills bear scant resemblance to a pterosaur.

In contrast, the shoebill's candidature was enthusiastically received by another notable zoologist—Dr Maurice Burton, who discussed its merits as a prospective pterodactyl mimic within the *North Rhodesia Journal* in 1961, and also in the then-weekly British wildlife magazine *Animals* (18 February 1964). Within this latter account, Burton confidently opined:

> The 'pterodactyls' are, beyond question, shoebill storks, which look reptilian at the best of times and, in flight, can very well be mistaken for pterodactyls, especially by someone who is not on his guard.

This is all very well, but according to 'pterodactyl' eyewitnesses their mystery beasts have teeth (specifically commented upon by J.P.F. Brown) and bat-like membranous wings—features singularly lacking in the shoebill. Equally, whereas the shoebill is feathered, the kongamato is not supposed to have any feathers at all. In response to these morphological discrepancies, Burton argued that the shock of seeing a creature as grotesque in appearance as the shoebill (especially when revealing its impressive 8-ft wingspan in flight) would be enough for an observer to convince himself that he had seen such distinctly non-avian attributes as those. I am not all satisfied by this explanation, however, as I doubt very much that eyewitnesses could unconsciously delude themselves so profoundly, simply through being startled by the sight of an unusual bird.

In an article to accompany Burton's within the *North Rhodesian Journal*, C.W. Benson, a renowned expert on Africa's avifauna, also expressed doubts as to the shoebill's synonymity with the kongamato. Recalling the two incidents recounted here featuring native Africans speared in the chest by a swamp-inhabiting 'bird' with a huge beak and who associated the image of a pterodactyl with their attacker, Benson ruled out a shoebill on the grounds that its beak was not the correct shape to accomplish such a deed, and that its species was not noted for such acts of aggression anyway.

Incidentally, I cannot help but wonder whether those two supposedly separate incidents are really one and the same. Apart from the actual locations, only the manner in which the pterodactyl identity was aired differs between the two. In the Southern Rhodesian version, the man's terror at seeing the picture of a pterodactyl created the link; in the Northern Rhodesian version, the man sketched a pterodactyl-like beast. The latter reputedly occurred more than 20 years after the former—but was it simply an inaccurately-remembered version of it, rather than a separate incident in its own right?

Although acknowledging that shoebills have definitely been sighted at Lake Bangweulu (albeit only rarely), Benson doubted that this species could be responsible for the Jiundu kongamato, if only for reasons of habitat—the Jiundu marshes consist principally of swamp-forest and ambatch (pith trees), which would not be environmentally compatible with *Balaeniceps*.

Saddle-billed stork, revealing its very long, pointed beak (Sengkang)

Benson had even less time for Colvin's photo of a 'pterodactyl' in the Zambezi Valley. According to various ornithologists who had examined it, the photo portrayed a shoebill, but this species is not known from the Zambezi Valley. When Smithers looked at it, how-ever, he felt sure that the creature was really a juvenile specimen of the saddle-billed stork *Ephippiorhynchus senegalensis*—a tall, elegant species with an extremely long, spear-like beak, and which is well-known from this area of Africa. True, it is not very pterodactylian in appearance, but its beak might well be capable of inflicting the type of injury suffered by the swamp-invading native hunter(s) further north.

From all of this, the conclusions to be drawn are that although the shoebill is decidedly strange in appearance, and may perhaps explain one or two pterosaur reports by eyewitnesses unfamiliar with it, the morphology of this species totally conflicts with those detailed native accounts of featherless bat-winged horrors with teeth-crammed beaks—whose Jiundu swampland is equally inconsistent with the shoebill's preferred habitat. As for the saddle-billed stork, although it sports the correct beak for inflicting injuries attributed to the kongamato and its kin, the remainder of its morphology is completely different.

Moreover, as emphasised by Melland in his book, the Jiundu swamp in particular would be a perfect hideaway for living pterosaurs:

> I must say that the place itself is the very kind of place in which such a reptile might exist, if it is possible anywhere. Some fifty square miles of swamp, formed by an inland delta . . . the Jiundu River spreading out into innumerable channels, and—after receiving several tributaries, reuniting farther down into a single stream of crystal-clear water. . . The whole of the swamp is covered with dense vegetation: big trees that grow to a great height, tangled undergrowth with matted creepers and beautiful ferns. The soil is moist loam and decaying vegetation, the main

> channels and lesser rivulets being reminiscent of the peat hags [bogs] on a highland moor. . . In one place just to the west of the area there is a big hole resembling a crater. Nowhere else on high, well-drained ground have I seen such a morass: nor could one conjure up a more perfect picture of a haunted forest. If there be a *kongamato* this is indeed an ideal home for it.

There are countless other swamplands of a similarly daunting, impenetrable nature elsewhere in tropical Africa, including those associated with pterosaur reports. Such creatures, therefore, if they do exist, are not short of effective havens—whose forbidding nature greatly assists in distancing natives who might otherwise grow less fearful of their winged inhabitants, and who may even find their dead bodies before the natural processes of decomposition can begin to take their toll.

The natives' terror of the creatures, and the predictable outcome that they refuse to show Westerners their habitats, also help to explain the absence of hard evidence supporting the kongamato's existence. After all, if only to see one is death (according to native tradition), it is hardly surprising that the natives do not hunt such animals—or do they?

In 1953, Volume 4 in a series of memoirs entitled *Loin des Sentiers Battus* was published, penned by the Marquis of Chatteleux under his famous pseudonym, 'Stany'—which he used when documenting his varied travels around the world. Unfortunately, Stany was prone to elaborating those facets of his journeys that were not inordinately exotic—and in this particular volume, he claimed not only to have learnt much about the kongamato while sojourning in Northern Rhodesia near the Jiundu swamp during 1920-21, but also to have met a woman whose husband Tshipeshi, the village chief, had actually killed the very last three specimens! Heuvelmans, however, placed little faith in Stany's account, noting that the descriptions of the kongamato in his book corresponded exactly with those in Melland's, published 30 years earlier. Also, sightings of kongamatos here and elsewhere long after the early 1920s indicate that they had not been exterminated.

If, therefore, the Zambian kongamato and its morphological counterparts in other parts of East-Central Africa do exist, and they really are living pterosaurs, what type(s) could they be? Even though the term 'pterodactyl' is invariably used when referring to such beasts, the pair seen at Fort Rosebery by engineer Brown were relatively small and had long tails—characteristics more typical (when present together) of the rhamphorhynchoids.

Modern reconstruction of *Rhamphorhynchus*, the most familiar fossil rhamphorhynchoid (GDB)

Melland's account does not mention a tail for the Jiundu kongamato—but, equally, there is no statement that it doesn't have one. Hence the category of pterosaur involved here is uncertain. As most lizards possess noticeable tails, however, and as native descriptions of the kongamato frequently liken it to a winged lizard, it may not be too presumptuous to suppose that the kongamato itself has a sizeable tail—which would thereby place it most comfortably within the rhamphorhynchoid group too.

With regard to the giant winged beasts of the Namibian desert, conversely, their sheer size alone places them within the pterodactyloid group (although there were small pterodactyloids too, e.g. *Pterodactylus*, *all* of the *biggest* pterosaurs were pterodactyloids). I have deliberately refrained from mentioning these animals to any great extent here—because whereas their huge size and desert terrain sets them well apart from the moderately-proportioned swamp-dwelling forms reported elsewhere in Africa, these same features correspond closely with those described for creatures documented in the next section of this chapter. Hence it would be more logical to deal with the Namibian beasts there, rather than here.

Something that initially seems to set the kongamato and kin well apart from bona fide pterosaurs is a highly unexpected ability not mentioned here so far. Many natives have claimed that these winged terrors can dive into rivers, plunging directly underwater—only to surface underneath a native boat and capsize its hapless occupants into the water. Indeed, the name 'kongamato' is derived from 'nkongamato', which translates as 'boat breaker'. Such behaviour hardly calls to mind the ineffably aerial pterosaurs of the Mesozoic.

Yet as far back as 1929, Dr L. Döderlein proposed that some rhamphorhynchoids may have dived down into the water and hunted fishes beneath the surface, rather like modern-day gannets. Even their feet were webbed. The principal problem with this notion is that pterosaurs could not press their wings as close against their bodies as birds can, which means that their large flight membranes may have interfered with their swimming. Also, one would expect creatures that exhibit a fair amount of aquatic activity to possess at least a certain degree of specialisation for such activity, but fossil pterosaurs do not appear to do so. If the kongamato really is a resurrected rhamphorhynchoid, however, we can only assume that the 90-odd million years of continuing evolution separating the most recent fossil species on record from the living kongamato has satisfactorily dealt with this situation.

The concept of at least one species of 21st-Century pterosaur thriving in the heart of tropical Africa must surely be among the most thrilling cryptozoological subjects of all time. Nevertheless, the redoubtable combination of inaccessible habitat, superstitious/justified native terror, and aerial aptitude is likely to thwart for a long time to come any Westerner hoping to solve the mystery of these secluded swamp-dwellers. Having said that, it may already be too late.

In October 1997, cryptozoological field investigator Brian Irwin emailed me to say that during the previous month he had visited and even trekked through part of the Jiundu swamp, but none of the local villagers questioned by him there, not even the village chief, had any first-hand knowledge of the kongamato. An elderly villager said that he remembered hearing about it when he was younger, but didn't claim to have seen it himself. As a result, Irwin concluded that even if the kongamato had existed in the past, it had probably been exterminated because it was such a menace. Yet unless a body, or at least a skull or skeleton, of a kongamato is made available for formal scientific scrutiny, its reality will be discounted and its history rejected as fanciful legend.

In his classic children's book *The Water-Babies*, Charles Kingsley wrote:

> People call them pterodactyls: but that is only because they are ashamed to call them flying dragons, after denying so long that flying dragons could exist.

In view of traditional zoology's current contempt for the kongamato, it would be appropriate to reword Kingsley's lines as follows:

> People call them kongamatos: but that is only because they are ashamed to call them pterodactyls, after denying so long that pterodactyls could still exist.

A LIVING PTERODACTYL IN GREECE?

If pterodactyls living today in Africa seems unlikely, how much more so in Europe? After all, Greek mythology tells of many winged monsters, including the harpies and the Stymphalian birds, but I don't recall any mention of pterodactyls.

Nevertheless, Crete was the setting for the appearance of just such a creature—allegedly! At 8.30 am one morning in summer 1986, three young hikers (Mannolis Calaitzis, Nikolaos Chalkiadakis, and Nikolaos Sfakianakis) were hunting by a small river in the Asterousia Mountains when they saw a bizarre creature flying overhead. They described it as resembling a giant dark-grey bird but with membranous bat-like wings that sported finger-like projections at their tips, long sharp talons, and a pelicanesque beak.

It reminded all three of them of a pterodactyl (though it is true that boys tend to be more clued-up about dinosaurs and other prehistoric monsters than about birds), and certainly their description sounds more pterosaurian than avian. Conversely, it hardly need be said that a colony of modern-day pterosaurs on Crete would surely have been uncovered by science long ago—wouldn't it? Also, in fossil pterosaurs the entire wing was supported by the forearm and just a single digit (the fourth), not by more than one digit (the first three digits were all positioned together near the proximal upper edge of the wing, not at its tip as claimed by the eyewitnesses of the supposed Greek pterosaur).

PTEROSAURS IN TEXAS?

A reclusive race of pterodactyls undisturbed by the local populace and undiscovered by the scientific community amid the primeval swamps of the Dark Continent may not seem totally beyond the realms of reasonable probability. The concept of flying reptiles soaring through the open skies of modern-day Texas, conversely, is quite another matter.

Nevertheless, although fewer in number, some sober reports of pterosaurian lookalikes have indeed been filed from this part of the world. And, as only to be expected from the Lone Star State, they are apparently far bigger in size than any reported elsewhere.

The year 1976 was a major time for such sightings, whose epicentre was a small triangular area in the southeastern tip of Texas bordering with Mexico. The classic encounter took place on 24 February, and featured three schoolteachers driving along an isolated road southwest of San Antonio. Suddenly, a huge shadow fell over the road, and after glancing up to see what was responsible for it, Patricia Bryant spied a monstrous creature gliding overhead—no higher than the telephone wires, but with a wingspan of 15-20 ft. Its body was encased in a greyish skin, and when fellow teacher David Rendon observed it, he was struck by the distinctly bat-like appearance of its huge wings, which resembled great membranes stretched across a framework of bones.

Unable to put a name to what they had seen, the trio spent some time afterwards looking through encyclopedias in search of a comparable creature. Eventually they found one—its name was *Pteranodon*, a huge pterosaur from North America's late Cretaceous, with a long toothless beak, a lengthy bony crest, and a massive wingspan of up to 30 ft.

Unbeknownst to the teachers, just a few weeks earlier and only a little further south two other eyewitnesses had nominated a *Pteranodon* as the animal most similar to the mysterious winged beast that they had seen. This occurred after Libby and Deany Ford, in mid-January, had sighted what they initially referred to as a big black bird, but with a bat-like face, near a pond several miles northeast of Brownsville—just inside Texas's border with Mexico.

Life-sized model of a *Pteranodon* in flight (© Dr Karl Shuker)

A week before that, an even stranger beast had been spied, which seemed to combine features from both of those already described here. During the evening of 7 January, after hearing a loud crash, as if something had collided against his trailer home near Brownsville, Alvérico Guajardo decided to step outside and investigate—a decision he was soon to regret.

When he switched on the headlights of his station wagon to expose the intruder-cum-vandal, he was confronted by a terrifying apparition—glaring at him infernally with blazing eyes of scarlet. Garbed in inky-black plumes, and standing 4 ft high, it raised its 2-4-ft-long beak, and a pair of dreadful bat-like wings could be seen wrapped around its shoulders. Uttering a spine-chilling shriek, the monster backed away to escape the headlights' penetrating beam, while a greatly-frightened Guajardo fled to his neighbour's home. The next morning, still trembling, he recounted his amazing experience to a local news reporter, who was greatly impressed by what he felt to be Guajardo's evident sincerity—and fear.

Even more horrific was the attempted abduction of Armando Grimaldo, at Raymondville, just north of Brownsville—by a man-sized *beakless* entity with dark, leathery, unfeathered skin, the face of a bat or monkey, large flaming eyes, and a 10-12 ft wingspan. No sooner had the sounds of bat-like wings flapping and an unusual kind of whistling alerted Grimaldo to the presence of something near, as he sat in his mother-in-law's backyard on the evening of 14 January, than the afore-mentioned creature allegedly swooped down and snatched at him from above with big claws. Happily, it did not succeed in carrying him away, and soon flew off—leaving a shocked but uninjured Grimaldo to wonder forever afterwards just what it was that had attacked him and had torn his clothes on that nightmarish evening.

On 14 September 1983, ambulance technician James Thompson was driving along Highway 100 to Harlingen, midway between Raymondville and Brownsville, when a remarkable airborne object with an eyecatching tail flew across the road about 150 ft in front of him. Thompson was so astounded by what he saw that he pulled up and stepped out of his ambulance in order to obtain a better view of it as it skimmed over the grass.

Could giant winged cryptids exist in the Brownsville area of Texas? (© Dr Karl Shuker)

> Its tail is what caught my attention. . . I expected him to land like a model airplane. That's what I thought he was, but he flapped his wings enough to get above the grass. . . It had a black, or greyish rough texture. It wasn't feathers. I'm quite sure it was a hide-type covering. . . I just watched him fly away. . .

According to Thompson, the creature measured 8-10 ft in total length, with a thin body whose tail terminated in what he referred to as a fin, and had a wingspan that was at least as wide as the ambulance (5-6 ft). He also reported that whereas its neck was virtually nonexistent, the back of its head bore a hump, and there was also a structure near to its throat that reminded him of a pelican's pouch. Initially, he could only assume that it was some form of odd bird, but after reading up on the subject he revised his identification, describing the creature as a "pterodactyl-like bird".

In response to this record, staff at the Laguna-Atacosta Wildlife Refuge suggested several more conservative identities for consideration, including an ultralight aircraft, and an American white pelican *Pelecanus erythrorhynchos*. Needless to say, however, aircraft (ultralight or otherwise) do not flap their wings, and pelicans have neither a humped head nor a tail fin. What is particularly interesting about this report is that whereas the earlier ones were linked with the giant tailless *Pteranodon*, the tail fin of Thompson's flying mystery beast is a feature of the smaller, tailed rhamphorhynchoids.

A remarkably similar beast was sighted during much the same time period, flying roughly 120 ft away at a height of about 50 ft off the ground, by Richard Guzman and a friend, Rudy, one early evening in Houston. A sketch produced by Guzman appears in American cryptozoologist Ken Gerhard's book *Big Bird!* (2007), and depicts an indisputably pterosaurian entity—complete with a prominent bony head crest and a long finned tail (crests are traditionally a pterodactyloid characteristic, but at least two fossil

Is this what James Thompson and Richard Guzman independently sighted? (© William M. Rebsamen)

rhamphorhynchoids are known to have been crested too).

In his book, Gerhard noted how, after interviewing Guzman personally (on 9 October 2003), he took out his copy of this present book of mine's original edition, *In Search of Prehistoric Survivors* (1995). He then read out loud from it my documentation of Thompson's sighting to an enthralled Guzman who, inexplicably, had not seen my book before (what?!!), and had not previously known about Thompson's encounter.

Texas is not the only state in the U.S.A. to feature reports of 'neo-pterosaurs'. Carl Larsen documented a notable incident from Pennsylvania (*Pursuit*, fall 1982). It was around 9.30 am on 8 August 1981, and Leverne and Darlene Alford were nearing the town of Ickesburg after driving for two hours from Carlisle towards Lock Haven along Highway 74 crossing Tuscarora Mountain. Suddenly, the noise of their oncoming car disturbed two enormous winged creatures, each about 3 ft tall, perched on the ground up ahead.

To the Alfords' great alarm, the two creatures promptly began running down the road directly towards them, with wings fully outstretched (spanning the road's 15 ft width)—attempting to gain sufficient speed in order to become airborne. A mere 6 ft away from what seemed destined to be a major collision, the nearer of the two creatures turned to one side, and both succeeded in taking off into the air. For the next 15 minutes, the Alfords were able to follow them as they flew ever higher directly above the road, but eventually they dwindled to the merest specks in the sky. A sketch of one of these creatures by Darlene, depicting it as viewed from below while in flight, revealed that its 'tail' was nothing more than a slight outward curve beyond the posterior perimeter of the immense wings—the only other structure visible from below was its beak, characterised by its rounded tip. In a second sketch, showing the head and neck in profile, Leverne disclosed that they had each borne a prominent backward-pointing crest.

Were they birds? Not as we know them, judging from the Alfords' description:

American cryptozoologist Ken Gerhard (© Ken Gerhard)

> They didn't seem to have any feathers. They looked like they were covered with skin. I immediately thought of those prehistoric birds. They were both about the same size, with long necks curved in an S-shape. They were dark gray in colour, with rounded beaks.

Taxonomic error notwithstanding, the 'prehistoric birds' were presumably pterosaurs—and except for their round-tipped beak and long neck, the Tuscarora Mountain flyers certainly bore far more resemblance to *Pteranodon* than to any avian species. Furthermore, as we shall see shortly, there is another giant pterosaur from North America's Cretaceous that even shared their long neck.

Scott T. Norman (© William M. Rebsamen)

Prior to his untimely death in 2008 aged just 43, Scott T. Norman was one of America's most respected investigators of mystery beasts, and had participated in various cryptozoological field expeditions to remote regions across the globe. He was also a longstanding friend of mine whose honesty and objectivity as a chronicler of cryptids I can wholeheartedly vouch for, thereby making his own first-hand sighting all the more notable.

During the early morning of 22 July 2007, Scott, who until then had been an avowed sceptic of alleged pterosaur sightings, had been sitting outside a shed at a property on California's West Coast, where he and some colleagues had been investigating claims of pterosaur occurrence in this area. Suddenly, at around 2 am, while gazing out over a nearby pasture, he saw a remarkable creature glide just above the 18-ft-high shed and into the pasture before he lost sight of it in the darkness (he remained sitting there for a further 2 hours in case it returned, but it didn't do so). Scott only saw its silhouette against the night sky, but he could readily discern that its body was around 6 ft long, its neck about 1-2-ft long, its head approximately 3 ft long bearing a 2-ft crest which matched that of a *Pteranodon* in his opinion, and its wings spanned around 10 ft. He did not see any tail. Scott's sighting lasted 15-20 seconds, so he had a good look at the creature, and was very perplexed by it, his abiding memory being just how big it was, far larger than any bat.

One of the most notable recent sightings was that of Prof. Steven Watters, made while looking out of the open back door of his home in Crestview, Florida, just before noon on 14 November 2012. That was when he spied a huge featherless winged creature fly overhead at a closest height of around 100-150 ft, with an estimated wingspan of 8-12 ft, plus a readily-discerned tail as long as the creature's body that bore a diamond-shaped bulge at its tip. It was whitish-grey in colour, sported a long pointed beak, and eventually landed on the very top of a tree about 0.5-0.75 mile away from where Prof. watters was standing, before flying off again in a northeasterly direction. Watters was previously unaware of alleged pterosaur sightings in North America, only discovering that others had seen comparable creatures when subsequently researching his sighting online.

Another case worthy of inclusion here emanated from southeastern Arizona. If genuine, it is certainly the most important ever recorded, but it sounds almost too good to be true. The case was published by the *Tombstone Epitaph* on 26 April 1890, which claimed that six days earlier, two ranchers riding through the Huachuca desert between the Whetstone Mountains and the Huachuca Mountains outside Tombstone had encountered an incredible monster with featherless leathery wings, writhing on the ground as it made futile attempts to become airborne. If the report is to be believed, the creature was of truly gargantuan proportions—over 92 ft in total length, with a wingspan of about 160 ft! Faced with such a horror, most observers would have been more than content to redirect their journey elsewhere with all speed, but the ranchers were clearly made of sterner stuff. Unfazed by its enormity, not to mention its lengthy beak crammed with sharp teeth, they allegedly opened fire with their Winchester rifles, killing the creature outright.

Not surprisingly, it soon occurred to them that the people back home might find their story a little difficult to believe, so they cut off the tip of one of its wings and took it back with them as proof of the monster's reality. What became of this remarkable trophy, however, is unknown—and indeed, the entire tale might have been written off as a far-fetched yarn long ago, were it not for Harry McClure. In 1910, he had been a young man living in Lordsburg, New Mexico, roughly 97 miles northeast of Tombstone—and in 1969, after reading a magazine article on monsters that mentioned this incident, he came forward to announce that he not only remembered hearing all about it at the time, but also had known the ranchers themselves, who were well-respected citizens. Moreover, the true facts of the case as recalled from his youth were much more sober than the *Epitaph*'s account.

According to McClure, the beast's wingspan had only been 20-30 ft, and it had twice succeeded in becoming temporarily airborne, travelling some distance through the air before crashing back to earth. In addition to its wings, it had a single pair of horse-like legs, and its eyes were extremely large, as big as saucers. The ranchers had indeed shot at it with their Winchesters, but, contrary to the newspaper's claims, the beast was not killed, and the ranchers eventually left it alone, still striving to become airborne.

In recent years, many alleged vintage photographs have been (and still are) circulating online depicting the supposed carcases of large dead pterosaurs killed by US Civil War or other early American soldiers and displayed alongside them. Every one of these images that has been investigated, however, has been exposed as a hoax, involving skilful (or sometimes not so skilful!) photo-manipulation and photo-interspersion techniques with stock images of pterosaur models and genuine vintage pictures of war veterans, ranchers, or hunters.

Petroglyphs of the Piasa

No coverage of North America's winged reptilian monsters would be complete without mentioning the piasa of Illinois. On or around 1 August 1673, French Jesuit priest Father Jacques Marquette was journeying down the Mississippi River when, upon reaching the section that nowadays passes by Alton, he encountered two terrifying monsters. Fortunately, they were quite harmless—because they happened to be a pair of enormous petroglyphs, meticulously carved and painted about 80 ft

The replacement, 20th-Century piasa petroglyph (CC-SA Burfalcy)

above the river upon the side of the limestone cliff on which Alton would one day be built. The portion of the cliff that had borne the body of one of the petroglyphs had crumbled away with the passage of time—so only the monster's head remained. The other petroglyph, conversely, was still present in its entirety, and depicted what would have been an animate nightmare, had it ever lived.

According to the description included by Marquette in his published journal of 1681, the beast was as large as a calf and had a man's face, but its head and horns were those of a deer, it had a tiger's beard, red eyes, and a tail ending like that of a fish but so long that it passed over the beast's scale-covered body and between its legs under its body. It was painted in three colours—red, black, and bluish-green.

Later descriptions by other observers of this spectacular representation provided some significant additional details. The precise size of the depicted creature was 30 ft long and 12 ft tall. It possessed a pair of large wings, raised upwards as if it were about to take flight, with a span of around 16-18 ft. And its legs were bear-like, but terminated in the talons of an eagle. It is very fortunate that the creature's appearance had been documented in detail—because in or around 1856, some nearby quarrying disrupted the cliff face, as a result of which the petroglyph shattered and cascaded down into the river.

In the meantime, however, information regarding the identity of the creature portrayed by it had been obtained from the local Illini Indians, who referred to it as the piasa—'the bird that devours man'—and recounted various legends in which these creatures had terrorised the Indians' ancestors until eventually slain by them. Just a fanciful example of native folklore, or a distant, distorted memory of a real animal? And who created the petroglyphs anyway? The Illini of long ago? In his book *The Piasa, Or the Devil Among the Indians* (1887), Perry Armstrong wondered if this beast was based upon memories of a living pterosaur (specifically a rhamphorhynchoid)—a notion that has also been raised by several other investigators.

The answers to all of these questions are still unknown, but one thing is certain—anything that inspires a work of such size and elaborate composition as the piasa petroglyphs is evidently of great significance to the artist(s) responsible. It hardly seems likely that they would have invested such time and effort in portraying a wholly imaginary monster.

Incidentally, although the original piasa petroglyph has been lost, a newer version, created during the 20th Century, and based partly upon 19th-Century sketches and lithographs, is present upon a bluff at Alton, several hundred yards upstream of the earlier one. Unfortunately, however, it has to be regularly restored, due to the unsuitability of the limestone rock quality there for retaining an image painted upon it.

Staying with petroglyphs of a quasi-pterosaurian persuasion: for decades, cryptozoologists, creationists, and iconography researchers have been trading views as to the likely identity of the tantalisingly pterosaur-like creature, bright red in colour, depicted in ancient artwork decorating Black Dragon Canyon in Utah. True, there is indeed a resemblance to a pterodactyl with outstretched wings and even a possible crest upon its head like some latter-day *Pteranodon*, which had led some cryptozoologists to suggest that it offered proof of modern-day pterosaur survival in North America. However, new research has effectively torn apart this visual testimony.

Dating back to the agrarian Fremont culture (c.1-1,100 AD) but remaining undiscovered in modern times until 1928, in August 2015 this ambiguous artwork was revealed in an *Antiquity* paper authored by researchers co-led by freelance archaeologist Paul Bahn to be a composite petroglyph, not a single one as previously assumed. In fact, the 'pterosaur' is actually a combination of five separate petroglyphs, respectively depicting a sheep, a dog, a tall person with protruding eyes, a smaller person, and a snake-like entity. But how had this composite come to be?

In 1947, a certain John Simonson had traced over what he believed to be the outline of the one, single petroglyph with red chalk, yielding the pseudo-pterodactyl image, but this artefact was recently exposed by Bahn and company using a portable x-ray fluorescence device and a very special program/tool called DStretch. This enables researchers to photograph a petroglyph, upload it onto a computer, and then highlight its original pigments (even if invisible to the naked eye) while also distinguishing pigments that have been added later. So when DStretch removed the confusing effect caused by Simonson's red chalk, the true, five-piece artwork was duly revealed, with the pterodactyl of Black Dragon Canyon unceremoniously jettisoned into the dustbin of historical howlers.

Rebirth of *Quetzalcoatlus*

Notwithstanding the above accounts, it is likely that the mere suggestion of *any* modern-day pterosaurs (let alone giant ones) existing in Texas, not to mention elsewhere in North America too, would have received short shrift

The author alongside a statue of *Quetzalcoatlus* (© Dr Karl Shuker)

even among the more open-minded members of the cryptozoological community—had it not been for 1975, that is. For this was the year in which such a remarkable discovery was made that suddenly anything seemed possible.

It was then when palaeontologist Dr Douglas Lawson, from the University of California (Berkeley), announced the discovery of a colossal new species of pterosaur from the late Cretaceous. He had personally discovered the first remains in 1971, while still a postgraduate student at Texas University and undertaking excavations in the Big Bend National Park, Brewster County, in western Texas. During the next three years, the partial skeletons of three separate specimens were unearthed here.

Based upon these remains, which included a single radius (forearm bone) measuring over 2 ft long, calculations of the species' likely wingspan gave an initial estimate of up to 51 ft, but later fossil finds reduced and refined this to the currently accepted figure of 33-36 ft. Nevertheless, that is still more than sufficient to crown this revolutionary pterosaur one of the biggest flying creatures ever recorded by science—effortlessly exceeding fellow American pterosaur *Pteranodon sternbergi*, with a wingspan of around 30 ft.

Another of its claims to fame is that it is also one of the most recent ('youngest') of pterosaur species, because its fossils are only 68-66 million years old, thereby dating from the very end of the Cretaceous. Morphologically, it was quite different from the familiar *Pteranodon*—quite apart from its far greater wingspan, it also sported a much longer neck and limbs, but only an exceedingly short tail, plus a crest of uncertain size and shape. Emphasising how dramatic this winged wonder must have appeared in life, Lawson formally dubbed it *Quetzalcoatlus northropi*, after the Aztec deity Quetzalcoatl, whose symbol was a huge flying serpent.

Its lifestyle probably differed from *Pteranodon*'s too—for whereas the latter's fossils occur in marine strata and imply a largely piscivorous diet, those of *Quetzalcoatlus* were excavated from the flood plain of a meandering river that during Cretaceous times was about 250 miles inland. From this, two quite different hypotheses concerning its diet have arisen. According to one school of thought, it may have utilised its long toothless beak for probing out crabs and molluscs from the mud at the bottom of ponds and rivers. In contrast, others picture it as a terrestrial, quadrupedal stalker of small dinosaurs and other vertebrates that lived on land or in streams, thus sharing certain lifestyle aspects with long-necked storks.

When is a Pterosaur Not a Pterosaur?

From the episodes recounted in this section, it is clear that the eyewitnesses believed the creatures that they saw resembled pterosaurs more closely than anything else known to them. Nevertheless, certain of these beasts were evidently not pterosaurs. Alvérico Guajardo's attacker, for instance, had bat-like wings, but black plumes—thereby eliminating a pterosaur. The wings' bat-like appearance might simply have been an illusion created by the way in which they were being held, rather than a direct allusion to a naked, leathery look.

Similarly, the bat-faced winged creature spied by Libby and Deany Ford was initially referred to by them as a big black bird—only after seeing pictures of pterosaurs did they ally it with *Pteranodon*, whose slender face, bony crest, and extremely long pointed beak instantly differentiate it from any type of bat. (Making matters even more complex when attempting to analyse reports of apparent pterosaurs in North America is the abundance of accounts featuring what appears to be a still-undiscovered species of gigantic flying bird—a situation dealt with in its own right a little later in this chapter.)

Another bat-faced monster with wings was the would-be abductor of Armando Grimaldo. Here, however, its skin was specifically described as unfeathered, and it had no beak—thus ruling out birds and pterosaurs alike. Is it possible that the description of this creature's face was more than just comparative—that the animal really was a giant bat? In my book *The Beasts That Hide From Man* (2003), I discussed the plausible existence of such beasts in Java and parts of western Africa, but Loren Coleman clearly feels that this identity should be applied to certain of the cases documented in this chapter too—in his book *Curious Encounters* (1985), he referred to the Texas terrors from 1976 as ". . . unknown giant bats seen along the Rio Grande, near Brownsville". Who knows—this identity may even be relevant to the man-faced piasa, whose resemblance to a pterosaur is remote in the extreme.

Certain other accounts given in this section do recall pterosaurs—but the situation here is complicated by the fact that more than one type of pterosaur appear to feature in them. Whereas, for instance, the fin-tailed glider spotted by James Thompson corresponds with a rhamphorhynchoid, the creature spied by the three teachers near San Antonio and the Tuscarora Mountain twosome encountered by the Alfords recall a giant pterodactyloid, such as *Pteranodon* or *Quetzalcoatlus* (the latter is particularly reminiscent of the Tuscarora creatures). The pterodactyloid versions are very thought-provoking, as they approach in size the giant desert-dwelling pterosaur-like beasts described from Namibia earlier in this chapter.

If, somehow, a lineage of giant pterosaurs has persisted to the present day, *Quetzalcoatlus* seems to offer a more plausible ancestry for it than *Pteranodon* or any other genus. Its occurrence inland ties in with the habitats reported from the U.S.A. and from Namibia for their winged colossi. True, no *Quetzalcoatlus* remains have so far been unearthed in Africa, but those of a related giant, *Arambourgiania* [formerly *Titanopteryx*] *philadelphiae*, have been found in Jordan, in the Middle East.

Of course, whereas the concept of winged monsters soaring over the little-explored deserts of Namibia may not seem outrageously romantic, attempting to generate a similar degree of credibility for the scenario of such creatures existing undiscovered in a land as ostensibly familiar as the U.S.A. is another matter entirely—until we consider that only 200 miles east of Brownsville, Texas (a heartland of pterosaur sightings) is Mexico's Sierra Madre Oriental, which is actually one of the least explored regions in the whole of North America. A suitable home and hideaway for a viable quota of *Quetzalcoatlus*?

The likelihood that *Quetzalcoatlus* subsisted upon small vertebrates is also consistent with what one would expect the dietary preferences of the modern-day monsters to be (especially in Namibia's deserts), and its general morphology and size correspond very closely indeed with theirs. In addition, the massive crests upon its humeri (upper arm bones) for muscle attachment show that it must have possessed very powerful flight muscles. This indicates that in contrast to early speculation, *Quetzalcoatlus* was capable of powered, flapping flight as well as passive gliding—thereby yielding an identity equally consistent with reports of flapping and gliding pterosaurian mystery beasts.

Nor should we forget that it is the youngest pterosaur currently known from the fossil record, with remains dating right up to the very close of the Cretaceous, and that these selfsame

Two artistic restorations of *Quetzalcoatlus* in flight (top, LeCire; bottom, GDB)

remains eluded scientific detection until as recently as 1971. If there is any pterosaur that could conceivably be represented by still-undiscovered post-Cretaceous fossils, and even evolved modern-day representatives, *Quetzalcoatlus* must surely be the one.

Cryptozoology is full of curious coincidences—but few are more curious (or contentious) than the undeniable fact that modern-day reports of giant pterosaurian lookalikes just so happen to emanate from the very same region that was once home to a bona fide creature of this type. In addition, the eyewitnesses of the mid-1970s versions are not likely to have been aware of the near-synchronous discovery of *Quetzalcoatlus*—because reconstructions of it did not appear in popular-format books and newspaper articles for quite some time afterwards.

RESURRECTED 'SERPENT BIRDS' IN LATIN AMERICA?

Pterosaur reports have also emerged from Mexico, South America, and even the Caribbean. The late J. Richard Greenwell, onetime secretary of the International Society of Cryptozoology, had a Mexican correspondent who claimed that there are living pterosaurs in Mexico's eastern portion and was (still is?) determined to capture one, to prove beyond any shadow of doubt that they do exist. Worthy of note is that certain depictions of deities, demons, and strange beasts from ancient Mexican mythology are decidedly pterodactylian in appearance.

One particularly intriguing example is the mysterious 'serpent-bird' portrayed in relief sculpture amid the Mayan ruins of Tajin, at Totonacapan in the northeastern portion of Veracruz—noted in 1968 by visiting Mexican archaeologist Dr José Diaz-Bolio, and dating from a mere 5,000-1,000 years ago. Yet all pterosaurs had officially become extinct at least 66 million years ago. So how do we explain the Mayan serpent-bird—a non-existent, imaginary monster; a beast of legend inspired by sightings of pterodactyl fossils by laymen; a misidentified known creature; or a living creature lingering long after its formal date of demise but still eluding official scientific discovery? Although none of these solutions would be unprecedented, only one (if any) is correct—but which one?

Conversely, after hearing about a report from early 1985 concerning the supposed filming of some pterodactyls flying over Mexico's Yucatan Peninsula, Loren Coleman made some enquiries and drew a total blank from the locals.

Indeed, when he showed one man a sketch of a pterodactyl, he had no hesitation in identifying it as an eagle!

In his book *Strange Prehistoric Animals and Their Stories* (1948), explorer A. Hyatt Verrill discussed the theory of fossil-inspired illustration with regard to the Aztecs' remarkably pterosaurian depictions of one of their demons—a huge flying monster called Izpuzteque. A mere artefact of imaginative artistry—or something much more deliberate, based upon palaeontological precedence? Verrill favoured the latter option, commenting: "... such a pure invention would have been a most amazing coincidence". He adopted the same view in relation to some similar depictions present on a series of Code pottery that he had unearthed in Panama, noting:

> Not only do the drawings show the beak-like jaws armed with sharp teeth, but in addition, the wings with two curved claws are depicted. Included also are the short pointed tail, the reptilian head crest or appendages, and the strong hind feet with five-clawed toes on each.
>
> It is wholly unreasonable to assume that the pterodactyl-like figures were intended to represent birds; for even if artistic licence and imagination permitted the Indian artist to add teeth to the bird's beak and claws to the wings, he most certainly would not have given a bird feet with five toes.

On a beautiful clear sunny day in around mid-March 1971, having been in Cuba for about 4 months stationed to 2nd Battalion, 8th Regiment (Reinforced), H&S Co., 106 mm Recoilless Rifle Platoon, U.S. soldier Eskin C. Kuhn was at his platoon's barracks, looking out towards the ocean, when he saw an amazing sight. What he claims were two pterosaurs, each with a 10-ft wingspan, flew by at low altitude, estimated by him to be perhaps 100 ft, and at very close range, so close that he could see that the struts of their wings emanated from fingers. Eskin ran inside and asked his sergeant to come out and corroborate his sighting, but by the time that he did so, it was too late—the creatures had gone. As noted earlier in this section, however, the wings of pterosaurs were each supported by only a single strut—the highly elongated fourth finger—not by several. Consequently, Kuhn's claim that the struts of the creatures' wings emanated from fingers (i.e. plural) contradicts a pterosaur identity.

Reiterating what I mentioned in Chapter 1 in relation to the stoa, when writing his famous novel *The Lost World* Sir Arthur Conan Doyle was greatly inspired by local Waiká Indian reports of huge monstrous creatures said to inhabit a lofty tepui called Kurupira on the Venezuela-Brazilian border that the famous lost explorer Lt-Col. Percy Fawcett had spoken to him about after having visited similar plateaux during his Bolivian expedition to the great sandstone tepuis known as the Franco Ricardo hills. According to the Waiká, moreover, one such creature, called by them the washoriwe, would sometimes swoop down from Kurupira's high summit into the jungle at its base, skimming through this Indian tribe's territory on huge wings that boasted a span of 20 ft or more. In addition, it bore a long bony backward-pointing crest upon its head, and sported a very long pointed beak.

Waiká lore attests that this terrifying entity is the immortal forefather of all vampire bats. Yet whereas the immortal forefathers of all other beasts in their lore closely resemble their respective descendants (except for the much greater size of the forefathers), the long-beaked, bony-crested washoriwe bears scant resemblance to the short-faced, crestless vampire bats. Moreover, whereas these latter bats are strictly nocturnal, the washoriwe reputedly flies only during the daytime.

After highlighting these significant morphological and behavioural discrepancies in his 1997 cryptozoological encyclopaedia *Svet Tajemných Zvírat* ('*The World of Mysterious Animals*'), Czech zoologist Jaroslav Mareš pointed out how, in stark contrast, the washoriwe seemed to be very similar in form and lifestyle to pterosaurs. He also commented upon the curious coincidence of how frequently the finding of complete, perfectly-preserved fossil pterosaurs by palaeontologists had occurred in this same region in modern times. Might the Waiká's belief in the washoriwe have been inspired, therefore, by their own possible finding of fossil pterosaur remains here from time to time? Or might it even be, as again pondered by Mareš, that the abundance of such remains in this region lends support to the possibility that a pterosaurian lineage has persisted here right into the present day, currently undiscovered by science but well known to the local Indians, who refer to these airborne prehistoric survivors as washoriwes?

As noted in Chapter 1, Mareš organised an expedition to Kurupira in 1978, but unfortunately he did not find any cryptids there, and he also failed to ascend this tepui's high vertical sides in his bid to reach its summit's plateau. During spring 1997, however, Mareš met in Boa Vista, Roraima (Brazil's northernmost state), a gold-prospector identified by him in print only via the pseudonym 'Reginald Riggs' whom he had first encountered during his expedition to Kurupira 19 years earlier, and he learnt that, near a waterfall at Kurupira, Riggs had caught sight of a mysterious flying creature that a Waiká tribesman called Retewa had identified as a washoriwe. Moreover, in his cryptozoological encyclopaedia, Mareš stated that other gold-prospectors in this same area have claimed to have seen such creatures here, flying high above the jungle's tree tops, and some have even sworn that they were attacked by them.

Around February 1947, J.A. Harrison from Liverpool was on a boat navigating an estuary of the Amazon River when he and some others aboard spied a flock of five huge birds flying overhead in V-formation, with long necks and beaks, and each with a wingspan of about 12 ft. According to Harrison, however, their wings resembled brown leather and appeared to be featherless. As they soared down the river, he could see that their heads were flat on top, and the wings seemed to be ribbed. Judging from the sketch that he prepared, however, they bore little resemblance to pterosaurs, and were far more reminiscent of a large stork. Moreover, no less than three such species—the jabiru *Jabiru mycteria*, the maguari *Ciconia maguari*, and the wood ibis (aka wood stork) *Mycteria americana*—are indeed native here.

Jabiru (1000Faces)

Frigate birds in flight (CC-SA B. Navez)

An unequivocal example of pterosaur misidentification occurred in Argentina during the 1800s. This was when a supposed pterosaur was shot near Lake Nahuel Huapi—only for its carcase to be subsequently identified as that of a Patagonian steamer duck *Tachyeres patachonicus* (the only steamer duck species—of which there are four—that can fly).

Relevant here is a first-hand experience of my own. One early evening in 2007, I was standing at the top of Sugarloaf Mountain in Rio de Janeiro, Brazil, when, looking upwards, I was startled to see a number of superficially pterosaurian creatures circling high above in the sky overhead. Raising my trusty bird-watching binoculars to my eyes, however, I was swiftly able to disperse this illusion, because these putative prehistoric survivors were instantly exposed as seabirds known as frigate birds (in this particular instance the magnificent frigate bird *Fregata magnificens*).

These long, angular-winged relatives of gannets and cormorants do appear positively primeval on first sight, and might well deceive ornithologically-untrained eyes into believing that they had truly witnessed a phalanx of flying reptiles from the ancient past. Indeed, this is also the identity that Coleman believes to be the correct one for the supposed pterodactyls allegedly filmed in 1985 flying over the Yucatan Peninsula.

The most bizarre report of an alleged South American pterodactyl supposedly featured a (very) close encounter with one by a small commuter aeroplane carrying 24 passengers that was attempting to land in Brazil after having flown over this vast country's mountain jungles. According to a report of the incident that appeared in an issue from 1992 (precise date unknown to me) of *People*, an Australian fortnightly tabloid/celebrity-type men's magazine, a huge pterodactyl appeared and flew alongside the plane, causing the pilot to veer sharply to one side in order to avoid colliding with it—but this is not all. One of the passengers, a Dr George Biles, referred to in the magazine report as an American anthropologist, was quoted as having stated:

> This was a classic case of a white pterodactyl with a giant wingspan. Of course, I've heard the rumours for many years that these prehistoric creatures still roamed the Amazon. But I was sceptical like everybody else. But that wasn't an airplane of a UFO flying beside us. It was a pterodactyl.

A nothing if not uncompromising statement, always assuming, of course, that this case is genuine—but is it? After all, if it were, surely it would have attracted headlines worldwide, yet it didn't. And when I attempted to uncover the whereabouts, or even the reality, of Dr Biles himself via online searches, the only references to anyone of that name listed as an American anthropologist were a handful of website coverages of this very same case! In other words, 'American anthropologist Dr George Biles' would appear to exist only within the confines of this alleged incident.

In the *People* report, a second eyewitness, air stewardess Maya Cabon, is quoted as having said:

> Here was this giant monster flying right next to the plane. He was only a few feet away from the window—and he looked right at me. I thought we were all going to die.

With a mounting sense of impending doom, I conducted an online search for Cabon too, and, not to my great surprise, the only references found were ones to this present incident yet again.

Not to be outdone by Biles and Cabon, the aeroplane's pilot also reputedly offered up a soundbite: "He was coming straight at us and he was mighty big!" But as the pilot was not named in the *People* report, I was unable to check him out, although I strongly suspect that even if I had been able to do so, the result would have been the same as for Biles and Cabon.

Yet if, just for the sake of argument, we accept that this incident really did happen, the creature was clearly too near and far too big for there to be any realistic likelihood that it was simply a large bird—Biles's unequivocal identification of it as a pterodactyl is sufficient by itself to scupper that option anyway. But there, tantalisingly, is where the trail, such that it is, goes cold, leaving us to ponder futilely over whether it was all simply a tabloid invention, or whether it was a bona fide living pterodactyl. Whichever is correct, however, it was undeniably of monstrous proportions!

PTEROSAUR NEWS FROM NEW GUINEA, NEW ZEALAND, AND AUSTRALIA

Several notable cryptids have been reported from New Guinea and its offshore islands over the years, including alleged dinosaurs, the thylacine-like dobsegna and the tapir-like devil-pig (see Chapter 4), the enigmatic freshwater shark of Lake Sentani, and what may be an undescribed bird of paradise species from Goodenough Island. However, it is the most recent major cryptozoological riddle from this enormous yet still poorly-explored island—a winged mystery beast known as the ropen—that attracts the greatest international publicity.

I first heard about the ropen during a series of communications in late 1999 and early 2000 from veteran cryptid-seeker Bill Gibbons, who was hoping to search for it as part of a planned television documentary (but as yet unmade). According to Gibbons, who also later provided data to the cz@egroups online cryptozoological

Representation of ropen in flight (© Karl G Rose-Enigmatic Static)

discussion group, two different kinds of unidentified flying mystery beast allegedly exist on or around New Guinea, but the name 'ropen' is often, confusingly, applied to both of them. The 'true' ropen supposedly frequents Rambutyo (=Rambunzo), a small island in the Bismarck Archipelago off the northeastern coast of Papua New Guinea (PNG), and Umboi, a slightly larger isle sited between eastern PNG and New Britain. With a wingspan of 3-4 ft, a long tail terminating in a diamond-shaped flange, and a long beak brimming with sharp teeth, its description is startlingly reminiscent of *Rhamphorhynchus*.

Feared greatly by the PNG and Solomon Islands people, the ropen hides or sleeps during the day in caves on Rambutyo and Umboi, but is on the wing at night. Attracted to the stench of decaying flesh, it has reputedly been known to attack human funeral gatherings here—including one instance when western missionaries were present. Missionaries have also spied this creature on nocturnal coastal fishing trips, and it will attack native fishing vessels by snatching fishes from the nets as they are being hauled in by the fishermen. According to Gibbons, moreover, one day during the 1990s the residents of an Umboi village called Gumalong watched as a ropen flew from nearby Mount Bel, down over the jungle valley, and directly over Gumalong as it headed out to sea.

An even more formidable flying cryptid supposedly exists on PNG's mainland, where it is referred to as the duah ('ropen' and 'duah' come from different native languages, but both apparently translate loosely as 'demon flyer'). With huge leathery wings spanning up to 20 ft, a fairly long neck, and a bony crest, this aerial mystery beast is said to resemble *Pteranodon*. Intriguingly, however, locals state that the duah has glowing underparts. Gibbons claims that in 1995, a missionary saw one as it flew by a lake, and that another missionary more recently claimed to have seen several in a mountain cavern.

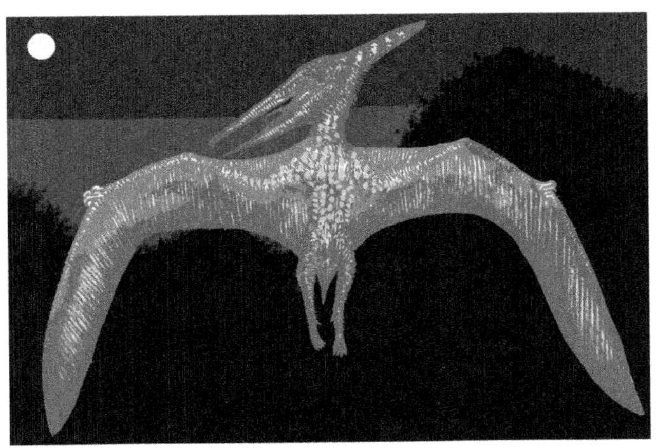

Duah sketch, showing its luminous underside at night (© Tim Morris)

The next important source of ropen news that I obtained was a letter by Robert F. Helfinstine, of Anoka, Minnesota, published in issue #33 of *Ancient American* during autumn 2000. In his letter, Helfinstine noted that he was acquainted with four of the eight members of an expedition (which included Dr M.E. Clark, a retired professor from Illinois University) that in 1994 sought "a large flying creature . . . called 'Ropen'" on PNG (presumably the duah, therefore, not the smaller, true ropen of the offshore islands). Interestingly, Helfinstine revealed that the natives claim that this creature's mystifying light derives from glowing patches on its underside, which can be actively turned on and off, like a firefly's bioluminescence.

He also reported a bizarre-sounding story of how one local had encountered a sleeping ropen, and had tied a nearby log to one of its legs to prevent it from escaping—only to see, upon returning with some companions, the fully-awake ropen flying away with the log hanging from its leg! According to another story, one of these powerful creatures had lifted a woman up off the beach and had transported her aloft to a mountain about four miles away. Helfinstine also noted that this cryptid is so attracted to decaying flesh, digging up freshly-buried human corpses to carry them away, that graves here have special coverings to prevent it from desecrating them.

The problem with such claims is that in all fossil pterosaurs documented so far, the feet were only very weakly muscled, and did not possess opposable digits. Consequently, even the biggest species would have been incapable of lifting and carrying off an adult human. So unless the duah is a modern-day pterosaur whose feet have become far more powerful and have developed opposable digits during 66 million years or more of continuing evolution, the latter claim is not realistic.

As for the duah's alleged bioluminescence, might this be due to glowing fungi from tree branches or cave interiors that have adhered to this creature's underside when pressed against such surfaces? A similar phenomenon is known to have been responsible for various cases of supposed luminous owls. True, this wouldn't account for the duah's reputed ability to turn the glow on and off at will, but that may be mere exaggeration.

Although the 1994 expedition spent some time seeking a ropen cave in the mountains, none was found. On a second expedition, several ropens were allegedly seen, but not photographed. Helfinstine claimed that missionaries and Peace Corp workers in New Guinea report that these animals fly from one offshore island to another. In January 2001, Helfinstine reiterated all of these details during a telephone conversation with Mark K. Bayless, a herpetological expert from Berkeley, California.

At much the same time, I discovered that during two expeditions to PNG, Dr Carl E. Baugh of the Creation Evidence Museum had learnt from the locals about a glowing flying beast termed the ropen. He also succeeded in spying such a creature himself at night, using a monocular night scope, and he photographed a strange print in the sand the following morning.

Later, in September 2001, Australian cryptozoologist Brian Irwin revealed on the cryptolist@yahoogroups.com discussion group that while in New Guinea during July-August 2001 he had visited some offshore islands allegedly home to the true, smaller ropen. On Rambutyo, locals claimed that it was quite easy 30 years or so earlier to see up to three of these flying at night, glowing (like the bigger PNG duah), on the island's uninhabited eastern side, but only one is seen at a time there nowadays. According to an eyewitness called Ralph (interviewed by Irwin), one night about 12 years previously, a ropen had dive-bombed his boat while he and a friend had been fishing on Rambutyo's east coast. After hitting the boat, the ropen had fallen into the water, where it splashed about for a time before flying off.

On nearby Manus Island, the local school's headmaster informed Irwin that he had once seen one of these glowing beasts sitting upon a branch of a tree on Goodenough Island, in the Milne Bay Province. And on Umboi, many locals claimed to have seen the ropen at night,

Artistic rendition of luminescent pterosaurs flying over a boat (© William M. Rebsamen)

including a policeman and a government employee during Irwin's own short stay there. It is thought to live near the top of Mount Bel, and is most frequently seen in the early morning on the island's western side. Sadly, Irwin did not spot one of these entities himself.

Fellow investigator Jonathan Whitcomb interviewed some eyewitnesses on Umboi who claimed to have spotted a huge pterosaur (a duah?) while hiking near Lake Pung as boys in or around 1994. This and many other alleged sightings are documented in his book *Searching For Ropens: Living Pterosaurs In Papua New Guinea* (2006), which is the first of several authored by him on the subject of putative living pterosaurs there and elsewhere around the world.

New Zealand, a land of many indigenous birds but few reptiles and even fewer mammals, is surely the last place one might expect to encounter any kind of present-day pterosaur, let alone a multicoloured one. Nevertheless, according to Whangarei resident Phyllis Hall, some time prior to the early 1980s she had been walking along a new motorway on North Island that had not yet been opened to traffic when a strange creature that looked to her like a pterodactyl flew "out of nowhere". Its under-wings were blue, but the rest of it was red, and it flew with an undulating motion. This description does not fit anything known to exist anywhere in New Zealand.

Even Australia is not immune from pterosaur sightings, as demonstrated by the following example, collected by Whitcomb, which featured a husband and wife who experienced their remarkable encounter while walking down a hill along a major thoroughfare in Perth, the capital of the state of Western Australia, at around 10.30 pm one evening in December 1997. Looking into the distance, the husband saw an indistinct object in the sky and continued watching as it soared closer. Within a minute, it was approximately 300 ft above them, and as the area was fairly well-lit he could see that it was a reddish-brown creature with leathery, seemingly unfeathered wings that spanned about 30-45 ft, and never flapped as it silently glided on through the sky. By now, his wife was watching it too, and she added that it had a lizard-like appearance, a long tail, and was definitely alive (a popular if implausible option offered by sceptics of reputed pterosaur sightings is that what are being observed in far-flung places right across the globe are man-made, remotely-controlled models).

Finally: on 16 January 2002, longstanding mystery beast investigator Todd Jurasek informed me that during the 1980s: ". . . a small pterodactyl was killed and displayed in a store front in Queensland until the rotten carcass was tossed out". He didn't have any further details, and I've been unable to uncover any either, so I've no idea what—if anything—this intriguing creature was. Might it have been one of those so-called devil fishes or Jenny Hanivers—exotic-looking but fake winged monsters created artificially by skilfully modifying the bodies of dead skates or rays before drying them out? Certainly, I can well imagine how a winged, superficially pterodactyl-like beast could be manufactured in this manner.

In any case, if this specimen did indeed exist, I wonder if some photos were taken of such a curious entity before it was discarded? If so, I'd like to see them, and any additional information concerning this brief but intriguing report.

EMBALMED, MUMMIFIED, AND STUFFED PTEROSAURS?

Speaking of manufactured monsters: Based upon surviving illustrations of the specimens in question, some cryptozoological researchers

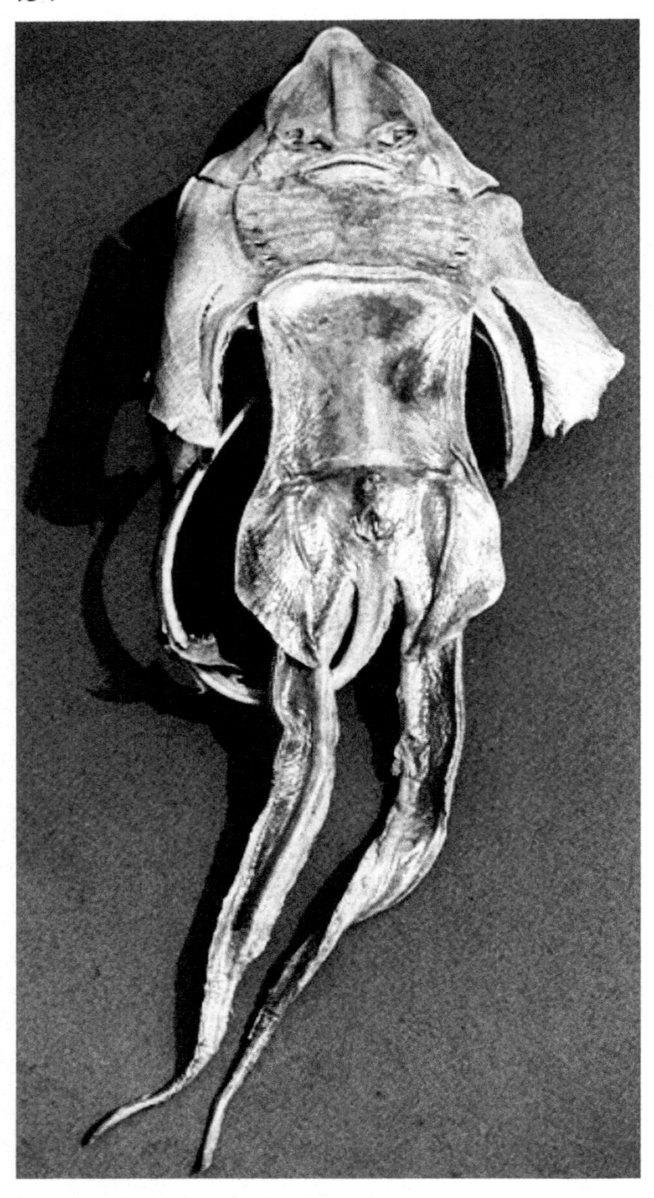

A so-called devil fish, in reality a modified dried skate (© Dr Karl Shuker)

have boldly speculated that certain preserved 'winged dragons' variously seen or owned by the likes of French naturalist Pierre Belon (1517-1564), Italian naturalist Ulisse Aldrovandi (1522-1605), and Cardinal Francesco Barberini (1597–1679), nephew of Pope Urban VIII, were actually recently-deceased pterosaurs. Three in particular have attracted particular interest in relation to this proposal.

The first specimen was one of several alleged winged dragons that Belon claimed to have seen in embalmed form while visiting Egypt, and a drawing of which he published in 1557, with variations upon it subsequently appearing in many works by other writers. The second one was supposedly a mummified African dragon presented to Aldrovandi as a gift by Francisco Centensis and subsequently depicted directly by Aldrovandi's painters in colour. And the third one was also a gift, but this time from King Louis XIII of France to Cardinal Barberini, and was depicted in a detailed drawing by Lyncean anatomist Giovanni Faber (accompanied by Faber's equally meticulous description of it) published in the multi-authored work *Rerum Medicarum Novae Hispaniae Thesaurus* (*'Thesaurus of Medicinal Treasures of New Spain'*), on the natural history of Mexico, which was published in 1651.

Although none of these specimens still exists, many of the illustrations of them do. Moreover, the extensive degree of morphological detail captured in these images has recently enabled Fayetteville State University biologist Dr Phil Senter working with Indiana University comparative literature expert Dr Darius M. Klein to analyse and identify with a high degree of confidence these specimens' true natures—and, in so doing, confirming that none of them was any form of pterosaur, modern-day or otherwise.

So what did they discover? Quoting from the abstract for the *Palaeontologia Electronica* paper in which they documented their investigation, the eye-opening findings of Senter and Klein are as follows:

> Comparison with extant animals reveals that Belon's and Aldrovandi's dragons are decapitated snakes with attached mammal heads. Their wings are the

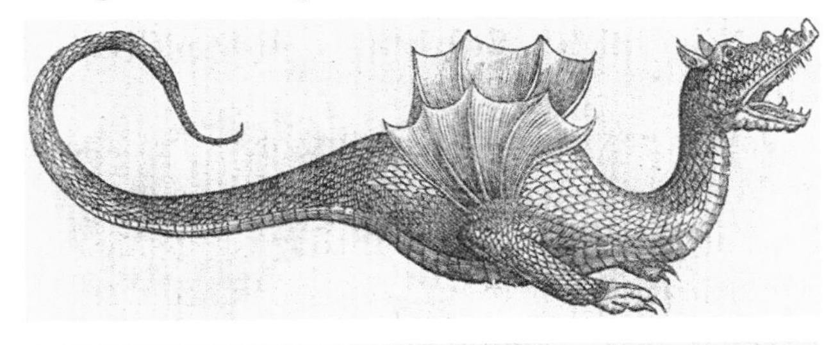

Illustrations of three preserved 'winged dragons' sometimes claimed to have been modern-day pterosaurs,—Belon's dragon (t); Aldrovandi's dragon (m); Barberini's dragon (b)

pectoral fins of flying gurnards *(Dactylopterus volitans)*. Their "legs" are the forelimbs of rabbits or canids in reptile-skin sleeves. The dragon illustrated by Faber and owned by Cardinal Francesco Barberini includes the skull of a weasel *(Mustela nivalis)*, the belly skin of a snake, the dorsal and lateral skin of a lizard, and the tail skeleton of an eel *(Anguilla anguilla)*. These hoaxes now join the list of discredited "proofs" of human-pterosaur coexistence.

In summary: fakes and frauds notwithstanding, there seems little doubt that very unusual flying beasts that cannot be readily dismissed as either birds or mammals are being seen in various disparate regions of the world. Whether any of them is truly a living pterosaur is another matter entirely. After all, there are no pterosaur fossils on record from beyond the end of the Cretaceous Period, 66 million years ago. Then again, as previously noted, many modern-day reports come from areas such as tropical forests where fossilisation is rare, or from inaccessible mountain ranges where fossils have not been sought.

Another problem with sightings of supposed living pterosaurs is that their descriptions tend to recall classic but nowadays outmoded reconstructions of what prehistoric pterosaurs may have looked like (rather than more recent but possibly less publicly-familiar palaeontological reconstructions). This intriguing situation is what one might expect if layman eyewitnesses are being influenced by more accessible yet less accurate pterosaur depictions. Then again, if by any chance a lineage of these flying reptiles has indeed survived into the present day, during the continued 66 million years of evolution that such creatures would have experienced, bridging the gap between the youngest pterosaur fossils on record and the present day, who can predict how their morphology might have changed? Perhaps a modern-day pterosaur may indeed have evolved into a form comparable to classic reconstructions.

Ultimately, only physical evidence can confirm just what these winged wonders being reported actually are. Yet in view of what happened to the brave native investigator who sought one such creature amid Southern Rhodesia's nightmarish swamplands, re-emerging with a serious chest injury, such an undertaking is clearly not for the faint-hearted.

As veteran cryptozoologist Dr Bernard Heuvelmans once wrote: "The trail of unknown animals sometimes leads to Hell".

IS DRACULA ALIVE AND WELL AND LIVING IN SOUTH AMERICA?

There are three modern-day species of vampire bat, all exclusively American, whose collective distribution range stretches southward from southern Texas to northern Argentina and southern Brazil. Despite their infamous (and erroneous) reputation as blood-suckers (they merely lap blood flowing out of wounds surreptitiously incised with their razor-sharp canine teeth), they are decidedly unremarkable in appearance—no more than 3.5 in long, with a wingspan not exceeding 6 in. And then along came a mega-vampire, called Dracula!

Within the *Proceedings of the Biological Society of Washington* (7 December 1988), researchers Drs Gary Morgan, Omar Linares, and Clayton Ray described a new species of vampire bat, 25 per cent larger than *Desmodus rotundus* (largest of the three known living species), and based upon two incomplete skulls and skeletal remains found in Venezuela's famous Cueva del Guácharo—home of the amazing echolocation-adept oilbird *Steatornis caripensis*. Dubbed *Desmodus draculae*, this giant vampire's remains date from the Pleistocene.

However, within a *Mammalia* paper from 1991 in which they described the partial skull of a Pleistocene-dated *D. draculae* bat from a cave in southeastern Brazil's Ribeira Valley, Brazilian zoologists Drs E. Trajano and M. de Vivo noted that there are reports of local inhabitants in the Ribeira Valley referring to attacks upon cattle and horses by large bats.

Trajano and de Vivo cautiously wondered if these could be living specimens of *D. draculae*—but despite extensive recent searches of caves in this area, none has been found . . . so far?

THUNDERBIRDS, BIG BIRDS, AND MONSTER BIRDS

According to Amerindian mythology in Canada and the U.S.A., the thunderbird was a massive bird of prey that soared through the skies on gigantic wings, bringing rainstorms and thunder to the lands below, and as befits a creature of such importance its effigy appears on the totem poles of many different tribes. Their names for it vary—to the Maliseet Nation of the Saint John River in New Brunswick and Maine it is the cullona, the Comanche of the southern Plains know it as the ba'a', Wisconsin's Potawatomi call it the chequah, omaxsapitau ('big eagle') is the name given to it by the Blood tribe of Blackfoot in Alberta, and so on—but they are all clearly referring to the same type of bird.

Moreover, although thunderbirds were once presumed to exist only within the confines of these legends, several anthropological studies of Amerindian lore have unearthed remarkable accounts ostensibly based upon real events. And some of these date back to less than 200 years ago.

One of the most detailed and dramatic examples featured a Cree member (by marriage) of the Blood Blackfoot tribe, called White Bear, and occurred during a very harsh winter around 1850. This was when he and three fellow hunters, in a desperate quest for meat with which to feed their starving kinsfolk, set out to Devil's Head Mountain, northeast of Banff in Alberta, Canada. Eventually they set up a base, and each of the men then journeyed forth in a different direction in search of suitable quarry. After a time, White Bear met with success, killing a deer, whose precious carcase he tied securely to his back before beginning the return journey to their base—but he never reached it.

An enormous shadow appeared overhead, and almost instantly he was hoisted up into the air, whereupon he saw to his horror that he was being carried off by a gigantic bird of prey—its huge talons having grasped the deer carcase strapped to him! His avian abductor effortlessly transported White Bear and his deer for some distance through the sky, finally arriving at a lofty cliff. Here was this monster's nest, into which it unceremoniously dropped its still-living but terrified human cargo.

When he had recovered from his precipitous descent, White Bear took the opportunity to take a closer look at the giant bird, only to learn to his even greater horror that it was an omaxsapitau—a colossal creature resembling an eagle but far larger than any living bird of prey presently known to science, yet well-known to (and greatly feared by) the Blackfoot.

Equally disturbing was his discovery that he was sharing the nest with two juvenile omaxsapitaus—for whom he and his deer were evidently intended as their next meal, as testified by the assortment of bones, including some recognisably human remains, readily visible inside the nest. Clearly White Bear was not the first human to have been carried away by the chicks' parent; but as he had no intention of sharing his predecessors' terrible demise, he lost no time in formulating a plan of action with which to achieve his escape.

Without warning, he grabbed the two young birds' feet, and immediately threw himself out of the nest, hauling them after him as he plunged down through the air. In a bid to halt their rapid, uncontrolled descent, the two birds naturally opened their wings—and, as a result, acted as a pair of parachutes, enabling White Bear to land safely on the ground. As soon as his feet made contact, he released his hostages, who promptly flew away—but not before he had pulled two long feathers from their tails, which would confirm the reality of his harrowing experience when recounting it back home to his fellow Blackfoot tribesmen.

This story is reminiscent of an Arabian Nights story featuring the abduction of Sinbad the sailor by an immense carnivorous bird called the roc, but, unlike Sinbad, White Bear was a real person—a professional eagle trapper who had died in 1905 at the age of about 83. His grandson, Harry Under Mouse, was the source of this account, which he gave to Dr Claude Schaeffer from Montana's Museum of the Plains Indians, in Browning—who published it, along with a number of others, in June 1951.

A Northwest Coast styled Kwakwaka'wakw totem pole depicting a thunderbird perched on the top (CC-BY Dr Haggis)

It might be easy for the more sceptically-minded to dismiss such reports as modern-day Amerindian fables inspired by traditional thunderbird legends—if it were only Amerindians who were encountering these creatures. In reality, however, numerous Americans of European descent have also reported seeing them—particularly in the skies over the

Pursued by a thunderbird! (© William M. Rebsamen)

southern United States—and most have been likened by their eyewitnesses to huge vultures or eagles, with immense wingspans. As a consequence, the media habitually refer to them as 'Big Birds'—although a certain star of the American children's TV series *Sesame Street* may also have served as inspiration!

The profound significance of these birds' alleged wingspans can be readily emphasised by keeping in mind the following data. Notwithstanding its relatively small body, the wandering albatross *Diomedea exulans*, a southern oceanic seabird, has the greatest wingspan of any known species of bird alive today—when spread fully, its narrow wings yield a span of up to 12 ft. The world's largest bird of prey, the Andean condor *Vultur gryphus*, whose huge distribution range stretches southward from Venezuela to Tierra del Fuego, is second in line, with mighty wings reaching up to 10.5 ft. And the largest bird of prey native to North America, the California condor *Gymnogyps californianus*, has a wingspan of up to 10 ft.

A typical sighting of a Big Bird was that of Clyde Smith, his wife, and Les Bacon on 10 April 1948 at Overland, Illinois, when they spied a huge object flying overhead that seemed to be a dark-grey pursuit plane—until it flapped its wings! As revealed by Loren Coleman in *Curious Encounters* (1985), this is just one of several similar reports of a cryptozoological UFO (i.e. an unidentified flapping object!) that emanated from Illinois and the Illinois-Missouri borderlands in that same month.

During the evening of 24 April, for instance, two Missouri patrolmen from the St Louis Police Department—Francis Hennelly and Clarence Johnson—spied an enormous bird silhouetted across the moon that they estimated to be the size of a small aeroplane. Earlier that day, E.M. Coleman and his teenage son had observed a gigantic bird flying at a height of about 500 ft above Alton, Illinois, whose body resembled a naval torpedo and which cast a shadow as big as the one that a Piper Cub would yield if flying at the same height.

Perhaps the most breathtaking sighting reported during this outbreak of Big Bird activity, however, was that of St Louis chiropractor Dr Kristine Dolezal. On 26 April, she had been watching an aeroplane passing by from the porch of her second story apartment when suddenly a gigantic bird appeared, and seemed set on a direct collision course with the plane. Fortunately, however, this did not occur, and both cruised out of sight into the clouds.

It is possible that these giant creatures are either curious about the nature of aeroplanes, or may even feel threatened by them, because there are other cases on file featuring confrontations between small planes and Big Birds. In 1947, for example, a pilot claimed that while flying over Arizona he had narrowly avoided a collision with a gargantuan bird whose wingspan exceeded 30 ft in his opinion. One day in May 1961, a businessman from New York was piloting his light aircraft along the Hudson Valley River when he was 'buzzed' by an immense bird gliding past him without flapping its wings ". . . like a fighter plane making a pass", as he was later to describe it to the press. And in his *Encyclopedia of Monsters* (1982), Daniel Cohen briefly reported what may have been an even more sinister encounter.

In November 1962, a United Airlines plane crashed into a wooded area of Maryland. All passengers and crew were killed, and on the wreckage were traces of blood and feathers, from which the official enquiry concluded that the plane had collided with a flock of birds, probably geese, of which a few must have been sucked into the engine. There have been several tragic occurrences of this nature worldwide, so there seemed no reason to doubt such an explanation here—except for one disconcerting discrepancy. The plane's tail assembly apparently bore a number of gouges and slashes—but would geese be capable of perpetrating such a violent assault? According to Cohen, there were those who voiced the opinion, based upon these marks, that the plane had been attacked and brought down by an angry thunderbird, i.e. a belligerent Big Bird that may have deemed the craft to be a rival.

Envisaged close encounter between a light aircraft and a Big Bird (© William M. Rebsamen)

Pennsylvania has hosted its fair share of Big Bird appearances down through the years. In 1892, lumberman Fred Murray reputedly spied a flock of 'giant buzzards' with 16-ft wingspans near his lumber camp in Cameron County. Even more sensational, however, was the specimen observed in 1898 by Potter County School superintendent Arch P. Akeley.

According to his testimony, the bird was 4-6 ft tall when standing upright, sported grey plumage, a feathered head and neck (but without a ruff around its neck), and a wingspan of more than 16 ft. For a Big Bird report, this is an unusually detailed description, but there is a very good reason for that. Instead of being little more than a silhouette soaring at a great height through the skies, at the time of Akeley's sighting this particular individual was being held captive in a secure cage! What may be the only Big Bird ever captured and exhibited alive had been trapped and successfully caged by a farmer in Crawford County after he had spotted it pecking at the corpse of a dead cow at the far end of his farm, near Centerville.

The farmer had assumed that it was either a California condor or an Andean condor, but Akeley knew that it could not be either of these species. Apart from its much greater size, its grey plumage and feathered head instantly distinguished it from both species (which are black and bald), and its lack of a neck ruff further differentiated it from the Andean condor. What became of the farmer's unique zoological exhibit is not known.

Stretching from Harrisburg in Pennsylvania down as far as Lake Ontario in New York, and covering 200 square miles, the Allegheny Plateau is a rugged wilderness in which many Big Bird sightings have occurred, dating from antiquity right up to the present day. One of the most noteworthy of these was made around 1940 by naturalist Robert Lyman, in the Black Forest region, which he documented as follows in his book *Amazing Indeed* (1973):

> I saw a huge bird which I am certain was a thunderbird. It was on the ground in the center of Sheldon Road, about two miles north of Coudersport, Pennsylvania. It was brownish in color. Legs and neck were short. It was between three and four feet tall and stood upright like a very large vulture. When I was about 150 ft away it raised to fly. It was plain to see its wingspread was equal to the width of the road bed, which I measured and found to be 25 feet. I will concede it may have been 20 feet but no less. The wings were very narrow, not over one foot wide. . . Other local reports claim the thunderbirds are grayish in color. As they mature they may change from brown to grey.

The skies of western Canada and the northwestern U.S.A. are also traversed by Big Birds, with Washington's Olympic Mountains among the most frequent localities for sightings of such creatures. Of particular interest are the claims and traditions of several different Amerindian tribes from this region of North America that its seas are fished by thunderbirds for whales, which they carry aloft to their nests high in the mountain peaks. According to the Twana Indians of the Hood Canal region, their hunters sometimes found these nests, which are said to be simple structures composed of sticks and leaves.

For the most significant and controversial Big Bird episodes of modern times, however, it is to the state of Illinois that we must return, which experienced a major spasm of sightings in 1977—including what is alleged to have been the horrific near-abduction of a 10-year-old child.

It was about 8.30 pm on 25 July 1977, when three boys playing in a back yard at Lawndale, in Illinois's Logan County, saw two enormous black birds swooping down from the sky. They had evidently spotted the children, because they immediately pursued one of them, Travis Goodwin—but before they could reach him, he had dived into the swimming pool. Without further ado, they switched their attention to a second member of the trio—10-year-old Marlon

Artistic representation of the attempted avian abduction of Marlon Lowe (© William M. Rebsamen)

Lowe. He too began to run away, but unlike Travis he was not so lucky.

According to Marlon's subsequent testimony, even as he was still running he felt the talons of one of the birds seizing the shoulder straps of his singlet-like sleeveless shirt, and was promptly hauled up about 2 ft into the air by his winged abductor. Marlon screamed at the top of his voice for his mother, and his cries alerted not only his parents but also two of their friends, cleaning out a camper nearby. All four adults came running straight away, but his mother Ruth was the first to arrive on the scene—a spine-chilling scene that seemed to owe more to surreality than reality.

Right before their eyes was the horrific spectacle of a black vulture-like bird, with a large curved beak, a white ruff around its long neck, and a wingspan estimated at 8-10 ft, carrying Marlon a distance of 30-40 ft from the Lowes' back yard into their front yard. During this brief but terrifying journey, Marlon was frantically punching up at the bird, which seemed to be experiencing no difficulty in transporting his 65-lb weight (given as 56 lb in some accounts). Happily, however, one of his punches evidently hit home, because the bird suddenly dropped him, and continued flying towards and over the camper, accompanied by its partner.

By this time, the other three adults had also reached the spot, and together with Ruth they were able to watch the birds fly off towards some large trees along Kickapoo Creek. Marlon, meanwhile, had picked himself up and had fled for safety into the camper. As the bird had carried him via his shirt's straps, he was not scratched, but his shirt was frayed, and he suffered nightmares when he attempted to sleep. Ruth phoned the police, who duly searched the area that night and the following day, but nothing was found.

Marlon Lowe and his mother Ruth, following his attempted abduction by a giant mystery bird (© Range Pictures/Bettmann Archives/United Press International)

As with the Illinois Big Bird 'flap' of 1948, however, for quite a time afterwards birds fitting the description of Marlon's abductor and partner were spied elsewhere in the state. Tragically, however, this did not prevent the Lowes from becoming the focus of considerable ridicule, as well as the targets for anonymous, sick-minded cranks—who would periodically deposit dead animals upon their doorstep, including the carcase of an eagle.

Over the years, people reporting sightings of many different types of mystery creature unrecognised by science have often run the gauntlet of mockery and disbelief from neighbours and colleagues—such occurrences are a sad but all-too-common side-effect of cryptozoological confrontations. Yet few have been as unpleasant and devastating as the evil, merciless harassment experienced by the Lowes. Only a year later, the stress stemming from the memory of his abduction, and compounded by the vile hate campaign, had left its mark for all to see—Marlon's hair, once coppery-red, had turned grey.

Readers may well have wondered why the North American birdwatching fraternity has not featured in reports of Big Birds. Surely, out of the countless numbers of binocular-toting 'twitchers' compiling their life lists, someone somewhere has spied one of these creatures—but, if so, why have they not come forward? I think that we need look no further than the Lowes' experiences for an answer to this mystery. There is a universe of difference between the thrill of reporting a vagrant European warbler in Illinois or a Californian finch turning up in Missouri, and the stigma long associated with confessions of the cryptozoological kind.

By mid-August 1977, publicly-released reports of Big Birds in Illinois had ceased—but not before a onetime Marine combat cameraman had succeeded in filming what he believes to have been the two specimens responsible for the Lowe incident. On the morning of 30 July, John Huffer (known locally as 'Texas John'), from Tuscola, was fishing with his son at Lake Shelbyville when they spied two exceedingly large birds (but one notably bigger than the other) roosting in a tree, and depositing baseball-sized droppings around its base. Having his cine-camera to hand, Huffer decided to film them, so with the aid of his boat horn he flushed them out of the tree, and began filming the larger of the two birds as it took to the air, obtaining more than 100 ft of film before it settled in another tree out of camera range.

Judging from excerpts shown on American TV and stills reproduced in two newspaper accounts, however, intervening foliage had partially obscured the bird for much of the time, rendering impossible any attempt at precise identification or scaling. According to Huffer, it had a 12-ft wingspan and an 18-in-long neck, and the plumage of both birds was black. Huffer was familiar with the appearance of America's known birds of prey, particularly turkey vultures, and was greatly surprised when ornithologists dismissed the bird in his film as the common red-headed turkey vulture *Cathartes aura*. In response to this claim, he swiftly pointed out that his bird lacked the turkey vulture's red neck, and he also considered that the shape of its head and legs were different.

Sightings of Big Birds in North America have continued to the present day, and many additional examples are included in the newest, 2004 edition of Mark A. Hall's indispensable book *Thunderbirds! The Living Legend of Giant Birds*—not only the most comprehensive work ever published on this subject but also a cryptozoological classic of modern times. A chapter on Pennsylvanian Big Birds also appears in the revised edition of his book *Natural Mysteries* (1991). Resulting from these extensive studies, Hall has acquired a detailed picture of the Big Bird's general appearance and lifestyle, which can be summarised as follows.

Excluding a handful of sightings that featured readily identifiable birds clearly unrelated to the Big Bird phenomenon (e.g. great blue heron, cranes, storks, pelicans, peacock!), the creatures in question are, indisputably, birds of prey. They have been variously compared by their eyewitnesses with giant vultures, eagles, and one or other of the two species of condor; their plumage has been given as grey, black, or brown, which is sometimes said to be much paler underneath than on top; their

Turkey vulture with wings outstretched (Cameron64)

beaks are long and hooked; their heads and necks variously feathered or unfeathered; and their wingspans immense—estimates of 10-30 ft being most common—with occasional references to white wingtips.

Of great significance is their consumption of living prey, aided by the apparent ability and strength of their taloned feet to carry off victims as heavy as children (Marlon Lowe's is not the only such case on file)—and even, according to Amerindian lore, adult humans and whales.

They build nests, albeit of simple construction, on lofty cliff faces and crags, and during the year they undergo a marked pattern of migration. During the colder winter months, they remain in the warm climes of the southern U.S.A., but at the onset of the spring they journey northwards to spend the spring and summer in the Pacific Northwest. In so doing, their appearance coincides with the seasons of thunderstorms, which explains how these birds became associated with them by the early Amerindians. Travelling for much of the time over lofty, remote mountain ranges, they have succeeded in eluding scientific detection, but there is one expanse of territory along their route that provides little cover—the extensively-populated areas of the Midwest. Presumably, then, it is no coincidence that there are a notable number of Big Bird sightings on file from Illinois and Missouri.

Such information as this is already of great interest, but what makes it even more so is the unequivocal fact that when viewed *in toto* it does not correspond with the natural history of any species currently known to science in the living state.

A Verisimilitude of Vultures, a Comparability of Condors, and an Equivalence of Eagles

Excluding species such as buzzards, hawks, falcons, and kites, which are far too small to warrant serious consideration, there are only four birds of prey native to North America that could have any bearing upon the Big Bird—all of which are regularly nominated by ornithologists as the solution(s) to any Big Bird case that is brought to public attention.

The most popular of these is the turkey vulture, Big Bird Candidate #1, distinguished from the next two by its brown plumage. The skin of its bare head and neck is bright red in colour, it has a total length of 2 ft, and a wingspan of 6 ft. When airborne, it soars in wide circles, its wings held in a broad v-shape, rocking rapidly from side to side. Abundant throughout most of the U.S.A., and regularly spied in fields and along roadsides, it is a very familiar bird—so much so that it seems unlikely (though not impossible) that anyone could fail to recognise it, and mistake it for something much larger and less familiar.

Candidate #2 is the black vulture *Coragyps atratus*—just under 2 ft long, with a 4.5-ft wingspan, bare black face but feathered head and neck, shorter tail but longer neck than the turkey vulture, and mostly black plumage offset by distinctive white wingtips. Often invading settlements in search of garbage and small animals, this species inhabits the southern and southeastern portions of the U.S.A., including Illinois and Missouri. When in flight, which tends to be a rather weak, low-powered affair, it flaps its wings deeply and then glides for a short period, during which time the wings are held horizontally.

Black vulture (© Chad Arment)

Except for its smaller wingspan, the black vulture's morphology and flight strongly recall some Big Bird descriptions—to the extent that I consider it plausible that certain of these reports can be explained by unexpected encounters with big, bold individuals of this species lurking in closer proximity to human habitation than usual, with exaggeration of their wingspan resulting from the eyewitnesses' surprise at meeting them.

The third candidate is by far the best for explaining some of the larger Big Birds from fairly long ago. As noted earlier, its 9-10-ft wingspan earns the California condor the title of North America's largest bird of prey. Almost 4 ft in total body length, this magnificent species has predominantly black plumage but with white under-wing coverts that yield a striking triangle of white beneath each wing when viewed in characteristic soaring flight from below. Its head and neck are bare, exposing bright reddish-pink skin (bluish-grey as a juvenile), except for a small patch of black feathers on its forehead. Its hooked beak is very large and heavy, and when in flight its long, wide wings are held in a straight line.

This species traditionally boasted a very appreciable distribution range—stretching along the Pacific Coast from British Columbia and southwards through Washington, Oregon, and California to northern Baja California in Mexico. During the Pleistocene, moreover, it even extended eastwards across the southern U.S.A. into Florida. Following a merciless programme of persecution during the 19th Century, however, its numbers plummeted horrendously everywhere. By the second half of that century, it had vanished throughout the entirety of its former range north of San Francisco and Baja California.

By 1980, just over a century later, less than 40 birds survived—consisting of two tiny populations in California, and the species seemed doomed to extinction. Consequently, after much deliberation and following successful attempts to hatch California condor eggs in captivity and rear the fledgelings to adulthood, in 1987 the last six specimens known still to exist in the wild were themselves captured, and were introduced to their captive-raised counterparts maintained at San Diego Zoo and Los Angeles Zoo. For several years thereafter, the California condor was officially extinct as a free-living species, but in 1992 it re-entered North America's ornithological checklist when five specimens hatched at Los Angeles Zoo were released in Ventura County—the first step in what is hoped will be this species' successful revival in its native homelands. Following further releases, they now exist elsewhere in the wild too, including Arizona's Grand Canyon (where I was fortunate enough to observe a couple when I visited it in 2004), and Utah's Zion National Park.

California condor, with identification numbering, in Utah's Zion National Park (Phil Armitage)

Clearly, therefore, it is unlikely that the California condor, despite its great size, could be responsible for many (if any) Big Bird sightings from the past few decades—but it probably featured in some from pre-20th Century times and must have profoundly influenced Amerindian legends of the thunderbirds. Indeed, Schaeffer suggested that it might even have inspired the Blackfoot stories of the omaxsapitau. It is true that Alberta is outside the official boundaries of its former distribution range, but who can be certain of what these boundaries were, during those distant ages when the first thunderbird legends arose?

This trio of species belongs to a taxonomic family of birds termed the cathartids or New World vultures—on account of their exclusive American distribution in modern times (though fossils of some long-extinct Old World species are also known). They share a number of characteristics that readily differentiate them from the accipitrids—i.e. the large raptorial taxonomic family that contains the eagles, hawks, buzzards, harriers, and the true, Old World vultures.

Indeed, although traditionally assumed to be closely related to the latter birds, initial DNA analyses coupled with anatomical studies seemed to suggest that the cathartids were sheep in wolves' clothing—or, to be zoologically precise, storks in vultures' plumage! That is, it appeared that they must have descended from the stork lineage, but went on to pursue a flesh-eating lifestyle, gradually transforming via convergent evolution into species bearing a superficial external appearance to the true, Old World vultures.

More recently, however, multi-locus DNA studies have argued against a taxonomic affinity between cathartids and storks, allying the cathartids with the accipitrids after all, though not closely. (Additionally, the falcons, formerly classed as close accipitrid relatives, are nowadays deemed to be only distant relatives of these latter birds too.) Consequently, they are assigned by some ornithological researchers to their own taxonomic order, Cathartiformes. Interestingly, two of the most notable anatomical and behavioural features that distinguish the cathartids from the true, Old World vultures and other accipitrids are of particular relevance to the Big Bird issue.

Unlike accipitrids, the feet of cathartids are very weak and ill-designed for grasping, so that they are unable to carry anything but the smallest of prey. Consequently, apart from devouring diminutive items of living prey such as rats, lizards, and fledgeling birds, and consuming unattended eggs in nests, they thrive exclusively upon carrion. Also, whereas accipitrids construct nests, cathartids do not—depending upon the species, they lay their eggs on bare ground, ledges, in caves, hollow logs, or hidden amid vegetation or boulders.

From this, it hardly seems likely, therefore, that any of the three cathartids discussed here can be responsible for the thunderbird nests reported by various Amerindian tribes. Nor would they appear to be capable of abducting humans or whales as attested by eyewitnesses in Big Bird incidents of this nature.

The only native bird of prey in North America that offers any prospect of solving these latter mysteries is Candidate #4—the golden eagle *Aquila chrysaetos*. With a total body length of 2.5-3.5 ft and a very impressive wingspan of up to 8 ft, this is not a bird of inconsiderable stature. Moreover, although it is present throughout the year in the western half of the U.S.A. and occurs during the winter months in most of the eastern half, it is rarely encountered by humans, unless they penetrate its forbidding realms of remote mountain peaks and barren deserts. Consequently, its appearance is not as well-known as those of the more accessible cathartids. Its dark brown plumage, feathered head and neck, and dark undersurface (only in the juvenile are white under-wing patches present) collectively set this species apart from the three large cathartids too. Also worth noting is that several Amerindian names for the thunderbird incorporate 'eagle' or 'big eagle'.

Golden eagle (George Hodan)

Most significantly, however, as a member of the accipitrids the golden eagle builds a nest (eyrie), usually on a high mountainous crag, and has extremely powerful grasping feet, with which it can seize and haul aloft prey as large as lambs and fawns—and perhaps, in exceptional cases, even larger victims. As outlined by John Michell and Bob Rickard in *Unexplained Phenomena* (2000), there is some controversial—

and, in at least one case, seemingly incontrovertible—evidence that certain eagles have been powerful enough to lift and carry away small children.

One such case was that of 5-year-old Marie Delex, dating from 1838 and occurring in the Valais region of the Swiss Alps in central Europe. She had been playing with a friend on a mossy mountain slope, when suddenly a huge eagle swooped down upon her and bore her away, ignoring the screams of her terrified companion. The screams, however, did alert some peasants working nearby, who ran to the spot, but found nothing other than one of her shoes, at the edge of a precipice. She had clearly not fallen over it, however, because two months later a shepherd found her corpse, horribly mutilated, lying upon a rock about 1.5 miles away from where she had vanished.

19th-Century engraving depicting Marie Delex's tragic abduction by a golden eagle

This was one of two incidents recounted by Felix Pouchet in his book *The Universe* (1873). The second, which took place in autumn 1868, was quoted directly from the account of an unnamed teacher in Missouri's County Tippah, and read as follows:

> A sad casualty occurred at my school a few days ago. The eagles have been very troublesome in the neighbourhood for some time past, carrying off pigs, lambs, &c. No one thought that they would attempt to prey upon children; but on Thursday, at recess, the little boys were out some distance from the house, playing marbles, when their sport was interrupted by a large eagle sweeping down and picking up little Jemmie Kenney, a boy of eight years, and flying away with him. The children cried out, and when I got out of the house, the eagle was so high that I could just hear the child screaming. The alarm was given, and from screaming and shouting in the air, &c., the eagle was induced to drop his victim; but his talons had been buried in him so deeply, and the fall was so great, that he was killed—for either would have been fatal.

The most famous case of avian abduction on record, however, which seems to have been proven beyond any reasonable doubt, featured Svanhild Hartvigsen (née Hansen) (1929-2010). Her incredible experience was investigated in such painstaking detail by Steiner Hunnestad that in 1960 he published an entire book on the subject.

As a 3.5-year-old girl on the fateful day in question—the morning of 5 June 1932—

Svanhild had been playing in the yard of her parents' farm on the Norwegian island of Leka when a very large white-tailed sea eagle *Haliaeetus albicilla* swooped down, and lifted her bodily into the air, its huge talons firmly seizing her dress as it carried her through the sky for over a mile until it reached Haga Mountain at the valley's end. Here it dropped her onto a narrow 800-ft-high ledge, less than 70 ft below its nest. Happily, the child's 42-lb weight had defeated its attempt to deposit her directly in it.

In the meantime, her absence had been noticed and swiftly acted upon by the people of Leka, who had sent out a search party to trace her. Looking upwards, some of the searchers spotted an eagle hovering above a ledge at the valley's farthest point—and when they scaled the ledge to investigate, they discovered Svanhild, very frightened but still alive and unharmed. As a souvenir of her terrifying experience, throughout her life she retained the checked, talon-torn dress that she had been wearing on that scary morning in 1932.

Consequently, if an unusually large (and strong) golden eagle, not the most familiar of sights for many people in the U.S.A., were to swoop down unexpectedly from the skies, not only would the shock of its appearance assist greatly in exaggerating its size in the eyes of its observers but also it may be capable of accomplishing the acts of child (though not adult) abduction claimed for Big Birds. Here, then, is a promising solution to some Big Bird sightings.

Having said that, it needs to be emphasised that one piece of 'evidence' that has been put forward by some proponents of this avian-engendered child-abduction scenario is nothing of the kind. On 18 December 2012, a video was uploaded onto YouTube (at: https://www.youtube.com/watch?v=9feAQtgX1fw) by a user named Mirrorviewvids that showed a golden eagle swooping down from the sky, lifting a small child off the ground with its talons, then flying off for a few feet before dropping him. Not surprisingly, this dramatic film attracted considerable online attention, but the next day all was explained via the following official statement concerning it, posted by the University of Quebec's Centre NAD (National Animation and Design) on its website:

> The "Golden Eagle Snatches Kid" video, uploaded to YouTube on the evening of December 18, was made by Antoine Seigle, Normand Archambault, Loïc Mireault and Félix Marquis-Poulin, students at Centre NAD, in the production simulation workshop class of the Bachelors degree in 3D Animation and Digital Design.
>
> The video shows a royal eagle snatching a young kid while he plays under the watch of his dad. The eagle then drops the kid a few feet away. Both the eagle and the kid were created in 3D animation and integrated in to the film afterwards.
>
> The video has already received more than 1,200,000 views on YouTube and has been mentioned by dozens of media in Canada and abroad.
>
> The production simulation workshop class, offered in fifth semester, aims to produce creative projects according to industry production and quality standards while developing team work skills. Hoaxes produced in this class have already garnered attention, amongst others a video of a penguin having escaped the Montreal Biodôme.

Back in the real world, it is worth noting is that the bald eagle *Haliaeetus leucocephalus* is of similar size to the golden eagle, but its unmistakable white-headed adult form must surely eliminate it from contention as a Big Bird. Immature bald eagles, conversely, are all-brown, lacking the adults' characteristic white head, but they are smaller in size than adults.

Yet even surprise and terror cannot sufficiently 'stretch' a golden eagle in order for this species to explain some of the classic Big Bird reports—featuring truly enormous specimens that soar effortlessly like wraiths around and alongside light aircraft whose dimensions are little (if at all) larger than their own. To discover a plausible identity for these, we must turn away from the present—for only among the past has science ever encountered anything similar.

Turning to the Teratorns

During the Pleistocene, North America was home to some very large cathartids now extinct—including the La Brea vulture *Breagyps clarki*, and the great condor *Gymnogyps amplus* (the California condor's direct ancestor). None, however, could compare with two vulture-like species of such colossal size that they became known as teratorns—aptly translated as 'monster birds'.

With a wingspan of 11.5-12.5 ft, Merriam's teratorn *Teratornis merriami* was larger than any modern-day bird of prey, and its fossils have been found in quantity in California's famous La Brea tar pits. Yet even this spectacular creature was dwarfed by *Aiolornis* (formerly *Teratornis*) *incredibilis*—a well-deserved name for a species with a truly incredible 18-ft wingspan! This species is known from fossils uncovered in California and Nevada, but in 2000 the first teratorn species to be discovered north of La Brea was formally described, thereby greatly expanding these birds' known zoogeographical distribution in North America. Named *T. woodburnensis*, it was based upon a partial specimen uncovered at Legion Park, Woodburn, in Oregon, that dated from the late Pleistocene around 12,000-11,000 years ago. It was found in a stratum containing mastodont, ground sloth, and condor bones, and which bears evidence of human habitation (hence humans may well have seen this bird). Moreover, with a wingspan estimated to have been over 14 ft, the Woodburn teratorn would have been a spectacular, unforgettable sight.

It was back in 1979, however, when the monster bird to end all monster birds made its debut. Unearthed several years earlier as a fragmented partial skeleton at a site roughly 100 miles west of Buenos Aires, Argentina, by La Plata palaeontologists Drs Rosendo Pascual and Eduardo Tonni, and dating from the late Miocene (8-5.3 million years ago), this gargantuan teratorn was christened *Argentavis magnificens*—'magnificent Argentine bird'. It also necessitated a total rethink of the maximum size attainable by a flying bird—and little wonder, for its fossils revealed that *Argentavis* was around 4 ft long, 5 ft tall, 160 lb in weight, and boasted a scarcely-conceivable estimated wingspan of 23 ft. To express these dramatic dimensions another way, this colossal bird was approaching the size of a Cessna 152 light aircraft!

It scarcely needs to be pointed out that if *Argentavis* were alive today, in terms of dimensions it would correspond very effectively with those of the really large Big Birds on file. And although, contrary to initial speculation, further studies confirmed that *Argentavis* would have been able to become airborne (provided that it received sufficient assistance from strong winds), it is likely that for much of its time in the air it would have relied upon

Scale diagram of *Argentavis* alongside human (© Tim Morris)

thermals to support it in a lazy, soaring, gliding mode of propulsion, with only occasional, intermittent periods of flapping flight—thereby resembling the overland flight of condors, and comparing once again with reports of Big Birds.

It is hardly surprising, therefore, that an undiscovered descendant of this bird, or at the very least a modern-day version of *some* form of teratorn, is the favourite identity for Big Birds among cryptozoologists. And as far as sightings of giant soaring raptorial birds are concerned, I too agree that this identity provides the closest correspondence. Reports featuring Big Birds engaged in abducting children, adults, whales, or any other large prey victims, conversely, pose a very real dilemma.

For although teratorns are placed in a separate taxonomic family from the cathartids, they are more closely allied to this group of carnivorous birds than to any of the others, and share with them one key morphological feature that wreaks havoc with attempts to cast these giant species in the mould of avian abductors. Just like cathartids, their long-toed feet were extremely weak—structurally, they were wholly incapable of seizing and bearing aloft even very insubstantial prey.

If anything, their feet were even weaker than those of the cathartids, as they apparently did not use them for seizing prey at all but used them for walking instead. For as revealed from studies on teratorns by Tonni and Dr Kenneth Campbell from the Los Angeles County Museum, the structure of these birds' pelvic girdle and their relatively long, stout legs strongly indicates that they were probably more agile on the ground and favoured a more active mode of feeding than the cathartids, stalking prey on foot instead of seeking carrion from the air. In addition, the construction of their elongated hooked beak implies that they were predators that consumed their prey whole.

Cryptozoological sceptics who denounce the existence of Big Birds claim that such enormous creatures would require appreciable supplies of large prey in order to sustain their huge bulk. What is known of the teratorns, however, suggests otherwise—because in order for them to consume their victims whole, these giant birds evidently did not prey upon very large animals. As Campbell and Tonni pointed out, the skull dimensions of *Argentavis* (about 2 ft long and 6 in wide) suggest that this species could only engulf creatures up to the size of hares (i.e. about 1 ft long and 10 lb in weight), and Merriam's teratorn probably subsisted upon frogs, lizards, nestlings, and fledgeling birds—far cries from the scenario of megafauna predation envisaged by the Big Bird sceptics.

Certainly, amid the vastness of the North American continent a modern-day teratorn should have no problem in obtaining sufficient quantities of its preferred mini-prey.

Artistic representation of a teratorn (© Hodari Nundu)

How Many Types of Big Bird Are There?

From the latter discussion, it is clear that no single species can be satisfactorily identified as the modern-day cryptozoological Big Bird. On the contrary, its file constitutes a heterogeneous collection of reports featuring several different species of bird—and the same appears to be true for the traditional mythological thunderbird. I believe that, like many other legendary animals, the thunderbird is a non-existent composite, deriving from Amerindian observations of least three different real-life species—whose most striking attributes gradually became amalgamated in the telling and retelling of myths to 'create' this awesome entity. Its enormous size may have stemmed from sightings of a living species of teratorn, its general appearance was based upon the California condor as seen during those far-off times when it was abundant and widespread, and its formidable prey-seizing abilities were inspired by those of the golden eagle.

The Big Bird of the present day is a similar mosaic of separate species. Once again, if they are accurate, sightings of aircraft-sized monsters soaring high in the skies can surely only refer to a still-undetermined species of living teratorn. True, some such observations in this category may be based upon exaggerated size estimates on the part of the observers, who in reality have spied nothing more startling than vultures or eagles, or the California condor in earlier days. However, there are many reports of gigantic birds from eyewitnesses well-equipped for accurately estimating flying objects—notably pilots. Also, terrestrial encounters, such as those of Lyman and Akeley, who were close enough to their respective birds to obtain reliable assessments of their size (Lyman achieved this by measuring the width of the road that had been spanned by the bird's outstretched wings), are of great value in confirming that Big Birds of genuinely huge size are a reality.

Some of the smaller Big Birds are assuredly nothing more than native species spied unexpectedly or by observers unfamiliar with them,

Life-sized silhouette of *Argentavis magnificens* displayed at the Los Angeles County Museum, with teratorn researcher Dr Kenneth Campbell standing alongside it (© Natural History Museum of Los Angeles County)

and the abducting contingent of Big Birds must surely feature large golden eagles—as long as we limit our options to native species. However, among the category of persons who enjoy keeping potentially dangerous wild animals as pets (which include such formidable exotica as big cats, highly venomous snakes, and tarantulas), birds of prey are particularly popular. Considerable numbers, many belonging to non-native species, are currently owned in North America.

Inevitably, there will be escapes (but which will go unreported if the birds have been kept illegally), as well as deliberately-engineered releases by owners no longer so enamoured of their unusual pets. Consequently, I consider it likely that some of these have swollen the already-bulging dossier of Big Bird sightings.

One species that particularly impresses me in relation to this possibility is the harpy eagle *Harpia harpyja*—whose native distribution extends from southern Mexico to northern Argentina. Weighing up to 20 lb with a total length of up to 3.5 ft, and a wingspan of 6.5-7.3 ft, the female harpy (often as much as a third larger than the male) is widely recognised to be the world's most powerful eagle—and for good reason. With enormous talons thicker and only fractionally shorter than the terrible claws of a fully-grown Kodiak bear (one of the world's largest carnivorous land mammals), this imperious, grey-plumed assassin can lift heavier prey than any other modern-day raptor currently known to science.

Beautiful 19th-Century engraving of a harpy eagle

Traditionally, its largest prey was believed to be tree sloths, opossums, and rabbit-sized rodents called agoutis, but in 1991 California University biologist Dr Peter Sherman disclosed that while recently walking through Peru's Manu National Park he had encountered a harpy eagle preparing to devour a freshly-killed red howler monkey. As he drew nearer, the bird flew away, leaving its kill behind instead of carrying it away to its nest, and when he weighed the abandoned carcase he found it to be about 15.5 lb—almost as heavy as the bird itself!

This was by far the heaviest prey specimen ever recorded for the harpy eagle, and makes one wonder what the maximum size of prey victims could be for exceptional harpies approaching their species' upper size limit. Perhaps we already have a clue.

During his extensive travels in South America from the 1950s to the 1970s, film maker Stanley Brock kept two female harpy eagles, both called Jezebel. The larger of these had a 6.5-ft wingspan, and as he mentioned in his books, it had a disturbing propensity for swooping after toddlers belonging to the local Indians and attempting to take hold of them with its massive talons. The existence of an escapee harpy or two in North America might conceivably solve some controversial claims of avian abduction currently cluttering the Big Bird file.

Scarcely less daunting than the harpy eagle is the African crowned eagle *Stephanoaetus coronatus*, another popular species among raptor enthusiasts, which has a wingspan of up to 6 ft, and a verified record for attacking small children. As noted by Peter Steyn in his book *Birds of Prey of Southern Africa* (1982):

> One grisly item found on a nest in Zimbabwe by the famous wildlife artist D. M. Henry was part of the skull of a young African. That preying on young humans may very occasionally occur is borne out by a carefully authenticated incident in Zambia where an immature Crowned Eagle attacked a 20 kg seven-year old schoolboy as he went to school. It savagely clawed him on head, arms, and chest, but he grabbed it by the neck and was saved by a peasant woman with a hoe, who killed it, whereafter both eagle and boy were taken to a nearby mission hospital. The boy was nowhere near a nest, so the attack can only have been an attempt at predation.

In yet another instance, when assisting in the investigation of the disappearance of a four-year-old girl, Simon Thomsett came to believe she was the victim of a crowned eagle after the severed arm of a child was found in a tall tree that was inaccessible to leopards and known to be used as a crowned eagle cache.

African crowned eagle (CC Jon Mountjoy)

The crowned eagle, incidentally, is also a contender put forward by some researchers as the killer of the famous Taung Child. This juvenile *Australopithecus africanus*, a primitive hominin, is estimated to have been 3-4 years old, to have stood 3.5 ft tall and to have weighed up to 22 lb, but is known only from its fossil skull, which is approximately 2.5 million years old. This was excavated in 1924 at Taung, South Africa, amid a pile of other mammalian remains believed to have been killed by the same predator. Damage to the Taung Child's skull and eye sockets is consistent with what eagles inflict upon modern-day primates, but there has been much argument as to whether a crowned eagle could have lifted and carried off this particular individual—as its killer must have done, in order to explain why the Taung Child's skull was found in the pile of killed victims.

In the Lowe case reported earlier, however, the birds in question were certainly not harpies or crowned eagles. Indeed, the only species that bears any resemblance to them is another non-native, the Andean condor—whose size, plumage colouration, long neck, white neck ruff, and curved beak all feature in the description of Marlon's abductor. Yet even if we explain their origin as escapees or releases from captivity, there is still a major flaw with this identity—as a cathartid, this species should not be able to lift anything remotely as heavy as a 10-year-old boy.

The literature occasionally concedes that in times of desperation, condors will attack livestock and prey upon newborn deer while remaining on the ground. (They will also do the same when attacking beached whales spotted around the South American coastlines—a noteworthy snippet of information in view of the whale-devouring tendencies attributed to the legendary thunderbird.) Yet this behaviour cannot compare with abducting a struggling, 65-lb boy, who would pose major lifting problems even for an oversized eagle, let alone a weak-footed cathartid.

If the Lowe case is genuine, there are only two options open. The more conservative of these argues that the birds were a pair of escapee or released Old World accipitrids, of which the best fit is the mighty griffon vulture *Gyps fulva*—a species with a conspicuous neck-ruff of white, and a wingspan that can slightly exceed 9 ft. Its plumage is dark brown rather than black, but the evening viewing conditions during the Lowe incident might have made the birds seem darker than their true colour.

The second, much more radical option is that there is an unknown species of giant accipitrid alive and well in North America—at least as large as a condor, and perhaps as immense as a teratorn. Indeed, a teratorn-proportioned accipitrid would, uniquely, explain virtually *all* Big Bird reports single-handedly—uniting the teratorn's dimensions with the grasping capabilities of the accipitrids. Such an identity could also explain why, according to traditional Amerindian lore, the thunderbird's numbers fell precipitously following the demise of the great herds of bison (hunted into near-extinction by Western settlers), because this lore claims that bison calves were the thunderbird's favoured prey.

Alternatively, we could hypothesise that the giant Big Birds constitute an evolved teratorn, with feet lately modified for grasping. However, because even as relatively recent a species as Merriam's teratorn did not exhibit the slightest rudiment of such a development, and as it would not be needed anyway if a modern-day teratorn perpetuated its foot-stalking hunting technique, this is not a very realistic proposition.

Returning to the possibility of a teratorn-sized accipitrid: the reason why I have not given this option greater coverage here is that it is unsupported by the merest fragment of physical, tangible evidence. Whereas the concept of a living teratorn in the New World has a healthy palaeontological history to substantiate its credibility—Merriam's teratorn, for instance, is known to have survived until at least as recently as around 11,000 years ago, less than the blink of an eye away from the present day on the geological time scale—there is no evidence whatsoever from the fossil record to indicate that an accipitrid counterpart ever existed here. Naturally, the absence of any palaeontological precedent does not automatically deny the possibility of a creature's onetime existence—but when considering anything as revolutionary as a teratorn-sized accipitrid, until (if ever) some such evidence does come to hand I would much prefer to place its candidature on hold.

Having said that, I cannot deny that there are certainly some very intriguing modern-day eyewitness accounts on file that record sightings of mystery birds resembling truly gigantic eagles. There is also the anomaly of Washington's (aka the Washington) eagle, which I have documented in detail on my *ShukerNature* blog (so too has fellow writer Scott Maruna on his *Biofort* blog).

Named after the U.S.A.'s first president, George Washington, this highly controversial form was based not only upon sightings but also upon an absolutely enormous specimen (bigger than bald eagles and even golden eagles), which sported distinctive all-brown plumage. It was shot in Kentucky by the very famous 19th-Century American artist John James Audubon, who had previously had four sightings of similar birds in the wild. Audubon then painted his specimen and claimed that it represented a new species, separate from all other known eagles, but ornithologists later discounted this claim. Audubon's specimen is seemingly lost, tragically, so there is no physical evidence of Washington's eagle that can be examined anatomically and genetically to determine conclusively its taxonomic status.

As for the afore-mentioned eyewitness accounts of giant eagles, which some cryptozoological investigators have postulated may actually constitute Washington's eagles, here is a selection, excerpted from my *ShukerNature* coverage of this avian cryptid:

One such report was placed online by Maruna, and recounted a sighting of a huge raptor one winter's morning in 2004 by William McManus and his wife, spied across a small meadow between their cabin home (located roughly 15 miles north of Stillwater, Minnesota) and a river channel, as it perched

Washington's eagle as painted by John James Audubon

in a dead tree. They immediately discounted any bald eagle, adult or immature, and also a golden eagle, on account of the sighted bird's dark brick-red colour, and also because of its size, which they estimated to be well over 3 ft tall. After viewing it for some 2 hours at a distance of only 200 ft or so, they moved nearer but the bird flew away, revealing itself to be indeed an eagle (and not a condor, as they had begun to wonder), with an eagle's head and neck, and an enormous wingspan far exceeding a bald eagle's. After seeing Audubon's painting of his Washington's eagle in Maruna's blog, McManus believed that this is what he and his wife saw that day.

This communication received two interesting responses posted in Maruna's blog. One, posted by 'Kurt N' on 24 October 2006, stated that he had seen at the Hawk Mountain bird observatory in Pennsylvania a photograph taken there in or around 1993 depicting a large unidentified eagle that was clearly no bald or golden eagle but had elicited speculation that it was some kind of sea eagle. And in a post of 9 January 2007 (later removed), a reader with the username of dogu4 mentioned that a couple of years earlier (2005), a bush pilot and some passengers flying over a remote stretch of western Alaska claimed to have seen a gigantic bird.

As dogu4 also noted pertinently:

> If there were ever an area where it [Washington's eagle] could have survived as a small population un-noticed, the coastal areas out along the edges of the Y-K Delta would fit the description as far as inaccessibility and remoteness, not to mention the abundance of sea-life, rookeries and breeding grounds for a multitude of relatively un-disturbed populations of birds and sea-mammals. And the incredible solitude.

An equally thought-provoking report reached me in November 2009, courtesy of a long-standing correspondent called Mark, from Birmingham, Alabama, who emailed me with news of an intriguing report aired a month or so earlier on an episode of the popular North American late-night radio talk show entitled *Coast To Coast AM*, presented by George Noory. Apparently, on two separate occasions during the course of this particular episode, which focused upon cryptozoology, a man aged in his 40s called in, claiming a remarkable sighting made by him and two ex-army friends while in a harbour around the Aleutian Islands off western Alaska.

The caller stated that while standing not far away from some telegraph poles, they saw that perched on top of one of them was what he referred to as a gigantic bald eagle. He mentioned that there were other, normal-sized bald eagles flying around it and that it was therefore very easy to estimate its size—10 ft or so tall. This seems far too large, but perhaps the caller was thinking of wingspan rather than height? In any event, he told host Noory that he and his two friends looked at each other in amazement, hardly believing what they were seeing. Referring to it as a gigantic bald eagle implies that it had a white head. Yet, as already discussed here, only adult bald eagles have this, but near-adult, all-brown bald eagles are sizeable birds themselves and just as eyecatching, so we cannot say for sure that simply because it was likened to a bald eagle the Alaskan mega-eagle must have been white-headed. What we *can* say, however, is that if it was all-brown, it would bear much more than a passing resemblance to that most enigmatic of northern U.S.A. mystery birds, Washington's eagle.

Juvenile bald eagle, painted by John James Audubon, which lacks the adult form's characteristic white head

More recently still, beginning on 17 March 2011 with one that was forwarded by *Fortean Times* to me, I received a series of detailed emails from Mike E. Richburg of West Columbia in South Carolina, U.S.A., in which he described what he believes to have been a truly enormous eagle that he spied approximately 20 years earlier in his home state near the Combahee River. A letter from him (then using the pseudonym Mike Richards) in which he reported his sighting was later published in *FT* (June 2011).

But now, combining all of the much more comprehensive accounts contained in his various emails to me and presented here with his kind permission for me to do so using his real name, this is Mike Richburg's full description of what he saw:

> I witnessed a very, very large bird (6-7 ft. tall) that looked almost exactly like a Golden Eagle except for its extraordinary size. The avian flew off with a small deer in its grasp. Its wingspan was wider than one lane of US HWY 21. It lifted the deer from the roadway and cleared the treetops with three cycles of its wings.
>
> The deer's neck was clearly visibly broken, and not just a little bit. It had clearly very recently happened. I have spent much of my life outdoors, hunting, fishing, hiking, camping, etc, and have seen deer in all manner of states and speak from a position of knowledge

on the deer itself. I am also quite clear on the size and age of the deer, and stand firmly by my 50 lb estimate. Also not included was what first caught my eye, the movement of spooked (alerted) deer far ahead in the roadway. I saw the flashes of white on their hind ends as they bounded briefly about in the roadway, in an obvious attempt to get the heck out of there, apparently as afraid of this thing as I was. I am sure there are many more small details that may or may not be important.

Also, one thing that really stuck out to me about the whole thing is when the bird looked up. It didn't look at my vehicle. It looked straight at me. In my face. We made eye contact, which totally freaked me out, and convinced me that the creature was very intelligent. This has always stuck out as being very impressive to me. It looked straight into my eyes. Still kinda creeps me out.

In addition, I will inform you that I think about what I saw almost daily, and frequently dream about it. I often wake up when the animal looks at me. I still feel afraid of it. I was so afraid at the time of the sighting, it is hard to explain. Let's just say I knew a human would be no match for this thing.

I could actually see the wind created by the flapping motion of the bird's wings affect the vegetation (bushes, grass, etc) on the side of the road. This was most noticeable on the second cycle of the wings, which was the first "full cycle" while still low over the roadway. This effect was very similar to the "wave" of air created by a passing truck that can be seen in tall grass on the roadside. Also of note was the great ease in which the bird took flight with the deer in tow. Much like a Hawk with a mouse, or an Eagle with a 3 lb rabbit. It looked very easy for the bird, and I am sure it was not the first time he flew away with very large prey.

I have extensively researched both the Haast's Eagle [*Harpagornis moorei*, the giant eagle of New Zealand that officially died out in c.1400 AD] and Washington's Sea Eagle, both thought to be extinct. I believe what I saw was similar to both in appearance with some differences worth noting. The bird I saw clearly had a yellowish beak, most depictions of Washington's Eagle report a dark beak. The overall coloration however, was not that dissimilar. The underbelly coloration was similar to Haast's eagle, but appeared to be a different pattern. The overall size of the bird I saw was also even larger than either of these great birds. The bird I saw was very, very Eagle like, as I previously stated, looking almost exactly like a Golden but much larger. Its feathers were also a little more "ruffled" or out of place, for the most part, except for the head itself where the feathers clung closely and neatly, giving a "smooth' appearance. This greatly contrasted with the "rough" look of the rest of the bird, which started at the neck.

The bird I saw also had very impressive legs. The top of the leg was very large and apparently muscular. Comparable to a great dane with feathers. The lower legs looked very much like tree trunks, and were very noticeably stout. The feet themselves were also very impressive and noticeable, not as yellow as I would have thought, more like dirty

Mike E. Richburg's drawing of the bird that he saw (© Mike E. Richburg)

dishwater grey and had the look of leather that was excessively cracked and peeled. There ain't no lizard in the bird I saw. It was 100% Eagle, and could have lifted prey much larger.

As I have stated in the summary attachment, it was in the roadway directly in front of me at close range. I had a very good look at it.

Here is his summary attachment, received by me on 18 April 2011:

I wanted to thank you again for your interest in my encounter with a giant eagle type bird in SC in about 1990. Although I am not much of an artist, I took a shot at drawing what I saw that morning, and have enclosed the illustration with this correspondence. Albeit crude and somewhat primitive, it provides a visual aide. If the average picture is worth a thousand words, then this one ought to be worth a million. This is how the animal looked when it raised its head and looked at me, at which time I finally realized what I was looking at. The bird flared out his wings slightly and gave me a very intimidating glare, as if to say that the deer was his and he was ready to defend it. It was a very fierce and menacing look. Of particular interest was the very robust, stout girth of the body and the very wide stance. His right leg was on the deer's neck and the left on the hind end. His eyes also seemed very large, even for such a large animal. The bird also had very noticeably large feet. I did not notice the great length of the beak until he turned. Here is a brief list of observable physical attributes.

Size of animal:
6-7 ft from highest part of wings to roadway as pictured. Please note that highest part of wings was tilted toward vehicle, and wingtips pointed away from vehicle and noticeably laying on road, as were the tail feathers. Exact height of animal if upright is unknown.

Width of wingspan was undeterminable until bird turned around and took flight, was then estimated at least 16 ft, and probably around 18 ft. Broadness of each wing estimated at least 4-5 ft from leading edge to trailing edge.

Very wide girth and very wide stance. Distance between legs at least 2.5-3 ft.

Very large feet with large visible talons.

Color of animal:
Overall color was dark brown to charcoal gray.

The underside of the wings appeared lighter in color with white "spots" that were actually not round, but square-ish and rectangular and "L" shaped.

The underside of the tail which the bird presented to me after turning around was also much lighter in color as it got closer to the bird. Tips of tail feathers were dark on underside as well as the top.

Yellow beak.

Yellow to gold eyes.

Grayish looking feet with black talons.

In a subsequent email, sent to me on 3 May 2012, he included two further descriptive accounts, both of which offered some important additional insights into the appearance and nature of the giant eagle that he had encountered:

> I awoke from a dream this morning with the proverbial light bulb over my head. I have now realized why the bird's feet looked like they did. In a word, it was mud. To fully understand why I am so sure, one must appreciate the geography of the area. This is a flat almost featureless coastal plain, except for the Carolina Bays (a unique unexplained phenomenon in their own right), that was once the sea floor. Fresh water is in abundance, but everywhere there is water, and in a lot of places there is not, there is also a very greasy black to charcoal gray mud. This is the only kind of mud around and there is plenty of it. This mud, of course, dries to a lighter shade of gray. The bird had to drink fresh water, and to do that I am 100% sure he would have had to stand in that kind [of] mud, or fly at least 50 miles. I am now very confident what I saw was half dried mud that was flaking off, thus explaining the coloration and extreme texture on the bird's feet.
>
> Several years after the sighting I confided in a friend and told him what I had seen. The conversation turned to the pattern under the wings and my curiosity about it. I insisted that it must be for camouflage purposes, to help conceal the large animal. In my attempt to describe the pattern I took some white spray paint to a piece of plywood I had. One thing lead to another, and with the help of some other colors, I had quickly made a rough imitation of the pattern over the whole side of the 4′ x 8′ sheetof plywood. A debate about the effectiveness of the camo ensued immediately.
>
> To settle the debate we tried a very unscientific study, by placing the plywood in different surroundings, and stepping back to take a look. It did not work against a clear sky, nor around a stand of Pine trees and a few other places. To make a long story short it sucked in everyplace we put it except for one. When we placed the wood up in an Oak tree behind my house it became

virtually undetected. There were other hard wood trees around, behind and in front of the wood. There were Persimmon, Gum, Sparkleberry, Maple etc. It became astonishingly clear that the white "blocks" closely imitated the small "bits" of sky one can see when looking up thorough the trees.

I had hoped to conduct the same experiment again just to send you pics, but I was much younger then and the chances of me climbing 30′ up in a tree with a 4′ x 8′ sheet of plywood are slim to none, and I think I just saw Slim leaving town.

Conclusions: The camouflage system is very area specific and only works well in growths of deciduous trees, where it excels in the extreme. Noteworthy is the fact that this is also where the bird was seen. This pattern works very well half way up a mature 50′-70′ tree, where the bird might likely hunt from at times. The pattern was so effective that a deer, much less the average person, would never detect the presence of the great bird unless the bird moved or made noise, even with its wings open and just above its prey. In fact, if the bird was on a large branch next to the trunk of the tree, much like a large Owl might sit, it would be virtually invisible even with the wings fully opened, as long as it was facing towards its prey. I believe the bird would in fact hunt this way due to its many advantages. Humans hunt deer from tree stands in similar fashion to gain these same advantages. The predator's scent is not at ground level, better vision, etc. It would only make sense that this creature would use these advantages as well.

The combined accounts from Mike Richburg quoted here constitute the most extensive eyewitness description of an unidentified giant eagle in North America that I have ever read, added to which is his hand-drawn illustration. Although not identical, this mysterious mega-eagle certainly recalls the equally mystifying Washington's eagle, and makes me wonder anew as to whether such a form was not only truly distinct from all other North American species of eagle but still survives on this continent today, albeit assuredly in very small numbers, yet very effectively camouflaged, thus preserving the secrecy of its existence from all but a scant few, highly fortunate observers

Alternatively, is it conceivable that the individual birds documented here, including the specimen upon which Audubon established his Washington's eagle species, in reality constitute freak, over-sized specimens of the golden eagle, possibly exhibiting genetically-induced gigantism?

Of particular interest is Richburg's statement that the eagle carried off a small deer whose weight was estimated by him to be approximately 50 lb. Adult male golden eagles living in the wild weigh around 8 lb on average, and adult females living in the wild weigh around 11 lb on average, though females weighing up to 15 lb have been confirmed. Golden eagles are normally able to lift and carry off prey roughly half their own weight—which would be around 4 lb for an adult male, and 5.5 lb for an adult female, although prey victims up to 9 lb in weight have been recorded. Yet even this is far short of the 50-lb weight claimed by Richburg for the deer abducted by the giant eagle seen by him. So either he over-estimated the deer's weight, or there are some phenomenally powerful—as well as gargantuan—eagles out there.

Another eyewitness report lending support to this daunting prospect is the following one.

It was posted in two sections on my *ShukerNature* blog by a reader identifying themselves only as 'Micah and Kim' on 28 April 2014, in response to my article on Washington's eagle and other mystery mega-eagles:

> I had to comment before I even finished reading the article. I grew up in the northern great lakes region (specifically Michigan's UP [Upper Peninsula]) and saw a lot of eagles growing up. We always assumed a brown eagle was an immature bald eagle. It was relatively common for my parents and I to spot large (8 foot wingspan) eagles. One incident sticks out, however. When I was about 12 traveling with my parents by car through the deep woods of the western UP, my father saw a bald eagle on the side of the road. We passed it, stopped and reversed to get a closer look. It was massive, easily the biggest eagle any of us had ever seen. It was scavenging a roadkilled whitetail doe. We got as close as 20 feet to it before it decided that it didn't appreciate our presence. It sunk its talons onto the doe, opened its wings to its fully nine foot wingspan and flew up and away—carrying the doe with it! It carried the doe at least twenty feet up and thirty feet away into the woods before dropping it. We assumed that it decided that it wasn't able to clear the trees with it. It was a smallish doe, but a small whitetail doe will easily weigh 60 pounds.
>
> My previous comment was submitted before I finished the article. Now that I am done reading, I am struck by the similarities between my story and Mike Richburg's. The posture of the bird feeding on the deer, the three great flaps of its wings and the wind they generate specifically. I need to clarify that the bird I saw was very much a mature bald eagle. It had a white head, bright yellow beak and talons and deep solid mahogany brown elsewhere. The only thing unusual about my encounter was the bird's massive size (9 foot wingspan for sure and approximately 3 foot height) and the fact that it carried off the deer it was eating. I believe my sighting was near Watersmeet, Michigan, although I would have to ask my dad for sure. It was around 1986 or 87.

Whatever the explanation for such birds might be, it would appear that eagles much bigger than science presently believes they have any right to be may indeed exist in the U.S.A.

Finally, and as if matters were not already complex enough, in 1988 the Big Bird subject gained yet another source of confusion. This was the year in which the first three specimens from a planned series of 17 Andean condors were released in California. The reason for their introduction was so that the scientists involved in the captive breeding of the gravely endangered California condor could test the techniques that they would be using when they eventually released captive-bred specimens of this latter species (a programme that began in 1992).

Although a very worthy project, it does pose problems when assessing reports of Big Birds—especially those resembling Andean condors. From now on, we will never be sure whether such birds are genuine Big Birds, or simply some of these latter introduced condors.

It is not the first time that what had been traditionally thought of as a single type of mystery creature has ultimately been unmasked as a veritable jigsaw of many different, entirely separate pieces. I revealed a similar scenario for the mystery cats of Great Britain in my book *Mystery Cats of the World* (1989), and Heuvelmans did the same in various of his books for the Nandi bear and the great sea serpent. Moreover, in the case of the Big Bird there is a particular urgency in obtaining an understanding of the correct situation.

As emphasised by Mark Hall, the truly gigantic type of Big Bird appears to have decreased dramatically in numbers over the centuries, judging from Amerindian testimony and records dating back to North America's pioneer days. Although, as revealed here in this present book, teratorns are unlikely to pose any physical threat to humans, their outward similarity to vultures would be sufficient for them to be persecuted and destroyed. There is a great need, therefore, to obtain conclusive evidence for the reality of these mega-birds with all speed—in order to ensure that they do not suffer the fate of their smaller relative, the California condor. The same applies if these birds are, alternatively, an exceptionally large species of eagle or other accipitrid still awaiting formal scientific recognition.

Humanity's savagery almost destroyed one variety of thunderbird—the California condor. Will science's indifference and dismissal condemn a second, even more spectacular version to the same tragic fate?

Eagle and bear depicted in painted wooden wall plaque, of Canadian Haida or Kwakiutl style (© Karl Shuker)

The Curious Case of the Absent Photograph

Even by cryptozoological standards, the Big Bird case is rather unusual, inasmuch as it has been dominated lately not by high-profile sightings or studies, but by a continuing search for an item of evidence more than a century old —and which may never have existed anyway.

In 1886, an Arizona newspaper called the *Tombstone Epitaph* allegedly published a photograph of a dead Big Bird belonging to the gigantic thunderbird variety, nailed to a barn with its wings fully outstretched and its somewhat pterodactylian beak open. In the photo, it was apparently displayed alongside a row of six men who served as a scale for accurately estimating its wingspan—which was revealed to be an incredible 36 ft!

In some accounts referring to this photo, the men are more specifically but variously described as American Civil War soldiers in uniform, or farmers, or as rifle-toting hunters; and the pterosaurian similarity of the bird's beak is greatly exaggerated—to the extent that the entire creature is likened to a huge pterodactyl. However, it is clear that those accounts citing it as a pterodactyl have confused it with the subject of a totally separate *Tombstone Epitaph* report—namely, one dating from 26 April 1890, and featuring the supposed giant pterosaur discovered floundering in the Huachuca desert by two ranchers, as detailed earlier in this present chapter.

Naturally, this unique photograph from 1886 would be invaluable in yielding support for the giant thunderbirds' existence and in determining their taxonomic identity—but the sad fact of the matter is that it has since proven to be as elusive as the bird that it reputedly portrays. Despite extensive searches through the *Epitaph*'s files, and also those of many other publications believed or suspected by investigators to have reproduced it at some stage, the photo has never been located.

One of the most promising leads in recent years came when Fortean writer W. Ritchie Benedict recollected that the photo had been shown on an episode of a Canadian television programme called *The Pierre Berton Show*, transmitted during the early 1970s. As far as Benedict could recall, the episode featured an interview with the late Ivan T. Sanderson, who supposedly held up a copy of the photo to the TV camera, yielding a close-up shot of it. Sadly, however, despite conducting extensive searches Benedict was unable to locate an existing copy of this episode.

Out of curiosity, during some of my own investigations concerning this missing photograph I contacted the Audio-Visual Public Service division within the National Archives of Canada, enquiring whether they could locate this episode. I subsequently learnt from research assistant Caroline Forcier Holloway, however, that she had not been able to uncover it and needed a precise production or release date for it in order to continue looking, because there were 597 episodes in this series still in existence, each of which contained more than one guest. Moreover, there were others that seemed to have been lost, so there was no guarantee that the episode containing Sanderson was among the 597 preserved ones anyway. Unfortunately, however, the whole crux of the problem is that we simply do not know when it was produced or released—hence this particular avenue of investigation seems to have reached an impasse.

Moreover, one of my correspondents, Prof. Terry Matheson, an English professor at Saskatchewan University with a longstanding interest in the thunderbird photo, claimed in a letter to me of 22 September 1998 that Sanderson appeared on *The Pierre Berton Show* not in the early 1970s, but actually no later than the mid-1960s. This is because Prof. Matheson vividly remembered seeing this episode and talking about it afterwards with a friend with whom he was working on the Canadian Pacific Railway as a summer job, and he only worked there from 1965 to 1967. Here is what he wrote:

> The particular episode of the programme . . . did not take place in the early 1970s. I remember watching the segment dealing with the thunderbird—part of an extended interview Pierre Berton had with Ivan Sanderson—from my home, when I was an undergraduate student at the University of Winnipeg in the mid-1960s. By the 1970s I was in graduate school in Edmonton. I know the programme could not have aired much later than 1965, because I recall discussing it initially with my

> mother and grandmother, who had also watched the show; with college friends, who made me the subject of much good-natured ridicule; and sometime later with a friend from Calgary whom I had met while employed on the Canadian Pacific Railway, as I was (over the summer months) from 1965 to 1967. I cannot recall the precise date of this conversation with my railroad friend, nor can I recollect the date I watched the programme with pinpoint accuracy, but would guess that it aired the winter before my first summer on the railroad, that is, 1964-65; at the very latest, the following year (1965-66). That might be a good place to start.

Prof. Matheson's confident placing of his well-remembered conversations concerning the latter TV show within the mid-1960s, coupled with his precisely-dated period of employment on the railway in the 1960s, as well as his undergraduate studies also occurring exclusively in the 1960s, would certainly seem to disprove previous assumptions that this particular show was not screened until the early 1970s—unless, perhaps, it was re-screened at that time, following its original screening in the mid-1960s? However, his letter also contained another notable challenge to traditional assumptions regarding this show:

> To the best of my recollection, the photograph was not shown, at least not on this particular programme. I definitely recall Sanderson's allusions to the photograph, which he described vividly and with great precision. Although I can envision Sanderson's description as if it were yesterday—the bird nailed to the wall of the barn, the men standing in a line spanning the wingspan, etc—he did not, however, have the photograph in his possession when the interview took place, although he certainly claimed to have seen it. Incidentally, some time after this, Sanderson set up a society for the investigation of paranormal phenomena [SITU—the Society for the Investigation of The Unexplained]. I joined, and in response to my inquiry about the photograph, was told that they did not have a copy. Receiving this news led me to wonder at the time if the photograph might be an example of an urban myth or legend.

If, as would now seem to be the case, the thunderbird photo was not shown by Sanderson on that elusive mid-1960s episode of *The Pierre Berton Show* after all, one of the most promising avenues for tracing it—by seeking an existing copy of this specific episode—has gone . . . or has it? Might Sanderson have also appeared on a second, later episode of the show, this time armed with the photo? Yet according to SITU, which contained Sanderson's extensive archives, they did not have a copy of it.

Another possibility that seemed worthy of pursuit was that the photo had featured in the American magazine *Saga*, well known for regularly publishing articles of a Fortean nature. Fellow investigators had suggested that those issues spanning the years 1966-68 were likely ones to examine for its presence—so I duly transmitted this information to the Library of Congress in Washington D.C., whose research specialist Travis Westly very kindly instigated a search on my behalf. Sadly, however, after checking through each issue from January 1966 to March 1969, he was unable to discover any photo or article relating to any type of enormous bird—another dead end.

Is this photograph of a marabou stork responsible for many people believing that they have seen the missing thunderbird photo?

Over the years, numerous people have claimed to have seen this evanescent image—which makes it all the more frustrating, and mystifying, that no-one seems to have retained a copy of it.

Perhaps, therefore, as sceptics have often suggested, the thunderbird photo has never existed at all, and should therefore be dismissed as nothing more than an unusual example of urban folklore. Alternatively, there are others, including myself, who wonder whether at least some of those people who claim to have seen it have actually seen a superficially similar, 'lookalike' picture, depicting some large but known species of bird with wings outstretched, and years later have misremembered what they saw, erroneously believing that they had actually seen the thunderbird photo. Such an event would be a classic case of false memory syndrome.

Interestingly, one image that could certainly have inspired people to believe that they had seen the genuine thunderbird photo is a certain photograph of a large African marabou stork *Leptoptilos crumenifer* (syn. *crumeniferus*) held with its beak open and its massive wings outstretched by some native men. Tellingly, this picture appeared in a number of popular books worldwide during the early 1970s, including the *Guinness Book of Records* edition for 1972, which at that time was second only to the Bible as the world's bestselling book, so was certainly seen by a vast number of people all around the globe.

I first proposed the marabou stork picture as a possible false memory trigger in relation to the real thunderbird photo (always assuming, of course, that the latter image really does/did exist!) way back in 1993—in a letter sent to Bob Rickard at *Fortean Times* on 15 February

1993 and in one sent to Mark Chorvinsky at *Strange Magazine* on 2 July 1993. My letter to Mark was subsequently published by *Strange Magazine* in its fall-winter 1993 issue (for a comprehensive *Strange Magazine* article of mine on this same subject, check out its December 1998 issue). Here is what I wrote in my letter:

Numerous people around the world believe that at one time or another they have seen the notorious "missing thunderbird photograph," allegedly published within a *Tombstone Epitaph* newspaper report in 1886 (see *Strange Magazines* #5, 6, 7, 11). In view of its extraordinary elusiveness, however, in many cases it is much more likely that their assumption is founded upon a confused, hazily recalled memory of some other, superficially similar picture instead—i.e. a "lookalike" photograph. A particularly noteworthy "lookalike" for the missing thunderbird photograph appeared on p. 35 of the British version of the *Guinness Book of Records* (19th edition, published in 1972), and is reproduced alongside this letter of mine. It depicts a large African marabou stork *Leptoptilos crumenifer* standing upright with its extremely large wings (which can yield a wingspan in excess of 10 ft.) held outstretched by some native tribesmen flanking it, and with its startlingly pterodactyl-like beak open wide. This picture thus incorporates a number of features supposedly present in the thunderbird photograph—a very big bird with a pointed pterodactyl-like beak, and an extremely large wingspan, whose wings are outstretched, and flanked by various men. Bearing in mind that the photo is a very old one (possibly dating back to the first half of this century [i.e. the 20th Century]), and also that the *Guinness Book of Records* is a worldwide bestseller, and that this photo might well have appeared not only in the English version but also in many (if not all) of this book's other versions around the world [as far as I am aware, the same picture layout does indeed appear in all versions worldwide within any given year], it is evident that countless people will have seen it over the years, of which some may well have been unconsciously influenced by its striking (indeed, archetypal) image when contemplating the issue of the missing thunderbird photograph.

Expanding upon what I wrote there, in addition to the *Guinness Book of Records* edition for 1972 this photograph also appeared in the first edition of the late Gerald L. Wood's bestselling *Guinness Book of Animal Facts and Feats* (1972); and I learnt from his widow, Susan Wood, that it had been in a collection of pictures and postcards that he had already amassed by the early 1940s (which is why I stated in my *Strange Magazine* letter that it may date back as far as the early years of the 20th Century).

It is not even the only such photo of a marabou stork in existence either. Below is a second, albeit slightly less evocative one, which appeared in a book by Richard Tjader entitled *The Big Game of Africa*, published in 1910:

Marabou stork photograph from Tjader's book

Of course, one might argue that neither marabou stork photo could have influenced peoples' belief that they had seen the thunderbird photo because in both photos the men standing alongside the stork were native African tribesmen, which are surely not going to be falsely remembered in subsequent times as American Civil War soldiers, ranchers, or farmers by these photos' viewers. In reality, however, the human mind has a remarkable capacity for 'filling in' details that are not actually there. The sight of the very striking marabou stork image may well be more than sufficiently memorable for viewers, years later, to be convinced (albeit wrongly) that the human figures around it were indeed Civil War soldiers rather than African natives. The same phenomenon explains how sightings of inanimate objects have in the past been deemed to be mystery cats, man-beasts, ghosts, and various other animate entities.

In his letter to me, Prof. Matheson raised another thought-provoking but very different point concerning false memory syndrome and the thunderbird photo:

> Although your suggestion that people's memories of a similar photograph might have been confused with that of the thunderbird is entirely possible, as I'm sure you know, Sanderson was a great raconteur, a man whose verbal gifts could cause anyone to imagine that they had actually seen something he had only described in words. Indeed, many years after watching the programme, I met an individual who had also seen the Berton interview and was initially positive that the picture had been shown.

Yes indeed, the power of verbal suggestion. Wars have been instigated as a result of the mesmerising oratory skills of certain leaders, let alone belief that a picture had been shown on a television programme when in reality no such picture had appeared (unless of course, as suggested earlier, Sanderson had appeared on *The Pierre Berton Show* not once but twice, and had shown the thunderbird photo on the second occasion, but which Prof. Matheson had neither seen nor was even aware of).

Incidentally, the December 1997 issue of *Fortean Times* not only contained a detailed account of modern-day thunderbird reports by Mark Hall but also included a succinct account of my suggestion that the marabou stork photo in the *Guinness Book of Records* 1972 edition may have influenced some people in their belief that they had seen the missing thunderbird photo. Deftly combining our separate contributions to the subject, this issue's front cover duly sported a breathtaking illustration by artist Steve Kirk of a marabou stork-inspired thunderbird!

Returning to Prof. Matheson's letter: elsewhere in it, he mentioned a line of investigation of his own that he had conducted in relation to the thunderbird photograph, and highlighted a fascinating and extremely pertinent fact, but one that seems to have attracted little or no attention from other investigators. What he did was to go right back to the starting point of the entire mystery—by writing directly to the *Tombstone Epitaph*, and enquiring whether such a photo had indeed ever appeared in their newspaper:

> In an interesting reply, they both denied any knowledge of the picture and also pointed out that the reproduction of photographs in newspapers was at that time—the late nineteenth century—not common anywhere in North America. In checking our local newspaper—the *Winnipeg Free Press*—to see if this was the case, I found that photographs

rarely if ever appeared before the early 1900s, at least in that newspaper.

This is true, but engravings, drawings, and other artwork did commonly and widely appear in newspapers at that time. Moreover, I learnt from correspondent Adam Selzer that such illustrations as these were often referred to colloquially as photographs anyway, even though they weren't actual photos. So perhaps there really was a thunderbird illustration, but in the form of an engraving or something similar rather than a photograph, yet which has nonetheless been confusingly referred to since then as a photograph. Even if so, however, the mystery remains as to why no-one seems able to locate a copy of it anywhere.

So is the thunderbird photograph (or artwork) fictitious, illusive rather than elusive, nothing more than a fable of our times, perpetuated into the present day by false memory syndrome—inspired in turn by visual lookalikes and seductive verbal suggestion, not to mention an ever-increasing number of faked versions circulating online and elsewhere?

Or, against all the odds, might it truly be real? Could there actually *be* a missing thunderbird illustration—one that originated in 1886 but has subsequently been reprinted elsewhere—concealed in some old, yellowing magazine somewhere?

Next time that you clean out your attic and find a pile of dusty magazines there, have a look through them before you throw them out—just in case. You never know what you may discover inside! And needless to say, if you do find the thunderbird illustration, be sure to contact me and let me see it!

Preserved Thunderbird Feathers?

Cryptozoologists are familiar with the long-standing mystery of the missing thunderbird photograph, but what about an alleged thunderbird feather?

Interviewed recently by Tucson-based freelance writer Craig S. Baker for an online article on unsolved Wild West mysteries (at: http://mentalfloss.com/article/56759/9-unsolved-mysteries-wild-west), veteran Wild West author/investigator W.C. Jameson made a claim of considerable potential significance to cryptozoology regarding the legendary thunderbirds.

Jameson stated that a Cherokee treasure hunter he once knew told him that while looking for a long-lost cache of Spanish silver in a Utah cave, he had dug up several huge feathers, each one over 18 in long and with a quill of comparable diameter to one of his fingers. Above the cave's mouth, moreover, was an ancient pictograph of an enormous horned bird. Could this have been a piasa?

Jameson has also claimed that he actually owns the stem (i.e. quill) of one of these remarkable mega-plumes, albeit broken and incomplete, thus 'only' measuring 18 in long, and that its species had not been positively identified by any of the several (unnamed) ornithologists who had seen it. A photograph of Jameson's alleged thunderbird feather quill can be viewed in Mark Turner's *Mysterious World* blog (at: http://markturnersmysteriousworld.blogspot.co.uk/2011/06/thunderbird-legends-sightings-evidence.html)

Assuming that Jameson's story is accurate, could this giant feather be a bona fide thunderbird plume? Tangible, physical evidence for cryptids is, by definition, a rare commodity, so such a specimen could be of great scientific worth, thanks to the considerable power of modern-day DNA analysis in ascertaining taxonomic identity or kinship.

For by subjecting the feather to such analysis (using samples of dried blood if present at its base, or viable cells collected from the calimus—the portion of the quill that had previously been imbedded underneath the bird's skin), biotechnologists might succeed where the ornithologists have reputedly failed, and duly unveil the hitherto-cryptic nature of its avian originator's taxonomic identity.

Let us hope, therefore, that someone will be able to persuade Jameson to submit his giant mystery feather for formal DNA testing—always assuming of course that it really is a feather . . .

After all: during medieval times, crusaders returning home to Europe from the Middle East often brought back with them as unusual souvenirs what they had been told by unscrupulous traders were feathers from an immense fabled bird known as the roc or rukh—said to be so enormous that it could carry off elephants in its huge talons. Even its plumes were gigantic, up to 3 ft long. In reality, however, when examined by naturalists these were swiftly exposed as the deceptively feather-like leaves of a species of raffia palm tree (genus *Raffia*).

A giant leaf from a raffia palm tree, masquerading as a roc feather (© Dr Karl Shuker)

A Stuffed, Taxiderm Thunderbird?

In 1998 I received the following fascinating information of possible relevance to North America's ongoing thunderbird/Big Bird mystery. And this time it involves something much more substantial than a missing thunderbird photograph—nothing less, in fact, than what may be a missing stuffed thunderbird!

In an email to me of 19 October 1998, the afore-mentioned Prof. Terry Matheson from Saskatchewan University stated:

> Years ago, a friend of mine who had lived in northern Ontario told me that in the town of Spanish, Ontario, there is a stuffed specimen of a huge bird that no one has ever been able to identify. The bird had (I presume) been sighted locally and killed. I wrote the town hall asking about this, but received no reply, and, although I vowed that I'd investigate whenever I happened to be in the vicinity, I've never had occasion to be there. Who knows? Although I got nowhere, an inquiry from a person with

> genuine credentials, an acknowledged expert such as yourself, might elicit a response. It might be worth pursuing.

Indeed it might, which is why I lost no time in following Prof. Matheson's lead in contacting the town hall in the Ontario town of Spanish noted by him, but I received no reply. And email searches for additional info failed to uncover anything either.

Then, on 3 August 2012, I received the following email message from Facebook friend Rebecca Tosh Xayasith, who had been very kindly investigating this case on my behalf:

> An update on the stuffed Thunderbird in Spanish, Ontario. Couldn't find much info. I did however find out, through a friend, that the nearby Massey Area Museum had never heard anything about it. He e-mailed the museum, and they did respond back. Also, the local library has heard nothing about it either. Spanish is a small town, population around 650+ people, and it is on the decline. They lose more people every year. If indeed there IS a stuffed bird, I'm wondering if it is a bird that is known to the world, but, maybe not known to the people of that area. And this town is slowly dying. I would imagine, if they had such a thing as a stuffed Thunderbird, they would use it to the town's advantage. They would attract MANY tourists if they had such a thing, and it just might save the town. So, I seriously doubt there is anything there, or anything that is unknown to the world.

She makes some very valid, pertinent points. After all, there is little doubt that a stuffed thunderbird, or any spectacularly large bird, would help in attracting tourism to any town possessing one. So, sadly, it appears that the bird has flown—at least figuratively!

Is This Mystery Mega-Bird a Canard?

Finally: a mystery bird larger than an ostrich and spied by hundreds of observers before being killed and placed on display is surely not the easiest of creatures to vanish from the pages of history. And yet this is precisely what seems to have happened to the extraordinary-sounding entity documented in the following *Columbus* [GA] *Daily Enquirer* newspaper report from 2 September 1868, which I recently found reprinted on various websites, including Chris Woodyard's excellent *Haunted Ohio*. Here is the report:

> A Singular Bird
>
> On Sunday of last week a novelty in the bird line was killed in Kentucky, opposite Mound City, Ill., by a man named Jim Henry, of that city. The *Cairo Democrat* [a newspaper in Illinois, not Egypt!] says:
>
> It is larger than the ostrich, and weighs one hundred and four pounds. The body of this wonderful bird is covered with snow white down, and its head is of a fiery red. The wings of deep black, measure fifteen feet from tip to tip, and the bill, of a yellow color, twenty-four inches. Its legs are slender and sinewy, pea green in color and measure forty-eight inches in length. One of the feet resembles that of a duck, and the other that of a turkey. Mr. Harney [this name is interchangeable with Henry, depending upon which website's version of this report is consulted] shot it at a distance of one hundred yards, from the topmost branch of a dead tree

> where it was perched, preying upon a full sized sheep that it had carried from the ground.
>
> This strange species of bird, which is said to have existed extensively during the days of the mastodon, is almost entirely extinct—the last one having been seen in the State of New York during the year 1812. Potter has it on exhibition in his office at Mound City. Its flight across the town and river was witnessed by hundreds of citizens.

Fascinating though it is, there is so much to doubt in this bizarre report that it is difficult to know where to begin. A bird bigger than an ostrich eating a full-grown sheep in a tree, sporting a 15-ft wingspan plus legs of differing appearance, and once-common many millennia ago but now reduced to this single, last-known, and soon-afterwards-demised specimen—it all sounds far more like the fanciful product of a journalistic hoax than anything engendered by Mother Nature, that's for sure.

Who, moreover, is Mr Potter, and why are there no follow-up reports or pictures, especially as this feathered behemoth was supposedly exhibited dead by Potter? Even its slayer's name changes within the report (and also between different websites' versions of it). Nevertheless, just on the very slim off-chance that there really is more (rather than less) to this strange story than meets the eye, if any readers have extra information regarding it, I'd love to hear from you!

THE MONSTROUS MAKALALA

Speaking of carnivorous birds larger than an ostrich: No one doubts that the tallest species of bird alive today is indeed *Struthio camelus*, the ostrich—no-one, that is, except for the Wasequa people (most probably an alternative, kiSwahili name for the Zigua, according to one of my correspondents, whom I shall refer to here by his Blogger username, Pat the Plant). They inhabit an unspecified interior region of mainland Tanzania eight to nine days' journey from the coast of Zanzibar (the Zigua do live directly inland from Zanzibar).

Reconstruction of the makalala's possible appearance in life (© Markus Bühler)

According to a report by a Count Marschall (*Bulletin de la Société Philomatique*, 1878-9), as recently as the 1870s these people averred that their territory harboured a monstrous bird even taller than the 8-ft-high ostrich, equipped with very long legs, the head and beak of a bird of prey (which it puts to good use when feeding on carrion from animal carcases), and the ability to take to the air in sustained, powerful flight. Also, each of its wingtips bears hard plates composed of a horny, compact substance, and when it strikes its wings together they produce a very loud noise, earning this bird its local name—makalala ('noisy').

Marschall claimed that the makalala is said by the Wasequas to be very fierce, but can be killed if the correct strategy is employed.

Engaging upon an extremely hazardous version of 'playing possum', the would-be assassin has to lie on the ground and feign death, until the makalala approaches close enough to seize the supposed human carcase—whereupon the latter must reanimate himself instantly and deliver the fatal blow before the makalala realises its mistake!

So far, this could all be discounted as fanciful native folklore—but physical remains of the makalala may have been recorded too. Marschall mentioned a Dr Fischer, who saw in Zanzibar an object that he identified unhesitatingly as a rib from some form of gigantic bird. Narrowing from one end to the other, this alleged rib had a width of 8 in at its widest end, and was just under 1 in at its narrowest end. Unfortunately, Marschall did not record whether Fischer sent it to a scientific institution for conclusive identification and retention.

However, Marschall did record another possible source of makalala remains—because he noted that native chiefs placed makalala skulls on their heads, using them as helmets! Could any of these bizarre examples of protective headgear still be owned today by Wasequa tribesmen?

Thanks to my afore-mentioned correspondent Pat, I now have a copy of a second makalala document from the same time period—namely, the published account by the Dr Fischer alluded to by Marschall in his own report. He was Dr Gustav A. Fischer, and his account of the makalala was part of a much longer report co-authored in German with Dr A. Reichenow, which was published in 1878 within the *Journal für Ornithologie*. Interestingly, in his own account Fischer described the makalala as being very shy (rather than very fierce as claimed for him by Marschall in his report), and stated that he was reluctant to believe that the rib-like structure came from a bird (whereas Marschall claimed that Fischer readily identified it as such), but otherwise the two descriptions correspond well with one another.

Assuming, against all the odds, that the makalala is real—that the frightening scenario of a *carnivorous* bird taller than the ostrich surviving into historical times somewhere in mainland Tanzania's interior is not a grotesque fantasy but a sober fact—what could it be? Several interesting, albeit mutually-exclusive, lines of speculation compete for attention.

The first of these to be discussed here was kindly brought to my attention by German cryptozoologist Markus Bühler. Breeding throughout much of sub-Saharan Africa, sporting an immense wingspan of up to 10.5 ft (even greater spans have been claimed but presently not verified), standing up to 5 ft tall, and weighing as much as 20 lb, the well-known, previously-mentioned marabou stork is certainly an extremely impressive, potentially formidable bird. Indeed, when specimens are scavenging from a carcase, they will sometimes even ward off vultures once the latter birds of prey have torn chunks of flesh from the carcase with their hooked beaks (which marabous lack). Even so, it seems unlikely that such a familiar species could have somehow been converted by local myth and superstition into a mystery bird.

However, during the Pliocene, Africa was also home to an even bigger marabou stork, *L. falconeri*, Falconer's marabou. Like *L. crumenifer*, it was widespread across northern and eastern Africa but stood around 6.5 ft tall (taller than an adult human of average height) and weighed up to 44 lb (as heavy as a small child). In comparison to *L. crumenifer*, Falconer's marabou exhibited a slight reduction in wing size, therefore possibly being more terrestrial than its modern-day relative, but it was still fully able to fly. As birds often look much bigger than they actually are, due to their plumage and pneumatic internal system adding substantial volume to their forms, this already-huge species would have been truly monstrous in appearance, added to which its possibly greater terrestrial lifestyle means that it may possibly have been able to kill and eat bigger creatures than *L. crumenifer*.

Based upon fossil evidence, Falconer's marabou stork had become extinct by the end of the Pliocene 2.58 million years ago, but if it had somehow survived into historical times (with what would be its more recent fossils not having been uncovered so far), there is no doubt that it could have been a thought-provoking makalala candidate (albeit one lacking the raptorial beak claimed by the Wasequas for the makalala). Even the latter's supposed wing-clapping sounds might in reality have been a confused memory of the beak-clapping sounds often produced by storks, and which would have been very loud if made by Falconer's marabou. However, there is currently no scientific evidence that the latter species did survive into historical times.

Another very large and intriguing species of bird that once inhabited Africa is *Eremopezus eocaenus*, which, as its name indicates, lived during the Eocene (specifically the late Eocene, between 36 and 34 million years ago). Its fossil remains, which have been obtained from Jebel Qatrani Formation deposits around the Qasr el Sagha escarpment, north of the Birket Qarun lake near Faiyum in Egypt, indicate that this was a very large, flightless, and quite possibly predatory bird, probably as tall as a small emu or large rhea but bulkier in form. Its taxonomic position has incited much debate, and it has yet to be confidently allied with any existing avian lineage, but the enigmatic *Eremopezus* does possess certain interesting and quite specific anatomical similarities with the secretary bird—a highly distinctive African species that will feature a little later in this discussion of potential makalala identities.

Could *Eremopezus* itself, however, be linked to the latter mystery bird? It seems implausible that this species could have lingered on into the present day or given rise to modern-day descendants without some geographically intervening remains having been found somewhere between Egypt and Tanzania's portion of East Africa. Then again, as previously noted, the fossil record is famously incomplete.

The author with a life-sized model of the North American terror bird *Titanis walleri* (© Dr Karl Shuker)

With flagrant disregard for zoogeographical dictates, the makalala readily recalls the phorusrhacids or terror birds. These were an aptly-named taxonomic group of huge flesh-eating birds known predominantly (but not exclusively) from the New World, and which attained their awesome zenith with a truly gigantic, spectacular species from Argentina's Patagonia region called *Kelenken guillermoi*.

Sporting a massive 28-inch-skull armed with an enormous hooked beak, this 10-12-ft-tall horror died out approximately 15 million years ago during the mid-Miocene, whereas *Titanis walleri* (originally thought to have been 10-12 ft tall too until further finds led it to be downsized to a still-daunting 5-6 ft) not only reached North America but lived there in Texas and Florida until as least as recently as 2.5 million years ago, making it the youngest terror bird species currently known. However, these fearful birds were flightless, as their wings were vestigial. Moreover, although confirmed terror bird fossils have been discovered in the Americas and also Antarctica, the only known fossil evidence for their erstwhile existence in Africa is a single femur from an individual that had lived during the early or early-to-mid-Eocene (i.e. between 52 and 46 million years ago) in what is today southwestern Algeria. In 2011, this mysterious species was named *Lavocatavis africana*.

Even so, could the makalala be an undiscovered modern-day species? There is one notable precedent for such speculation, because some zoologists consider it plausible that a *living*, flying species of phorusrhacid-related bird is already known from Africa—namely, that strange, stork-like bird of prey called the secretary bird *Sagittarius serpentarius*. Although it is commonly classed as an aberrant accipitrid based upon molecular analyses, egg albumen comparisons have suggested in the past a closer taxonomic allegiance between this species and a pair of South American birds known as seriemas—which in turn constitute the last surviving members of a phorusrhacid-allied taxonomic family.

In any event, the secretary bird affords a compelling correspondence to the makalala's

19th-Century engraving of a secretary bird, from Charles Orbigny's *Dictionnaire Histoire Naturelle*

morphology (albeit on a rather more modest scale). Standing up to 4.5 ft tall on notably long, crane-like legs, and endowed with strong wings that support a powerful, soaring flight, plus the head and hooked beak of a bird of prey, the secretary bird constitutes a very acceptable makalala in miniature. Furthermore, when attacking snakes (an important part of its diet) it frequently shields itself from potentially fatal strikes with its outstretched wings, which are equipped with horny tips—i.e. claws on the tips of its 'finger bones' (phalanges), instantly recalling those of the makalala.

This last-mentioned correspondence is particularly telling, because there are very few species of bird alive today that are equipped with these wingtip claws. Indeed, other than the secretary bird, the only ones presently known are the three species of crane-allied birds called finfoots or sun-grebes, plus three vaguely grouse-like relative of waterfowl known as screamers, native to South America, and including the black-necked screamer *Chauna chavaria*, the cross-sectional shape of whose wing spurs is such that they are particularly noisy when clapped together. In addition, a strange pheasant-like bird known as the hoatzin *Opisthocomus hoazin*, again from South America, produces curiously reptile-like offspring able to crawl along tree branches by virtue of two large, mobile claws on each wing, but these are lost as the chicks mature. Over the years, the hoatzin has been classified with numerous different avian groups, including the galliforms, cuckoos, touracos, mousebirds, waders, sand-grouses, and many others, but it is currently deemed to represent the oldest living avian lineage, discrete from all others alive today.

Certain other birds, like the jacanas or lily-trotters, the spur-winged goose *Plectropterus gambensis*, the spur-winged plover *Vanellus spinosus*, and a pair of Antarctic endemics called sheathbills, possess horny spurs on their wings, used in combat. However, these are variously sited on the 'wrist bones' (carpals) or 'hand bones' (metacarpals), not on the finger tips.

Out of all of these species, moreover, only one—the secretary bird—is predominantly carnivorous. Could the makalala, therefore, be some form of extra-large secretary bird—not necessarily as tall as the Wasequas state (their fear of it could certainly have inflated their estimate of its height), but much bigger than today's single *known* species? If so, a suitable scientific name, based upon the morphological description for it given here, would be *Megasagittarius clamosus*—'the noisy, giant secretary bird'.

Staying with the secretary bird line of speculation, is it conceivable, alternatively, that the makalala was a false secretary bird, i.e. some other raptorial species, possibly another accipitrid, that had assumed via convergent evolution a form outwardly comparable to *Sagittarius*? Although this is another suggestion with no tangible evidence to support it directly, there is actually an interesting confirmed precedent for such an ostensibly unlikely premise.

In 1989, Drs Alan Feduccia and Michael R. Voorhies formally described a remarkable new species of North American fossil accipitrid from the late Miocene whose tarsometatarsal structure was nearly identical morphologically to that of the secretary bird. Indeed, the convergence was so striking that they christened this species *Apatosagittarius terrenus*, which translates as 'terrestrial false secretary bird', because they considered it likely that just like the true secretary bird, it had exhibited a predominantly terrestrial hunting lifestyle. In fact, it was only because the tarsometatarsus bore some attached phalanges whose structure was very different from those of the secretary bird that Feduccia and Voorhies were able to confirm that *Apatosagittarius* was not a true secretary bird, but was merely an anatomical impersonator.

Finally, a sizeable bird native to western Tanzania but possibly venturing eastward occasionally into the region supposedly inhabited by the makalala is the shoebill *Balaeniceps rex* (discussed earlier in this present chapter with regard to alleged living pterosaurs reported in Africa), which sports not only a huge hooked beak but also a very impressive 8.5-ft wingspan. However, the shoebill's wings do not possess horny tips, so it could not make the loud wing-claps characterising the makalala. In addition, being principally piscivorous it doesn't scavenge carcases, it is shy of humans, and as its overall appearance is so singular that it seems unlikely the Wasequa would confuse such an unmistakeable species with anything else or convert it into a much larger, quite different mystery bird, this would seem to rule out the shoebill from further consideration concerning the makalala—unless, of course, there is a still-undiscovered species of giant shoebill out there. . . ?

All of the lines of speculation discussed here —with identity contenders ranging from marabou storks, shoebills, and terror birds to secretary birds, false secretary birds, and even the anomalous *Eremopezus*—are certainly absorbing and thought-provoking, but even if any of them is valid, it is scarcely likely to yield a living makalala, sadly. After all, a bird as large and as visually distinctive as this one would surely be hard-pressed indeed to remain undiscovered by science for long, regardless of the geographical locality involved—yet there do not appear to be any post-19th-Century reports of its existence.

Consequently, even if the makalala was a reality in the 1870s, presumably it no longer survives—but that does not mean that its former existence cannot be verified. As noted earlier, among the valued possessions and relics of present-day Wasequas there may still be one or more of the revered helmets worn by long-departed chiefs. Should one of these tribal heirlooms pass into the hands of an ornithologist, the lucky recipient could well find himself holding a bona fide makalala skull!

GIANT BIRDS FROM THE TOMBS OF THE PHARAOHS

The phoenix may be the most famous feathered mystery associated with ancient Egypt—but it is not the only one. Equally mystifying is the bennu bird, which frequently appeared in the Books of the Dead—the hieroglyphic texts that accompanied the deceased in their coffins during the Fifth Dynasty (c.2466-2322 BC) at Heliopolis. Each text was a long roll of papyrus, summarising in picture writing the life of the deceased person alongside whose corpse it had been placed within the coffin. It also offered advice on gaining acceptance for entry into the next world, and revealed that upon acceptance the person would encounter the holy bennu bird, which would bear his or her soul to Ra, the supreme god.

Within the picture text, the bennu is unambiguously depicted as a gigantic heron—taller than a man, with long legs and pointed bill, a slender curved neck, and a pair of very elongate plumes on its head—resembling those of the Eurasian grey heron *Ardea cinerea*.

Archaeologist Dr Ella Hoch, from Copenhagen University's Geological Museum, became curious to discover the inspiration for the noble bennu. Perhaps it had originally been based upon the grey heron (which does exist in Egypt), but had been enlarged in the depictions to emphasise its sacred status and significant role? Alternatively, it might have been inspired by travellers' reports of the genuinely lofty goliath heron *A. goliath*, standing up to 5 ft tall, from the Arabian peninsula and sub-Saharan Africa. By far the most compelling possibility, however, had still to be disclosed.

Depiction of the bennu (GFDL Jeff Dahl)

In 1958, extensive archaeological studies began on the island of Umm an-Nar (aka Umm al-Nar), which lies adjacent to Abu Dhabi of the United Arab Emirates (UAE), in the lagoon complex off the Trucial Coast. Great quantities of animal remains were eventually unearthed, which Hoch documented during the late 1970s in her museum's *Contributions to Palaeontology* journal. To her surprise but delight, these included some fragments constituting the distal end of the left tibiotarsus (lower leg bone) of an enormous heron—one that was probably even taller than the goliath heron, the world's tallest living species. Indeed, some estimates give its height as having been up to almost 7 ft, i.e. taller than an average human, and with a wingspan approaching 9 ft. Other remains from this newly-revealed colossus were later found, some from separate sites in the Umm an-Nar settlement, and one fragment from Failaka Island, offshore from Kuwait.

The Umm an-Nar material dates from 2,600 BC to 2,000 BC (i.e. the third millennium BC), whereas the Kuwait bone is more recent, from c.1,800 BC, thereby collectively yielding a span of time that wholly encompasses the Fifth Dynasty of Egypt (c.2,494-2,345 BC)—during which period the bennu appeared in Books of the Dead illustrations. The model for the bennu may therefore have been this now-extinct giant heron—a possibility not lost upon Hoch, who christened its species *Ardea bennuides*.

However, there is also evidence to suggest an even more startling prospect—that this giant heron was still alive as recently as 200 years ago.

In an *American Journal of Science* report from 1845, its author, a Mr Bonomi, disclosed that between 1821 and 1823 traveller James Burton had chanced upon three enormous conical nests, all within the space of a mile, at a place called Gebel ez Zeit (aka Gebel Zeit), situated on the Red Sea's Egyptian coast, opposite the Mount Sinai peninsula. They had been constructed from all manner of materials, ranging from sticks, weeds, and fish bones to fragments from what had apparently been a very recent shipwreck, and which included a shoe, strands of woollen cloth, a silver watch, and the ribcage of a man—a victim of the shipwreck, whose remaining bones and tattered clothing were spotted by Burton a little further along the coast.

The nests were colossal in size, with an estimated height (and also a basal diameter) of about 15 ft, and an apical diameter of 2.5-3 ft.

Understandably, Burton was thoroughly bemused concerning their origin—until he began to question the local Arabs. They stated that the nests were those of a huge type of stork-like bird that had deserted the area not long before his arrival there.

Bennu depicted on ancient Egyptian papyrus

No such bird is known here today; but as his curiosity had been aroused by this episode, Bonomi undertook some investigations of his own, and uncovered persuasive, independent evidence in support of the Arabs' claim. Delving through the documents of early writers describing Egypt in the far-off days of the pharaohs, he came upon the description of a giant stork once native to the Nile Delta region, and which, in the form of a painted bas-relief, is sculptured upon the wall within the tomb of an officer belonging to the household of Pharaoh Cheops (also called Khufu or Shufu), from the Fourth Dynasty.

According to these sources, it was a bird of gregarious habits, with white plumage, long tail feathers, and a long straight bill; the male additionally bore a tuft on the back of its head, and another upon its breast. However, these features all more readily recall a heron, an egret or even certain cranes rather than a stork, and they compare well with representations of the bennu. Is it plausible, therefore, that this 'stork' was actually *Ardea bennuides*?

Whatever its taxonomic identity, however, at the time of Pharaoh Cheops (c.2,600 BC) it still survived in the vicinity of the Delta, because specimens were occasionally trapped by the region's peasantry. The nests spied by Burton were probably the product of several generations of these birds (thereby explaining their immense size) rather than just a single pair—and their most recent contributors may have been this species' last representatives.

How tragic, and how ironic, if the bennu, the bird that bore the souls of humans to Heaven, ultimately met its own death at the hand of humans.

SYLVIORNIS, AND THE DYNASTY OF THE DU

The following quote is from Rev. J.G. Wood's *Illustrated Natural History* (1863):

> [In Port Essington, northern Australia] great numbers of high and large mounds of earth exist, which were formerly thought to be the tombs of departed natives, and indeed, have been more than once figured as such. The natives, however, disclaimed the sepulchral character, saying that they were origins of life rather than emblems of death; for that they were the artificial ovens in which the eggs of the Jungle Fowl were laid, and which, by the heat that is always disengaged from decaying vegetable substances, preserved sufficient warmth to hatch the eggs.

The bird to which Wood was referring is the Australian jungle fowl *Megapodius freycinet tumulus*, belonging to a taxonomic family of extraordinary gallinaceous birds from south-east Asia and Australasia called megapodes—which are renowned for incubating their eggs not by sitting upon them like other birds, but by utilising various sources of external heat instead. Some megapodes, such as the Australian jungle fowl, erect huge mounds composed of dead plant material, into which they then deposit their eggs and allow the heat generated inside the mounds by the rotting vegetation to incubate them. Others excavate pits in the ground, within which they then bury their eggs in order for them to be hatched once again by whatever external sources of heat are present—including the sun's rays, subterranean steam, and even volcanic activity.

Today's megapodes are little more than dwarfs in comparison with certain fossil species. If, as currently believed, the ancestors of today's aboriginals reached Australia no later than 48,000-40,000 years ago, they would have been familiar with the Australian giant megapode *Progura* [=*Leipoa*] *gallinacea*. Although generally similar in overall shape, this very robust species was approximately half as large again as today's biggest megapodes. These are Australia's closely-related but only 2-ft-long mallee fowl *Leipoa ocellata* and the slightly larger brush turkey *Alectura lathami*—this latter vulture-headed species creates mounds up to 9 ft in height. Leg bones of an even larger, hitherto-unknown form of megapode have been discovered in a Polynesian midden on one of Fiji's tiniest islands, demonstrating that it too was clearly contemporary with early humans. Yet even these sizeable species pale into insignificance in comparison with a truly spectacular relative, recently exposed as a bona fide prehistoric survivor.

To the east of Australia is the island of New Caledonia, politically a special collectivity of France, whose southernmost tip looks out towards a very much smaller island nearby—the Isle of Pines. In 1976, the *Bulletin de la Société d'Etudes Historiques de Nouvelle Calédonie* published an extremely interesting paper by historian Dr Paul Griscelli, who revealed that according to oral traditions of the Houaïlou people (one of the isle's native Melanesian tribes), this minuscule speck of land once harboured an enormous and thoroughly extraordinary bird.

Artistic representation of the du (© Tim Morris)

Their ancestors called it the du, and described it as a huge, aggressive bird with red plumage and a star-shaped bony casque on its head. Although unable to fly, it could run very swiftly, usually with its wings outstretched. It laid a single egg, which took four months to hatch (from November's close to April's onset), but it did not incubate the egg itself.

As expected with any item of mythology, there were certain aspects that were unquestionably fictional. In particular, it was asserted that the du practised a most unlikely cuckoo-like deception to avoid brooding its egg—by laying it in a hollow banyan tree used as a lair by some form of giant lizard, in order for the lizard to incubate it instead!

In stark contrast to such tall tales as this, however, the fundamental, mainstream details regarding the du are noticeably precise and sober, indicating that these may have stemmed from sightings long ago of some real creature. Even so, bearing in mind that many fabulous birds of monstrous form with no claim whatsoever to a basis in reality do occur in numerous myths and folktales around the world, it is nonetheless very possible that the du would have been dismissed as nothing more than just another example of this latter type of fantasy animal, especially as there appeared to be no way of pursuing the matter further. Fortunately, however, the du's seemingly imminent descent into obscurity was not to be.

While preparing the final draft of his paper, Griscelli received some staggering news. Excavating fossils at Kanumera, on the Isle of Pines, in 1974 Paris researcher Dr Jean Marie Dubois had disinterred some bones that had since been identified by Prof. François Poplin of Paris's Natural History Museum as limb portions from a gigantic bird! Relics of the du? In addition, when the bones were dated by radiocarbon techniques, they were found to be no older than 3,500 years, and as humans were known to have reached the Isle of Pines prior to this period, they would have been well-acquainted with the species from which these bones had originated.

The dimensions of this vanished bird's remains have shown that it was as large as the modern-day emu (up to 6 ft tall), and flightless. Indeed, after studying the bones, in 1980 Poplin published a formal scientific description of this dramatic species (*Centre National de la Recherche Scientifique, Paris*), named by him *Sylviornis neocaledoniae*, in which he announced that it may have been a ratite. That is, one of those widely-dispersed species of giant flightless bird that include the ostrich, rheas, emus, and cassowaries, as well as the extinct moas and elephant birds.

Three years later, however, following further researches, Poplin changed his mind—announcing that *Sylviornis* was actually an immense megapode (*Comptes Rendus*, 1983). Moreover, as he and co-worker Cécile Mourer-Chauviré went on to discuss in another paper (*Géobios*, February 1985), there was an intriguing supplementary source of evidence favouring this identity, contained within the Houaïlou's corpus of mythology. For according to their oral traditions, the du did not incubate its egg. Unfortunately, however, if we exclude from serious consideration their lizard-hatching legend, their folklore offers no clue as to the resulting fate of the egg.

Nevertheless, as Mourer-Chauviré and Poplin pointed out just a few months later (*La Recherche*, September 1985), researchers seeking the answer to this riddle on the Isle of Pines might not have to look very far for it. For if their hypothesis is correct, the riddle's answer is readily visible in many parts of the isle, and also on New Caledonia.

Both of these islands bear a great number of very large mound-like structures, measuring up to 150 ft in diameter and as much as 15 ft in height. Constructed from materials present in their immediate surroundings (soil, particles of iron oxide, coral debris, sometimes black silicon as well), their precise nature has never been conclusively ascertained—despite a century of archaeological investigations. In the meantime, it has simply been assumed that they are ancient tumuli (burial mounds) erected by the islands' early human occupants. However, no human remains have been found within them. Hence Mourer-Chauviré and Poplin postulated that they may really be egg-incubator mounds of *Sylviornis*, the last visible evidence of this mighty species' former existence. If ultimately verified, this identification would not only solve a longstanding topographical mystery but also demonstrate that *Sylviornis* existed on New Caledonia itself, not just on the Isle of Pines.

Speaking of New Caledonia, it is possible that a much smaller megapode once hailed from this island too. When Captain James Cook's second voyage reached New Caledonia in 1774, one of its members recorded seeing a small, bare-legged form of gallinaceous bird that could well have been a megapode. Certainly, fossil bones of a comparably-sized megapode of relatively recent date (geologically speaking) have been unearthed here; their species has been named *Megapodius molestructor*.

Returning to the du/*Sylviornis* issue, there is no doubt that a mound-erecting bird the size of an emu would have been a stunning sight, and it is one that may have persisted for much longer than originally believed. According to Dr Jean-Christophe Balouet in his book *Extinct Species of the World* (1990), this species is now thought by some researchers to have survived into early historical times—until at least the 3rd Century AD (though more conservative estimates date its extinction at around 150 BC).

Even so, it had vanished long before its island domain was reached by Europeans—but why? After all, during his researches into Houaïlou mythology, Griscelli found that for at least a time the du had been deemed by their ancestors to be little less than an avian deity—a sacred entity of reverence. The punishment for killing a du was death, and an insignia in the shape of its casque was a symbol of great esteem, worn only by the most powerful of tribal chiefs—but, like all things, customs change, especially when eroded by changing priorities.

On as tiny an island as the Isle of Pines, food is scarcely plentiful, particularly for entire human tribes—not surprisingly, cannibalism was prevalent in those early days of human occupation. It could only be a matter of time, therefore, before the awe-inspired taboos protecting the du would be weakened and violated by the more practical realisation that these enormous, robust birds were actually an abundant, readily-accessible source of top-quality meat. And so it was that this colossal bird's once-exalted status plummeted precipitously, and prosaically—from the du as demi-god, to the du as dinner!

There could only be one outcome. Long before Westerners could ever receive the opportunity of witnessing this magnificent bird, *Sylviornis* had been exterminated. For many centuries it would survive merely as a curious native legend, its morphology and lifestyle preserved only by verbal transmission through successive generations of its annihilators' descendants—until the day when its bones would finally be disinterred, and the reality of the Isle of Pines' bygone feathered deity at last be confirmed beyond any shadow of doubt.

In 2005, Mourer-Chauviré and Balouet published a detailed update paper on *Sylviornis neocaledoniae* in which they reclassified this extinct species, deciding that it was sufficiently distinct from all other megapodes to warrant the creation of an entirely new taxonomic family to house it—Sylviornithidae. And in March 2016, a *PLoS ONE* paper authored by a team of researchers including Dr Trevor H. Worthy from Flinders University in Adelaide, South Australia, which presented the first extensive documentation of post-cranial *Sylviornis* remains, stated that this species lacked the megapodes' specialised digging adaptations, so was certainly not a megapode, but rather a stem-galliform. In consequence, the team deemed it improbable that *Sylviornis* created the mounds present on New Caledonia. Yet it has long been established that these edifices do not contain human remains, so it is clearly time that they were thoroughly examined to determine whether they contain any *Sylviornis* eggs or shell fragments. After all, someone—or something—created these tumuli; they did not arise spontaneously of their own accord . . .

From mythical du to extirpated *Sylviornis*—clearly not a resurrection . . . or is it? Tantalisingly, while visiting New Caledonia in 1991, Danish zoologist Lars Thomas discovered that the native people speak of the du as if it were still alive today, and describe it accurately. Tenacious racial memories—or *Sylviornis* survival in some remote New Caledonian outback?

AND FINALLY . . .

Should *Sylviornis* be found to exist still, this would indisputably be a major revelation—as would the discovery in living form of any of the other winged mystery beasts documented in this chapter. So keep your eyes on the skies—perhaps one day an ardent birdwatcher will succeed in adding a teratorn (or even a pterosaur?) to their life list!

CHAPTER 3

Monsters From the Ancient Waters—Living Leviathans of Sea and Lake

Leviathan that crooked serpent . . . the dragon that is in the sea.

THE HOLY BIBLE—*ISAIAH* (27:1)

The famous Lough Ness, so much discours'd for the supposed floating island, for here it is, if anywhere in Scotland.

RICHARD FRANCK—*NORTHERN MEMOIRS* (1658)

Judging from the enormous number of unclassifiable sea creatures reputedly spied over the years, the ocean depths must surely be among the most plausible places on this planet to harbour undiscovered prehistoric survivors—should any such creatures truly exist. By far the most (in)famous marine monster on record is the 'great sea serpent'—for despite being reported from the earliest times and from all around the world, its very existence, let alone its likely zoological identity, has incited heated discussion and dissension within the scientific community for centuries. Yet with a bulging dossier containing hundreds of separate eyewitness accounts, many from highly responsible and respected observers, there is clearly a case to answer.

Those zoologists willing to accept the sea serpent's reality agree that more than one species is involved, because the eyewitness descriptions are far too disparate for any single species to explain them. This scenario is equally applicable to the subject of lake monsters—unidentified beasts of many forms reported from inland bodies of freshwater worldwide.

Based upon the multi-identity theory, several different sea serpent classification systems have been devised by researchers down through the years, of which the most famous is that of Dr Bernard Heuvelmans, which was first published in his book *Le Grand Serpent-de-Mer* (1965), and again three years later in the English-language version *In the Wake of the Sea-Serpents*. (Incidentally, this latter tome was not a direct translation of the former, but rather a revised adaptation, and also included a greatly-abridged version of another of his French-language books, dealing with the giant squid and colossal octopus.)

In his grand scheme of sea serpent classification, Heuvelmans conceived no fewer than

Artistic representation of the nine sea serpent types postulated by Dr Heuvelmans (© Tim Morris)

nine distinct categories. Each constituted a different, scientifically-undiscovered species, and which (after discarding a sizeable number of accounts as unreliable or as outright hoaxes) he believed collectively explained all of the major sea serpent sightings reported from around the world down through history.

These hypothesised species were: a giant yellow tadpole-like creature of indeterminate taxonomic affinities; a gigantic 'super eel' (and/or a very elongate shark form); a marine reptile resembling a prehistoric mosasaur or a flippered crocodilian; an immense sea turtle; a many-humped serpentine zeuglodont (or zeuglodont-like) cetacean; an armoured anomaly that he considered to be another zeuglodont due to his mistaken belief that armoured zeuglodonts were known from the fossil record (see later); an exceedingly primitive stem cetacean of superficially otter-like form but much greater size and still possessing four limbs (his so-called 'super-otter'); and two separate types of pinniped, both of which were either tailless or near-tailless, like all modern-day species.

One of these pinnipeds, with a shorter neck, huge eyes, and a very noticeable mane, was dubbed by him the merhorse. The other, which combined the body and limbs of a typical otariid or eared seal (i.e. fur seals and sea-lions, possessing external ears) with an exceedingly long, giraffe-proportioned neck, he dubbed the long-necked (nowadays shortened to long-neck or longneck), and proposed for it the binomial name *Megalotaria longicollis* ('long-necked big otariid'). (In their 2003 book *The Field Guide to Lake Monsters, Sea Serpents, and Other Mystery Denizens of the*

Traditional restoration of plesiosaurs in life

Deep, veteran American cryptozoologists/Fortean writers Loren Coleman and Patrick Huyghe merged the merhorse and long-neck into a single sea serpent type, which they dubbed the waterhorse.)

Sea serpent classification systems notwithstanding, a sizeable proportion of reports describing ostensibly mysterious water monsters most probably constitute unusual or mistaken observations of known creatures—otters or sea lions swimming in a line, certain rarely-spied species of whale or fish, colonial invertebrates of larvacean or thaliacean (salp) identity with which few non-specialists would be familiar, and various non-animate phenomena such as waterspouts, floating algal mats, seiches (standing waves moving back and forth between two boundaries in enclosed water basins such as lakes), or temperature-generated optical illusions.

Sometimes, however, the creatures involved appear strikingly similar to certain animals officially believed to have died out millions of years ago. As these particular water monsters may therefore be prehistoric survivors, it is with them that this chapter is concerned, and they represent several (but not all) of Heuvelmans's hypothesised sea serpent categories.

PLESIOSAURS AND PERISCOPES

[Caveat lector: As explained in this book's new introduction, the publisher's brief for the original edition was for me to concentrate specifically upon those proposed cryptid identities that involve postulated prehistoric survivors. In the case of long-neck water monsters, the principal such identity is of course the plesiosaur, so this is the identity focused upon here.

As the principal alternative identity, a postulated long-necked seal famously suggested by Heuvelmans, is not a prehistoric survivor per se, far less discussion concerning it is included here, but it does receive extensive coverage via a major chapter devoted entirely to it in my recent book *Here's Nessie!* (2016), which should be consulted by anyone interested in this particular line of zoological speculation.]

The most famous of all prehistoric aquatic beasts are undoubtedly the plesiosaurs, which, although dominating the seas during the Mesozoic Era to much the same extent as the dinosaurs ruled the land, were not themselves dinosaurs (despite their frequent media appellation of 'aquatic dinosaurs'). They were instead the predominant members of an entirely separate reptilian superorder, the sauropterygians. During an early stage in its evolution, the plesiosaur lineage diverged to yield two morphologically distinct groups—the pliosaurs, with short necks and long heads, and the 'true' plesiosaurs, with long necks and short heads. The pliosaurs are dealt with in their own section later in this chapter; so for the purposes of convenience, from here onwards until the pliosaur section is reached, it should be understood that all mentions of plesiosaurs refer specifically to the true plesiosaurs (unless otherwise stated).

A surviving species of plesiosaur is unquestionably the most popular identity for one of the major morphological categories of water monster—generally termed the long-neck, whose basic outline as seen above the water surface is often likened by eyewitnesses to a periscope or umbrella handle. Such creatures include the following selection.

Morgawr and Other Long-Necks of Sea and Shore

A typical example of a marine long-neck was seen at close range one sunny day in August 1910 by Orkney wildfowler W.J. Hutchinson, his father, and his cousin, while sailing to a series of exposed rocks in Meil Bay called the Skerries of Work. To their surprise, a shoal of whales suddenly leapt out of the water, fleeing the area in such uncharacteristic haste that the men were worried that the resulting disturbance would capsize their boat. Once the whales had departed, Hutchinson's father, who was steering, looked ahead, hoping to catch sight of the Skerries, and spied something else instead—something so extraordinary that all thoughts of wildfowling were instantly forgotten. As Hutchinson later recalled:

> I heard him say, 'My God boys, what's that?', pointing ahead. I looked up and saw a creature standing straight up out of the sea—with a snake-like neck and head, like a horse or camel!

Hutchinson was so frightened by this apparition that he reached for his gun and pointed it at the creature's head, but his father wisely prevented him from shooting. Instead, they sat quietly and watched it, from a distance of less than 100-150 yards. They estimated the length of its neck above the surface to be at least 18 ft. Dark brown in colour with lighter bands, it widened gradually towards the water surface until it was roughly as thick as a man's body, but in overall appearance it was so slender that the dark-coloured head perched on top seemed disproportionately large. After about five minutes, the beast slowly submerged—disappearing vertically beneath the water, but so gently that it made no splash at all.

A sea monster swimming with its head and tall vertical neck above the water surface like an animated umbrella handle was observed through binoculars on 20 April 1920 by Thomas Muir—third officer aboard the Royal Mail

Steam Packet Company's ship, *Tyne*, travelling near St Paul's Rocks (500 miles off the coast of South America) towards the end of its journey from Casablanca to Rio de Janeiro. According to Muir, the beast turned and looked at the ship, then moved closer until it was about 400 yards away. Still watching the ship, it cruised alongside on a parallel course for about five minutes, its head raised 30-35 ft above the surface, before curving its neck down like a swan and diving out of sight under the waves.

At around 4.30 pm on 31 October 1922, an unidentified animal was spied in the Gulf of Mannar between India and Sri Lanka by various crew members aboard the steamer *Bali*. According to the fourth officer, P. Kruyt, its head and neck recalled those of a giraffe but were even larger. The neck did not taper but was of uniform thickness throughout, which he estimated to be about 18 in, and the head terminated in a blunt point. The portion of its neck that emerged above the water surface was roughly 15 ft long and grey-green in colour. The creature remained visible for about two minutes before diving head first back into the water.

On 28 November 2008, Kent-based mystery beast researcher Neil Arnold passed on to me a remarkable, previously-undocumented long-neck sighting from the late 1950s, with kind permission from its source to publish it (which I duly did shortly afterwards, originally in one of my Alien Zoo cryptozoology news columns for *Fortean Times*). Arnold's source was a correspondent called Nick (whose email address I have on file), and one of the eyewitnesses in question was his father. Nick was anxious that the sighting be preserved, so here once again is his account of it as written by him in 2008:

> I am 39 years old with three children and my father is now 72. He was one of the last people to have been conscripted to national service, serving as navigator on RAF Shackleton aircraft, in 1957, flying missions over Europe and much of South America. The official job of the Shackleton was as a spotting aircraft, and as such the pilot and crew were highly trained in this respect; their job was to search for submarines and relay the positions back to base. One of the tell-tale signs they used to look for high above was a dark shape underwater, which could often be a shark or whale or other creature but occasionally they would get lucky and find a submarine near to the surface. On one occasion, having seen such a dark shape near the surface, they flew down to investigate, and as they flew by this shape, every crew member on board all saw the same thing—the neck and head of a large sea creature protruding from the surface. The nearest thing they could think of at the time was a plesiosaur. They were all in agreement but of course the pilot on the way home forbade them absolutely from talking about it; if they had done so they would all be immediately grounded on returning home and accused of being drunk on duty. So anyway, my father decided to research what the creature might have been, and found that it was almost definitely an 'elasmosaurus' [long-necked plesiosaur], as the one thing that was very obvious was the huge length of the animal's neck. My father was forced to conclude, and of course he'd read all about the coelacanth, that there were indeed much larger 'extinct' creatures out there, undiscovered to science. When I reached adulthood he was to tell me of the incident. Twenty years on, he still will

swear to his grave that this is what they all saw on the flight. . . The year was 1957, the location was "about 500 miles north of the Canary Islands, in open sea". The RAF Squadron was 228 Squadron, flying out of St Eval in Cornwall. (This air base has since been closed but I believe remnants of the site remain.) I don't know who the other crew members were and I have been thinking about making an official enquiry to the RAF archives. However what I hope is important is the location (vague, sorry), species and date. I am more than aware that most scientists consider the possibility of any large dinosaur [sic] still alive as astronomically small, and my father himself also thinks now with sadness that he might have seen one of the last ones; his own reasoning is that at this point, not ever so long after the war, there really was not much shipping around; indeed, the area where they had the sighting was entirely devoid of shipping and maybe this was the reason the creature had considered it safe to surface, far from the throb of ship's propellers.

One zoologist who read Nick's account suggested that perhaps they had seen a tentacle from a giant squid rise above the water surface, but Nick's father totally discounted any such possibility.

One of the most intriguing long-neck reports is that of the *Fly* sea serpent—due to the exceptionally good sighting of the creature afforded by the transparent quality of the water through which it was allegedly seen by its eyewitness.

To quote Edward Newman (*The Zoologist*, February 1849):

Artistic rendition of an encounter with a cryptid like the *Fly* sea serpent (© William M. Rebsamen)

> Captain the Hon. George Hope states that, when in H.M.S. *Fly*, in the gulf of California, the sea being perfectly calm and transparent, he saw at the bottom a large marine animal, with the head and general figure of the alligator, except that the neck was much longer, and that instead of legs the creature had four large flappers, somewhat like those of turtles, the anterior pair being larger than the posterior: the creature was distinctly visible, and all its movements could be observed with ease: it appeared to be pursuing its prey at the bottom of the sea: its movements were somewhat serpentine, and an appearance of annulations or ring-like divisions of the body was distinctly perceptible.

A sketch of the creature also included a long tapering tail.

In his book *The Case for the Sea-Serpent* (1930), Lt-Commander Rupert T. Gould rightly lamented the absence in this report of such fundamental information as the time, date, and a more precise location relative to the alleged sea serpent encounter (as he also noted, the Gulf of California occupies an area of some 40,000 square miles), as well as the fact that no official account of Hope's sighting appeared in the *Fly*'s log.

Even so, Gould did concede that the logs of certain maritime vessels from which multiple (as opposed to single) eyewitness sightings of sea serpents had occurred—and which had even been reported formally to the Admiralty—did not contain such accounts either, so this is not as unusual as it might otherwise seem. In fact, sightings like these were often not logged simply because the crew feared that they would not be believed and may even be reprimanded by officialdom for making such entries. It is a great tragedy that such a detailed sighting as Hope's was not substantiated by the presence of other eyewitnesses, but, to quote Gould: "With all its defects, the story is too interesting to omit".

The crystal-clear waters off Aegina, one of the Saronic Islands in Greece's Saronic Gulf, were the setting for a sighting of what may well have been a young long-neck underwater. Occurring in 1907 (*not* 1928, as often claimed—see below), it was made by none other than the creator of Sherlock Holmes and The Lost World—Sir Arthur Conan Doyle (and also his wife). He described their sighting as follows in his book *Memories and Adventures* (1924):

> We were steaming past Aegina on a lovely day with calm water around us. The captain, a courteous Italian, had allowed us to go upon the bridge, and we—my wife and I—were looking down into the transparent depths when we both clearly saw a creature which has never, so far as I know, been described by Science. It was exactly like a young ichthyosaurus [sic], about 4 feet long, with thin neck and tail, and four marked side-flippers. The ship had passed it before we could call any other observer.

Needless to say, Conan Doyle had muddled his Mesozoic sea reptiles—from his description, he clearly meant a plesiosaur, not an ichthyosaur, which was short-necked and very fish-like in superficial outward form—but that is not the only muddle connected with this case, as I discovered when investigating it.

In his sea serpent book, Heuvelmans stated that Conan Doyle's sighting had occurred in 1928, but as it had already appeared in 1924 within the latter's above-cited book, he was clearly incorrect. Moreover, he then attributed to Conan Doyle a quote that had appeared in Harold T. Wilkins's book *Secret Cities of Old South America* (1952), and which described the creature that he and his wife and seen, but which was fundamentally different from the description given in Conan Doyle's own book. Here is the supposed Conan Doyle quote that appeared in Wilkins's book:

> I saw, swimming parallel to the ship, under water, a curious creature around four feet long with a long neck and large flippers. I believe, as did my wife who also saw it, that it was a young plesiosaurus.

Wilkins stated that it was in 1929 when Conan Doyle (who died in 1930) had written the above lines (in which he had amended his identification of the mystery animal from an ichthyosaur to a plesiosaur) but did not give a published source for them. As for this sighting's correct date: in a July 1996 *Fate* article,

American Fortean researcher Mark Chorvinsky revealed that on 5 May 1969, Conan Doyle's son, Adrian Conan Doyle, wrote to Heuvelmans correcting his sea serpent book's 1928 date for it, stating that his parents had in fact witnessed the beast in 1907. Dates notwithstanding, however, it remains a noteworthy long-neck sighting, especially in view of the main observer.

Yet another very valuable sighting of a long-neck seen in full while underwater took place in or around July 1965 (not 1969, as erroneously stated in some publications and websites). This was when the U.S. research submarine *Alvin*, captained by Bill Rainnie and piloted by Marvin McCamis, was submerged at a depth of almost 1 mile in the Tongue of the Ocean off the Bahamas, surveying the underwater listening array *Artemis*. Suddenly, they spied an extraordinary creature, which possessed a long neck, a somewhat snake-like head, and two sets of flippers that propelled its thick body. Before the submarine's cameras could be activated, however, the creature swiftly ascended, swam away, and was not seen by them again. Nevertheless, in this particular case the sighting *was* duly entered into the craft's log, and both men not only were experienced submarine observers, with hundreds of dives to their credit, but also went on to receive a number of commendations for expertise and bravery in their field, thus cementing their status as persons whose word could be trusted as being reliable and credible.

Artistic rendition of the *Alvin* submarine crew's underwater sighting of a plesiosaur-like long-neck (© Thomas Finley)

Moreover, American cryptozoologist Scott Mardis corresponded with McCamis regarding this remarkable sighting (which is how its correct year was confirmed) prior to McCamis's death in 2004, and sent him a number of illustrations, including plesiosaur reconstructions. In reply, McCamis stated that these resembled the mystery beast that he and Rainnie had seen on that eventful occasion back in 1965. Having observed this creature in its entirety underwater (as opposed to a mere head-and-neck surface sighting), and as their description of its morphology is undeniably very reminiscent of a plesiosaur, their report remains one of the most compelling ones in favour of the reality of a marine cryptid potentially allied to such creatures.

A particularly famous—and infamous—marine long-neck is Morgawr (Cornish for 'sea giant'), the frequently-sighted sea monster of Falmouth Bay in Cornwall, England. Reports of such a creature date back many years, but Morgawr did not make headline news until 5 March 1976. This was when the *Falmouth Packet* newspaper published two extraordinary photos sent in by a pseudonymous reader who signed her attached letter 'Mary F'.

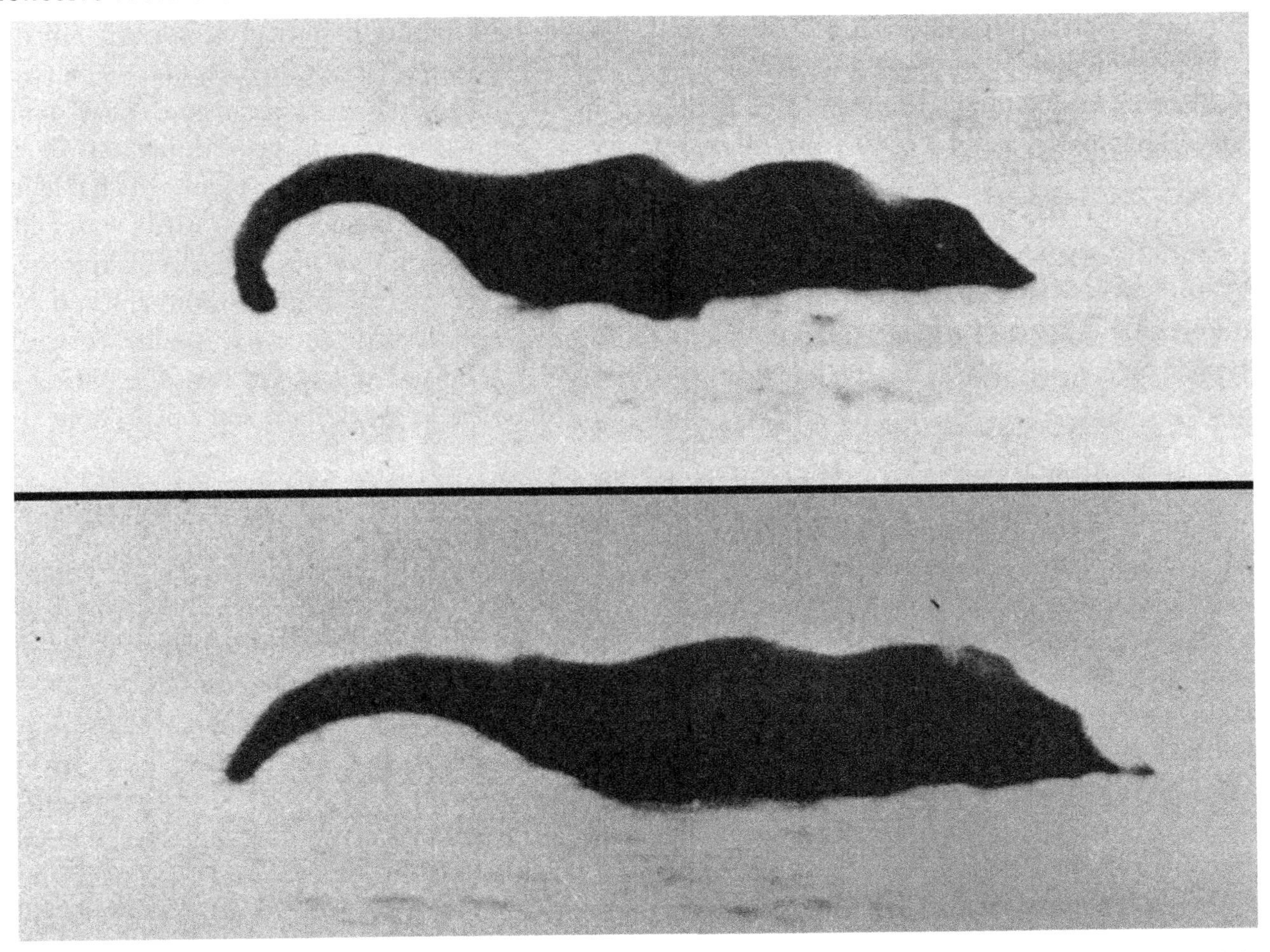

Morgawr, the Falmouth Bay sea serpent, as photographed by 'Mary F' (© Fortean Picture Library)

The photos depicted what seemed to be a dark, bulky sea monster with humps along its back and a long curving neck bearing a very small head at its tip; no background details were visible. In her letter, published alongside the photos, 'Mary F' claimed that they were of a sea monster that she had spied about three weeks earlier from Trefusis, Falmouth, and that the absence of background detail was due to the sun shining into the camera and a haze upon the water. The portion of the beast visible above the sea's surface was 15-18 ft long, and was black or very dark brown, but had only remained there for a few seconds before submerging. She described it as looking:

> . . . like an elephant waving its trunk, but the trunk was a long neck with a small head on the end, like a snake's head. It had humps on the back which moved in a funny way . . . and the skin seemed to be like a sealion's. . . As a matter of fact the animal frightened me. I would not like to see it any closer. I do not like the way it moved when swimming.

For many years, these remained the most spectacular, if enigmatic, of all sea serpent photos, especially as the identity of 'Mary F' was never formally uncovered. In 1990, however, after an exhaustive investigation of these and certain other mysterious water monster photos, Maryland's renowned Fortean researcher Mark Chorvinsky, a university-trained expert in cinematic and other photographic special effects, expressed serious doubts as to their authenticity. Moreover, in a detailed published account (*Strange Magazine*, fall 1991), he considered that the cryptic 'Mary F' might actually be Tony 'Doc' Shiels, an Irish mystic and self-confessed trickster.

Irrespective of these photos' source and nature, there are also many reports from independent eyewitnesses attesting to the spasmodic occurrence off the Cornish coast of a classic long-neck sea monster. In January 1976, for instance, Truro dental technician Duncan Viner had been walking along Falmouth's coastal path observing sea birds when he saw what he assumed at first to be a whale—a dark hump rising up through the water. As he watched, however, a long neck suddenly appeared, turned around, then sank back beneath the waves. The creature seemed to be 30-40 ft long, and was several hundred yards off the beach.

Later that same month, London holidaymaker Amelia Johnson was walking near Rosemullion Head when she saw emerging from Falmouth Bay's waters ". . . a sort of prehistoric dinosaur thing, with a neck the length of a lamp-post". And a very detailed Morgawr sighting was made on 10 July 1985, off Portscatho, by Dorset writer Sheila Bird and her scientist brother Dr Eric Bird, visiting her from Australia's Melbourne University.

It was about 8 o'clock in the evening, and they were relaxing on the cliff top to the west of Portscatho, when Dr Bird leapt to his feet in amazement—having spied an extraordinary creature propelling itself smoothly and rapidly in an easterly direction just offshore. Estimated at 17-20 ft in length, with grey, slightly mottled skin (ascertained with the help of two passers-by with binoculars), it had a long neck, small head, and large hump protruding out of the water, as well as a long muscular tail visible just beneath the surface. The two Birds watched it for several minutes, its head held high as it swam gracefully through the water—then suddenly, without any warning, it sank vertically beneath the water and vanished, creating no disturbance to mark its former presence.

Lakes, Lost Worlds, and Long-Necks in South America

Freshwater long-necks have been reported from well over a hundred different lakes throughout the world, and include the following examples among their number.

The most publicised lake-dwelling long-neck from South America was the so-called Patagonian plesiosaur, which at the height of its fame even received coverage in the august journal *Scientific American*. In January 1922, Dr Clementi Onelli, Director of Buenos Aires Zoo in Argentina, received a letter from a Texan adventurer called Martin Sheffield, who had spent a number of years as an itinerant prospector living off the land in Patagonia. In his letter, Sheffield claimed that some nights previously, after pitching his hunting camp close to a mountain lake near Esquel, he had encountered a strange animal:

> . . . in the middle of the lake, I saw the head of an animal. At first sight it was like some unknown species of swan, but swirls in the water made me think its body must resemble a crocodile's.

Not surprisingly, Sheffield's description conjured up images of plesiosaurs in Onelli's mind, and also reminded him of a somewhat earlier report. In 1897, he had spoken to a farmer living on the shores of Patagonia's White Lake, who informed him that a strange noise was frequently heard there at night, resembling the sound that a cart would make if dragged over the lake's pebbly shore—but that was not all. On moonlight nights, a huge beast could be seen in the lake, with a long reptilian neck that would rise high above the water, unless disturbed—whereupon it would instantly dive and disappear into the depths.

Heartened by these and other reports, Onelli organised an expedition to follow them up, which duly set forth on 23 March 1922, led by José Cihagi, superintendent of Buenos Aires Zoo. It eventually reached the lake where Sheffield had experienced his sighting, but with the approach of winter further explorations were abandoned and the expedition returned to Buenos Aires. Interestingly, Sheffield had also previously contacted former American president Theodore Roosevelt concerning the swan-necked beast that he had seen in the mountain lake. As a result of this, Roosevelt, who was famed for his hunting skills, had apparently pondered over whether to launch a search for it himself, but he never actually did so, and he died in 1919, three years before Onelli's expedition set out.

And so it was that apart from a jaunty tango entitled *El Plesiosaurio* (composed in 1922 by Rafael D'Agostino, with lyrics by Amilcar Morbidelli, and sheet music depicting on its cover a caricature of Onelli riding a plesiosaur) plus a brand of cigarettes also named after it, nothing else of note emerged regarding the putative plesiosaur of Patagonia for many years—until the 1980s. Since then, however, numerous reports have been aired by the media concerning a similar water beast, nicknamed Nahuelito, which is said to inhabit Nahuel Huapi, a 204-square-mile Argentinian

Sheet music cover for the *El Plesiosaurio* tango

lake ensconced amid the Andes winter-sport resort of Bariloche.

Auyan-tepui is a lofty tepui (table mountain) in Venezuela, one of the inspirations for Sir Arthur Conan Doyle's classic cryptozoological novel *The Lost World*, populated by dinosaurs, pterosaurs, plesiosaurs, and other prehistoric survivors. In 1955, however, during an expedition to this isolated plateau, naturalist Alexander Laime allegedly sighted some creatures that gave the more optimistic zoologists reason for believing that the theme of Conan Doyle's novel may not be wholly fictitious after all. While searching for diamonds in one of the rivers at the summit of Auyan-tepui, he spied three very strange beasts sunbathing on a rocky ledge above the water. Superficially seal-like, closer observation revealed that they had reptilian faces with disproportionately long necks, and two pairs of scaly flippers. Drawings that he made of them at that time are reminiscent of plesiosaurs. There is, however, one very unexpected feature—none of them was more than 3 ft long.

Could they have been young specimens? Laime believed that they were adults, but belonging to some pygmy species of plesiosaur, whose small size has enabled it to persist into the present day without disturbing the ecological balance of this enclosed system. More conservative opinions favour some long-necked type of otter as a more plausible identity, whereas others have even likened them to a crocodile.

In 1990, Auyan-tepui played host to an expedition led by biologist Fabian Michelangeli and including scientific reporter Uwe George, for whom this was his sixth exploration of a South American tepui. During their visit, Michelangeli and his brother Armando spied a silhouette of a beast closely resembling those reported by Laime, but as they drew nearer to investigate, the beast plunged into the river and disappeared from view. As for various German TV reports claiming that one had actually been captured, these were inspired by the procurement of nothing more spectacular than a common species of lizard.

'Giant Newts' of Siberia

One of the most famous and readily-recognisable cryptozoological images is the much-published sketch of Siberia's Lake Khaiyr (aka Khainyr) monster prepared by its principal eyewitness, Nikolai Gladkikh, back in 1964, during Moscow State University's Northeastern Expedition (surveying mineral deposits), as led by Dr G. Rukosuyev. It was published later that same year in a *Komsomolskaya Pravda* report that also contained an interview with Rukosuyev.

Since then, this very striking picture has appeared in numerous cryptozoological works, my own included. Due to an extraordinary revelation, however, published once again by *Pravda* (6 August 2007), the whole episode featuring the creature so depicted has been exposed in a new and very shocking light, as will now be explained. But first, permit me to set the scene: In the original edition of this present book of mine, what was assumed at that time to be the truth concerning the Lake Khaiyr

Nikolai Gladkikh's illustration of the Lake Khaiyr monster

monster sighting was concisely documented by me as follows:

> One of the most distinctive long-necks ever reported was spied in 1964 by Soviet biologist Dr Nikolai Gladkikh, while visiting Lake Khaiyr, a remote body of water in the east Siberian province of Yakutia and largely ignored by science until then. Standing at the lakeside one early morning, Gladkikh caught sight of a very big and extraordinary beast on the opposite shore [apparently there to eat the grass, according to some versions of this sighting]. Its huge body was bluish-black in colour, with two pairs of ill-defined limbs, a long sturdy tail and an elongate gleaming neck carrying a small head at its tip. Its most eye-catching feature, however, was the low, triangular dorsal fin, containing vertical rays, that ran from the base of its neck to the beginning of its haunches.
>
> By the time Gladkikh had alerted his colleagues to come and see it, this unidentified mystery beast had vanished, but he was able to prepare a good sketch. Before the end of their visit the expedition's leader and two other members were able to confirm the sketch's accuracy, when they briefly saw the head and dorsal fin of the monster appear above the surface at the centre of the lake, and watched its long tail furiously lashing the water. Its dorsal fin sets this animal well apart from any species currently known to science, but it is not the only time that such a feature has been reported for long-necks.
>
> In 1942, two Soviet pilots surveying this lake reported seeing two bizarre animals that were later likened to giant newts. As some newts, exemplified by males of the familiar *Triturus cristatus* [great-crested newt], bear a conspicuous dorsal crest, this lends support to Gladkikh's story.

The 6 August 2007 *Pravda* report, conversely, tersely stripped his story of all support.

To begin with, this radical reassessment of the Lake Khaiyr monster episode claimed that far from being a Soviet biologist, Gladkikh was nothing more than a migrant worker hired by the expedition. Moreover, it alleged that in addition to confirming Gladkikh's account of the monster by way of his own sighting, team leader Rukosuyev had made some unusual extra allegations—asserting that there were no fishes in the lake, that birds never landed on its surface, and that the locals often heard "muffled sounds and splashes of water" coming from the lake.

An expedition was duly dispatched to Lake Khaiyr to investigate all of these claims (a fact not generally mentioned in Western reports of this case), and it swiftly refuted all of them. Yes, there were fishes in the lake, and yes, birds did land upon it. Most significant of all, however, was this new expedition's discovery that none of the locals had ever seen any strange creatures in it. Moreover, despite combing the bottom of this lake (which, incidentally, is actually a very shallow body of water, only about 23 ft deep), scuba divers dispatched by the expedition found nothing unusual.

Consequently, the expedition had what the *Pravda* report described as "a heart-to-heart talk" with Gladkikh, who confessed that he had invented the entire story but was unable to explain clearly why he had done so. Accordingly, *Pravda* concluded: "He concocted it either to entertain himself and his friends or as an excuse for shirking his duties at work".

Yet in spite of *Pravda*'s ostensibly decisive dismissal of Gladkikh's testimony, and, in turn, the whole monster encounter, there are some glaring omissions to consider. What about Gladkikh's colleagues who also saw the creature—if any of them are still alive, why weren't they interviewed by the new expedition, especially Rukosuyev? And if they are dead, surely that should have been noted in the *Pravda* report, if only to avoid the very questions being posed here? Ditto for the two Soviet pilots from 1942.

Nevertheless, the damage has been done—sufficient doubt has been cast upon this whole episode by *Pravda* for Gladkikh's Lake Khaiyr monster sketch to be viewed forever more with scepticism by all but the most trusting of investigators. Exit another item of tantalising mystique from the panoply of mystery beast memorabilia previously assumed to constitute valid cryptozoological evidence untainted by the shameful shadow of fraud.

Notwithstanding this outcome, however, also on record from Yakutia is a long-necked, snake-headed monster reputedly inhabiting Lake Labynkyr. According to Anatoly Pankov, a chronicler of mysterious happenings in this part of the world, the creature supposedly raised its neck above the lake's surface in full view of a team of geologists sometime during the 1950s, and lunged upwards to seize a flying bird in its jaws while watched by a number of astonished reindeer hunters.

Storsjöodjuret—Sweden's Big-Eared Water-Cow

A conspicuous tourist attraction at the mountain town of Östersund in central Sweden is a monument (nowadays daubed with graffiti, unfortunately) depicting the storsjöodjuret—the long-neck of Lake Storsjön, which is the deepest lake in Sweden and has a surface area of 176 square miles. With sightings dating back to 1635, extensively reported during the 19th Century, studied by zoologist Dr Peter Olsson, and spasmodically occupying the cryptozoological news headlines in more recent times too, it is one of several aquatic monsters referred to by cryptozoologist Ivan T. Sanderson as Swedish water-cows.

According to Sanderson's description, the Storsjön example is ". . . very large with greenish looking shiny skin, a long thin neck at least 10 feet long, small head, the usual bumps, and a capacity for rushing headlong about the lake at the speed of a motor-boat, leaving a huge wake, and making sudden sharp turns." Its eyes, moreover, are said to be very large, and it has two pairs of large flipper-like limbs.

Graffiti-defaced statue of Lake Storsjön monster at Östersund, Sweden (© Lars Thomas)

One of the most peculiar characteristics of this long-neck is its 'ears'—a somewhat inadequate name for the pair of strange head-borne projections reported by several eyewitnesses. One investigator, the then curator of Östersund's Jämtlands läns Museum, wrote in 1965 to Loch Ness monster researcher Tim Dinsdale, describing the storsjöodjuret's 'ears' as ". . . great fins of the back or of the head, possibly ears, described as little sails, which can be laid tight on to the neck. These fins or ears have an estimated diameter of 1 metre [3 ft]."

They would therefore be extremely large ears! More likely is that they are exposed portions of a much larger dorsal crest, whose remaining portions are hidden beneath the water surface.

In 1894, an attempt was planned by a local company to capture the storsjöodjuret, using a very large ovoid spring trap (measuring 5 ft by 3 ft) baited with piglets, and watched by a Norwegian harpooner, but nothing transpired. Today, the catching tools are on display in the Jämtlands läns Museum. In summer 1987, a much more scientific (but, alas, no more successful) hunt for the monster took place, launched by this same museum.

Champ-ion Long-Neck of the U.S.A.

Looming large on the long-neck front in North America is Champ, the monster of Lake Champlain—an immense body of freshwater 109 miles long, up to 11 miles wide, and a water surface area of 440 square miles, straddling the borders of Vermont and New York, but also nudging its northernmost tip into Quebec, Canada. Reports of a huge horse-headed monster (the chaousarou) inhabiting this massive lake stretch back into Amerindian lore, and it has been spied for decades by local Westerners, but only since the 1960s has it attracted international interest. Those sightings featuring a head and neck as well as the obligatory hump(s) account for about 37 per cent of all Champ sightings, according to veteran Champ investigator Joseph W. Zarzynski during the 1980s.

One afternoon in 1976, for example, while at Treadwell Bay, north of Plattsburgh in New York, at a distance of 300 yards Orville Wells saw what he later called ". . . a huge animal or whatever swimming very steadily and slowly across our bay. . . . It appeared as something out of the past or a prehistoric monster". According to his description, it was 20 ft long, and brownish in colour, with a long neck, upright head, and two humps.

Just after 7.00 pm on 14 June 1983 and at a distance of 550 ft, Tim and Dick Noel had a sighting of Champ off Vermont's North Hero Island. Once again, a long neck was clearly spied, followed by at least two (possibly three) humps, as the dark-coloured animal moved through the water. They estimated its total length at 20-25 ft, and its height at 3-4 ft.

And on 27 August 1983, at a distance of 900 ft, Eva Gauvin and a friend caught sight of Champ from Camp Marycrest, Grand Isle, Vermont. Before it submerged, they were able to discern a long neck, two humps, and an estimated total length of 20-25 ft.

The most important Champ sighting to date, however, must surely be that of Sandra Mansi and her husband Anthony, which took place just after mid-day on 5 July 1977 somewhere around St Albans, Vermont, at a range of 100-150 ft. Believed by them to be 15-20 ft long, with a head and neck raised 6-8 ft above the water surface and very dark, fairly smooth-looking skin, what makes this particular appearance so important is that it was actually photographed—Sandra Mansi successfully snapping a colour Polaroid of it before it disappeared again. Although far from lucid, it had captured the image of what does appear to be the head, upright neck, and back of some form of large, brown-coloured creature.

Prof. Paul H. LeBlond, from the University of British Columbia's Department of Oceanography, is a renowned expert in the field of wave dynamics, and after carefully studying the

Sketch of Sandra Mansi's Champ photograph (© Thomas Finley)

Mansi photo in an attempt to gauge the likely dimensions of the object depicted in it he announced in 1982 that the lower and upper limits for its water-line dimension range from 16 ft to 57 ft 4 in. Moreover, we must remember that this size range is only for the *photographed* portion of the object—it does not include whatever portion remained invisible beneath the water surface. Consequently, whatever it is, it is certainly very large.

In September 1993, a Japanese team was on site at Lake Champlain, recording a TV documentary and cruising the lake armed with video cameras and sonar equipment, but nothing emerged—except, that is, for some new publicity for a very unusual Champ identity. According to an article detailing the team's search in *USA Today* (8 September 1993), some Champ believers are contemplating the credentials of *Tanystropheus longobardicus*—a superficially lizard-like, *terrestrial* reptile from Central Europe's mid-Triassic Period (245-228 million years ago). However, its disproportionately long neck is the only feature that allies it to Champ sightings. Moreover, whereas Champ's neck is apparently quite flexible, *Tanystropheus*'s neck was extremely rigid, because although this was over 9 ft long (longer than the combined body and tail length), it only contained 12-13 hyper-elongate, rod-like vertebrae. Nevertheless, the *Tanystropheus* identity contender has inspired the coining of *Champtanystropheus americansus* as a potential taxonomic name for Champ (should it exist).

One of the most frequently-cited of the early Champ sightings was made in July 1609, and by none other than the lake's European discoverer—Samuel de Champlain. According to a report by historian Marjorie L. Porter (*Vermont Life*, summer 1970), de Champlain had described the beast as about 20 ft long, serpent-like but as thick as a barrel, with a horse-like head. In other words, a description that compares well with those given by modern-day eyewitnesses, and which has since been

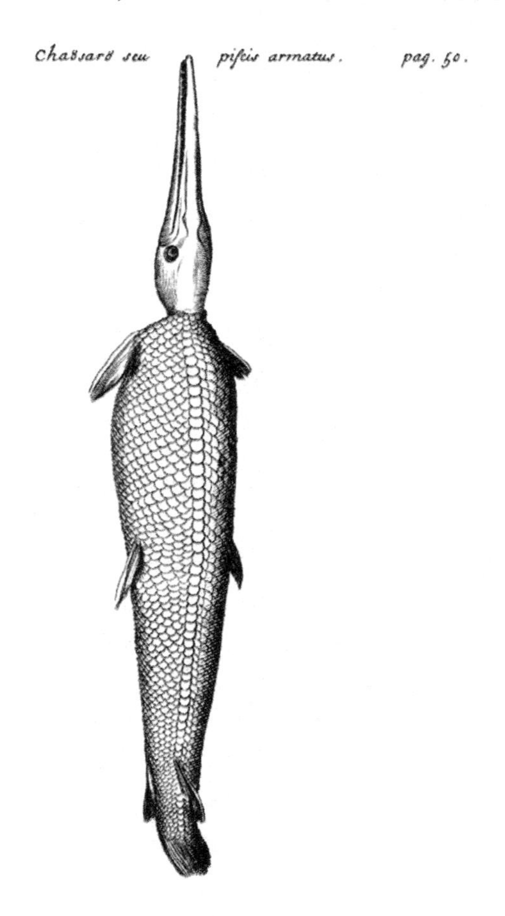

Chaousarou, from *Historiae Canadensis*

Tanystropheus statue (© Dr Karl Shuker)

requoted in many other publications. More recently, however, a very different account of his sighting has emerged—contained in no less authoritative a source than *The Works of de Champlain* (edited by H.P. Biggar). According to this, which contains de Champlain's own words, he saw some chaousarou:

> . . . about five feet long, which were as big as my thigh, and had a head as large as my two fists, with a snout two feet and a half long, and a double row of very sharp, dangerous teeth.

Whatever these were, they bore no resemblance, and hence no relationship, to Champ. From de Champlain's description, they were most probably specimens of a voracious, 4-6-ft-long fish called the longnose garpike *Lepisosteus* (aka *Lepidosteus*) *osseus*—a taxonomically significant creature, as it belongs to a group of archaic, armour-plated fishes called ganoids, which mostly died out many millions of years ago. Only a few modern-day species are known, all of which inhabit North American rivers and lakes. Moreover, an illustration of a chaousarou in François Du Cheux's *Historiae Canadensis* (1664) leaves no doubt whatsoever that it is indeed a longnose garpike.

How ironic for one of the most celebrated Champ sightings to have featured a bona fide prehistoric survivor—but the wrong one!

Issie—A Lesson in Long-Neck Deception

Nowadays, the image of the long-neck is so familiar in the context of water monsters that,

Statue of Issie near Lake Ikeda, Japan, depicting it as a long-neck (© Shin-ichiro Namiki/Fortean Picture Library)

outside cryptozoological circles, it has become something of a stereotype—applied to all such creatures regardless of whether or not this action is justified by the eyewitness accounts on file. Issie, the monster of Japan's Lake Ikeda, is a case in point.

Sightings of this water beast generally feature at least two black humps moving swiftly through the water, often accompanied by noticeable water disturbance—as with the 60-90-ft-long black object observed by about 20 people on 3 September 1978; the multi-humped shape photographed by a Mr Matsubara on 16 December 1978; the 15-30-ft black form videoed during the early afternoon of 21 October 1990 by Kazuo Kawano; and what would seem to be two separate Issies filmed together on videotape by Hideaki Tomiyasu on 4 January 1991. A long neck, however, is conspicuously absent from such reports. Yet a statue of Issie erected near the lake on Kyushu depicts the monster as a dragonesque long-neck—more akin to a certain Caledonian loch monster than to anything reported from Ikeda.

Bessie and Baby Erie

A similar identity crisis to that of Issie is also suffered by Ohio's Bessie, the monster of Lake Erie in the Great Lakes. Although its many sightings generally describe an extremely long, decidedly serpentiform creature, Bessie is often represented in artwork and other visual portrayals as a plesiosaurian long-neck. Perhaps the most intriguing of these portrayals is Baby Erie.

On 12 November 1998 I was emailed by Canada-based field cryptozoologist Bill Gibbons, who had received some potentially electrifying news, which he then proceeded to tell me about in confidence (and in which it stayed until after the news broke). In or around 1992, an Ohio-based taxidermist had apparently found dead on the shore of Lake Erie an extraordinary creature, 3.5 ft long, that resembled a miniature plesiosaur, complete with a long neck, two pairs of flippers, and a long tail that bore a serrated dorsal fin and also a small terminal caudal fin.

Not surprisingly, he had taken its carcase back home with him, and after removing its soft internal tissues had preserved it as a taxiderm specimen. He had then placed it in his shop's window, where it was seen and purchased by collector Dr Carl E. Baugh. Gibbons promised to send me a video of this potentially stupendous find, but when I received and watched it I could see that the creature was simply some form of fish that had been artificially modified and moulded into a plesiosaurian form.

Three major clues alerted me to this fact straight away. A large pouch-like structure was visible just behind the base of the creature's jaws, which some cryptozoologists who had also seen the video were calling a gullet extension, but which I knew full well to be an operculum—the large gill-cover possessed by actinopterygian (ray-finned) fishes and certain others. The rayed structure of its long dorsal fin and its caudal fin was also characteristically actinopterygian in form; and it bore a readily-perceived lateral-line sensory system running horizontally along its flank from head to tail—a familiar feature of fishes but not of any reptile.

Once the specimen's existence was officially made public a little later, it became popularly known as Baby Erie, and was thought by some mystery beast investigators to be a genuine plesiosaur—in spite of its unequivocal, diagnostic fish characteristics. It was also examined both externally and via CAT scans to determine its internal structure at various scientific institutions. Oddly, however, the results were mostly inconclusive, although they did establish that its flippers were fake, and that it was a single animal, rather than the kind of cleverly-constructed composite gaff incorporating parts from more than one type of animal that are so often exhibited as curiosities at sideshows (the so-called Feejee mermaids are prime examples of such gaffs, usually combining the body of a large fish with the head, arms, and thorax of a monkey).

During the mid-2000s, this enigmatic specimen was investigated by the *Paluxy* website's owner, Glen J. Kuban, who presented a detailed report of his findings on his site. He had successfully traced and spoken by phone to the taxidermist responsible for finding and preserving it (who proved to be Larry 'Pete' Petersen, of L & D Bait and Tackle in Lakewood, just outside Cleveland, Ohio), as a result of which the true nature of Loch Erie's 'baby plesiosaur' was finally exposed. Quoting from Kuban's report:

> I telephoned Petersen, whose shop happened be not far from where I lived at the time. As I inquired about the "monster" Petersen related that he did indeed find the carcass washed up on the shore of Lake Erie, and that he sold it to Baugh during Baugh's visit. Petersen was surprisingly candid about the manner in which he had processed the carcass. He said that when he found it, it was already decaying and had been "pecked at" by birds, but was evidently some kind of fish, with a hook still in its mouth. He said that he decided to stuff it and fashion it into a sea-serpent like creature as an attention-getting display for an upcoming taxidermy trade show. He said it had a long fin along the rear part of the body, which he notched into triangular shapes to make it appear more dragon like. He also bent and

The "Baby Erie" specimen with taxidermist Larry Petersen, courtesy Dr. Ben Scripture

When Dr. Carl Baugh acquired the specimen in 1998, it was soon examined by Dr. Ben Scripture, of Warsaw, IN, who notes: "This was my evaluation: the taxidermist did a so-so job on trying to make it look like some odd unique creature. It had pelvic paddles instead of fins, however, they were very obviously fake—made of a kind of composite/plastic material, and attached rather poorly. The dorsal fin which ran the whole length of its body was also obviously trimmed to make it look undulating instead of a straight edge. Doing some research which included being allowed to look at specimens at the Chicago Museum of Natural History, I determined it was an American eel. They are common in Lake Erie, and can get up to 3 feet or more in length. I did try to get enough tissue to extract some DNA to do PCR, however, the tanning chemicals made it very unlikely that I would be able to obtain any useable DNA, and honestly I didn't give it a lot of effort as it was obvious to me that the specimen was not authentic in the way it was presented/tampered with, and pretty much seemed to be an eel."

> sewed the neck into an S-shape to foster the same impression, and finally, sewed little pieces of skin to form little flippers. . . I asked if he knew what species of fish it was. Despite being a taxidermist, Petersen said that he was unable to identify it.

Baby Erie's precise taxonomic identity remains unconfirmed. However, Kuban feels that it is probably a burbot *Lota lota*.

Up to 4 ft in length, the burbot is a long-bodied freshwater relative of the cod, is somewhat eel-like in outward appearance, and is already known to exist as a not-uncommon fish in Lake Erie. After comparing photos of burbots and Baby Erie, I think it likely that this identity is correct, because the two do correspond very closely (allowing for the modifications wrought upon the latter specimen by Petersen during his transforming of it into a mock long-neck). Sadly, however, we will never know for certain, because Baby Erie was recently destroyed in a fire that swept through Dr Baugh's office where it was kept.

A Long-Neck Called Nessie

The most famous mystery beast of *any* type, and the archetypal long-neck lake monster, is undeniably the elusive creature allegedly inhabiting Loch Ness, which is situated on the Great Glen Fault that runs southwards from Inverness to Fort William in the Scottish Highlands. Nessie's distinguished history supposedly dates back at least to the 6th Century AD, when a mysterious water beast was encountered by St Columba (though in the River Ness, not in Loch Ness, and at about 580 AD, as revealed by Celtic history expert Prof. Charles Thomas in a *Cryptozoology* paper, rather than 565 AD, as erroneously given in many other works).

Since 1933, which saw the creation of a new motoring road (the A82) overlooking the northern shoreline of this immense but hitherto-secluded lake, Loch Ness has hosted countless sightings of what many people believe to be huge but unremittingly reclusive water beasts with long necks, small heads, and humped backs. And prior to the A82's arrival, there had always been local beliefs in the existence within the loch of strange, sometimes frightening animals generally referred to as kelpies or water-horses.

Unless they are not resident here but migrate to and fro from the sea via the River Ness, and/or unless they habitually 'loch-hop' from one lake to another (both of which possibilities are by no means unlikely), if long-necks do occur in Loch Ness on a long-term basis there must be a viable population.

Loch Ness is approximately 23 miles long, up to 1.5 miles wide in places, and has a maximum claimed depth of 813 ft recorded at a spot called Edwards' Deep (although an unconfirmed sonar reading obtained by Loch Ness tour-boat operator Keith Stewart using 3-d sonar equipment and made public in January 2016 claimed a depth of 889 ft within a supposed, hitherto-unreported loch-bottom trench sited about 9 miles east of Inverness). It also boasts a volume of about 263 billion cubic feet (Scotland's biggest loch by volume, and containing more water than all of the lakes in England and Wales combined), a surface area of 22 square miles (Scotland's second-largest loch by surface area, beaten only by Loch Lomond), and a diverse biota with substantial quantities of fish readily able to sustain such a population, despite what critics and sceptics might claim (see Nessie investigator Roland Watson's thought-provoking calculations at http://lochnessmystery.blogspot.co.uk/2012/02/is-there-enough-food-for-nessie_12.html for more details). Consequently, problems of adequate space and food are easily dismissed—unlike the riddle of these beasts' identity.

As the eventful history of the Loch Ness monster (or LNM for short) has been thoroughly examined in many published full-length Nessie studies, this necessarily brief coverage will concentrate upon those aspects of the beast most relevant to its identification.

The most popular image of Nessie (or a Niseag in Scottish Gaelic, which apparently translates as 'pure') is that of a long slender neck and one or more humps rising briefly up above the loch's surface, thereby assigning it to the long-neck category of water monster—and which, with minor variations, has been reported by numerous eyewitnesses over the years, as shown by the following selection of sightings. (Over 1,000 alleged Nessie sightings have been recorded in total.)

On 12 July 1934, naval officer Captain F.E.D. Haslefoot was walking along a disused railway line leading from Fort Augustus to Inchnacardoch Point when he saw two oval black humps in the loch, moving slowly towards the shore. They were roughly 8 ft apart, and 1 ft or so above the water surface, but as he watched they suddenly turned and travelled away up the loch for about 200 yards at a speed of about 10 knots, one in front of the other with the anterior hump generating a feather-like wash, before slowing down again to around 3 knots. As he watched, a head and neck emerged in front of one of the two humps, black in colour and roughly 4 ft out of the water. Haslefoot described the head as being spade-shaped, but only saw it for a moment before the creature flicked it sideways back down into the water. After swimming leisurely until about 50 yards from the loch's shore, it submerged.

In October 1936, from a spot about 3 miles outside Foyers, Marjory Moir and some friends enjoyed a 14-minute view of a long-necked beast with three humps, which was apparently resting on the water surface. It was estimated by them to be approximately 30 ft long, with humps of differing sizes (the middle one was largest, the anterior one smallest) and a slender neck whose head had no discernible features and frequently dipped back and forth into the water. Then, without warning, it sped away towards Urquhart Bay, throwing up a huge wash, before returning to its original resting site.

A view out over Loch Ness

One day in early August 1946, Mrs Norah Atkinson and her husband, holidaying in Scotland, were taking a drive northwards on the A82 alongside Loch Ness. Abruptly, one of their car's tyres developed a puncture, so they pulled up for Mr Atkinson to change the tyre. While he did so, his wife gazed across the tranquil splendour of the loch, but it didn't remain tranquil for very long. To quote the description that she gave in 1959 to British zoologist Dr Maurice Burton:

> Suddenly, there was a terrific upheaval of water and up came a long swan-like neck with a small head. Then the whole

> body appeared, elephant grey in colour, two humps and very long and powerful. . . . We both watched until it disappeared underneath the water just below the edge of the bank. What a terrifying sight it was.

She had been so amazed by this sight that she had called to her husband to come over and see it too, which he did, both of them watching the creature in awe until it submerged. Indeed, they were so shocked by what they had observed that they abandoned their holiday forthwith and drove back home to Inverness with all speed to tell their friends all about their extraordinary encounter. Although he had become famously sceptical of Nessie after initially considering the likelihood of its reality quite favourably, Burton was so impressed by Mrs Atkinson's testimony that he included it in his book, *The Elusive Monster* (1961).

On 11 May 1962, from the window of a cottage at Alltsaigh, Edith Christie spied a single-humped long-neck with an egg-shaped head about 160 yards away, generating a tremendous wash as it swam rapidly northward up the loch at the speed of a fast motor boat. Its neck was about 4-5 ft long, and a 2-ft-long greenish-black hump followed about 6 ft behind it.

During a 12-minute sighting by Fort Augustus resident Katherine Robertson and her friend, a Lutherian nun visiting from Germany, two of these creatures were seen together, and for part of that time they were no more than 300 yards away. The date was 18 August 1971, and the two women were walking along the road behind Bolum Bay, when Robertson spied an extremely large animal swimming near to the shore, but away from the Abbey, on their left. The two friends observed it crossing the bay to the loch's far side, and Robertson estimated it to be about 45 ft long, with two low humps, and a small squarish head perched upon an erect, swan-like neck. As it was progressing across the bay, Robertson's friend spotted a second, smaller long-neck, also swimming from the vicinity of the Abbey and travelling quite speedily in the direction of the larger one until it finally caught up with it on the opposite side. This second animal had the same form of head and neck as the larger one, but only one hump was visible, and the creature's length was estimated to be a mere 15 ft or so. After a few moments together, the two beasts submerged.

Sceptics regularly dismiss the LNM on the grounds that there is no tangible, physical evidence for such a creature's existence, evidence that could be subjected to formal scientific examination in order to determine its originator's taxonomic identity. On one very notable occasion, however, some such evidence was indeed obtained, nothing less, in fact, than sizeable samples of flesh from an apparent Nessie—only for them to be carelessly thrown away! Here's what happened.

In 1978, a holiday cruiser owned by truck driver Stan Roberts, rented out to a family with an elderly grandfather, collided heavily with a substantial unknown object while sailing on Loch Ness near Urquhart Castle. As later recalled by Roberts in a *Daily Record* interview:

> The propeller stopped turning. The family were very alarmed. The old man had a heart attack and seemed to have died. There was no radio on board so they let off distress flares to get a tow back to Fort Augustus. The grandfather was taken by ambulance to hospital where he was found to be dead.

Roberts was duly informed by the rental managers. However:

> They simply told me there had been an accident. It was only later that I learned more—what had been found on the underside of the boat when they pulled it out of the water.

Boatyard workers who examined the cruiser had found:

> . . . flesh and black skin an inch thick along the propshaft. [However,] the workers chiseled the flesh away and threw it into the Caledonian Canal. I said you stupid b—s. It would have proved that Nessie was here.

Indeed it might. Certainly, to quote Adrian Shine of the Loch Ness Project when told of this incident:

> Very frustrating. With modern DNA techniques we could have learned a lot about exactly what had caused the damage.

In fact, this was quite possibly the single greatest lost opportunity in the entire LNM history to conduct a direct scientific examination of Nessie, because there is no *known* animal species resident in the loch that is big enough to have caused such a collision.

One morning in January 1980, while the loch's waters were totally still, fisherman Donald MacKinnon viewed a long-neck for almost 10 minutes with binoculars, noting three humps: ". . . and a long neck which reminded me of the submarine periscopes I'd seen in the war".

A surprising number of reports are on file concerning mysterious, unidentified submarines appearing in lakes and stretches of ocean throughout the world, whose presence has been made known only via brief glimpses of their periscopes projecting up above the water surface. Scandinavian fjords seem particularly prone to such appearances, and there has been much discussion regarding unexplained appearances and disappearances during World War II of craft assumed (but never proved) to have been periscope-bearing U-boats.

Two very different lines of thought have been aired as to their identity. One is that they are spy vessels, covertly observing while remaining largely unobserved themselves—except for their tell-tale periscopes. The other is that they are not submarines at all—but rather that they are long-necks, which makes MacKinnon's description particularly noteworthy.

Reports of long-necks in Loch Ness and re Nessie in general continue to the present day, as does international interest in this cryptozoological mega-star.

In June 1993, for example, Edna MacInnes and David McKay claimed to have watched a huge beast with light-brown body and giraffe-like head and neck for 10 minutes before it dived out of sight.

Seven years later, 'Loch Ness 2000' was a major new exhibition based at Drumnadrochit, dealing with the 'official' natural history of this famous lake, but inevitably featuring Nessie too. Designed by veteran Ness naturalist Adrian Shine, 'Loch Ness 2000' was formally opened in June 1999 by renowned explorer Sir Ranulph Fiennes.

During late March 2000, Swedish explorer Jan-Ove Sundberg led a week-long GUST (Global Underwater Search Team) survey at Loch Ness, dubbed 'Nessie 2000', using hydrophones day and night in the hope of recording sounds of the cryptozoological kind within its deep waters. The hydrophones had a range of 5 miles in ideal conditions, and unlike sonar equipment, which might disturb any creatures that may exist here, they do not create any disturbance when recording. The outcome of this survey was that some mysterious sounds were indeed recorded, including a series of pig-like grunting noises whose frequency of 741-751 Hz was comparable to sounds produced by various very big, known aquatic species such as killer whales, walruses, and elephant seals.

July 2001 marked the tenth anniversary of Nessie seeker Steve Feltham's vigil at the lochside of Loch Ness. For the past decade he had diligently albeit unsuccessfully scoured the loch's waters for a sign of its legendary inhabitant, and had lived throughout those ten years in nothing more luxurious than an old library van. However, he hoped to swap it for a double-decker bus, converting the bottom deck into a Loch Ness exhibition, and was already planning his next decade here. (And at the time of writing, May 2016, he's still there.)

In a British survey conducted during autumn 2006, over 2,000 adults across the UK were asked to give their opinion as to Scotland's most famous figure, present or past. Obviously, the surveyors expected someone like Sean Connery, Robert Burns, Ewan McGregor, William Wallace, or Robert the Bruce to top the poll. But no, for although those names did indeed appear in the list they were all trounced by a most unexpected name—Nessie, the Loch Ness monster! One can only assume that a fair few of those questioned had cryptozoological leanings!

Nevertheless, due to the dearth in recent years of noteworthy reports, Nessie seekers were beginning to fear that Scotland's most famous figure was no more—always assuming, of course, that she had ever existed to begin with! In December 2015, however, Gary Campbell, Keeper of the Official Loch Ness Monster Sightings Register, was delighted to announce that 2015 had been a "vintage" year for such reports. Needless to say, many claimed sightings *are* reported each year, but as Campbell points out, the vast majority can be readily explained and are therefore eliminated from consideration for addition to the register, but in 2015 the register had formally accepted no fewer than five as being unexplained by normal phenomena, the most for 13 years.

As for photos of Loch Ness long-necks, even the most striking examples have failed to sustain universal acceptance. In December 1975, a team from the Academy of Applied Science (based at Concord, New Hampshire), led by its founder Dr Robert H. Rines, released two underwater photos obtained during their research at the loch in June 1975 that incited a media sensation. One seemed to show the long neck and body of a plesiosaur-like beast, and the other resembled a close-up of a horned head (this was dubbed the 'gargoyle' photograph).

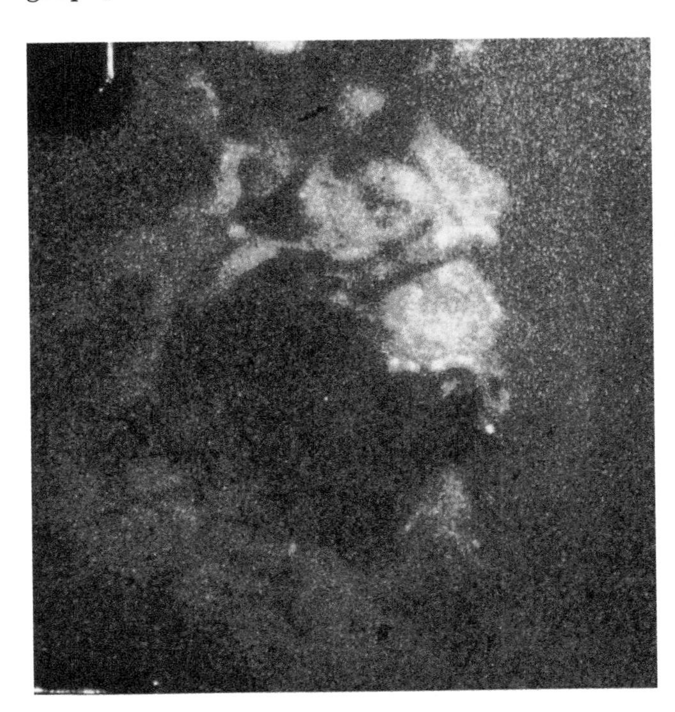

The 'gargoyle' photograph (© Dr Robert H. Rines/Academy of Applied Science)

The 'Surgeon's Photograph' (Fortean Picture Library)

Some LNM authorities, including Nicholas Witchell, later opined that the 'gargoyle' may simply be a decayed tree stump that was located and eventually dredged up in October 1987 from the area of the loch bottom where the photo was taken. Leading on from this are suggestions from some researchers that the 'body and neck' photo merely depicts sediment swirling up from the loch bottom around the edge of the 'gargoyle' tree stump, whose features must have been concealed by shadow. Yet in the opinion of various others, such 'solutions' as these appear more contrived than convincing.

The greatest controversy, however, is reserved for the Loch Ness long-neck photo *par excellence*—the 'Surgeon's Photograph', whose image of a head and slender neck protruding up through the water is everyone's immediate mental picture of Nessie whenever the subject of Loch Ness is raised. The picture was one of two that Lt-Col. Robert K. Wilson (1899-1969), a London gynaecologist, claimed to have snapped after he had seen something on the loch from a spot about 2 miles north of Invermoriston during a car journey along the A82 on the morning of 21 April 1934.

Sceptics have variously attempted to dismiss the photo's 'head-and-neck' image as that of a bird, or the tail of a diving otter, or even as the dorsal fin of an out-of-place killer whale or a giant sturgeon. During the 1990s, however, two rival allegations were made that sought to expose the photo as a blatant hoax—but as these allegations are mutually exclusive, only one (if any) can be correct; and, to be perfectly frank, I am by no means persuaded by the claims for either of them!

In 1992, the Danish weekly magazine *Hjemmet* published an article in which musicologist Prof. Lambert Wilson, a former conductor of the Aberdeen Symphony Orchestra, claimed that he had constructed a 'sea serpent' swimming mask, featuring the famous head and neck configuration in the 'Surgeon's photo', with tiny eyeholes enabling him to see, which he had worn while swimming in the loch on a day in late summer 1934. According to his story, this is what the London surgeon, coincidentally sharing the professor's surname, had seen and photographed. However, all other accounts concerning the taking of this photo placed the date in question in April, not during the summer. In any case, the very concept of someone swimming around in the loch with the model of a Nessie perched on top of their head is one for which I personally would be most reluctant to stick my neck out (so to speak!).

The second allegation received widespread publicity during March 1994, when researchers David Martin and Alastair Boyd announced that, only shortly before he died, in November 1993, expert model-maker Christian Spurling had confessed to them that the photo's image was actually that of a 1-ft-tall 'head-and-neck' model—constructed by him from plastic wood in January 1934, then attached to a clockwork toy submarine. This was secretly photographed (and afterwards sunk) in the loch. Surgeon Wilson (apparently fond of practical jokes, and supplied with full details concerning the model) was then recruited to publicise the photos. Martin and Boyd subsequently documented their investigations that had led them to put forward this startling claim in an absorbing book, *Nessie: The Surgeon's Photo Exposed* (1999).

Of especial note at this point is that a long-overlooked article appearing in London's *Sunday Telegraph* newspaper back on 7 December 1975 and written by 'Mandrake' (the pen-name of one Philip Purser) had actually claimed that Spurling's stepbrother Ian Wetherell had admitted to 'Mandrake' that he, while in the company of his father Marmaduke ('Duke') Wetherell and an insurance broker friend called Maurice Chambers, had taken some photos of a small toy submarine bearing the 'head-and-neck' made of rubber tubing in an inlet of the loch before sinking it, with one of these photos becoming the Surgeon's photo. Remarkably, however, this potentially significant article signally failed to be mentioned in any Nessie-related publication for almost two decades, which is a notable mystery in itself. More about this article a little later. Meanwhile, both of the Wetherells (and Wilson too) were dead by the time that Spurling made his own alleged confession, so they could not be questioned about it following its public release.

Spurling claimed that he had been requested to produce this model by his stepfather Duke and stepbrother Ian (who, Spurling alleged, also recruited Wilson as the fourth member of their conspiracy)—in order to avenge the recent public humiliation suffered by Duke when he rashly identified as genuine some footprints found on the shores of Loch Ness that were soon afterwards exposed as crude fakes. They had been produced by someone using an ashtray—not an umbrella stand, as claimed in many other publications—made from the foot of a hippopotamus! (Bearing in mind, however, that the Wetherell family actually owned such an object, it is by no means unlikely that Duke himself was the footprint hoaxer, but had not anticipated the ridicule that would rebound upon him.) As far as the quartet of hoaxers were concerned, there could be no better way of seeking retribution than to discredit the monster itself.

Sadly, the media's coverage in 1994 and since then of this supposed hoax was largely uncritical (indeed, it is still widely if mistakenly assumed to have been conclusively proved). So it was left to some astute newspaper readers to point out its array of inconsistencies, which can be summarised as follows.

The type of clockwork toy submarine available at that time could not have supported such an unwieldy structure as a 1-ft-tall head-and-neck model without being in serious danger of overbalancing; the only way to counter this would have been to place ballast inside the submarine—which would have promptly sent it plunging down beneath the water! True, in fairly recent times an ITN film crew and a Japanese film crew have successfully floated what they considered to be replicas of Spurling's 'head-and-neck' submarine. But as no-one has ever seen the latter toy (always assuming of course that it did actually exist!), having to rely instead entirely upon Spurling's verbal description of it and a picture of a toy submarine shown to him by Martin and Boyd that he claimed to be similar to the one that he had used, such accomplishments by the two teams clearly are by no means as impressive as they might outwardly seem to be.

The photo shows the head-and-neck surrounded by ripples. Yet this is not to be expected if the craft were moving at the time of being photographed.

The 'head-and-neck' actually looks to be rather more than 1 ft tall. Indeed, based upon a comparison of the length of adjacent wind waves (with the wavelength estimated from modern results on wind waves and contemporaneous weather information), oceanographers Prof. Paul LeBlond and Dr Michael Collins calculated in a *Cryptozoology* paper of 1987 that the neck's height above the water level was 4 ft.

Also, how can the 'model submarine' identity explain the second 'Surgeon's photo', whose 'head-and-neck' image has a very different outline from that of the first photo? Two contrasting explanations have been proposed—either the entity photographed was indeed alive, and had moved its position and orientation between the taking of the two photos; or the second photo was not depicting the same entity as the first, famous one (and may not even have been taken at the same time and/or location). Unfortunately, there seems no way of determining which (if either) of these explanations is the correct one.

Moreover, there are various noteworthy discrepancies between Ian Wetherell's confession as contained in the 'Mandrake' article of December 1975 and Spurling's own confession, which do not appear to have been highlighted anywhere before. According to Wetherell, the hoaxers were himself, Duke, and Chambers, whereas according to Spurling they were himself, Wetherell, Duke, and Wilson. So what had happened to Chambers? In the article, Wetherell is quoted as saying: "I took about five shots with the Leica", but if so, where are the others? Did they not come out? Only the Surgeon's photo and the second photo are known. Also in the article, the head-and-neck is said by Wetherell to have been made from rubber tubing, whereas Spurling claimed that it was made from plastic wood. Wetherell also claimed that the toy submarine was "only a few inches high", whereas Spurling stated that it was 1 ft high. How can these discrepancies be explained—a succession of memory lapses, or the failure of two brothers to synchronise their stories?

In addition, I am always very wary of 'death bed' confessions. Why wait so long, especially as the 'Mandrake' article had already been published way back in 1975 anyway? Even the faking of one of the world's most mysterious photos is not a crime (and releasing a supposed confession to having done so certainly hadn't troubled Ian Wetherell). Indeed, if anything, the release of the 'truth' while the hoaxers still lived might well have guaranteed them instant fame and lucrative financial gain (or at least it might have done for Wetherell had the article containing it not instantly sunk into absolute oblivion for two decades, by which time he had died).

A final curiosity is that in their book, Martin and Boyd also put forward as 'proof' that the Surgeon's photo was faked a very odd claim made in letters written during November 1970 by a Major Egginton, one of Wilson's friends, to Nessie researcher-author Nicholas Witchell. Egginton alleged that Wilson had told him that this photo had been faked by a keen amateur photographer friend of his (presumably Ian Wetherell), who had first of all snapped a photo of the loch and then, once back home, had superimposed a LNM model (presumably Spurling's?) onto the photographic plate. Martin and Boyd have claimed that these letters, retained by Witchell, constitute the most important evidence against the Surgeon's photo that has ever been uncovered, but in reality it is only anecdotal. In addition, few photographs have ever received such in-depth scrutiny as the Surgeon's photo, so if it had indeed been created by superimposing one image upon another back in the 1930s, when photographic techniques were far less sophisticated than they are today, I feel sure that this would have been exposed by now.

Moreover, the most surprising element of this particular aspect of the case, yet one that, strangely, Martin and Boyd seem not to have realised, as they have apparently not focused upon it during their investigations, is that if Egginton's claim *is* true, then a crucial component of the much-vaunted 'confessions' of Wetherell and Spurling is false. This is because the Egginton and the Wetherell/Spurling testimonies do *not* support one another; on the contrary, they fundamentally contradict each another. After all, if the Surgeon's photo were indeed the result of a Spurling-constructed 'head-and-neck' model borne upon a toy submarine being photographed by Wetherell directly at the loch and then sunk there (the proposal made throughout their book by Martin and Boyd), then simple logic dictates that it couldn't have been created by Wetherell at home by superimposing the model upon the plate of an earlier-snapped photo of the loch.

My own personal opinion regarding this entire episode is that the real hoax was not the 'head-and-neck' submarine at all, but rather the *claim* concerning the 'head-and-neck' submarine. In short, the craft never existed—only the story of it, invented (with careless variations by them in their respective tellings of it) by two sons seeking revenge upon Nessie for their father's humiliation (and in so doing conveniently ignoring the fact that if, as seems highly likely, he'd hoaxed the footprints anyway, he was the author of his own downfall).

After all, why go to all the trouble of constructing such a craft (one that, by its very nature, is highly improbable) and perpetrating a hoax with it (with all the attendant risks of being caught in the act), when all that you need to do is to release many years later (in 1975 and 1993 respectively) a superficially plausible *story* of a hoax—one that cannot ever be conclusively proved or disproved by anyone else afterwards, yet which will nonetheless cast for ever more a deep shadow of doubt upon a photo that has become over the years one of the most famous pieces of evidence in support of the

Loch Ness monster's existence? Such a scenario would achieve the desired effect to the maximum extent yet via the minimum of effort. Consequently, this is much more likely to be the one that actually did occur.

Yet even if the craft did exist, how very convenient that it sank (or was sunk) directly after being photographed—thereby ensuring that it could never be found. Once again, therefore, as there is absolutely no physical evidence to confirm its original existence (no photographs of it before it was placed on the loch are known, nor any preliminary sketches of it with head-and-neck attached or of the head-and-neck itself, nor even any preparatory notes about how the latter might be produced), why should anyone believe that this craft ever did exist?

After all, the ultimate irony here is that because there is no physical evidence to confirm its existence, only anecdotal, the LNM is dismissed out of hand by sceptics; and yet we are expected to believe unquestioningly in a Nessie-head-and-necked toy submarine for which there is no physical evidence either, only anecdotal once again.

A hoaxed photograph, or a hoaxed hoax? With Wilson, the Wetherells and Spurling all long gone now, no-one will ever know for sure. Meanwhile, the media and numerous online websites largely accept entirely uncritically that this iconic photo was a hoax, thereby unscientifically destroying the credibility of the LNM's single most significant piece of evidence, which is precisely what Spurling had hoped for. All in all, therefore, an excellent result for Spurling but a sad result indeed for modern-day cryptozoology.

There is no doubt from reading their book that Martin and Boyd conducted sterling detective work in pursuit of answers concerning the Surgeon's photo, in particular bringing to belated widespread attention the long-forgotten 'Mandrake' article. Yet somehow I fear that it is Spurling and the Wetherells who were the ones that truly destroyed the credibility of the most iconic cryptozoological image of all time—and without having to do anything more tangible than spin a couple of very fanciful, inconsistent yarns.

In 1960, an amazing piece of cinefilm was shot by aeronautical engineer Tim Dinsdale at Loch Ness. After spending five unsuccessful days seeking the monster, it was now the morning of 23 April—the sixth, and final, day of his long-planned search. Driving along the loch's Foyers Bay stretch of road, he suddenly spotted a mahogany-coloured hump-like object protruding up through the water, roughly 1,300 yards away and oval in shape, with a conspicuous dark blotch upon its left side. By then he

Tim Dinsdale in 1979 (© Prof. Henry H. Bauer)

was out of the car and standing by the loch, observing the object through 7x binoculars (so his view of it was very detailed)—and that was when it began to move.

Dinsdale began to film the object at once, shooting about four minutes of black-and-white film with his tripod-mounted 16-mm Bolex cinecamera as what he now believed to be some form of immense living creature swam away towards the far shore, generating a distinctive v-shaped wake and submerging slowly—before suddenly changing direction, swimming southwards parallel with the shore but still submerging until it had almost vanished beneath the water. Following its painstaking analysis of this remarkable film, the Royal Air Force's Joint Air Reconnaissance Intelligence Centre (JARIC) released an official report in 1966 announcing that in its opinion the object in the film "probably is an animate object"—in other words, part of a living creature, rather than a submarine, or surface craft such as a rowing boat. If so, then the creature is certainly huge, because JARIC estimated the hump to be 12-16 ft long, a cross-section through it to be not less than 5 ft high and 6 ft wide, and moving at a speed of 7-10 mph.

Although a major piece of evidence supporting the reality of an animal of truly monstrous proportions, the featureless nature of the hump filmed by Dinsdale cannot shed any light on its owner's taxonomic identity—an all-too-frequent problem with Nessie data. In recent years, moreover, sceptics have sought to cast doubt upon the validity of the JARIC analysis's conclusion, by speculating that it was a boat after all, and even that what appeared to be the image of a larger body portion staying submerged beneath the water surface was merely a shadow of the portion above the water surface. To my mind, however, this latter claim in particular seems spurious (and how could a boat submerge in the manner seen in the film?). Perhaps the best summary of the respective claims offered in recent times by the film's critics and supporters can be found in Gareth Williams's book *A Monstrous Commotion: The Mysteries of Loch Ness* (2015). The most pertinent excerpt from that summary reads as follows:

> In 1999, Richard Carter, Adrian Shine and Dick Raynor tried to recreate the sighting under similar lighting conditions, aided by a 16mm clockwork Bolex camera and a dark wooden 15-foot boat that followed the same zigzag course across the Loch. They believed that the result looked 'remarkably like' Dinsdale's 'Monster'. This is in line with some alternative interpretations of the film. Shine and others have suggested that the 'Monster' can be changed into a man in a dark-hulled boat simply by watching the film on television with the contrast turned up. Shine later superimposed magnified images of the object taken from 170 frames using an 'image-stacking' method which removes the grain in the film and random artefacts. According to Shine, this reveals a 15-foot boat with one and possibly two passengers—and a lighter spot on the prow, where a circular licence-number plate would have been carried during the 1960s. His conclusion has since been endorsed by commercial image processing experts, including some of the JARIC personnel who produced the original report in 1966.

Others beg to differ. Angus, the Dinsdales' youngest son, has also seen a modern reanalysis of the original 1960 footage. No boat magically creeps out of the grain; instead, a dark 'shadow' is revealed beneath the surface, showing a body, tail and a distinctive diamond-shaped flipper.

Also, a very comprehensive refutation of the 'man on a boat' identity for the monster was presented by Prof. Henry H. Bauer in a detailed communication posted on his behalf by Loren Coleman to several cryptozoology and water monster Yahoo-based online chat groups on 4 December 2003, as well as later on Bauer's own website (at: http://henryhbauer.homestead.com/LochNessFacts.html) and makes fascinating reading.

By virtue of its objectivity, sonar evidence is considered by the scientific community to be much more persuasive than eyewitness accounts, and even photographic evidence—which is open to all manner of interpretation (not to mention increasingly frequent claims of hoaxing, due to the sophisticated techniques of digital photo-manipulation readily available nowadays). From the late 1960s onwards, many research teams—including those from Birmingham University in England, the British Museum, the Academy of Applied Science, and the Anglo-American 'Operation Deepscan' of October 1987—have obtained important, reproducible traces of solid, seemingly animate objects encountered in midwater or near the sides or bed of the loch. They were swift-moving, exhibited profound diving abilities, measured around 20 ft long, and were readily distinguishable from shoals of fish or inanimate objects such as vegetable mats.

These thereby provide new insights into Nessie biomechanics and spatial distribution underwater. Once again, however, they have offered few morphological clues with which to identify its species—except, that is, for a remarkable occurrence in August 1972, which, in my opinion, yielded the most important evidence currently obtained in support of the Loch Ness monster's reality as a huge water beast of a still-undiscovered species.

With sonar apparatus and an underwater camera positioned in Urquhart Bay, Dr Robert Rines's team from his Academy of Applied Science were monitoring their equipment during the early hours of 8 August 1972 when a flurry of movement was suddenly detected by the sonar. From the readings obtained, this appeared to be a shoal of fishes swimming rapidly away from something solid and much larger that was following closely behind—something that moved purposefully rather than passively drifting, and which measured 20-30 ft long.

Moreover, while the sonar had been obtaining traces of the body, the underwater camera had been photographing it. When its film was developed, two consecutive frames were found to depict something that computer-enhancement techniques revealed to be a diamond-shaped, pointed-tipped object resembling a hind flipper—whose orientation had changed slightly between the two photos, suggesting movement. Estimated to be 4-6 ft long, it was attached to a much larger body, and possessed what looked like a median keel or stiffening rod.

Intriguingly, it bore little resemblance to the flippers of pinnipeds, cetaceans, or other aquatic mammals, and did not correspond to the fins of eels or other ray-finned fishes either. Conversely, its external outline recalled the rhomboidal flippers of plesiosaurs.

A *third*, less-publicised photo on the 'flipper' film (but reproduced in this present book's original 1995 edition) may portray *two* Nessies—depicting one lying on its side with two flippers visible, and the tail of a second close by. Correspondingly, the sonar traces showed

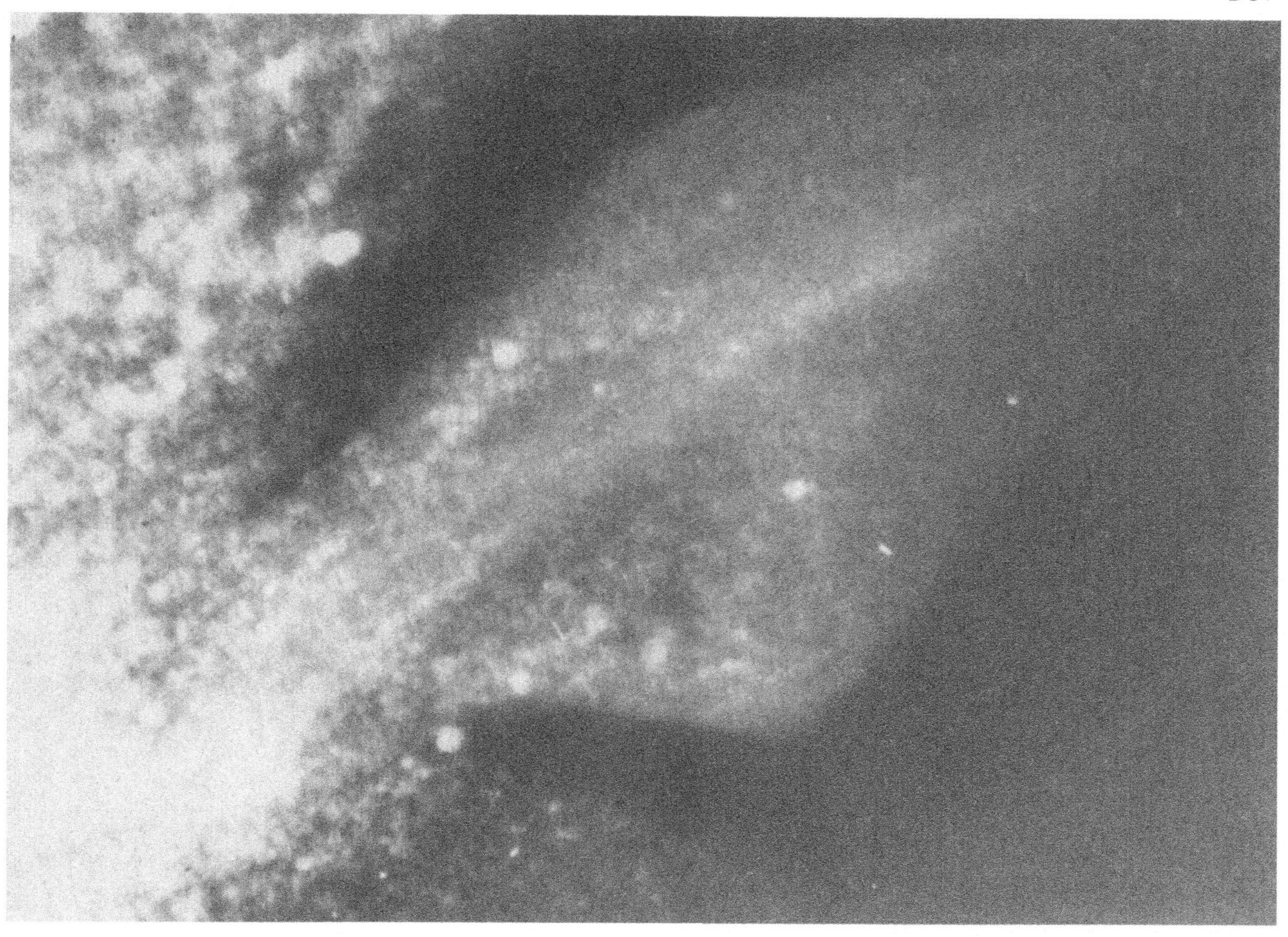

One of Dr Robert Rines's 'flipper' photographs (© Dr Robert Rines/Academy of Applied Science)

that a second 20-30 ft object had indeed been present for a time within the vicinity of the sonar equipment and camera.

What makes these results so significant, and possibly unique at present, is that whereas eyewitness reports are subjective (and thus are open to criticism and doubts regarding their reliability and interpretation), here were two independently-obtained pieces of data (sonar traces and photographic film) that convincingly supported one another, and, of particular importance, were obtained by totally objective, disinterested witnesses—machines.

It was the distinctive image captured by the two 'flipper' photos that led on 11 December 1975 to the Loch Ness monster's formal scientific christening by Rines and British naturalist Sir Peter Scott—dubbing it *Nessiteras rhombopteryx* ('monster of Ness with diamond fin').

In addition, its shape corresponded perfectly with the dark, hairless, rubbery flipper that briefly broke the surface when, in September 1970, veteran American cryptozoologist Prof. Roy P. Mackal and two colleagues aboard a boat on Loch Ness spied a huge beast about 30 ft away, rolling underwater in Urquhart Bay.

(As with Dinsdale's film, however, there has been much sceptical discussion and claims regarding the sonar results and especially the 'flipper' photographs, and whether the latter may in fact owe their distinctive appearance to 'retouching' techniques applied to genuine photos of loch bottom sediment. Some of the sonar and photography-related issues are far too lengthy and technical to be presented here, but I recommend interested readers to consult various online sources, including http://www.lochnessinvestigation.com/flipper.html and http://blogs.scientificamerican.com/tetrapod-zoology/2013/07/10/photos-of-the-loch-ness-monster-revisited/ plus http://henryhbauer.homestead.com/LochNessFacts.html for further details.)

Nevertheless, there is still a heartfelt wish among Nessie believers for the creature to be spied in its entirety once in a while, rather than offering nothing more than fleeting glimpses of humps and long necks, or tantalising sonar traces of sizeable but anonymous bodies. In fact, such a wish may have occasionally been granted, because there are a few detailed eyewitness reports on file that claim sightings of Nessie as seen *out* of the water, moving around on land.

One of the earliest noteworthy instances occurred one afternoon in September 1919 on the shore of the marshland opposite Inchnacardoch Bay. A 15-year-old girl (the future Mrs Margaret Cameron), was playing on the bay's beach with her sister and two brothers when they heard a loud crackling noise coming from the marshland's trees—and then saw a huge

Reconstruction of Nessie incorporating the diamond fins of the 'flipper' photographs (© Jeff Johnson)

beast emerge from among them. Pointing directly towards its amazed observers, it lurched down to the shore and into the water, humping its shoulders and twisting its small camel-like head from side to side upon a long neck. It seemed to be at least 20 ft long, with two short, round feet at the front of its body, a second pair further back, and a shiny grey skin.

During the 1930s, another group of children claimed to have seen a monster, this time in Urquhart Bay's bushy swamp. When later shown pictures of animals in a book, the creature selected by them as being similar to what they had seen was the plesiosaur.

The 1930s also yielded the Spicer sighting, one of the most dramatic Nessie reports ever recorded. On 22 July 1933, between 3.30 and 4.00 pm, Mr and Mrs George Spicer—visitors from London with no prior knowledge of Nessie—were driving along the road linking Dores to Foyers when they were confronted by a most incredible sight.

Emerging from the bushes about 200 yards further up the road, and raised several feet above its surface, was a horizontal object resembling an elephant's trunk—but it quickly resolved itself into a neck when followed swiftly by a thick, grey-coloured body estimated to be about 5 ft high. Its feet, if it had any, were not visible to its two eyewitnesses because the lower portion of its body was hidden in the slope of the road. Something protruded from the area of its shoulder, which the Spicers later deemed to be the tip of its tail, curling forwards along the side of its body facing away from them.

Its body moved via a series of jerks and its neck arched upwards as it progressed rapidly across the road and into the bracken separating it from the lochside. Judging from the width of the road, the beast's total length was put at over 25 ft, but moments later it had vanished, leaving no trace in the water. The Spicers were repulsed by the beast's appearance: "It was horrible—an abomination . . . a loathsome sight"—forceful descriptions that one would not expect if its species was in any way familiar. Mr Spicer also referred to it as the "nearest approach to a dragon or prehistoric animal that I have ever seen in my life".

William McCulloch, a cyclist to whom the Spicers spoke soon after their sighting, swiftly rode back to the spot where the creature had emerged, and he was able to confirm that the bushes on both sides of the road and leading down to the loch were extensively flattened, as if a steamroller had been driven over them.

Another memorable land sighting occurred 5.5 months later, and featured an exceptionally qualified eyewitness. At 1.30 am on the bright moonlit morning of 5 January 1934, 21-year-old veterinary student Arthur Grant was riding home on his motorbike, stone-cold sober and travelling from Inverness to Glen Urquhart, when he saw something dark move in the shadow of some bushes along the road's right-hand edge at the turn for Abriachan. What happened next is so startling that Grant's own account of it deserves to be quoted in full:

> I was almost on it when it turned what I thought was a small head on a long neck in my direction. The creature apparently took fright and made two great bounds across the road and then went faster down to the loch, which it entered with a huge splash. I jumped off my cycle and followed it but from the disturbance on the surface it had evidently made away before I reached the shore. I had a splendid view of the object. In fact, I almost struck it with my motorcycle. The body was very hefty. I distinctly saw two front flippers and there

Artistic reconstruction of Arthur Grant's land sighting of Nessie (© William M. Rebsamen)

> seemed to be two other flippers which were behind and which it used to spring from. The tail would be from 5 to 6 feet long and very powerful; the curious thing about it was that the end was rounded off—it did not come to a point. The total length of the animal would be 15 to 20 feet.

He gave further descriptive details at a meeting of Edinburgh's Veterinary Society:

> Knowing something of natural history I can say that I have never seen anything in my life like the animal I saw. . . . It had a head rather like a snake or an eel, flat at the top, with a large oval eye, longish neck and somewhat longer tail. The body was much thicker towards the tail than was the front portion. In colour it was black or dark brown and had a skin rather like that of a whale. The head must have been about 6 ft from the ground as it crossed the road, the neck 3.5 to 4 feet long and the tail 5 or 6 feet long. Height from belly to the back would be about 4.5 feet and overall length 18 to 20 feet . . .

Inevitably, Grant attracted ridicule. However, a search made shortly afterwards, led by a Fellow of the Zoological Society of Scotland called H.F. Hay, discovered some large flipper-like tracks measuring 24 in long and 38 in across amid shingle—some 70 yards further up the beach from where Grant had seen the beast go down the road's steep bank towards the loch. The team found some flipper tracks there too, spaced about 5 ft apart.

At 4.30 pm on 5 June 1934, a maid called Margaret Munro was looking out of the window of her employers' house overlooking Borlum Bay when she saw a very large beast about 300 yards away, turning back and forth mostly out of the water on the bay's shore. Watching it through binoculars for 25 minutes before it re-entered the water and swam away, she discerned a disproportionately small head perched upon the end of a giraffe-like neck, a huge dark-grey body with grey skin like an elephant but white underparts, and a pair of short front flippers. According to Munro, it was ". . . able to arch its back into large humps"—but whether the back physically arched itself, or whether it bore protuberances that could be raised or lowered to yield humps, is uncertain from this description.

A more recent land sighting occurred on the afternoon of 28 February 1960, when Torquil MacLeod spied an immense animal half ashore on the beach at Horseshoe, about 2.5 miles south of Invermoriston. With grey elephant-like skin and a long neck resembling an elephant's trunk, virtually throughout MacLeod's 9-minute-long, binocular-assisted sighting the creature faced directly away from him, but he spotted two paddle-like rear flippers—and once, as it turned around to re-enter the water, a front flipper also came briefly into view.

What makes this particular sighting so intriguing is MacLeod's estimate of the beast's size—using the graticulations on his binoculars' lenses, he calculated that even without the tail, hidden in the water, the animal was 40-60 ft long! This is much bigger than anything reported by other eyewitnesses of terrestrial Nessies—twice as long, in fact. A plausible explanation of this anomaly, as put forward by Nicholas Witchell in his own coverage of these reports, is that perhaps it is the smaller, younger members of this species that come ashore most often—with the larger, fully mature members (like the MacLeod-observed specimen) spending most if not all of their time in the water. This is a logical deduction, as their much heavier weight would make the larger individuals less adept on land than the smaller, lighter ones. (Alternatively, MacLeod may simply have made an error in his calculations.)

Is this what Nessie might look like on land?

After studying some of these accounts, even the afore-mentioned Nessie supporter-turned-sceptic Dr Maurice Burton was sufficiently impressed to suggest that it may actually be more rewarding to look for Nessie on land than in the water. Notwithstanding this, what remains unanswered is the question of *why* such a species should venture onto land in the first place.

Surely, creatures inhabiting a lake amply supplied with fish and other edible livestock scarcely need to seek prey out of the water. Reproduction would seem on first sight a more promising line of investigation—whether to lay eggs or to bear live young, many aquatic animals do come ashore for this purpose. However, if Witchell's suggestion concerning the size of Nessie specimens generally reported on land is correct, then it would seem that these are juvenile individuals rather than adult, sexually-mature ones.

Yet even if the reason *why* such animals come ashore is still obscure, the apparent fact that they *can* move overland provides important support for eyewitness claims that similar beasts inhabit a number of other Scottish lochs—including Morar, Shiel, Oich, Lochy, Quoich, and Arkaig. Perhaps this species actively migrates between lakes on a regular basis, or perhaps from one original inland home it has progressively invaded several others.

Ending this account of the LNM is a truly remarkable underwater encounter between eyewitness and enigma. One day during the 1880s-1890s, diver Duncan Macdonald was lowered into Loch Ness at Johnnies Point, near the loch's Fort Augustus entrance to the Caledonian Canal, in order to seek a sunken ship there. Once beneath the water, however, his eyes soon beheld a sight that swiftly dispelled any thoughts of sunken ships far from his mind.

At a depth of about 30 ft, Macdonald had seen a creature at least as big as a goat lying upon a shelf of rock—a creature that in his opinion recalled a huge frog, and which was staring directly at him! Not the most pleasant of experiences, surely. After all, an underwater eyeball-to-eyeball confrontation with what could well be an exceedingly large carnivorous beast of uncertain identity and even more uncertain temperament would be a decidedly unnerving experience for anyone! The most extensive documentation ever published of this fascinating case and also of a second, lesser-known underwater 'monster' sighting at Loch Ness can be found in my recent book *Here's Nessie!* (2016).

If we assume that Nessie does indeed exist (and I fully realise that this is certainly a big assumption), what could it be, zoologically-speaking? Numerous contenders have been put forward by LNM researchers down through the years, but by far the most popular and persistent Nessie identity is a prehistoric survivor, in the form of an evolved, modern-day plesiosaur.

IN PRAISE OF PLESIOSAURS

During the 19th Century, when considering the morphology of plesiosaurs, an English palaeontologist called Dean Conybeare likened them to "snakes threaded through the bodies of turtles". If we exclude the turtles' carapace from any such comparison, it is quite an apt description, for whereas plesiosaurs' bodies and their two pairs of paddle-shaped flippers are reminiscent of those belonging to the great marine turtles, their very long, slender necks and small heads are certainly rather ophidian in superficial appearance—exemplified by the elasmosaurs. In these, the last known group of plesiosaurs, which apparently died out alongside the dinosaurs at the close of the Cretaceous, the neck was disproportionately long—in *Elasmosaurus* itself, it measured 26 ft in comparison with the 20-ft length of the entire remainder of the animal, and contained 71 vertebrae.

Plesiosaurs fed upon fish and water-dwelling invertebrates, and again like turtles they were almost exclusively aquatic, coming onto land only to lay their eggs. It is now believed that they swam like turtles (and penguins!) too, moving their flippers up and down like great wings, so that they literally flew through the water (discussed in further detail later here).

In the light of this plesiosaurian précis, the long-neck cases documented above indicate that people are seeing creatures that *look like* plesiosaurs—but *are* they plesiosaurs?

There is no doubt that the external correspondence between plesiosaur and long-neck

Early plesiosaur illustration, from *Water Reptiles of the Past and Present*, 1914

is much more convincing than is true with most of the other cryptozoological identities on offer —giant eels, enormous worms or slug-like molluscs, specialised whales, huge sturgeons, or much-modified amphibians (this last-mentioned contender is rendered even less likely on account of the fact that hardly any marine amphibians are known, either living or prehistoric).

Nevertheless, there is a great deal more to consider than outward similarity when seeking to obtain a satisfactory zoological identification of a mystery beast. This concept was encapsulated by marine biologist Dr Forrest G. Wood from the U.S. Naval Ocean Systems Center, when commenting upon the nature of the object portrayed in Sandra Mansi's 'Champ' photo during a press conference held by the International Society of Cryptozoology on 22 October 1982 at the University of British Columbia:

> I will tentatively accept that it was a living animal. In appearance, it most closely resembles a member of the long-extinct group known as the plesiosaurs. That does not *make* it a plesiosaur. All I can say is that, in general appearance, it most closely resembles a plesiosaur, which was an aquatic reptile. I accept it was a living animal, but I can't say what kind.

Some of the issues facing anyone attempting to champion the plesiosaur's candidature as the long-neck can be effectively dealt with by considering the classic case of the modern-day Comoros coelacanth *Latimeria chalumnae* —whose discovery in 1938 (followed by that of a closely-related Indonesian species, *L. menadoensis*, in 1997) sensationally resurrected from extinction an entire lineage of prehistoric lobe-finned fishes, the crossopterygians.

Plesiosaurs and the Missing Post-Cretaceous Fossils

A major obstacle to the likelihood of plesiosaur persistence according to cryptozoological sceptics is the absence of verified post-Cretaceous plesiosaur fossils. Surely, they argue, if plesiosaurs had survived to the present day, we would have unearthed fossilised remains of them that bridge the gap between modern times and the close of the Cretaceous 66 million years ago. However, only quite recently has anyone identified any post-Cretaceous fossils of coelacanths either—yet *Latimeria* is conclusive proof that its lineage has indeed survived. There are also many other modern-day animals (so-called Lazarus taxa) with little or no fossil history linking them to prehistoric times (i.e. ghost lineages), as already noted in this present book's new introduction.

Interestingly, in 1993 American palaeontologists Drs Spencer Lucas and Robert Reynolds disclosed that since 1980 two samples of elasmosaur fossils (one a single cervical vertebra, the other a collection of about 40 incomplete cervical vertebrae) had been collected in California's Cajon Pass from a rock unit that may pertain to the San Francisquito Formation; some invertebrate fossils from this formation indicate a Palaeocene age. No direct association of known Palaeocene fossils and the plesiosaur remains, however, could be demonstrated. Hence Lucas and Reynolds decided to follow the more conservative option, and class the plesiosaur material as dating from the late Cretaceous. Yet a number of other supposed post-Cretaceous plesiosaur fossils have also been documented; these are traditionally discounted as so-called zombie taxa, i.e. deemed to have been reworked from earlier strata—but *have* they?

American cryptozoologist Scott Mardis has made a particular study of these controversial fossils, and a list summarising his findings made while searching through the palaeontological literature, presented in the 15 July 2014 post from Jay Cooney's excellent *Bizarre Zoology* blog, makes very interesting reading. Quoting from that list, here are its most intriguing entries:

1. Elasmosaurid plesiosaur fossils mixed with Paleocene microfossils in the Takatika Grit formation (Cretaceous-Paleocene) of the Chatham Islands of New Zealand.

2. The discovery of a specimen of *Plesiosaurus crassicostatus* in the Paleocene Waipara Greensand of North Canterbury, New Zealand.

3. The discovery of two sets of Elasmosaurid vertebrae (one articulated) allegedly associated with Paleocene microfossils in the Paleocene San Francisquito Formation near Cajon Pass, California. [Noted earlier.]

4. A plesiosaur tooth in the Aruma Formation (Paleocene-Eocene), of Saudi Arabia.

5. A plesiosaur vertebrae [sic], assigned to the now-discarded genus *Discosaurus vetustus*, allegedly from the Eocene marine deposits of Choctaw Bluff, Clarke County, Alabama, deposits that have also produced specimens of basilosaurine whales.

6. A set of fossil vertebrae from alleged Cretaceous deposits in Mullica Hills,

New Jersey, is acquired by paleontologist Richard Harlan in 1824. He describes one of the vertebrae and assigns it to the Plesiosauria that same year. In 1851, paleontologist Joseph Leidy mysteriously reassigns Harlan's plesiosaur vertebra to the dolphin genus *Priscodelphinus* and declares it to be from the Miocene epoch. What about the age and identity of Harlan's other vertebrae? Is the age and identity of the *Priscodelphinus* vertebra completely resolved?

The most startling putative post-Cretaceous plesiosaur fossil of all—or, if adopting the orthodox alternative explanation, the most dramatically reworked plesiosaur fossil of all—must surely be the isolated dorsal vertebra of an elasmosaur that was found in a Pleistocene erratic boulder (geschiebe) near Wisbar, northern Germany, in July 2008, and formally documented in 2011. According to lithological studies conducted upon this fossil, however, it can be correlated with the 'Köpinge' sandstone from the Ystad-Vomb area in southern Sweden (which dates from upper Lower Campanian to lower Upper Campanian within the late Cretaceous). Consequently, despite its Pleistocene setting it is officially deemed to be of late Cretaceous age.

Similarly, as revealed by Dr Brian Witzke from the Iowa Geological Survey in an *Iowa Geology* paper from 2001, and subsequently documented by Mardis, reworked and transported Cretaceous fossils, including plesiosaur bones, are sometimes found in the glacial tills and associated gravel deposits in Iowa, especially in the western part of the state. Moreover, in an email sent to Mardis on 21 January 2011, Witzke revealed that the University of Iowa's paleontological collections contain one such specimen, undocumented in the literature at the time of the email, consisting of a plesiosaur metacarpal derived from glacial gravels on the south edge of Iowa City. These gravels are largely reworked from pre-Illinoian glacial deposits in the area (about 1 million years old).

Yet even if any of the cases recorded here really do feature bona fide post-Cretaceous plesiosaur fossils, surely plesiosaurs could not have withstood the rapid diversification of aquatic mammals that occurred during the Cenozoic Era? Mardis feels that such a conclusion may not be warranted:

> If we accept the possibility that the fragmentary plesiosaur material spanning from the Paleocene to the Pleistocene may not be "reworked", then there may be no 70 million year ghost lineage for plesiosaurs. Would not the emergence of marine mammals such as cetaceans and pinnipeds have severely hampered any potential post-Cretaceous comeback the plesiosaurs may have made, had a few squeaked through the K/Pg extinction event? Perhaps not. Plesiosaurs seem to have persisted through such ecological shifts as the presumed extinction of ichthyosaurs, the rise of mosasaurs and the coming and going of metriorhynchid crocodiles. Would the introduction of cetaceans and pinnipeds be that different? Some would argue that warm-blooded, highly intelligent marine mammals would have had a distinct advantage over any reptilian competitors. There is some evidence to suggest that plesiosaurs may have been homeothermic, lived in gregarious social groups and cared for their young.

Moreover, biologist Dr Michael Woodley, author of *In the Wake of Bernard Heuvelmans*

(2008), a highly-regarded, in-depth assessment of Heuvelmans's classic but nowadays increasingly faulty sea serpent classification, has pointed out that the plesiosaurs' 'benthic grazer' ecological niche was not occupied by mammals following the latter reptiles' supposed extinction at the end of the Cretaceous. Surely, then, a niche remains for long-necked marine creatures (e.g. plesiosaurs) today?

The paradox inherent in the reworked plesiosaur fossils concept was masterfully summarised as follows by Jay Cooney in a *Bizarre Zoology* blog post of 9 May 2014:

> Looking at the litany of "reworked" plesiosaur fossils spanning from the Paleocene to the Pleistocene, I think the "no geologic evidence of plesiosaurs after the Cretaceous" argument is shot to hell. True, there is no unambiguous post-Cretaceous evidence that is not fragmentary. Nevertheless, there is some evidence. If only one isolated tooth, flipper bone or vertebra of a plesiosaur is found in a Mesozoic deposit, the paleontologists do not immediately invoke reworking to account for its presence in the strata where it was found. But let the same bone be found in a Cenozoic deposit, complete with other Cenozoic marine vertebrates, and it is immediately tossed into a refuse bucket labeled "reworked". Instead of marveling over this potential relict survivor, it is barely mentioned in the literature with disdain and then shoved into a drawer, hidden away to be forgotten or conveniently lost. If "reworking" will not get the job done, then one can always say it was mislabeled or even misidentified. . . There is a term for this. It's called "moving the goal posts". No doubt that the arch-nemesis of the Great Sea Serpent, Sir Richard Owen, would . . . approve.

Amen to that!

In view of the latter considerations, it is clearly time that the earlier-listed and other ostensibly anachronistic plesiosaur fossils received proper radiometric dating (until recently, direct dating of vertebrate fossils had been unsuccessful), and also that the strata containing them were comprehensively, directly dated, in order to discover unequivocally whether such fossils have indeed simply been reworked from older strata or whether they are truly much younger than any previously confirmed examples.

A Question of Habitat

Whereas many fossil coelacanths were freshwater species, the modern-day *Latimeria*'s physiology is modified for a marine existence. Its evolutionary adaptation for survival in a profoundly different habitat from that of its ancient predecessors is of great relevance to the question of present-day plesiosaur survival—because one of the arguments that has been frequently raised by sceptics in the past is that as the plesiosaur species known from fossils were marine, freshwater long-necks could not be plesiosaurs. Taking this line of thought even further: as the freshwater long-necks so greatly resemble marine long-necks that it seems reasonable to assume that they all belong to the same major taxonomic group of animals, if these freshwater versions are not plesiosaurs then surely their marine lookalikes cannot be either.

As pointed out as far back as the mid-1970s by Prof. Roy P. Mackal in his book *The Monsters of Loch Ness* (1976), however, some fossil plesiosaurs had been discovered under con-

Is Nessie an evolved freshwater plesiosaur? (© Wm Michael Mott)

ditions that implied a freshwater environment, especially rivers and estuaries—and, as he also noted, we could speculate that while pursuing fish, or escaping predators of an inflexibly marine nature, these reptiles travelled up rivers and perhaps into lakes, a theory that would be famously espoused in the palaeontological literature by Dr Alfred Cruickshank during the 1990s. Propelling this hypothesis to its logical conclusion, it would not be unreasonable to imagine a situation whereby, finding ample food in large bodies of freshwater and freed from the problems of predation, some plesiosaurs might have gradually transformed into resident freshwater creatures—a scenario already recorded for such noteworthy forms as the seals of Lake Baikal and the sharks of Lake Nicaragua.

And sure enough, in more recent years a number of smallish early Cretaceous leptocleidid plesiosaurs have been discovered that were indeed resident freshwater species, and were seemingly specialised for permanent existence within this specific habitat—rather than being there merely through retreating from marine environments into the relative safety of freshwater in order to avoid predation as postulated by Cruickshank. Of interest is that in 2012, a second plesiosaurian lineage, the pliosaurs, was also revealed to have a freshwater representative—*Hastanectes valdensis*—and other plesiosaurian groups are now thought to have evolved such forms too.

As the crocodilians never succeeded in invading many of the more temperate bodies of freshwater around the world, a vacant niche for a large reptilian predator in such habitats could conceivably be filled by plesiosaurs. And the seas are so unutterably vast that there would surely be more than sufficient space for a wide variety of aquatic mammals and reptiles to co-exist without competing excessively with one another—as noted earlier, the niche for a long-necked benthic grazer filled by the plesiosaurs during the Mesozoic seems not to have been occupied by any currently-known species during the Cenozoic.

Furry Endothermic Plesiosaurs and Giant Giraffe-Necked Seals

Another purported problem to be faced when seeking to reconcile long-necks with plesiosaurs is that according to various eyewitness descriptions, the former sometimes possess features not exhibited by any of the fossil plesiosaurs on record. These include whiskers and other furry accoutrements, horns or snorkel-like projections, one or more dorsal humps, an absence of any discernible tail, and an unexpectedly flexible neck.

In his book *In the Wake of the Sea-Serpents* (1968), Dr Bernard Heuvelmans cited these discrepancies with traditionally-conceived plesiosaur morphology as good reasons for jettisoning the plesiosaur identity in favour of a novel mammalian contender—an undiscovered species of giant giraffe-necked seal, which he christened *Megalotaria longicollis*. He proposed that it occupies the ecological niche left vacant by the plesiosaurs' extinction, and thus, via convergent evolution, had acquired certain of their characteristics, most notably a very lengthy neck. Peter Costello subsequently adopted this same identity for freshwater long-necks in his book *In Search of Lake Monsters* (1974). (See also my own book, *Here's Nessie!* (2016), for a comprehensive documentation and discussion of the merits and demerits of the long-necked seal's candidature as the identity of the long-neck category of water monster.)

Although such a beast could be responsible for certain long-neck sightings, from the cases documented earlier in this chapter and from the discussion that follows here, however, it will become apparent that this identity cannot satisfactorily explain many others—especially those, for instance, in which a very distinct (and sometimes relatively long) tail is spied, which is not a typical pinniped feature.

In any case, in my view it is unnecessary to invent an entirely new animal (i.e. one that has no palaeontological predecessors) in order to explain long-necked water monsters when there are reasonable explanations for the morphological differences between such creatures and fossil plesiosaurs. After all, not even the so-called swan-necked seal *Acrophoca longirostris* from Peru and Chile's late Miocene to early Pliocene bears any realistic resemblance to the postulated appearance of Heuvelmans's hypothetical *Megalotaria*.

For instance, the 'hair' and 'whiskers' allegedly exhibited by some long-necks may not be true hair at all. It could constitute sensory filaments, composed of soft tissue that might not leave impressions in fossilised specimens. Equally, the horns or snorkels may be protrusible breathing tubes. As for the dorsal humps: these could not be vertical undulations of the animals' bodies—unlike mammals, most reptiles cannot perform such movements (the long-extinct thalattosuchians or marine crocodilians were apparently a rare exception—see later), and the framework of the plesiosaurian body was in any case extremely rigid. More reasonable are the popular suggestions that they are expanses of fatty tissue, inflatable airsacs, or even exposed portions of a dorsal crest—and, as with sensory bristles, it is unlikely that the

Nessie portrayed as a long-necked seal (© Anthony Wallis)

outlines of breathing tubes, rolls of fat, airsacs, or membranous dorsal crests would be readily preserved in fossils. There is, however, another, much more important, conceivable possibility—that these structures are, in evolutionary terms, a recent innovation.

All of what we know regarding plesiosaurs is based upon fossils that are at least 66 million years old, and which indicate that those ancient species were indeed unfurred. If, however, any plesiosaurs have survived to the present day, 66 million years of intervening, continuing evolution might have yielded animals markedly different from their Cretaceous ancestors—animals that may, for instance, have developed fur, and/or snorkels, airsacs, and crests (not to mention more flexible necks and loss of their tails—see later).

I have seen speculation regarding continued post-Cretaceous evolution for plesiosaurs and other officially long-extinct prehistoric creatures discounted as special pleading by some crypto-sceptics, but in my view it is exceedingly myopic to display this attitude, i.e. the wilful refusal to countenance that if a given taxon has survived beyond the date of its most recent fossils right through to the present day, millions of years of continued evolution are likely to have worked all manner of anatomical changes. After all, this has happened with every major group of animals already known to be represented in today's fauna, so there are plenty of precedents for post-Cretaceous evolutionary change, including, once again, the coelacanths.

The living *Latimeria* is up to five times larger than many of its long-deceased fossil relatives, and, as noted earlier, is modified for a marine (rather than a freshwater) existence. True, during their Mesozoic reign, the plesiosaurs exhibited a relatively conservative degree of morphological transformation, at least macroscopically—but the changes that I have postulated here for an evolved, modern-day plesiosaur are not dramatically macroscopic either, as readily demonstrated by the fact of how outwardly similar long-necks seem to be to fossil plesiosaurs.

During the 1970s, a revolution occurred in dinosaur research—sparked by a dramatic theory promulgated by Dr Robert Bakker and given widespread public airing in Adrian Desmond's bestseller *The Hot-Blooded Dinosaurs* (1975). Namely, that at least some of the non-avian dinosaurs, hitherto looked upon as sluggish cold-blooded (ectothermic) creatures, might well have been active warm-blooded (endothermic) animals, just like mammals and birds.

Back in the late 1700s to mid-1800s, some naturalists deemed pterosaurs to be fur-covered bats (even marsupial bats, according to certain opinions!) rather than reptiles—a bizarre identification that was subsequently and discreetly discounted. However, following the remarkable discovery in Kazakhstan during the 1960s of a bona fide furry species of fossil pterodactyl, *Sordes pilosus* (whose 'fur' is actually composed of filaments termed pycnofibers that are not true hairs—see also Chapter 2), it became increasingly likely that a physiology based upon warm-blooded principles (and irrefutably evident that the presence of a furry outer body covering) had indeed evolved independently in wholly unrelated groups of vertebrates. Since then, moreover, some researchers have proposed that certain therapsids (mammal-like reptiles) could also have been hairy; and since my writing this present book's original edition, *In Search of Prehistoric Survivors*, it has of course been irrefutably confirmed by many palaeontological discoveries that certain dinosaurs, notably various theropods, possessed feathers, some of which were once again hair-like in form.

It is not impossible, therefore, that post-Cretaceous plesiosaurs (and perhaps even earlier ones) also evolved both of these traits, ultimately yielding endothermic furry species. What is definitely known is that certain Mesozoic plesiosaurs did inhabit very cold marine environments. In particular, a new species of Cretaceous plesiosaur that was officially named *Umoonasaurus demoscyllus* in 2006 lived in waters off the coast of what is today southern Australia. What makes this so interesting is that when this species was alive, Australia had not yet drifted north to occupy its present-day position but was still a southern polar continent, located at a latitude of approximately 70°S. In the words of Dr Benjamin Lear, leader of the Australian study team that formally described *Umoonasaurus* in a *Biology Letters* paper, as quoted in a *LiveScience* interview on 7 July 2006: "This is equivalent to the middle of the southern Antarctic Ocean today" (see also the 2010 French study regarding endothermic plesiosaurs later in this section).

In order to meet the high energy requirements for maintaining an endothermic metabolism, however, their habitat would need to contain an abundance of suitable food. Whereas this would not be a problem for marine plesiosaurs inhabiting the vast seas, freshwater versions might face problems if confined to relatively small, land-locked lakes—unless they were capable of travelling overland from one stretch of freshwater to another. For plesiosaurs frequenting giant biota-rich lakes, conversely, a small population could probably sustain itself without obligatory recourse to terrestrial migration.

In any case, also available for consideration is an alternative scenario that offers the benefits of endothermy without involving its endergonic burdens. One popular reason among cryptozoological sceptics for discounting the prospect of plesiosaur survival is that creatures resembling them have been reported in areas that are too cold to permit ectothermic reptiles to survive. Yet even if plesiosaurs never did acquire an endothermic metabolism, they could still exist in such localities—thanks to a physiological anomaly known as gigantothermy.

Ectothermic creatures weighing 1 ton or more lose body heat so slowly that to all intent and purposes they are functional endotherms (i.e. comparable to creatures that generate body heat metabolically, such as mammals and birds), and are thus termed gigantotherms. One marine vertebrate living today that has long and famously been thought to exhibit a gigantothermic physiology is the leathery turtle *Dermochelys coriacea*, which, with a verified maximum length of 7 ft, is far smaller than any plesiosaur potentially responsible for long-neck reports.

Of course, young leathery turtles would be too tiny to be gigantothermic, but as they are born and mature in tropical zones this would not be a problem. Only as an adult does this species migrate into colder, northern waters, and once again the same could be true of plesiosaurs—although as they would exceed the minimum size threshhold at a younger age than the leathery turtle, they could begin their northerly migrations earlier.

In recent times, however, further studies of the leathery turtle have revealed that its resting metabolism is not responsible for its body heat, so it is not gigantothermic after all. Instead, it derives its body heat from increased activity, constantly swimming but very rarely resting, thereby obtaining heat from muscle activity. Once again, however, the same principle might apply to young plesiosaurs of comparable size to adult leathery turtles.

In short, whereas a gigantothermy-type mechanism could satisfactorily facilitate the presence of adult plesiosaurs in low-temperature zones normally unfavourable to reptiles, it would not be sufficient in order to enable plesiosaurs to reside on a permanent basis in colder, northern waters, such as Loch Ness or the Arctic seas, because this mechanism could not be utilised by young, small individuals. However, an increased-activity lifestyle comparable to that of the leathery turtle might enable them to exist there.

And indeed, in 2014 a research team that included palaeontologist Dr Matthew J. Vavrek from the Philip J. Currie Dinosaur Museum in Alberta, Canada, published a *Cretaceous Research* paper in which they documented the discovery in Arctic Canada of early Cretaceous non-marine plesiosaur fossils that were primarily juveniles but which nonetheless "would have been living in a region that experienced at least seasonally cool temperatures". In other words, young plesiosaurs were indeed able to survive in colder, northern freshwater localities after all. The source of these significant fossils was the Hauterivian–Aptian Isachsen Formation of Melville Island in Nunavut, Canada.

Moreover, there is a second way of keeping warm in such circumstances that could also have been developed by boreal plesiosaurs. Phocids or earless seals (i.e. not otariids or eared seals, which constitute the sea-lions and fur seals) swimming in Arctic or Antarctic waters contract the tiny blood vessels (arterioles) in their skin, which enables the skin to cool down to a temperature only marginally above that of the surrounding water. This prevents heat from leaving their body, so that, whereas their skin is cold, the interior of their body remains warm—thereby effectively insulated from the low environmental temperature but without necessitating any metabolic rise in order to achieve this.

The leathery turtle also exhibits various anatomical specialisations for surviving in cold-water habitats. These include countercurrent heat-exchange mechanisms in their large flippers, fibrous fatty tissue and a thick skin saturated with oil, plus an insulative carapace.

Alternatively, the possibility of plesiosaurs evolving full, true endothermy during continued post-Cretaceous evolution has long stimulated speculation among palaeontologists and cryptozoologists alike. In 2010, however, a fascinating scientific study was published in the journal *Science* confirming that endothermy was in fact already present among Mesozoic plesiosaurs (as well as mosasaurs and ichthyosaurs), let alone any putative post-Cretaceous survivors. Including Dr Eric Buffetaut among

Leathery turtle engraving from 1896

the French research team responsible for this major breakthrough, here are the study's findings, as summarised in the paper's abstract:

> What the body temperature and thermoregulation processes of extinct vertebrates were are central questions for understanding their ecology and evolution. The thermophysiologic status of the great marine reptiles is still unknown, even though some studies have suggested that thermoregulation may have contributed to their exceptional evolutionary success as apex predators of Mesozoic aquatic ecosystems. We tested the thermal status of ichthyosaurs, plesiosaurs, and mosasaurs by comparing the oxygen isotope compositions of their tooth phosphate to those of coexisting fish. Data distribution reveals that these large marine reptiles were able to maintain a constant and high body temperature in oceanic environments ranging from tropical to cold temperate. Their estimated body temperatures, in the range from 35° ± 2°C to 39° ± 2°C, suggest high metabolic rates required for predation and fast swimming over large distances offshore.

In short, Mesozoic plesiosaurs were indeed able to maintain a higher body temperature than that of their external living environment. This in turn suggests a high metabolism adapted to predation and fast swimming over long distances, even in cold water. And to reduce body heat loss, plesiosaurs may even have developed blubber, just like whales did during the Cenozoic. So as plesiosaurs could live in cold, temperate locations after all, their alleged presence in Loch Ness, for instance, becomes less problematic from a purely physiological basis (always assuming of course that they originally made their way into it from the sea and became trapped there; geologically, this is a quite recent body of freshwater, only about 10,000 years old).

Moreover, in the case of putative evolved, modern-day plesiosaurs, the presence of blubber might explain the variable humps often reported with long-necks by eyewitnesses. Equally, their youngsters might not only be well-supplied with insulating layers of fat but also be profusely furred (like seal pups are). Although this is of course all entirely speculative, it offers up one very intriguing concept for consideration.

Sightings of otters are regularly reported from Loch Ness, and these lithe furry beasts have often been put forward as an identity for the Ness monsters—despite the fact that, if eyewitness reports and sonar traces are accurate, these latter beasts are very much larger than any form of otter. Who knows, perhaps in reality the exact *reverse* of this scenario is the truth—wouldn't it be supremely ironic if some supposed otter sightings actually featured juvenile hairy plesiosaurs?!

The Tale of a Plesiosaur's Tail

In their respective books on sea serpents and lake monsters, Heuvelmans and Costello claimed that long-necks appear to lack tails, and that in sightings that seemingly featured tailed long-necks the eyewitnesses had mistaken the paired rear flippers of such creatures

Reconstruction of an Antarctic cryptoclidid plesiosaur's possible appearance if sporting a thick layer of blubber (© Tim Morris)

for tails. In reality, however, long-necks with two pairs of flippers *plus* a well-delineated tail have been clearly observed on a number of occasions, in the sea and in lakes, as well as on land—which thereby points towards a plesiosaur more than a pinniped in the identity stakes.

The most likely explanation why many long-neck eyewitnesses do not report seeing a tail is quite simply that the tail is not rendered visible when swimming. In whales and other cetaceans, fishes, and the extinct ichthyosaurs and mosasaurs, the tail is the principal locomotory organ—explaining why it is or would have been readily apparent to observers of such creatures. In the case of the plesiosaurs, conversely, the tail's function as a rudder would not offer any reason for it to emerge above the surface. Only during those rare occasions necessitating a plesiosaur to come ashore—to lay eggs or (if a freshwater species) to move from one lake to another one close by—would the tail acquire an additional role, assisting the neck in preventing the terrestrially-inept creature from overbalancing. (Traditionally, it was believed that plesiosaurs lacked a sternum or breast-bone—which would have made forays onto land very difficult—but in 1991, zoologists Drs Elizabeth Nicholls and Anthony Russell presented new evidence to challenge this assumption.)

In any case, even if the tail did become visible during swimming it might not be recognised for what it was. Heuvelmans noted that in certain cases featuring sea serpents of types other than the long-necks, tails of enormous lengths are described—which led him to suggest that these tails might actually have been trails of bubbles. If true, I see no reason why the reverse explanation could not apply in cases featuring allegedly tailless long-necks—i.e. that eyewitnesses have wrongly identified a relatively nondescript, inactive tail as a bubble trail generated by the swimming animal.

The concept of evolved plesiosaurs can also offer a clue in this tale of the long-neck's tail (or lack of it). In the elasmosaurs, the last plesiosaurs known from the fossil record, the tail was extremely short—and as propulsion was accomplished exclusively via their flippers, its function was limited to that of a relatively dispensable rudder. Thus it would not be surprising to find a further 66 million years of evolution dispensing entirely with this appendage, yielding tailless plesiosaurs. Consequently, even if some long-necks really are tailless, this still does not eliminate a plesiosaur identity from contention.

Plesiosaur Flippers and Plesiosaur Flight (Underwater!)

When our knowledge of a creature is based solely upon fossilised remains, our perceptions of its appearance, movements, behaviour, and general lifestyle can change quite drastically as new fossil discoveries are made and research upon them is carried out.

It was traditionally assumed, for instance, that plesiosaurs swam using their flippers as oars, i.e. moving them backwards and forwards in a horizontal plane, with the action of drag upon them providing the propulsive thrust. However, from studies of the shape of their pectoral and pelvic girdles, and the presence of excellent areas for muscle attachment on the ventral portion of these girdles enabling these animals' flippers to achieve powerful downstrokes, researchers now believe that plesiosaurs swam by underwater flying. That is to say, they moved their flippers up and down like penguins, sea-lions, and marine turtles, with propulsive thrust being provided not by drag but by lift. This belief is also supported by the pointed tips of plesiosaur flippers—flippers used as oars gain no advantage from being pointed, but they do if they are being used as wings.

This fundamental change of opinion is of relevance to the question of whether long-necks are indeed living plesiosaurs, because underwater flying yields a much greater wake than rowing, and many long-neck eyewitnesses have remarked upon the great displacement of water occurring as the creature swam by. Once again, this would argue against a seal identity as these swim by rowing rather than flying. Furthermore, there is also an item of evidence regarding long-neck locomotion that would appear to eliminate sea-lions—an important matter, bearing in mind that, unlike seals, these creature do swim by underwater flying (and that Heuvelmans categorised his *Megalotaria* as an otariid). Namely, the remarkable 'flipper' photos obtained at Loch Ness in August 1972 by Dr Robert Rines's scientific team, because when closely scrutinised they ostensibly reveal a number of key clues regarding the taxonomic identity of the animal in question.

During his analysis of these photos for Witchell's book *The Loch Ness Story* (1989), veteran Nessie researcher Dr Denys Tucker, formerly of the Natural History Museum's Ichthyology Department, pointed out the presence of what seem to be two ribs visible beneath the skin of the animal's flank, just in front of the base of the flipper (which is apparently a hind flipper). Their position indicates that these are dorsal ribs, but whereas those of mammals do not extend as far back as the hind limbs, they do in plesiosaurs and other reptiles, implying a reptilian identity for the flipper's owner (thus eliminating otariids).

As for the flipper itself, it is ideally structured for underwater flying—relatively broad, flattened, pointed at its tip, and thin, but also rigid, thereby presenting a large, inflexible area of resistance to the water when held face-on during the propulsive downstroke, and enhanced by a novel non-skeletal flap at the trailing edge (yielding this flipper's distinctive diamond shape) to prevent turbulence. And during the passive upstroke, this flipper would provide first-class lateral stability, by functioning as a hydrofoil.

The fossil flipper of a *Rhomaleosaurus* plesiosaur (© Dr Karl Shuker)

The presence of the long ridge-like structure running down the central axis of the flipper has been highlighted by opponents of the plesiosaur identity, who note that no such skeletal component has ever been recorded from the flippers of fossil plesiosaurs. Indeed, the only known animal of any type whose flippers display a similar structure is the Australian lungfish *Neoceratodus forsteri*. In reality, however, there is no proof that this mystifying ridge *is* a skeletal component. On the contrary, it is far more likely to be a wholly external keel, a refinement for increasing the streamlining of the flipper by enhancing water flow across it.

All in all, the enigmatic Nessie flipper as depicted in those photos is precisely what one might expect from the limb of an aquatic reptile of evolved plesiosaurian persuasion.

Why the Plesiosaur Identity Wins by a Neck

Views regarding flipper locomotion are not the only ones with cryptozoological pertinence to have changed over the years in relation to plesiosaur biomechanics. When the first plesiosaur remains were excavated and studied in Victorian times, scientific reconstructions of the living animals depicted them swimming at the surface of the ocean with gracefully curving necks—held vertically erect like surrealistic reptilian swans, or even coiling in a highly flexible, serpentine manner.

Later, however, such images as those were replaced by ones which portrayed plesiosaurs with quite inflexible necks that could not be raised or even curved to any great degree. This change of view was succinctly expressed in 1914 by Chicago University researcher Prof. Samuel Williston:

> Textbook illustrations of the plesiosaurs usually depict the necks, like those of the swans, freely curved. . . But the plesiosaurs did not and could not use the neck in such a way. They swam with the neck and head, however long, directed in front, and freedom of movement was restricted almost wholly to the anterior part. The posterior part of the neck was thick and heavy, and could not have been moved upward or downward to any considerable extent and not very much laterally. From all of which it seems evident that the plesiosaurs caught their prey by downward and lateral motions of their neck.

This state of affairs persisted for many years, and for cryptozoology the inevitable outcome was that plesiosaurs could no longer be looked upon as serious contenders for the long-neck identity—because they were deemed incapable of swimming with their head and neck held high above the surface of sea or lake in the distinctive manner characterising this particular category of water monster.

Created by sculptor Benjamin Waterhouse Hawkins (1807-1894), two of the famous Victorian plesiosaur statues at the Crystal Palace Park in Bromley, London, sculpted with highly flexible necks (© Dr Karl Shuker)

That in turn enhanced the credibility of the plesiosaur's greatest challenger in the identity stakes, the hypothetical 'giraffe seal' *Megalotaria*, because pinnipeds can certainly raise their head and neck up above the water surface. More recently, however, opinions have changed yet again.

Based upon further studies and specimens, palaeontologists now believe that plesiosaur necks were considerably flexible after all. For example, according to an authoritative encyclopedia of prehistoric life from 1988, with Profs Brian Gardiner, Barry Cox, Robert Savage, and Dr Colin Harrison as consultants:

> The plesiosaurioids [i.e. the true, non-pliosaurian plesiosaurs] fed on modest-sized fishes and squid. Their long necks enabled them to raise their head high above the surface of the sea and scan the waves in search of traces of their prey.

Similarly, in the words of biophysicist Prof. R. McNeill Alexander, writing in 1989:

> If plesiosaurs had eaten worms or clams, we might suppose that they used their necks to reach down to the bottom, dabbling like ducks or swans, but their spiky teeth seem more suitable for catching fishes and squid-like animals which would probably have been too active to be caught easily that way. It seems likely that they darted at prey, extending their long necks to catch things as they swam by. The movement could have been fast, if the neck was held out of water. Herons use their long necks to dart at fish, though they stand in the water instead of floating as plesiosaurs presumably did.

If an evolved modern-day plesiosaur could hold its neck vertically upwards out of the water as portrayed by this model, it would compare morphologically with the 'periscope' head-and-neck sightings reported for Nessie and other long-neck water monsters—but could it? (© Dr Karl Shuker)

In Bristol University palaeontologist Dr Michael Benton's book *The Reign of the Reptiles* (1990):

> Elasmosaurs of the Cretaceous . . . took the long-necked adaptations to an extreme. Some had as many as 70 cervical vertebrae, and the neck could bend around upon itself two or three times. The elasmosaurs no doubt jabbed their snake-like necks rapidly among the scattering schools of teleost fishes. They could have darted the head in and out and seized several fish without moving the body at all.

Incidentally, although the presence of pronounced bony vertical processes (neural spines) upon the cervical vertebrae of plesiosaurs potentially decreased neck flexibility dorsally in these aquatic reptiles, the presence of cartilage and other soft tissue between the cervical vertebrae potentially increased such flexibility. Indeed, a *Comparative Biochemistry and Physiology* paper authored by Dr Maria Zammit and co-workers in 2008 claimed that certain elasmosaurs might have been able to raise their very lengthy necks dorsally by as much as 155°.

Also well worth bearing in mind in relation to the potential ability of plesiosaurs to poke their head and neck above the water surface is spy-hopping, in which the plesiosaur's body would be held vertically, not horizontally. So although the neck would project out of the water, it would in such instances still be in line with the body, not at right angles to the body, and therefore would not involve any dorsal flexion at all. This might explain the bold behaviour of the Canary Islands long-neck of 1957 documented earlier here, spied in open sea well away from any land (and therefore not expecting to be encountered?) with its head and neck well above the water surface by the crew of a surveillance RAF Shackleton aircraft flying overhead.

Moreover, as pointed out once again by Alexander in relation to plesiosaurs:

> If they swam under water with the long neck stretched out in front it would have been quite tricky for them to steer a straight course: if the animal accidentally veered slightly to one side, the water, striking the neck obliquely, would tend to make it veer more. . . . It seems possible that plesiosaurs often avoided this problem by swimming at the surface with their necks out of the water.

Clearly, then, there is a sound biomechanical reason for plesiosaurs to be able to raise their lengthy necks above the water surface. However, long-necks certainly do not spend all of their time with their necks out of the water—otherwise they would have been seen much more often—so most of their swimming must occur while completely submerged. Yet in view of Alexander's remarks, surely such behaviour conflicts with a plesiosaur identity?

Not necessarily. Bearing in mind that even when long-necks 'up-periscope' through the water their necks are nowhere near as long as those of elasmosaurs and other overtly long-necked fossil plesiosaurs, the simple conclusion would seem to be as follows. If plesiosaurs have indeed survived incognito into the present day, during their presumed 66 million years of post-Cretaceous evolution their necks have become shorter (at least by plesiosaur standards).

Alongside a flexi-necked plesiosaur statue (© Dr Karl Shuker)

Biomechanically, a shorter neck would be much less of a hindrance to swimming while wholly submerged—thereby reducing any necessity for a plesiosaur to spend most of its time with its neck held out of the water. This in turn would enable plesiosaurs to propel themselves faster than those ancestors currently known from the fossil record, whose notably long necks—either when held out of the water, or when submerged under it—would have greatly impeded rapid swimming. The fact that some long-necks are said to be very fast swimmers has, until now, posed something of a problem for proponents of the plesiosaur identity, but it can be seen that an evolved plesiosaur with a moderately but not dramatically long neck could well be capable of a very respectable turn of speed, even with its neck submerged. Also, if they do generate heat by increased muscle activity as postulated earlier in this chapter, that would tie in very neatly with rapid swimming behaviour.

Moreover, an evolved plesiosaur with a more flexible neck, one whose vertebrae perhaps possessed less pronounced neural spines than those of fossil ancestors but more intervertebral cartilage and other soft tissue (all of which would increase neck flexibility), might conceivably achieve the 'periscope' head-and-neck appearances so frequently reported for long-necks by eyewitnesses. Equally, such a plesiosaur, with a moderate neck but nothing like the extreme lengths known for fossil elasmosaurs, would not face the problems of generating sufficient muscular power to lift the neck and maintain its own balance in the water whilst doing so that an elasmosaur would have faced (and which therefore provides additional reasons why palaeontologists do not consider it likely that elasmosaurs could raise their necks to any significant extent vertically).

Monstrous Behaviour

Perhaps the classic sceptical response to the question of whether plesiosaurs have survived into modern times is the statement "If they had survived, science would have discovered them long ago". The informed cryptozoologist will soon point out, however, that this assumption is simply not warranted. As on so many other, previous occasions highlighted in this book, the coelacanth *Latimeria* constitutes a notable precedent in favour of plesiosaur persistence—by remaining unknown to science until 1938 in the case of the Comoros species and until as late as 1997 for the Indonesian species. Indeed, as I revealed in my book *The Encyclopaedia of New and Rediscovered Animals* (2012), these are just two of many spectacular species of aquatic animal to have been discovered and described by science since 1900.

Among these are: several new species of beaked whale; a creamy-white river dolphin from China called the baiji *Lipotes vexillifer* (1916); the pa beuk or Mekong giant catfish *Pangasianodon gigas* (world's largest freshwater bony fish; 1930); the huge and monstrously grotesque megamouth shark *Megachasma pelagios* (1976); a bizarre sea-bottom worm called *Riftia pachyptila* that inhabits 8-ft-tall tubes and flourishes huge scarlet tentacles (1977); and the Australian snubfin dolphin *Orcaella heinsohnri* (2005). If all of these can evade scientific detection for so long, how can anyone claim with certainty that there are no reclusive plesiosaurs in existence?

If such beasts do exist, however, there must be reasons *why* they are so elusive—which an examination of long-neck behaviour as reported by eyewitnesses may help to uncover.

One of the principal arguments put forward by sceptics of the giraffe-necked seal identity against treating seriously its candidature as the long-necks' taxonomic identity is that if these latter cryptids really were pinnipeds, then surely they would be seen far more frequently than is currently the case. This is because pinnipeds are exceedingly inquisitive and therefore very likely to make their presence readily apparent to anyone on or near the loch, and they also spend quite an amount of time ashore during the breeding season.

Aquatic reptiles, conversely, like crocodiles and water snakes, rarely show such interest in the presence of humans, and there is no reason to suspect that plesiosaurs would be any more inclined to do so. Moreover, as it is now known that some prehistoric plesiosaurs had become viviparous (bearing their young live), they need not even come ashore for reproduction.

Breathing air while staying unobserved would not pose any great problem for plesiosaurs either. Their nostrils' external openings are positioned very high on their heads (just in front of their eyes, in fact) rather than at the snout's tip. According to the traditional view concerning plesiosaur respiration, therefore, these beasts could lie directly beneath the water surface with only the tips of their nostrils protruding through it (and perhaps equipped with flaps or extrusible snorkels—the 'horns' noted by eyewitnesses?), thereby obtaining all the oxygen that they require while remaining undetected by all but the keenest-eyed observers.

In 1991, however, after studying a superb 200-million-year-old fossil skeleton of a 15-ft-long plesiosaur called *Rhomaleosaurus megacephalus*, Drs Arthur Cruickshank, Philip Small, and Mike Taylor revealed that it had possessed furrows that ran from the end of its snout and along its palate into each of its two internal nostril openings—which were positioned much further forward than those of such familiar aquatic animals as whales and crocodiles. This suggested that as the creature swam, water passing into the plesiosaur's mouth would travel along these furrows, into the nostrils' internal openings and out of them via their external openings. Consequently, the researchers speculated that these aquatic reptiles didn't use their nostrils for breathing, but rather for underwater sniffing—assisting them in their search for carrion and live prey.

How, therefore, did they breathe? Cruickshank, Small, and Taylor suggested that they accomplished this through their mouth—hence they would periodically come to the surface to gulp air. And sure enough, there are several eyewitness accounts of 'head-and-neck' sightings featuring Nessie and other long-necks in which the eyewitness has spied the creature making gulping sounds or movements, which would be entirely consistent with respiratory behaviour of this nature. Moreover, if these beasts chose merely to extrude the tips of their jaws above the surface, they could take in oxygen without being seen.

Plesiosaur behaviour also explains another initial anomaly of long-neck elusiveness—why are there no bona fide long-neck carcases on file? Today's advanced cetaceans are often stranded because of malfunctioning sonar and their inability to raise their necks out of the water in order to restore their sense of direction visually—whereas plesiosaurs, lacking sonar and potentially well-equipped with a flexible neck in evolved modern-day (and also quite possibly certain fossil) species, would not suffer such problems.

There is, however, another facet of plesiosaur behaviour that is also of great pertinence to this issue. Like certain modern-day marine vertebrates such as penguins and sea-lions—but unlike some cetaceans, seals, ichthyosaurs, and mosasaurs—plesiosaurs habitually swallowed large stones (duly dubbed gastroliths—'stomach stones'). For many years, zoologists were unable to suggest a reason for this behavioural dichotomy. The traditional explanation for stone swallowing was that they were utilised in grinding up food to facilitate digestion—but this is unsatisfactory, because, collectively, members of the one group ate much the same range of foods as those of the other group. Once again, it was Dr Michael Taylor, this time in 1993, who offered an alternative insight.

Taylor noticed that whereas the non-swallowers of stones all utilised their tail as the major organ of propulsion (the tailless seals use their hind flippers in an analogous fashion) and were fast swimmers, the swallowers often relied principally upon their limbs, achieving a type of relatively leisurely 'underwater flight' (as noted earlier here). From this, he proposed that the swallowers were ingesting (and, if no longer required, regurgitating) stones for ballast purposes—which would counteract the tendency of such creatures to rise upwards through the water while swimming slowly or remaining stationary (the much faster speeds attained by some non-swallowers when hunting would prevent this from happening to them). More recently, however, the buoyancy theory for gastrolith function has been replaced in the opinion of some researchers by one favouring these stones preventing the swallowers rolling from side to side instead. But whichever theory is true, it would also mean that a carcase containing a fair number of gastroliths is more likely to sink down to the depths than to be washed ashore.

Of course, it could be argued that sea-lions also swallow stones but many sea-lion carcases are nonetheless washed up, so if long-necks are plesiosaurs, why don't their carcases do the same? Ironically, however, exactly the same argument can be levelled against the *Megalotaria* identity too, yet this is strangely overlooked by plesiosaur sceptics. In reality, known sea-lion species are primarily coastal, which explains why their carcases are sometimes found, whereas if long-necks are indeed plesiosaurs (or long-necked otariids) but stay for the most part in mid-ocean rather than coastal zones, the missing evidence of stranded long-neck remains may no longer be a mystery.

In any case, there is a very significant but generally overlooked precedent involving large marine creatures not being stranded (or captured in nets, for that matter)—the megamouth shark *Megachasma pelagios*. As noted in various sections of the present book, this very large species (up to 18 ft long) remained entirely unknown not only to science but apparently even to local fishermen until 1976, when a specimen was accidentally hooked by a research vessel off Hawaii. Since then, more than 50 specimens have been recorded, from waters all over the world (thus revealing that it actually has a global distribution), yet almost all of these have been specimens found stranded on beaches or caught in nets. So why were no strandings or nettings reported anywhere, down through all the ages, prior to 1976? If the megamouth could—and did—avoid detection via stranded or netted specimens until a mere 40 years ago, how can we dismiss out of hand the existence of sea serpents simply because no confirmed stranded or netted specimens are known?

Although there do not appear to be any genuine long-neck carcases on file, there are certainly many records of *supposed* long-neck carcases, some decidedly plesiosaurian in appearance, but those that have been formally investigated have disclosed a very different identity—a rotting shark. On 25 April 1977, for example, the corpse of a huge outwardly plesiosaur-like beast was caught in the nets of the Japanese fishing vessel *Zuiyo Maru*, trawling in waters about 30 miles east of Christchurch, New Zealand. About one month dead, the carcase measured roughly 33 ft long, and was in an extremely advanced state of decomposition, smelling so badly that the crew very speedily cast it overboard—but not before its measurements were recorded, a few fibres taken from it, and some photographs obtained.

Biologists were very perplexed by the photos, and many identities were offered—including a giant sea-lion, a huge marine turtle, a whale, and a modern-day plesiosaur. For quite a time, its true identity remained undisclosed, until the fibres from it were meticulously analysed by Tokyo University biochemist Dr Shigeru Kimura. He discovered that they

The *Zuiyo Maru* carcase (Fortean Picture Library)

contained a very special type of protein, called elastoidin, which is found only in sharks—not for the first time, the scientific world had been fooled by a deceptively-shaped decomposing shark carcase. Indeed, this phenomenon is so well-known nowadays that cryptozoological writer Daniel Cohen has even coined an apt term for such a carcase—a pseudoplesiosaur.

Pseudoplesiosaurs generally turn out to be the decaying carcases of the basking shark *Cetorhinus maximus*, but as it looks nothing like a plesiosaur when alive, how do its dead remains acquire this appearance? In fact, the striking metamorphosis is easily explained.

When the carcase begins to decompose, the gill apparatus falls away, taking with it the shark's jaws, and leaving behind only its small cranium and its exposed backbone, which resemble a small head and long neck. The triangular dorsal fin also rots away, but sometimes leaves behind the rays, which can look a little like a mane—especially when the fish's skin also decays, allowing the underlying muscle fibres and connective tissue to break up into hair-like growth. Additionally, the end of the backbone only runs into the top fluke of the tail, which means that during decomposition the lower tail fluke falls off, leaving behind what looks like a long slender tail. The plesiosaur image is completed by the shark's fins, because the pectoral and sometimes the pelvic fins remain attached, yielding the plesiosaur's flippers.

Sometimes, however, the fins become distorted, so that they can (albeit with a little imagination!) look like legs with feet and toes, and male sharks have a pair of limb-like copulatory organs called claspers, which can thus

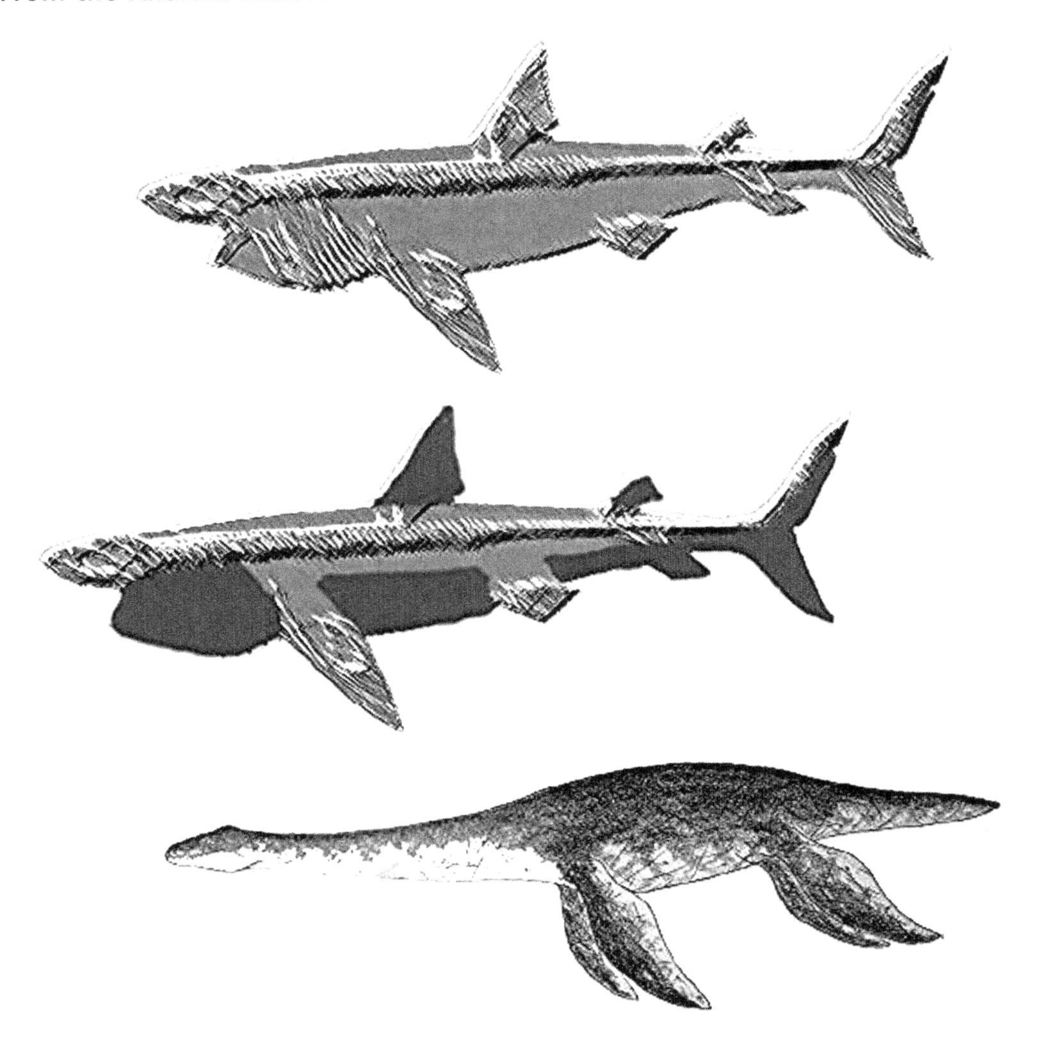

From top to bottom: basking shark carcase, pseudoplesiosaur, real plesiosaur (© Markus Bühler/ *Journal of Cryptozoology*)

yield a third pair of 'legs'—an extraordinary transformation from a dead basking shark to a hairy six-legged long-neck that explains the remarkable hexapod Stronsay sea serpent. A decidedly strange, giraffe-necked carcase, it was washed ashore in the Orkneys during October 1808, but its identity as a shark was confirmed by studying some of its preserved vertebrae, held at the Royal Museum of Scotland.

Yet another likely reason why long-necks are not encountered as often as one might expect is that they dislike the sounds emitted by the various mechanical contrivances in which humans put to sea or traverse the larger inland waters of the world; hence they actively avoid them. Nessie, for instance, has apparently reacted instantly on several separate occasions to the sound of a motorboat engine, and in a number of cases the sounds of sea-going vessels seem to have generated sightings of rapidly-departing marine long-necks.

The Course of True Extinction Never Did Run Smooth

So far, it has been shown that many objections commonly cited by sceptics for discounting the possibility of plesiosaur persistence into modern times are by no means as convincing as they seek to make out. However, there is one notable objection still to consider—the precise courses of supposed extinction taken by plesiosaurs and various other taxonomic groups.

Although coelacanth survival into the present day is commonly cited as a major precedent when considering prehistoric survival of other ancient lineages, and, as has already been discussed in various sections of this present book, it certainly has merit in this capacity, there is one area where it signally fails to provide the necessary degree of support.

Plesiosaurs and a number of other major groups of Mesozoic reptile (such as the ornithischian dinosaurs and non-avian saurischian dinosaurs, the pterosaurs, and the mosasaurs), as well as (albeit to a somewhat lesser extent) the ammonite molluscs, all exhibited high taxonomic diversity prior to a devastating collapse, an apparent mass extinction, by the end of the Mesozoic. In contrast, the coelacanths have always been far less taxonomically diverse and appear to have continued in this unspectacular but steady vein from the Mesozoic and onward through the Cenozoic into the present day, with no massive increase in diversity followed by a comparably massive collapse—or, at least, that is what the known fossil record indicates. But here is where the latter reveals its weakness as a basis upon which to make too substantial a judgement.

For although there are a fair number of Mesozoic coelacanth fossils on record, and two modern-day species alive today, the intervening 66 million years or more of coelacanth existence is represented entirely by the merest handful of remains, and which, in fact, constitute the only physical evidence preventing the entire coelacanth dynasty from exhibiting a ghost lineage spanning 66 million years or more. So where are all of their missing fossils? Even if the coelacanths have been taxonomically sparse throughout their Cenozoic persistence, they should still be represented by more than the most paltry smattering of palaeontological proof for their occurrence during that immense period of time—unless, as noted earlier when considering the fossil record's shortcomings, such fossils do exist but only in terrain where they can never be unearthed (for physical, political, and/or financial reasons); or the coelacanths died in regions where fossils never or only rarely formed anyway. What if all of the considerations presented here are also true for Cenozoic plesiosaurs and other undiscovered Mesozoic survivors?

Worth noting, incidentally, is that until the discovery of *Nichollssaura borealis*, an early Cretaceous leptocleidid species from Canada that was formally described and named in 2008, North America's fossil plesiosaurs had sported a ghost lineage of approximately 40 million years.

There is little doubt that if plesiosaurs have indeed survived into the present day (and again this same argument can be applied to pterosaurs, mosasaurs, and non-avian dinosaurs too), they are severely depleted taxonomically and numerically, the plethora of ecological niches that they once occupied now taken over for the most part by the explosion in mammalian and avian diversity that occurred during the early Cenozoic. However, certain specialised niches may still exist that they could occupy, albeit in reduced numbers and diminished diversity (in turn providing another reason why such creatures are not observed as often as one might otherwise expect). For instance, as noted earlier, Dr Michael Woodley has pointed to the 'benthic grazer' niche occupied by plesiosaurs during the Mesozoic seemingly remaining unoccupied by Cenozoic mammals. So although these long-necked reptiles' time as dominant alpha predators is undoubtedly long over, the possibility of contemporary plesiosaur survival in a more limited, specialised form is one that I find difficult to rule out entirely.

Finally: those readers who still consider a pinniped to be a more plausible long-neck than any plesiosaur might do well to recollect the Spicers' description of Nessie: "It was horrible—an abomination . . . a loathsome sight". Another eyewitness, Mrs H. Finley, exclaimed: "It was horrible! I wouldn't go to look at it if it was exhibited behind six-inch steel bars!". It is difficult to believe that any mammal, even one as extraordinary as a long-necked seal, would engender such profound feelings of revulsion.

Conversely, as Dr Denys Tucker so judiciously remarked, these are the very types of comment that we would not be at all surprised to hear from people visiting a zoo's reptile house. Just a coincidence?

Perhaps it is. Then again, however, as one Nessie researcher once drily observed: "If we regard everything as coincidence, we shall soon have to admit that Loch Ness has the highest 'coincidence rate' in the world!"

SERPENT-WHALES AND SUPER-OTTERS

The category of water monster most likely to be responsible for inspiring the 'sea serpent' name-tag consists of a number of unidentified marine and freshwater creatures with very elongate, serpentiform bodies, often of great length and flexibility, that can be thrown upwards into a series of distinctive vertical coils or undulations as the creatures swim.

Ironically, however, whatever their identity may one day prove to be, it will not be that of a snake, because snakes undulate horizontally—they cannot flex their bodies vertically. The only elongate creatures that may have been capable of achieving this mode of swimming are ones belonging to a specialised group currently believed to have died out many millions of years ago. Consequently, the serpentiform water monsters constitute another group of animals whose identity may involve a case of undiscovered prehistoric persistence, as highlighted via this next selection of cases.

Ogopogo—Many-Humped Monster of Lake Okanagan

Although many people have claimed sightings of water monsters, very few have apparently swum with one, and even fewer have actually touched one! Yet according to Mrs B. Clark, she did all of these things one warm, sunny morning in mid-July 1974, when swimming in British Columbia's Lake Okanagan.

Roughly 79 miles long, 2.5 miles wide, and 800 ft deep in places, this is one of North America's most famous 'monster' lakes, because there is a centuries-old tradition that Okanagan harbours a very long, serpentine creature, capable of flexing its body into vertical humps or coils. Westerners nicknamed it Ogopogo (after *The Ogo-Pogo*, an English music-hall song from 1924 featuring a banjo-playing water beast from Hindustan, whose ". . . mother was an earwig, his father was a whale"!), but the local Okanakane Indians called it the naitaka ('lake monster'). Today it is popularly dubbed Canada's Loch Ness monster.

At the time of her uniquely close encounter with Ogopogo, Clark was a teenage student, and had been swimming near the lake's southern shore at about 8.00 am. She estimated the creature to be 25-30 ft in length, 3-4 ft across, and very dull, dark grey in colour, with light stripes on its back and light round spots on its tail. Its skin appeared smooth and hairless, its body long and narrow, and she believed it to be very curious about her, but she did not see its head as the creature was moving away from her throughout her sighting—which she described to the International Society of Cryptozoology in 1987 as follows:

> I did not see it first. I felt it. I was swimming towards a raft/diving platform located about a quarter of a mile offshore, when something big and heavy bumped my legs. At this point, I was about 3 feet from the raft, and I made a mad dash for it and got out of the water. It was then that I saw it. . . .
>
> When I first saw it, it was about 15 to 20 feet away. I could see a hump or coil which was 8 feet long and 4 feet above the water moving in a forward motion. It was travelling north, away from me. It did not seem to be in much of a rush, and it swam very slowly. The water was very clear, and 5 to 10 feet behind the hump, about 5 to 8 feet below the surface, I could see its tail. The tail was forked and horizontal like a whale's, and it was 4 to 6 feet wide. As the hump submerged, the tail came to the surface until its tip poked above the water about a foot. . . . About 4 or 5 minutes passed from the time it bumped me until the time it swam from view. . .
>
> It was in the process of diving when I first saw it. It did not completely surface again, but it was so large and the water so clear that I could see it very well as it lazily swam north just a few feet below the surface. . . . It swam north in an undulating manner. Although it swam smoothly and well, it created a very large wake. It traveled about 3-5 miles per hour. It moved up and down. . . . After it was too far away to see any more, I could still see the large wake for several minutes. . . .
>
> At the time, I could not believe my eyes, so I told myself it was just a big fish. However, fish don't grow that big (in lakes anyway) or behave the way this thing did. This thing looked and acted more like a whale than a fish, but I have never seen a whale that skinny and snaky-looking before. Nor have I ever heard of any that fit that description. I have seen killer whales perform at the Vancouver Aquarium, and although the animal I saw had a whale-like tail, it didn't use it the same way as normal

Front cover of the sheet music for *The Ogo-Pogo—The Funny Fox-Trot*

> whales. By this I mean that it didn't seem to rely on it to the same extent as regular whales. Instead of using it as its main source of propulsion, it kind of 'humped' itself along like a giant inchworm. . . . I really have no idea what I saw, but it was definitely not reptilian, and I'm sure it wasn't a fish.

Although unquestionably the most memorable Ogopogo report on file, Mrs Clark's is far from being the only one. Over 200 sightings have been listed by the beast's leading investigator, Arlene Gaal, in her book *Ogopogo—The True Story of the Okanagan Lake Million Dollar Monster* (1986), and others have been reported each year since then. Some of the most notable examples on record include the following ones.

On 2 July 1949, Mr and Mrs L. Kerry were cruising upon the lake near the eastern lakeside town of Kelowna with visitors Mr and Mrs W. Watson when they all spied a long sinuous body, dark in colour and about 30 ft long, above the water surface, revealing five vertical undulations and what appeared to be a forked tail. The creature submerged and reappeared several times before finally swimming away, seemingly in pursuit of a shoal of fishes.

A five-humped monster, with a snake-like head raised 9 in above the water, was spied following about 250 ft behind their boat on 17 July 1959 by *Vernon Advertiser* editor Dick Miller and his wife. And a photograph depicting a series of dark humps above the water surface was one of five snapped on 3 August 1976 by Ed Fletcher during a thrilling 1.5-hour-long motorboat *pursuit* of Ogopogo—estimated by them to be 70 ft long, it surfaced and submerged more than a dozen times, briefly revealing a flattened, snake-like head at least 2 ft long.

No less startling was the observation made by Mrs and Mrs Rowden on 12 September 1976, for while relaxing that day at Okanagan's Peachland government lookout they were privileged to see *two* Ogopogos—swimming out across the lake to meet one another.

On 19 February 1978, Mrs McAdam from Penticton was just one of 30 eyewitnesses who viewed the astonishing spectacle of a 50-ft-long snake-like beast frolicking in the sun not far from shore at a campsite near Penticton. During its cavorting, which lasted for quite a time, several jet-black humps appeared, but eventually it swam out into deeper water.

During the evenings of 18 and 19 July 1989, former car salesman Ken Chaplin videoed an extraordinary object that he claimed to be a genuine Ogopogo. The prestigious *National Geographic* magazine was sufficiently impressed by the films to prepare computer-enhanced versions—and in the clearest section a low blackish shape with head raised can be seen moving towards the left, with what appear to be a vertical shoulder-like hump and a tapering tail also visible. The 'shape' then turns to the camera, turns once again, then swiftly flicks its tail upward before slapping the water in a downward motion as it rolls underwater.

During October 2000, American cryptozoologist Prof. Roy P. Mackal was seeking Ogopogo with a Japanese film crew. While at the lake, he interviewed a security guard who had had a summer sighting of Ogopogo. Based upon the description given, Mackal suggested that what the guard had seen may have been a juvenile Ogopogo, recently birthed, with possible foetal membrane attached. Mackal and the team did not have any sightings themselves, but they did make very strong sonar contact with what they believed to be two adult specimens.

On 18 April 2002, while a crew from Tripod Film and Video Productions was recording a scene reconstructing a 1978 sighting by Bill Steciuk of an alleged Ogopogo at the west of Okanagan Lake Bridge, with Bill's son Rob playing the role of Bill, three humps suddenly appeared about 200 yards away, above an otherwise totally calm expanse of water. They were undulating in and out of the water, at one point clearly revealing a space between one hump and the water, and were witnessed by a group of 14 observers, including Steciuk Snr himself, who was acting as an advisor for the film. Fortunately, the humps were filmed, yielding a clip lasting about 90 seconds and of excellent quality. One eyewitness, Renee Boucher, described what she saw as "something moving in a very slow and calm manner. It was moving up and down so gracefully. . . . It was huge, black and shiny, very slick looking. It was very, very long".

Nowadays, inevitably, Ogopogo has become a popular tourist attraction, and for those not fortunate enough to spot the real thing there is an impressive statue, replete with tall vertical coils and (unfortunately) an unrealistic dragon-inspired head, on display in Kelowna. In 1983, the local tourist association announced that a reward of 1 million dollars would be paid to anyone obtaining conclusive proof of the monster's existence before 1 February 1985—the fact that Ogopogo is still a mystery beast should be sufficient answer to those wondering whether the prize was ever awarded!

Another Canadian freshwater monster of the serpentiform variety is Manipogo of Lake Manitoba in the province of Manitoba. Sightings have been reported since at least as long ago as 1908, and traditional Native American lore includes centuries-old legends of huge snake-like water beasts here. However, events may have taken a seriously disturbing turn during the early part of summer 1997. For according to rumours swirling online and else-

Artistic representation of Ogopogo (© Richard Svensson)

where at that time, this was when a 45-ft-long serpentine monster with a horse-like head was allegedly shot here, then secretly smuggled away, and either sold to a reclusive purchaser for $200,000 or placed under guard with the Royal Canadian Mounted Police. Not surprisingly, the Mounties were inundated with enquiries and apparently investigated the stories, but no formal statement has been released, nor has any evidence in support of such claims been revealed.

Happily, however, even if such an event did occur, it clearly did not mark the end of Manipogo as a species (as opposed to a lone individual), because several notable observations have been recorded since. As recently as 2011, for instance, security personnel patrolling flooded cottages and home areas near Lake Manitoba reported many sightings of several humps emerging and then submerging again in this lake, as seen offshore at Laurentia Beach, Marshy Point, Scotch Bay, and certain other locations.

Huillia—An Ambiguity of Arches from Trinidad

The all-too-common mistake of assuming that serpentiform animals performing vertical undulations must be bona fide snakes might explain why, in January 1934, an enthusiastic expedition striving to capture the elusive huillia of Trinidad was singularly unsuccessful.

Known from media accounts emanating from the River Ortoire in the island's eastern portion, the huillia is said to be an elongate water monster measuring 25-50 ft long, which can flex its slender, scaly body into 'arches' that rise and fall as it swims—often at steamboat speeds. It is also said to come ashore and make its way across country, migrating overland from one stretch of water to another, and can roar like a lion.

A backbone capable of rising and falling in arches is not of ophidian origin—on the contrary, it implies a mammalian identity, as does its propensity for roaring. Sadly, the huillia-hunting party that set out in January 1934 was apparently unaware of this fact—because their lure consisted of a Hindu snake-charmer and flute, whose mesmerising melodies were expected to lull the beast into an obligingly motionless state of slumber, thus enabling them to secure it. If only. . . !

One book that is often cited in relation to the huillia is Edward L. Joseph's *History of Trinidad* (1838), but it does not actually contain any specific mention of the creature's alleged vertical undulations. What it does state is:

> On the eastern side of the island there is said to exist an amphibious kind of snake, which has been described to me as of the boa species. This is said to grow to more than 30 feet in length—it is said to be an inhabitant of the marshes and rivers of the Bande de l'Est, and is reported to have attacked men, or at least a man.

Similarly, in his own book *Five Years' Residence in the West Indies* (1852), Charles W. Day noted that in some of the rivers on the eastern coast, particularly the Oropouche, "enormous water boas, called Huilla, are found", which suspend themselves from overhanging trees and drop onto whatever prey passes underneath (note the slightly different spelling here of 'huillia'). In 1871, Charles Kingsley stated in his book *At Last: A Christmas in the West Indies* that the huillia was the common green anaconda *Eunectes murinus*. So is the huillia based merely upon sightings of very large boas (and whose claimed size is probably due more to exaggeration than accuracy of observation)?

Between November 1999 and January 2000, I exchanged letters with herpetological specialist Hans E.A. Boos from Port of Spain on Trinidad, who is extremely knowledgeable concerning the reptilian fauna of this island. Yet he has no knowledge of any cryptozoological huillia (nor even any published mention of it), only the very large ophidian constrictors known locally there as huillia/huilla/huille.

Consequently, Boos speculated that the huillia cryptid may have simply originated as a media invention to boost Trinidad tourism—and here's why he suggested this. He noted that a mock-up photo of a 'monster' huillia with vertical humps or coils moving across a river appeared prominently in an article concerning this creature on the front page of Trinidad's *Sunday Guardian* newspaper on 21 January 1934, which offered a reward to anyone who could photograph the real thing. Significantly, this article appeared just after the newspaper had received a report concerning Canada's vertically-undulating Ogopogo from a foreign correspondent, and also just after Loch Ness had seized headlines worldwide concerning Nessie. Indeed, in the article the fake huillia photo was aligned alongside a photo purportedly depicting Nessie.

Furthermore, what were referred to by this newspaper as "art photographs" of its huillia picture were even available to purchase (priced at 1 shilling) for a time in Port of Spain following the article's publication. In the subsequent days and weeks, various vague sightings of such a beast were duly reported (and as noted at the beginning of this present book's huillia section, even an expedition was launched to seek it), plus some maritime ones from the Caribbean—including an observation by the crew of a vessel called the *Mauritania*. But in view of the fact that the newspaper openly stated: "It is hoped that they will result in similar publicity for the island as has been received by Loch Ness", there can be little doubt, surely, that the huillia cryptid was truly a media-manufactured monster as opposed to a bona fide cryptozoological mystery beast.

The Marvellous Migo—A Composite Creation from New Britain

In January 1994, a Japanese TV crew arrived in Papua New Guinea during the rainy season to seek an elongate freshwater monster known to the local tribes as the migo or migaua. With Prof. Roy Mackal as scientific advisor, they travelled eastward to the island of New Britain in the Bismarck Archipelago and thence via sea vessel to a village sited near to a large, horseshoe-shaped lake. Variously referred to as Dakataua and Niugini, it had a diameter of 1,400 ft, a depth of around 30 ft, and contained three small islands plus a submerged volcano.

This lake is supposedly the home of the migo—but what allegedly sets it even further apart from other lakes is the strange claim that it contains no fishes. Consequently, the migo preys upon the abundant quantity of waterfowl that settle upon the water surface—which if true would be of great benefit to the team of investigators, because it would mean that migos would surface more often than is normally true for water monsters elsewhere.

After meeting the village chief and receiving eyewitness descriptions during the third day of their visit, the team actually succeeded in filming a migo, shooting a video of it from a range of about 0.7 mile. Judging from the video, the creature was over 33 ft long, swimming at a speed of 4 knots, and exhibiting the vertical undulations characterising many elongate water monsters. However, a detailed analysis of this video by Mackal as well as further observations made during subsequent visits to the lake revealed the surprising truth that the migo was not a single, individual creature at all. Instead, it turned out to be a composite, consisting of two male crocodiles and one female crocodile, all interacting together in a courtship display. Consequently, it may well be that the migo as a genuine cryptozoological entity does not exist.

Ireland's Elusive Horse-Eels

Horse-headed and serpent-bodied, what makes the horse-eels of Ireland more extraordinary than usual even for water monsters is their mystifying ability to inhabit lakes that seem too small to sustain them, only to disappear as soon as these lakes are investigated. Inevitably, such talents have attracted claims that they are paranormal entities, but it is far more likely (and reasonable) that they either are simply capable of moving overland from one lake to another, or don't exist at all.

A typical horse-eel case is that of the Lough Nahooin monster. Only 300 ft long, 240 ft wide, and 23 ft down at its deepest part, this tiny lake is situated close to Claddaghduff in County Galway, and on 22 February 1968 it hosted an important monster sighting. Events began when, while gathering peat in the early evening with his son and their dog, Stephen

Coyne spied a dark object in the water. When his dog began to bark, it revealed itself to be a large peculiar animal that swam towards them with its mouth open—until Coyne moved closer, whereupon the monster turned away and swam on further round the lake. Coyne's wife and their other four children joined him on the shore, viewing the beast in some detail, at as close a range as 15-30 ft for part of the time, before the evening grew too dark.

According to their description, it was about 12 ft long, with a smooth, black, hairless skin whose texture reminded them of an eel's, a slender body that was thrown up into a couple of vertical humps when it lowered its head beneath the water, a briefly-glimpsed tail, and a distinct neck with a horse-like head that bore a pair of horn-like projections on top (these have also been mentioned by a number of Ogopogo eyewitnesses). They did not notice its eyes, but when it opened its mouth the interior was pale in colour. Heartened by the Coynes' account and the ease with which he anticipated investigating such a small lake as this, veteran monster seeker Captain Lionel Leslie organised an attempt to trap the Nahooin beast during July 1968—leading a team of investigators (which included Prof. Mackal, armed with a tissue-collecting gun) who spent two days laying nets across the total breadth of the lake. Once the nets were in place, the investigators used an electronic fish-attractant and other devices to tempt the creature to show itself, but their methods neither cajoled it nor captured it.

A second attempt at Nahooin headed by Leslie was made in 1969, following another recent sighting of the creature, and involved dragging chains through the lake as well as netting it. However, this was unsuccessful too—leading to the obvious conclusion that the creature was not confined to the lake, but was sufficiently amphibious to be able come out onto land to avoid danger or to move to another locality if need be

In 1954, a fork-tailed, elongate, two-humped monster likened by its four frightened eyewitnesses to a giant worm, with a huge mouth and a head raised above the water surface on a slender neck, was spied swimming across Lough Fadda. This is another of County Galway's minuscule lakes, no more than 1.5 miles long and only 1,800 ft across its widest point.

On The Road—
An Overland Sea Serpent from the Firth of Tay

Just as there are freshwater and marine long-necks, so too are there freshwater and marine versions of the serpentiform water monsters—which, as with long-necks, are very occasionally seen on land. One of the most remarkable of these seldom-spied occurrences took place during the late evening of 30 September 1965 and was brought to cryptozoological attention by veteran monster hunter F.W. Holiday. It was 11.30 pm, and Maureen Ford (wife of amateur flyweight boxer David Ford) was driving with some friends along the A85 by car towards Perth, in northeastern Scotland. Close to Perth, Ford suddenly spied an extraordinary creature by the roadside, only a few yards from the banks of the River Tay, which enters nearby into the Firth of Tay—an inlet of the North Sea. She described it as: ". . . a long grey shape. It had no legs but I'm sure I saw long pointed ears".

Less than 2 hours later, it was seen again—but this time on the opposite side of the road, to where it had evidently crossed during the intervening period. At 1 am, Robert Swankie was driving along the A85 away from Perth towards Dundee, when his headlights revealed an amazing sight. As he later revealed in a *Scottish Daily Express* report (5 October 1965):

> The head was more than two feet long. It seemed to have pointed ears. The body, which was about 20 feet long, was humped like a giant caterpillar. It was moving very slowly and made a noise like someone dragging a heavy weight through the grass.

Swankie slowed down, and opened his window, but he could see another car not far behind, so he decided not to stop, and continued his journey. His testimony, and also that of Ford, were taped by an enthusiastic investigator, a Miss Russell-Fergusson of Clarach Recordings, Oban, and the police were also informed. In the *Express* report, one of their spokesmen commented that in the dark the headlights of a car could play tricks when they strike walls and trees—but as Holiday sensibly pointed out, if Swankie's sighting had merely been an optical illusion, why didn't he see monsters throughout his road journey? And how can an exclusively visual deception create a dragging sound?

Far more reasonable, surely, is the scenario of a reclusive sea creature emerging under the cover of darkness from the Firth of Tay, possibly via the River Tay itself, and, by sheer chance, being seen by two night-travelling eyewitnesses during its brief overland foray.

Marine 'Many-Humps' of North America

In his classification of sea serpents, those that are serpentiform in shape and given to yielding multiple vertical humps were referred to by Dr Bernard Heuvelmans as 'many-humped' sea serpents, and are most commonly reported in the seas around North America. He even gave their still-uncaptured species a formal scientific name—*Plurigibbosus novaeangliae*, emphasising the particular prevalence of sightings around the eastern U.S.A.'s New England coastline.

The Gloucester sea serpent, for example was reported by many eyewitnesses from 1817 to 1820 around Gloucester Harbor, Massachusetts, and its legacy included a description that summarises very succinctly the general appearance of the marine 'many-hump'. According to a news sheet on the monster, dated 22 August 1817, which was printed and sold by Henry Bowen of Boston, its elongate body ". . . resembles a string of buoys on a net rope, as is set in the water to catch herring. Others describe him as like a string of water casks." In other words, whenever it surfaced, its body yielded a series of small humps.

Dark brown or black in colour, with smooth scaleless skin, the creature allegedly measured 65 ft or so in total length, with a snake-like or turtle-like head the size of a horse's skull and held 6-12 in out of the water. Apparently attaining a maximum swimming speed estimated at 40-70 knots and capable of very agile turning movements, it was sighted by all manner of people at all times of the day under all sorts of viewing conditions.

Influenced, however, by the widespread belief at that time that sea serpents really were huge marine snakes, contemporary illustrations of the Gloucester beast generally portrayed it as just that—a gigantic scaly snake, with forked tongue and lateral coils (thus ignoring eyewitness testimony that its body was thrown into humps—i.e. vertical coils).

Unfortunately, its erroneously-claimed allegiance to the true snakes did not end there. In autumn 1817, a 3-ft-long 'baby sea serpent' with a series of humps along its back was captured in a field near the harbour of Cape Ann, north of Gloucester. When this odd-looking creature was observed by a committee of enthusiastic amateur investigators from the Linnaean Society of New England, it was deemed by them to be a young specimen of the

19th-Century engraving of the Gloucester sea serpent

Gloucester sea serpent, whose species was duly christened *Scoliophis atlanticus* ('Atlantic humped snake'). Unfortunately, once dissected and examined by fish researcher Alexandre Lesueur, it was found merely to be a deformed specimen of an embarrassingly common American snake called the black racer *Coluber constrictor*. Exit *Scoliophis atlanticus*!

North America's western coastline has seen its fair share of serpentiform but non-serpent water monsters too. One of the most important modern-day reports came from California's San Francisco Bay, featuring eyewitnesses William and Robert Clark. On the morning of 5 February 1985, they were sitting in their car enjoying a view of the bay just a few yards ahead when they noticed some seals swimming rapidly away from a wake following swiftly behind them—which, as it came closer, proved to be from an extraordinary beast resembling a huge black snake or eel, but swimming via vertical undulations of its body, readily visible in the bay's clear water.

With a maximum length of 60-75 ft, but an exceedingly slender body no more than 3 ft across, a 10-in-wide neck, a dorsoventrally flattened tail, and a thin snake-like head, the creature was dark brownish-green above, lightish yellow-green below, and although its skin appeared oily it also bore large round or hexagonal scales. Of particular interest was the brief sighting of a fin, when a portion of its attenuate body surfaced through the water—for according to the Clark twins, this fin was triangular in shape, so thin that it was translucent, and contained a series of ribs, yielding a fan-like appearance recalling that of ray-finned (actinopterygian) fishes' fins.

The Clarks were afterwards able to prepare a detailed description of the creature's strange method of propulsion, which is unlike that of any animal known to science:

> The shape was definitely that of a snake or an eel, and the head was visible as it swam perhaps only a foot under the water. Behind the head I could see what appeared to be four humps, and the animal appeared to be propelling itself by wriggling in a vertical manner. A series of four coils was created at the front half of its body, and these travelled backwards along the length of the neck where they would meet the middle body. At this point the undulations stopped abruptly and slowly dissipated along the remaining part of the body, which was dragged behind. There would be a short pause and another series of coils would begin . . . this was all happening at a very high speed.

This creature may well belong to the same unidentified species that had been spied at Stinson Beach, north of San Francisco, on 31 October 1983 by several different eyewitnesses, including five members of a construction crew repairing a highway overlooking the sea. According to one of these observers, Matt Ratto, it first appeared as ". . . three bends, like humps, and they rose straight up. Then the head came up to look around", after which the creature turned, lowered its head beneath the surface once more, and moved out to sea, the humps gradually submerging until it was lost to sight. There was agreement among all of the eyewitnesses that the creature was dark, slender, and approximately 100 ft long.

Ratto produced a sketch of what he had seen. Of particular note here is that he depicted the humps as vertical hoops—each with a definite space *between* its undersurface and the sea's surface.

Highly elongate, serpentiform aquatic monsters reported from British Columbia are not limited to Ogopogo and other freshwater examples—similar creatures have often been sighted around this Canadian province's coastlines too. On 4 February 1934, for instance, duck hunters Cyril H. Andrews and Norman Georgeson spied a long snake-like beast just below the surface of the water surrounding the rocks where they were standing near Gowland Head, South Pender Island.

As they attempted to retrieve a duck that they had shot and which had fallen into the sea nearby, their sinuous visitor raised its head and two vertical loops of its body out of the water only a few yards away. Its head was about 3 ft long and resembled a horse's in shape, but had neither ears nor nostrils, and its mouth contained a pointed tongue and fish-like teeth. Its loops were grey-brown in colour with a darker brown stripe running along the side of the body—which tapered towards the tail, and measured about 40 ft in total length.

The loops swiftly submerged again, but the head lingered above the surface just long enough to open its large mouth, seize the duck, and swallow it. Interest in its actions were not confined to the amazed duck hunters, for while it was ingesting its pilfered prey several gulls swooped down towards it, but, evidently accustomed to such behaviour, the animal kept them at bay by snapping its jaws from time to time until it finally submerged completely.

Determined to obtain official verification for their encounter, the hunters promptly phoned G.F. Parkyn, the local J.P., who arrived about 15 minutes later. Not long afterwards, he was rewarded with a good sighting of the beast, when its head and about 12 ft of a 2-ft-wide serpentine body (but undulating vertically) rose up out of the water about 20 yards from the shore. Several other local people saw it on this occasion too.

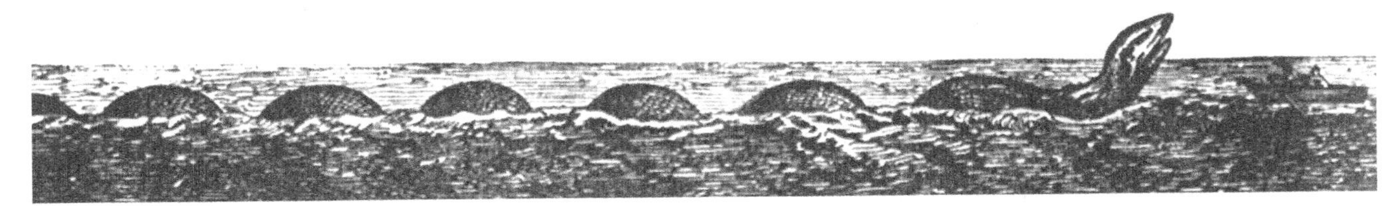

Hans Ström's Heroy sea serpent

The Great Sea-Worm of Norway

Another area of the world where serpentiform sea monsters have been sighted on many occasions is the icy waters around northern Scandinavia, particularly Norway—the veritable birthplace, in fact, of sea serpent reports. As long ago as 1555, Olaus Magnus, exiled archbishop of Uppsala in Sweden, had included an impressive chronicle of such reports within an exhaustive volume entitled *Historia de Gentibus Septentrionalibus*, dealing with the history of Europe's boreal lands. In this portion of the globe, these animals were generally termed the sœ-orm (sea-worm), but their description recalled the North American 'many-hump'.

A good example of this was the extremely serpentine but vertically sinuous specimen spied on several occasions at Heroy, Norway, by two neighbours of Pastor Hans Ström, called Reutz and Tuchsen, while travelling to church by boat. Ström prepared a sketch of the multiple-humped creature (reproduced above) based upon their description, and sent it to Erik Ludvigsen Pontoppidan, Bishop of Bergen, who included it in a major work of the 1750s that was published in English under the title *The Natural History of Norway*. He likened the creature's humps to "... so many casks or hogsheads floating in a line"—corresponding precisely with the "string of casks" description given for the Gloucester sea serpent.

Notwithstanding this, in his analysis of sea serpents Heuvelmans considered the Norwegian sea-worms and the New World many-humps to be taxonomically discrete entities, and dubbed the former type *Hyperhydra egedei*—'Egede's super-otter'.

This distinctive name was inspired by a sea monster roughly 100 ft long, sighted off Godthaab, western Greenland, on 6 July 1734 by a Greenland priest called Hans Egede, who documented it in his book *A Description of Greenland* (1745).

According to Egede, the monster rose up out of the water until its head was as high as his boat's masthead. Its body was 3-4 times as long as his ship, but no less bulky, and its skin was so rough and uneven that it looked as if it were covered in scales or a shell. Two features of especial interest were its snout and its forelimbs. Egede recorded that it spouted "... like a whale-fish" from its long pointed snout, and described its forelimbs as "... great broad paws". Yet despite its 'paws', it must have been distinctly serpentine, because Egede described its body's underparts as "... shaped like an enormous huge serpent". In addition, its tail—raised aloft and erect after the beast's forequarters had finally plunged back down beneath the waves—seemed "... a whole ship's length distant from the bulkiest part of its body". Another eyewitness, Pastor Bing, soon afterwards produced a sketch of this extraordinary animal, which closely matched Egede's verbal account.

Many of the eyewitness accounts of serpentiform water monsters given here are so

Hans Egede's sea serpent as depicted by Bing

precise and so compelling that there can be little doubt of such creatures' reality—and of their novel taxonomic status. Neither snake nor eel, nor indeed any other type of animal whose present-day existence is formally recognised by science, can offer a satisfactory identity for this perplexing category of mystery beast. When we turn to the fossil record for inspiration, however, a very different story emerges.

From Archaeocetes to Zeuglodonts—An A-Z of Serpent-Whales

On first sight, if asked to select a veritable sea serpent of the sinuous, near-limbless variety from the pages of prehistory we might be hard-pressed to come up with anything that corresponds more intimately than a basilosaurine—outwardly the personification of everything strange and serpentine that has ever been glimpsed amid the sometimes-tranquil sometimes-tempestuous seas, not to mention the elongate entities reported here from Okanagan and other large bodies of freshwater worldwide.

Basilosaurines belonged to an extinct sub-order of cetaceans known as archaeocetes ('ancient whales') or zeuglodonts ('yoked teeth'), and after thriving through much of the Eocene epoch they officially died out towards its end around 36 million years ago. Sometimes over 60 ft long (e.g. *Basilosaurus cetoides*), they were extremely elongate creatures, the most serpentiform mammals ever known to have existed—with a very small head, exceedingly long body containing an estimated 70 vertebrae, a pair of front flippers, two very reduced, tiny hind limbs (traditionally portrayed as little more than bony vestiges generally buried within the body wall, but see also later in this section), plus a short but very powerful tail flexed dorsoventrally and probably bearing a pair of horizontal flukes (as suggested by the presence of a 'baseball' vertebra).

Basilosaurines were fish-eaters, with slender jaws containing numerous teeth of two principal types—pointed and conical at the front, saw-edged at the back and each resembling two smaller teeth that have been yoked together (the derivation of 'zeuglodon').

The localities in which the fossilised remains of basilosaurines have been found (on all continents including Antarctica) indicate not only a widespread zoogeographical distribution for these creatures but also a variety of marine habitats—ranging from shallow offshore seas to the deeper and more open oceanic water. And as mentioned earlier with the plesiosaurs, if basilosaurines migrated into large stretches of freshwater in search of prey or to escape predators (or both) they could well have gradually adapted to survive on a permanent basis here too.

Moreover, because their vertebrae were marrow-filled (rather than solid as in modern-day cetaceans), it is supposed that basilosaurines were less well adapted for submerging than modern-day cetaceans. This is because pressure imbalance between the marrow in their hollow vertebrae and stronger pressure of deep water externally might conceivably have caused the risk of spinal injury with the vertebrae being crushed if the pressure became too much. In addition, being hollow, their vertebrae would have made these unusual whales more buoyant than those with solid vertebrae (another disadvantage for diving purposes). Consequently, basilosaurines probably swam and hunted only near the surface, not at depths, which in turn might help to explain the sizeable number of surface sightings on record for elongate water monsters.

As popularly claimed in the earlier literature, it was long assumed that basilosaurines swam via vertical undulations of what was widely thought at that time to be their exceedingly flexible backbone, thereby contrasting markedly with the horizontal flexures performed by elongate aquatic reptiles such as water snakes, or sinuous fishes like eels. If true, this would be of especial cryptozoological importance—for as seen from the selection of cases documented here, it is a characteristic behavioural trait of several sinuous water monsters on file, yet no *known* contemporary animals can perform this distinctive feat.

In more recent times, however, continued studies of basilosaurine fossils, in particular the fine structure of their vertebrae (as exemplified by the discovery in 2015 of the first-known complete *Basilosaurus* skeleton), have resulted in a starkly different palaeontological interpretation of their likely mode of (and limitations to) locomotion. For although it is still believed that they were capable of vertical arching undulations when swimming, their vertebral structure indicates that the spinal column was much more rigid than previously supposed, thereby seemingly ruling out the ability to produce the exceedingly flexible vertical coils and loops reported by eyewitnesses of serpentiform water monsters.

However, as noted earlier in this chapter with regard to the degree of flexibility in plesiosaur necks, the quantity of intervertebral cartilage present might have mitigated this ostensible inflexibility of the basilosaurine spine. Certainly, I find it difficult to believe that any creature so serpentine in outward form and with so many vertebrae would be as rigid and as limited in flexion as has been proposed in certain modern-day reconstructions of basilosaurines.

In any case, if a species of basilosaurine exists today, it will be an evolved form—36 million years of intervening evolution having no doubt wrought a number of morphological and physiological modifications, most notably potentially yielding a more flexible spine than in its prehistoric predecessors, and therefore much more likely to be able to achieve the marked vertical 'looping' and coiling reported for serpentiform water monsters. Moreover, certain elongate water monsters appear able to raise their necks above the water surface. Conversely, although the neck vertebrae of even the most advanced fossil basilosaurines were still unfused, the vertebrae themselves were only very short—thus the neck could not have achieved this feat. In contrast, the neck vertebrae of earlier forms were not only unfused but also quite long. Consequently, if this particular lineage had persisted, and continued evolution had acted in a manner to increase the length of the vertebrae even further, the result could have been a basilosaurine with a neck sufficiently flexible *and* long to be raised up through the water surface—precisely as reported by eyewitnesses of some elongate water monsters.

Equally, if the skin encompassing the digits (fingers) of the basilosaurines' forelimbs had gradually become more membranous during evolution, the outcome might well be a fan-like flipper, capable of being expanded and closed. To an observer untrained in zoology, this would greatly resemble the fin of an actinopterygian fish—thereby explaining how the San Francisco Bay sea serpent could undulate vertically like a serpentine mammal yet possess ostensibly fish-like fins.

Basilosaurines were probably the earliest cetaceans that were fully aquatic, but despite being highly modified for such an existence they retained various terrestrial characteristics that were lost in the two more advanced suborders of cetaceans, i.e. the toothed whales (odontocetes) and toothless whales (mysticetes).

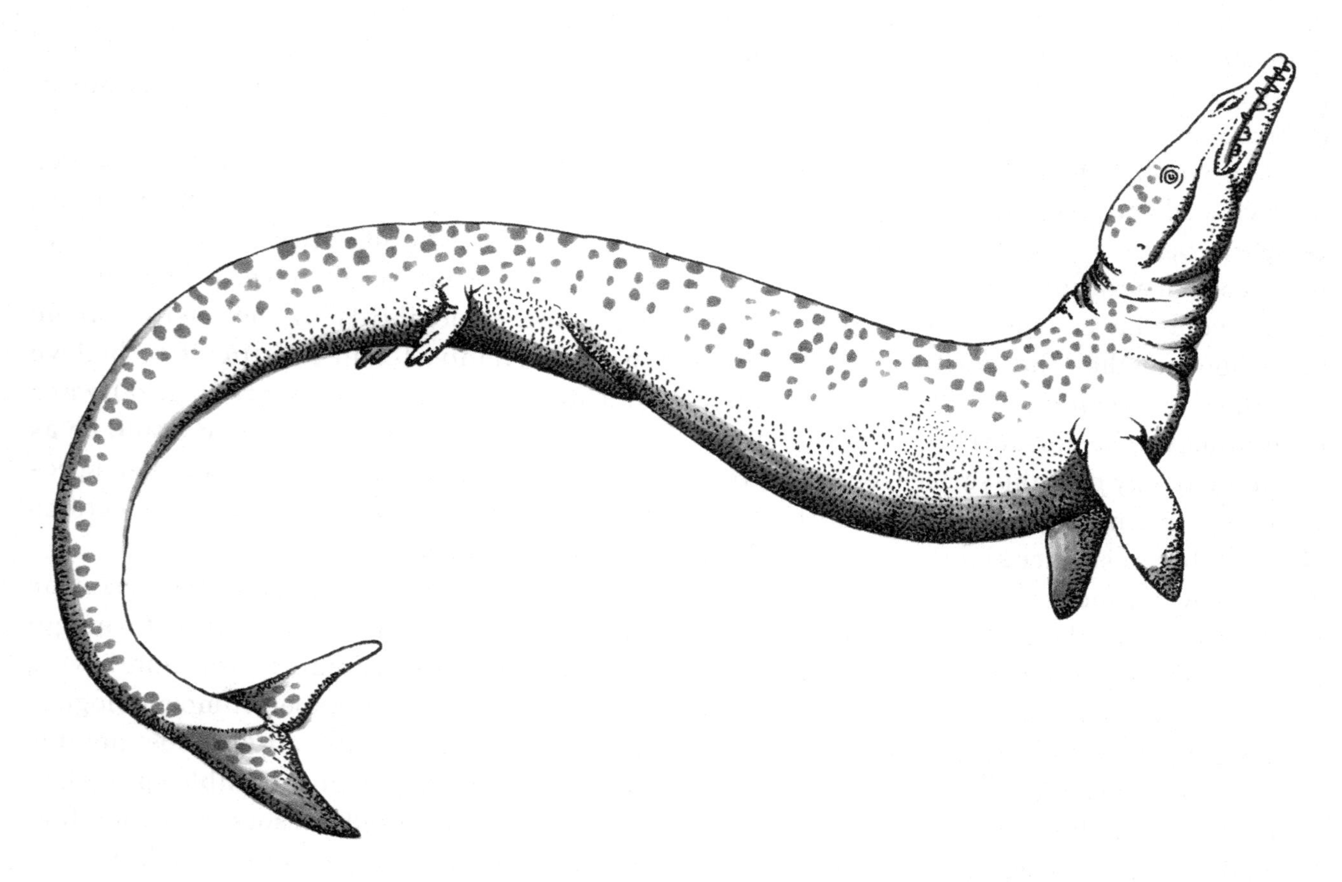

Restoration of possible appearance in life of a basilosaurine (© Tim Morris)

Artistic representation of an evolved, modern-day basilosaurine based upon eyewitness descriptions of serpentiform water monsters (© William M. Rebsamen)

These include: the forward position of the nasal bones; the limited degree of skull telescoping (compared with the extreme condition displayed by odontocetes and mysticetes); seeking prey by sight and smell rather than utilising sonar; as noted earlier, a short yet still flexible neck with unfused vertebrae (fused and thus immobile in more advanced cetaceans) that enabled them to turn their head; and moveable elbow joints (again fused and therefore rigid in more advanced cetaceans).

Indeed, whereas modern-day cetaceans are exclusively aquatic, some of the less specialised basilosaurines may have been capable of coming ashore for copulation and parturition, as their moveable elbow joints might have facilitated terrestrial locomotion. Such an ability could naturally explain the apparent migration of horse-eels from one lake to another in Ireland. And if, when on land, an evolved modern-day basilosaurine with a more flexible spine was able to move by humping its body vertically, rather like a giant caterpillar, this might explain the earlier-noted land sightings on the A85 near the Firth of Tay in September 1965. Obviously, as is so often the case in cryptozoology when there is no tangible, physical evidence to study, this is all speculation, nothing more, but it is by no means implausible nonetheless. Indeed, leading cetologist Dr R. Ewan Fordyce has even suggested that despite their aquatic specialisations, basilosaurines may have been able to wriggle across sandbars like giant pinnipeds.

On account of the compelling correspondence between basilosaurine morphology and their potential behavioural capabilities with the details reported by eyewitnesses for water monsters of the elongate, serpentiform variety, many cryptozoologists consider it likely that those latter unidentified mystery beasts constitute a surviving, evolved branch of this officially long-vanished cetacean lineage.

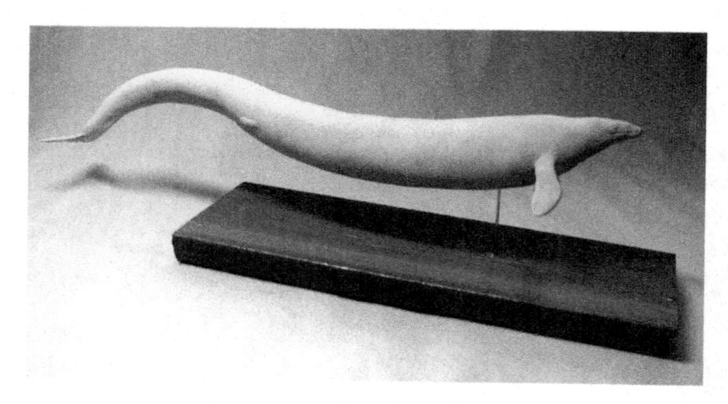

A model of a basilosaurine (© Markus Bühler)

Assuming that it is a genuine report (and not, as the more cynically-minded might suggest, a well-spun yarn inspired by careful reading of books on prehistoric animals that contain descriptions of basilosaurines), Mrs B. Clark's Ogopogo account, for example, is an uncannily-accurate description of what one might expect a living, evolved basilosaurine to be like, in terms of both its appearance and its behaviour. Everything in her report—from the creature's horizontal, forked tail fin and its body's sinuous, serpentine shape to its method of propulsion via vertical undulations and its typically cetacean curiosity regarding its human visitor—instantly recalls a basilisaurine while simultaneously discounting all other identities.

One of the observers of the Lough Fadda horse-eel during the 1954 sighting noted earlier was Clifden librarian Georgina Carberry. Observing it at a distance of only 60 ft, she considered the appearance of its body to be decidedly "... creepy ... [as it] seemed to have movement all over it all the time." Needless to say, this is precisely what one would expect from a vertically-undulating, serpentine basilosaurine.

Although outside the specific premise for this book's contents, a brief mention must also be made here of what I consider to be a plausible alternative identity for serpentiform water monsters that does not invoke a prehistoric survivor but is no less thought-provoking. Namely, an undiscovered species of highly-elongate, vertically-undulating, modern-day cetacean (or what British palaeontologist Dr Darren Naish once termed a "superweird elongate odontocete").

Various present-day taxa, notably certain species of delphinidan (e.g. the two right whale dolphins, genus *Lissodelphis*), some river dolphins, and also certain ziphiids or beaked whales, already exhibit quite an elongate body form and/or vertical undulation when swimming. Hence it would not involve a drastic degree of continued evolution along this morphological pathway to yield a decidedly serpentiform version. And bearing in mind that ziphiids are among the most elusive of all modern-day cetaceans anyway, if highly elongate water monsters are indeed of recent evolutionary origin a ziphiid ancestry in particular must surely offer the most likely candidature for these only briefly-glimpsed aquatic cryptids.

Humps or Hoops? That is the Question!

Returning to prehistoric survivors: Prof. Roy Mackal was a leading supporter of the basilosaurine identity for serpentiform water monsters, particularly Ogopogo and also a marine equivalent called Caddy (see later). This was the identity put forward by Heuvelmans for North America's 'many-humped' sea serpent too. However, he explained the creature's

A selection of ziphiids (© William M. Rebsamen)

humps as genuine dorsal projections—even considering a possibility proposed by Ivan T. Sanderson that they are inflatable hydrostatic organs used as reserves of air for prolonged dives, and for alleviating these immense animals' weight. This explanation has also been offered by supporters of the plesiosaur identity for long-necks, which often possess one or more humps (though never anywhere near as many as the serpentiform water monsters).

Yet as revealed, for instance, by Ratto's sketch of the Stinson beast, when raised sufficiently high out of the water the humps prove not to be humps at all, but rather hoops—because daylight can be seen between their undersurface and the water's surface. In other words, in the case of the elongate water monsters the humps are not exposed, solid projections borne upon the back of a much more substantial, submerged body (as would seem to be true for hump-yielding long-necks). Instead, they are exposed, vertically-undulating, hooped portions of the body itself—a body, furthermore, that is extremely slender and serpentiform.

First Ancestors and Walking Whales

As noted earlier, Heuvelmans believed the Norwegian sea-worm to be distinct from the American 'many-hump', referring to it as *Hyperhydra*, the super-otter, on account of its superficially otter-like appearance despite its much greater size. Nevertheless, he did believe it to be an archaeocete, like the 'many-hump', but one descended from a much more primitive stage in cetacean evolution.

Despite having only the merest vestige of hind limbs today, cetaceans are ultimately descended from quadrupedal land mammals. (Traditionally, these were believed to be a group of hyena-like condylarths called mesonychids, but recent molecular studies have shown that an artiodactyl—even-toed ungulate—ancestry is more likely.) Consequently, there was obviously some point during their evolutionary transformation into aquatic beasts when they still possessed hind legs.

Artistic rendition of the super-otter's putative appearance, based upon Heuvelmans's hypothesis (© Tim Morris)

In his book *Follow the Whale* (1956), Ivan T. Sanderson referred to these earliest cetaceans as 'first ancestors', and speculated that some might even survive today. Developing this intriguing prospect, in his sea serpents book Heuvelmans proposed that the super-otter may be a modern-day 'first ancestor'—equipped with a lengthy, highly flexible backbone and tail readily capable of vertical undulations, and two pairs of paw-like flippers (thereby explaining the unusual appendages described by Egede for the sea monster that he saw at Godthaab).

Although an undeniable prerequisite in the evolution of cetaceans, for a very long time the onetime existence of 'first ancestors' remained unconfirmed by the fossil record. It almost seemed as if the cetaceans had sprung into being in truest Athena-style, i.e. already fully-equipped with all of their sophisticated specialisations for an exclusively aquatic mode of life—and then along came *Ambulocetus*, the whale that walked.

In the journal *Science* (14 January 1994), Drs Johannes G.M. Thewissen from Northeastern Ohio Universities College of Medicine, Sayed T. Hussain of Washington D.C.'s Howard University, and Islamabad-based Mohammad Arif described a major new species of fossil mammal—represented by several specimens, including a skeleton and partly-articulated long-snouted skull, recently unearthed from river deposits in Pakistan's Kuldana Formation that are around 50 million years old (early Eocene).

Its skull, dentition, vertebrae, and forelimbs labelled it as a primitive cetacean, roughly the size of an adult male South American sea-lion (i.e. around 9 ft long, weighing 600-700 lb)—but what instantly distinguished it from any previously-recorded species was its well-formed pair of hind legs, with greatly-enlarged webbed feet. Also, its tail was probably much longer than in modern cetaceans, judging from the elongate shape of the sole caudal vertebra recovered with the remainder of the skeleton. Consequently, it may well have lacked tail flukes.

Its long backbone (the shape of whose lumbar vertebrae indicate that it was very flexible dorsoventrally), the structure and size of its hind legs (longer than the forelegs), and their well-developed feet collectively suggest that this species swam somewhat like otters. That is to say, it undulated its backbone and paddled up and down with its hind feet—in marked contrast to modern cetaceans, which rely upon

Could an undiscovered modern-day 'first ancestor', descended from the likes of *Ambulocetus* (pictured here), for instance, be the identity of the super-otter? (© Hodari Nundu)

powerful vertical strokes of their horizontal fluke-bearing tail as their principal means of propulsion. In addition, its hind legs would have enabled it to move around on land, albeit somewhat clumsily, in the fashion of sea-lions—something that no present-day cetacean can do.

Emphasising its unique capability, for a cetacean, to locomote both in the sea and on land, Thewissen and his colleagues dubbed it *Ambulocetus natans*—the swimming walking whale.

Needless to say, *Ambulocetus* has attracted enormous interest from palaeontologists and taxonomists alike ever since, but, less readily realised, its discovery was also a great boon for cryptozoology. For here was a creature that corresponded very closely indeed to Heuvelmans's hypothetical super-otter, the product of his attempt to visualise the likely morphology of the Norwegian sea-worm when seen in its entirety. A modern-day descendant of *Ambulocetus*, retaining its general shape, its two pairs of limbs, flexible backbone, and long tail, but greatly increasing its total length, could well be a perfect match for the super-otter.

Perhaps, therefore, a 21st-Century *Ambulocetus* does indeed exist, amid the Arctic waters of the far north. To my mind, however, there could well be a less radical solution than walking whales and super-otters to the mystery of the Arctic's elongate water monsters—not to mention an alternative, decidedly non-cryptozoological but unequivocally memorable explanation that for obvious reasons (as will be

Reconstruction of the likely appearance of *Ambulocetus* in life (© Tim Morris)

seen!) attracted considerable media attention when announced a few years ago.

Quadrupedal Basilosaurines?

To begin with: Egede's description of his monster's front limbs as paws need not be as literal as it sounds. As noted earlier with the San Francisco Bay sea serpent, the limbs might simply have been flippers with prominent digits (finger bones) encased in a fairly thin covering of skin, which could look rather like paws to the zoologically-untrained observer—or, indeed, to anyone unexpectedly confronted (like Egede) with the alarming spectacle of a massive, unfamiliar sea beast suddenly rearing up out of the water in close proximity to their boat!

Some sea-worm eyewitnesses claim that these animals have hind limbs as well as forelimbs—one reason why Heuvelmans proposed the super-otter identity, thus delineating the sea-worm taxonomically from all other vertically-undulating elongate water monsters. Thanks to a very relevant palaeontological find, however, I no longer see any reason for this delineation, because the identity that has already been put forward to explain all of these other beasts—i.e. a living basilosaurine—can readily deal with the sea-worm too.

In 1987 and 1989, Michigan University palaeontologist Dr Philip Gingerich and Dr Holly Smith, with Duke University colleague Dr Elwyn Simons, mapped out 243 partial skeletons of a very serpentine, 50-ft-long species of mid-Eocene basilosaurine, *B. isis*—excavated from an Egyptian site so renowned for its plentiful *Basilosaurus* fossils that it is known as Zeuglodon Valley. The 1989 digs yielded several nearly complete skeletons, which, for the first time, included *functional* pelvic limb and foot bones. Until then, the only known hind limb bones from these sinuous creatures were so incomplete that they were mistakenly thought to be functionless and vestigial—responsible in turn for the traditional belief that, just like modern cetaceans, zeuglodonts basically lacked external hind limbs.

Although the hind legs were only about 2 ft long, they were well-formed, rather than atrophied—as one might have expected if these limbs no longer served any useful purpose. Consequently, their trio of discoverers considered that they must have fulfilled some function—but what could it be? They were evidently too small to assist locomotion—either in the sea or on land—but when their investigators found that the ankles could still move, and the kneecaps could click up and down like an electric switch, a very different but plausible alternative scenario began to emerge. The ability to perform these movements could enable the hind limbs to act as guides in the correct positioning of basilosaurines alongside one another for successful copulation (*Science*, 13 July 1990).

As far as cryptozoology is concerned, a modern-day basilosaurine with externally visible hind limbs would be sufficient to explain the sea-worm. That in turn renders obsolete the need to invent super-otters or any other separate identity for this cryptid.

In his book *Bronze Age America* (1982), Prof. H. Barry Fell depicted several petroglyphs discovered at Peterborough, near Toronto in Ontario, Canada, and dating from around 1,700 BC. One of these could have great bearing upon the possible inter-relationship of identities between elongate water monsters and four-legged basilosaurines.

It depicts an extremely elongate beast, with a horse-like head and short horns (or pricked ears?) borne upon a curved erect neck, which merges almost imperceptibly into a long slender body possessing what appear to be two small pairs of limbs. Interestingly, the body curves upwards from the neck as far as a point beyond the hind limbs, whereupon it then executes a sharp downward turn, yielding a

Depiction of the enigmatic Peterborough petroglyph

discrete tail that terminates in a pair of flukes. Underneath this petroglyph is a strange inscription—W-A-L—that seems to be of Nordic origin, and Fell noted that hval is Old Norse for 'whale'.

Whatever the creature is meant to be, it surely cannot be an ordinary whale, however, because there is no known form of advanced, present-day cetacean that looks anything like it. In stark contrast, it is a strikingly accurate depiction of the vertically-undulating, elongate water monsters frequently reported from the lakes and seas of Canada—so much so that it could be easily taken to be a sketch made by one of these beast's modern-day eyewitnesses. And except for its horns or ears, it also bears a striking resemblance to what we now know at least some basilosaurines to have been like, i.e. possessing four external limbs instead of two.

Could the Peterborough petroglyph therefore lay claim to the world's oldest illustration of a cryptozoological water monster—a Bronze Age eyewitness's portrait of a still-undiscovered species of evolved, present-day basilosaurine? There are a number of other serpentine beasts depicted in various examples of ancient art around the world, but none can match the extent of detail exhibited by this particular one—which includes morphological features whose very existence was only unveiled by science as recently as 1989. A poor rendition of a modern cetacean? Or an excellent portrayal of a four-limbed living basilosaurine? Once Ogopogo finally makes itself available for study, we will then be able to decide for ourselves.

Returning to the concept of the 'super-otter' for a couple of brief digressions. Firstly: although it is associated nowadays with the question of sea serpent identification, a giant elongate beast that resembles an otter without necessarily being one had been proposed as long ago as 1961 as a possible explanation for some terrestrial encounters with Nessie—by zoologist Dr Maurice Burton, in his book *The Elusive Monster*.

Secondly: is it possible that the Norwegian sea-worm is more than just a super-otter in name only? Although such a theory is far more radical than those that propose a cetacean identity for this type of elongate sea monster, could the sea-worm actually be an undiscovered species of giant marine otter, much bigger and even more specialised for an aquatic existence than those otters currently known to science? The swimming behaviour of such a beast with a highly flexible backbone could correspond well with that of the sea-worm—and the sea otter *Enhydra lutris* confirms that the evolution of an otter well-adapted for a marine existence is not an unreasonable supposition. Similarly, the existence of a 6-7-ft-long species of freshwater otter in South America, the very impressive saro or giant otter *Pteronura brasiliensis*, shows that a tendency towards gigantism in this group of mammals is indeed present.

In fact, Heuvelmans, Mackal, and various other cryptozoologists have suggested that a huge freshwater otter exceeding even the saro in total length may explain the iemisch—a currently unidentified type of South American river monster.

Nevertheless, when comparing the claims of a taxonomically-genuine super-otter and a basilosaurine namesake as the sea-worm's identity, I prefer to adopt the same viewpoint as I did earlier in this chapter when assessing the relative merits of the hypothetical giraffe-necked pinniped *Megalotaria* and a modern-day plesiosaur in relation to the identity of the long-neck. The elegant simplicity of Occam's razor—no cryptozoologist should leave home without it!

And having said that, a more recent entry on the super-otter identity front may conceivably yield an equally valid if exceedingly different usage of this same principle—by offering as an explanation for Heuvelmans's elusive *Hyperhydra* a known whale engaging in known (if unexpected) behaviour. In a 2005 *Archives of Natural History* paper reappraising Egede's classic super-otter sighting, Drs Charles G.M. Paxton and Sharon L. Hedley from St Andrews University's Centre for Research into Ecological and Environmental Modelling, in conjunction with Norwegian cryptozoological researcher Erik Knatterud, proposed that what Egede had actually spied was a known species of baleen whale but exhibiting a state of sexual arousal, with its prominent genitalia misidentified by Egede as the flukeless tail of a sea monster! Not a super-otter after all, therefore, but merely another, albeit highly unusual, case of mistaken identity—a sexed-up cetacean rising to the occasion, in every sense!

Abandoning the Armoured Archaeocetes

It is both interesting and inconvenient to note that certain serpentiform water monsters, such as the San Francisco Bay sea serpent, the huillia of Trinidad (if it were ever anything more than simply a media invention), and Hans Egede's 'sea-worm with paws', were said to be scaly. Surely a scale-bearing water monster cannot be reconciled with a basilosaurine? Ironically, for quite a time this characteristic actually enhanced these monsters' degree of correspondence to serpentine whales. This was because some archaeocete remains have been recorded that were in association with plate-like scales, which were initially assumed to have become detached from their skin during the fossilisation process. These finds led to the concept of armoured archaeocetes, possessing dermal scales—from which the prospect of a living armoured archaeocete as the identity of vertically-undulating, scaly water monsters was a natural progression in cryptozoological thought, and duly featured in Heuvelmans's sea serpent classification system.

Today, however, armoured archaeocetes have quietly disappeared from the palaeontological textbooks, and for good reason. The consensus nowadays is that the skin of these primitive cetaceans was unscaled, with those scales discovered in association with various fossil specimens having been shown to belong to some other creature instead—a severe blow to cryptozoologists striving to identify those modern-day basilosaurine lookalikes that allegedly bear scales.

However, there is no certainty that such cryptids do bear scales, as a surface can appear scaly without actually being scaled. A rough skin may look scaly, for instance, or perhaps the skin bears a pattern of markings or ridges that superficially resemble scales—the latter state being a plausible solution to the San Francisco Bay sea serpent and (if real) the huillia.

Of course, scaly mammals are hardly unknown—as evinced by the pangolins and armadillos, belonging to two totally separate taxonomic orders. So we cannot, in any case, rule out entirely the possibility of an evolved basilosaurine with scales. Yet the relative scarcity of sightings featuring scaly elongate water monsters flexing vertically suggests that their scales may owe more to optical illusion than (epi)dermal reality.

Out of all of the prehistoric survivors mooted in this chapter as plausible identities for various types of unexplained water monster, I consider the existence of an undiscovered species of living evolved basilosaurine to be the least unlikely prospect. These veritable serpent-whales were ideally suited for an efficient aquatic existence, and by lacking the sonar-controlled mechanism of navigation developed by more advanced cetaceans they would not be vulnerable to the threat of beaching themselves if their sonar malfunctioned. Indeed, even if they were somehow cast ashore, their predicted flexible backbone, neck, and limbs may well have permitted them to clamber back into the water—or even to travel overland and seek out another aquatic habitat. Consequently, I cannot rule out of hand the possibility of at least one lineage of basilosaurines persisting into modern times. As the only serpentiform mammals in existence, they would be ecologically unique (hence avoiding direct competition with other aquatic mammalian taxa), and their exceptionally lithe mode of locomotion would be of great benefit in helping them to avoid unwelcome attention from humans.

Some sceptics say that the reason science has yet to discover a living basilosaurine is that such creatures do not exist. This may indeed be true. To my mind, however, the reason for science's failure may quite simply be that it has yet to devise the correct *means* of discovering them. That is the real problem facing today's seekers of elongate water monsters—the monsters themselves are, if anything, little more than a technicality, albeit a fascinating one, because whatever their identity may be, I sincerely believe that they are indeed real and out there. Consequently, I cannot help but wonder what an enterprising postgraduate research student might come up with, if offered such a challenge by way of a PhD project. University zoology departments everywhere, take note!

CADBOROSAURUS—A MARITIME CONUNDRUM FROM BRITISH COLUMBIA

In the original edition of this present book, I included the following highly-elongate sea serpent of British Columbia, Canada, within the section devoted to the serpentiform, many-humped category of water monsters and their possible identity as an evolved, modern-day basilosaurine. Since that edition's publication in 1995, however, a number of interesting developments have occurred regarding this particular cryptid's putative identity that merit it being treated separately.

Unquestionably the most famous British Columbian sea serpent is a creature known to the Chinook Indian nation as the hiachuck-aluck, and to the area's Westerners as Caddy. Short for Cadborosaurus, this affable name was coined during the 1930s by a local newspaper editor called Archie Wills for the exceedingly elongate water monster frequently spied around Cadboro Bay and the Straits of Georgia, separating Vancouver Island from the province's mainland, but also sighted as far north as Alaska's Pacific coast and as far south as Oregon's.

One of the earliest, but also one of the most detailed, Western eyewitness accounts of a Caddy is that of prospector Osmond Fergusson, who, with his partner, encountered one of these mystifying animals on the morning of 26 June 1897 near the Queen Charlotte Islands. After their sighting, Fergusson prepared two sketches—a small one (A) showing one of its vertical coils, and a larger one (B) showing the creature in its entirety, both of which are reproduced opposite.

These sketches were also referred to in his account of their encounter, which reads as follows:

> The boat was about 100 yards from shore. . . . I saw ahead of us what I thought was a piece of drift wood. On getting closer I noticed it was moving towards us. When within 50 yards I said to Walker (my partner), What is that?

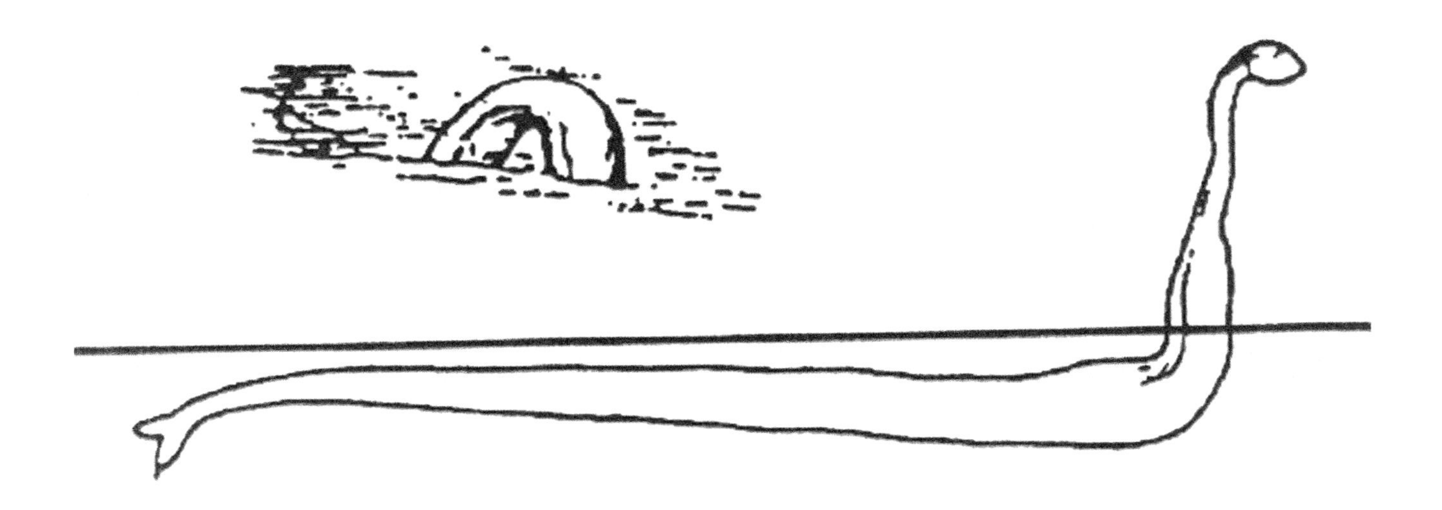

Fergusson's two sketches of the Caddy that he and his partner sighted

It seems to be moving this way (against the tide). What we could see was an object like sketch (A) sticking out of the water about two feet. When within a few feet of it the end uncoiled and raised a long neck about five feet out of the water with a head like a snakes [sic] on it. The arched portion making a broad flat chest like I have seen on the coba [cobra] I think.

When the serpent or whatever it was saw us it turned slightly towards land to avoid the boat. The head and neck were almost immediatly [sic] put under water again. As it passed the boat, at a distance, that with an effort, I could have thrown a[n] oar on it we could see a body about 25 feet long tapering with a fishlike tail and I think a continuous fin running the length of the body.

A slow undulating motion went along the body, while the tail part had a steady sweep from side to side of about six feet. A curious thing was the broad neck or chest part that formed the arch. . . . The only part out of water when the head was down was not exposed broadways in the direction the fish was going, but had a decided twist to the left allowing the water to flow through it.

Fergusson's sketches depict a species unlike anything known by science to be living today.

One afternoon in 1932, F.W. Kemp, an employee of the provincial Archives, was picknicking with his family on a tiny island off Victoria (British Columbia's capital city) when they spied at a distance of 1,200-1,500 ft but for several minutes in duration an enormous creature swimming up the narrow channel between two other small islands close by, creating a sizeable wake. According to an official statement subsequently prepared and signed by Kemp, the creature shot its head up out of the water, moving it from side to side as if taking its bearings, and then ". . . fold after fold of its body came to the surface", seemingly serrated towards its tail, like the edge of a saw, "with something moving flail-like at the extreme end". Kemp estimated this very elongate animal to be at least 60 ft long and about 5 ft thick. Interestingly, he also claimed that it bore around its head what appeared to be "a sort of mane, which drifted round the body like kelp". A mane has been reported by several other observers too in separate incidents.

Moreover, a year later, during October 1933, and this time in the company not only of his own family again but also of Major W.H. Langley (clerk of the Provincial Legislature) and his family, Kemp had a second Caddy sighting, in much the same location too. Not more than 100 ft away, they spied the dome of its back breaking the water surface. The creature was the size of a large whale, but was dark greenish-brown in colour, with serrated marks along its back and sides.

Several other notable Caddy sightings were reported during the 1930s and 1940s, and in my book *A Manifestation of Monsters* (2015) I documented an alleged Caddy carcase from Camp Fircom, British Columbia, whose image was captured on two vintage b/w picture postcards that claimed it had been found there on 4 October 1936. As I revealed, however, when the cards' photos are examined closely, this 'carcase' can be readily seen to be nothing more than a heterogeneous collection of detritus that had been artfully arranged by person(s) unknown to look superficially like a bona fide sea serpent carcase. In other words, a hoax.

In early July 1937, another alleged Caddy carcase came to light, this time recovered from the stomach of a sperm whale. Measuring 10-12 ft long, it was mangled and decomposed, but was still distinctive enough to warrant being photographed and afterwards displayed for a time at Naden Harbour whaling station in the Queen Charlotte Islands. According to long-standing Caddy investigator Dr Ed Bousfield, a zoological research associate at the Royal

As depicted here, Caddy is sometimes claimed by its eyewitnesses to be maned (© William M. Rebsamen)

British Columbia Museum, it had been a juvenile Caddy, and three coils behind its neck were still discernible. He and fellow Caddy investigator Prof. Paul LeBlond look upon this specimen as important scientific evidence for the reality of Caddy as a very distinct animal species. Looking upon it, however, is all that anyone can do now, quite literally—because although three photos of the carcase stretched out upon some packing cases still exist, the body itself was discarded long ago. Cryptozoological sceptics, meanwhile, have dismissed it variously as a foetal baleen whale or a highly decomposed shark.

What may (or may not—see later) have been yet another Caddy specimen was lost to science in 1968—but what makes this case very special even in the annals of cryptozoology, a subject liberally plagued with dematerialising evidence of the corporeal kind, is that the specimen in question was still alive. According to former sea captain William Hagelund, he had been fishing in Pirate's Cove, De Courcy Island, when he caught what some cryptozoologists

The two vintage picture postcards depicting
an alleged Caddy carcase from Camp Fircom

One of the Naden Harbour carcase photographs

believe to have been a baby 16-in-long Caddy in a dip net, and hauled it aboard his boat, placing it for safekeeping in a bucket of seawater where it remained overnight until he looked at it again the next morning. Describing his remarkable catch and the events that followed:

> His lower jaw had a set of sharp tiny teeth and his back was protected by plate-like scales, while his undersides were covered in a soft yellow fuzz. A pair of small, flipper-like feet protruded from his shoulder area, and a spade-shaped tail proved to be two tiny flipper-like fins that overlapped each other.
>
> . . . [Observing it the next morning] I felt a strong compassion for that little face staring up at me. I lowered the bucket over the side and watched him swim away quickly.

This action was not only kind-hearted but quite possibly of cryptozoological significance too. This is because such a response is more likely to be evoked by the sight of a mammal or even a fish than a reptile. Hagelund also prepared a sketch of the creature, which was subsequently published together with his account of this episode in his book *Whalers No More* (1987).

The sketch depicted it as elongate, with a long-muzzled toothy-jawed head, a large pointed front fin, heavy dorsal scaling along its entire back, ventral hairs, and sporting at the tip of its body two lengthy flippers yielding a very bifurcate tail.

During the late afternoon of 14 July 1993, two Caddys were spied in Saanich Inlet by pilot Don Berends and student pilot James Wells, who were practising landing a Cessna float plane. Each Caddy showed two loops above the water surface before diving, but when the plane landed on the area of the water where they had been, they reappeared about 20-40 ft further ahead, and began moving away at a rapid speed. The pilots could see daylight under the hoops (as with the Stinson sea serpent noted earlier), which were greyish-blue in colour.

This encounter's timing meets Bousfield's expectations, because his researches suggest that Saanich Inlet is used by Caddys from 15 June to 31 July each year as a breeding location, and that they move to shallow water with a sandy bottom or to deserted beaches to deliver their live young—normally at night, explaining why they are not seen on land more often.

Someone who claimed to have only narrowly missed seeing a Caddy ashore, however, was local resident Terry Osland. Early one morning in June 1991, she had been walking her dog towards the inlet's Ardmore Point beach when the dog became reluctant to go any further, tugging at the lead to go back as they turned by a cliff. Suddenly, Osland discovered the reason for her dog's apprehension—when she saw the face of a peculiar animal looking at her above the cliff's edge. The next moment, she heard a terrific splash, and when she peered over the cliff she could see the back end of a huge animal bigger than a killer whale plunging down beneath the surface of the sea. As she watched, however, the front end came up, dispelling any thoughts that the creature might simply have been a whale.

Smooth and hairless, its skin was grey-silver in colour, its neck was extremely long, and its tail ". . . was rounded like a lizard tail and it had, like little feet on the side back of the tail". This description suggests the presence of a pair of small flukes, like those reported by Fergusson in 1897 and by Hagelund for his putative baby Caddy.

In June 1996, a TV news bulletin broadcast in Oregon allegedly carried a report describing the recovery from waters somewhere off the American Pacific Northwest of a dead giant eel-like animal. It measured 25 ft long, and possessed genuine bones rather than mere cartilage. Could this have any relevance to Caddy sightings reported in this region?

Caddy sightings have continued into the present day, but the most newsworthy encounter took place in summer 2009, because not only did it feature as many as 15 individual Caddys being chased repeatedly by a pod of beluga whales *Delphinapterus leucas* no less, but also this remarkable incident was actually captured on video by one of its astonished observers. Kelly Nash, a Washington State fishing captain, was with his two sons on the deck of their boat, fishing in Nushagak Bay, Alaska, when they saw a herd of dark-coloured, saw-backed creatures that did not resemble any species known to them being rapidly pursued down a channel by a pod of belugas. One of the sons had the presence of mine to fetch their video camera, and proceeded to film three minutes' worth of this extraordinary spectacle, by which time the mystery beasts (which included juveniles as well as adults) had changed direction, swimming swiftly back up the channel towards the open sea again, with the belugas still closely pursuing them. The biggest specimens were longer than the belugas, i.e. in excess of

Size comparison of a Caddy alongside a human (© Connor Lachmanec)

around 18 ft (but not as long as 40 ft, as some commentators later claimed).

One of the adults, shielding a juvenile, pulled up next to the boat, and revealed a camel-like head. As described by British Columbia cryptozoologist John Kirk, who subsequently viewed the video footage: ". . . one of the creatures turns to look in the general direction of the camera and I must say I was stunned because it looked like a living breathing version of the famed Naden Harbour carcass obtained in 1937". Prof. Paul LeBlond also viewed the footage and he too likened the cryptid's head to that of the Naden Harbour carcase. Equally interesting is that one of the mystery animals was spouting, i.e. sending forth spouts of water into the air, an activity readily associated with cetaceans.

On account of their serrated backs, these creatures have been dismissed as sturgeons by some commentators, but Kirk is adamant that except for the serrations their backs did not match those of sturgeons, and also notes that beluga whales are not known to chase or hunt sturgeons. Nash sold their film's rights to Discovery Channel, who aired a very short section of it in July 2011 on its *Hilstranded* television show, in which its presenters, the Hilstrand brothers, viewed the segment and unsuccessfully attempted to locate one of the creatures. Two screen shots from the film can be viewed on the website *Cryptomundo* (at: http://cryptomundo.com/cryptozoo-news/caddy-screen/).

As with other water monsters, I have no doubt that a sizeable proportion of sightings are based upon misidentified known creatures—large seals, swimming deer, even fairly elongate but known cetaceans such as ziphiids all come to mind here. However, there are others, the 'true' Caddy sightings, replete with extremely long, serpentine, vertically flexing bodies and very lengthy necks, that are far less readily explained in this way.

Assuming (as ever, with any cryptid) that such reports as these are genuine, what could this most distinctive sea creature be? Prof. Mackal firmly believed that it was a basilosaurine, and if a very elongate, highly flexible evolved basilosaurine does exist, it would certainly offer a degree of comparison to Caddy—but even then, such a correspondence would be by no means a close one. Caddy's extremely lengthy neck, for instance, is at odds with what might be expected from even a very evolved basilosaurine, as is the manner in which it can apparently raise not just its neck but the entire front portion of its body vertically upwards, which instantly reminds me of certain pinnipeds, notably bull elephant seals. Consequently, if Caddy is indeed a living basilosaurine, it is by far the most morphologically specialised, radical version on record.

And speaking of pinnipeds: in late 2008 (and in June 2009 online), the scientific journal *Historical Biology* published a noteworthy paper authored by Michael A. Woodley (then a Royal Holloway, University of London, postgraduate biology student), afore-mentioned palaeontologist Dr Darren Naish, and Royal Holloway computer scientist Dr Hugh P. Shanahan. It was entitled 'How many extant pinniped species remain to be described?', and in it the authors examined the description record of the pinnipeds using non-linear and logistic regression models in an attempt to ascertain the number of still-undescribed species. Moreover, they combined that work with an evaluation of cryptozoological data, featuring certain marine cryptids that have been deemed by various mystery beast researchers at one time or another to be undiscovered pinniped forms. These included not only the long-neck category of sea serpent, Heuvelmans's merhorse, and the Bering Sea's leopard seal-like tizheruk, but also British Columbia's serpentiform Cadborosaurus. And there is no doubt that, looking closely at Fergusson's sketch of the Caddy that he and his partner Walker encountered back in 1897 off the Queen Charlotte Islands, it is indeed reminiscent of a pinniped, albeit an exceptionally elongate one.

Equally thought-provoking was a very different, non-mammalian identity, as presented by Bousfield and LeBlond in both a book (*Cadborosaurus: Survivor From the Deep*, 1995) and a scientific paper (*Amphipacifica*, 20 April 1995). In these publications, they proposed that Caddy was most similar to a plesiosaur, and, following a detailed morphological description of this cryptid as based upon the Naden Harbour carcase photographs, they formally christened its species *Cadborosaurus willsi*. A concise summary of their principal reasons for this taxonomic diagnosis can be found in their paper's abstract, and reads as follows:

> In general features of head, two pairs of flippers, and short tail, the animal appears least unlike some plesiosaurs of Mesozoic age. However, its large distinctive hind flippers are apparently webbed to the true tail to form a broad fluke-like propulsive caudal appendage. When swimming rapidly at the surface, the trunk region characteristically forms into two or more vertical humps or loops in tandem behind the neck.

They also postulated that it was probably poikilothermic (cold-blooded), like reptiles, claiming that such an elongate body's surface area/volume ratio would be so great that if Caddy were warm-blooded, it would need to eat constantly in order to offset the significant, continuous loss of heat from its body.

A significant problem faced by cryptozoologists when dealing with serpentiform water monsters is explaining their remarkable mode of locomotion, in which humps or body coils are thrown up vertically when these beasts move (often very rapidly) through the water in a horizontal plane. Bousfield and LeBlond, however, proposed a very interesting mechanism:

> Swimming propulsion in *Cadborosaurus* is almost certainly generated by the large and powerful fluke-like caudal region. The animal is able to reduce the very great drag effect of its snake-like body at rest by transforming the

trunk region into a series of vertical loops or humps in tandem. Such body transformation effectively results in a relatively short, rigid, longitudinal I-beam. . . . Friction is further reduced as the humps "draft" directly behind each other (as in highway trailer rigs in tandem), and part of the humps, neck and head are elevated into the relatively frictionless air medium. The thin, narrow front flippers presumably serve as hydrofoils in vertical elevation of the trunk, in rapid submergence, or in directional change, as required. Although the fluke-like tail motions are submerged and seldom visible, they are presumably rapid, as in the large thunniform [tuna-shaped] vertebrates, and provide the thrust required to drive the relatively frictionless humped body at the high speeds observed.

As described in a close sighting by Ken Kilner and Richard Smith at Roberts Bay, near Sidney, B. C. (personal communication), the loops move at exactly the same speed, each at a constant height above the surface, and at a constant distance between each. The musculature ripples in a continuous motion "like a travelling wave moving from head to tail", traversing all loops. No side-to-side (snake-like) undulations have been observed, and little wake is customarily created.

Vertical body flexing is generally confined to mammalian swimmers, but as Bousfield and LeBlond also pointed out in their paper, a notable reptilian precedent was provided by the thalattosuchians or marine crocodilians of the Mesozoic. Their limbs had become paddles, and their vertebrae each bore lateral spines

Artistic representation of a Caddy, emphasising its exceedingly elongate, flexible body (© Richard Svensson)

(hypophyses) that were strongly tilted upwards in a manner comparable to those of modern whales (rather than only very weakly upwards as in modern crocodilians), thereby allowing them to hold their body rigid in a vertical plane and propel themselves forward via rapid vertical movements (as in mammals) rather than lateral ones (as in other reptiles). Perhaps, they suggested, Caddy has comparable vertebral modifications, thus enabling it to flex vertically despite being (according to their proposal) a reptile.

In subsequent years, however, the notion that Caddy may be a living plesiosaur has incited appreciable controversy, including various criticisms by Naish in his *Tetrapod Zoology* blog, and also a detailed response in the ISC's scientific journal *Cryptozoology* by zoologists Drs Aaron M. Bauer and Anthony P. Russell. In their account, much was made by Bauer and Russell of the fundamental morphological differences between fossil plesiosaurs and the details reported for Caddy by eyewitnesses, although they failed to take into account that major changes may have occurred during 66 million years of continued, intervening evolution if Caddy were indeed a living plesiosaur. Particularly relevant, meanwhile, was their emphasis upon how any vertebrate capable of executing the pronounced vertical flexions necessary to engender Caddy's oft-reported coils must surely possess a spinal column containing numerous vertebrae (as opposed to the mere 26 estimated by Bousfield and LeBlond to have been possessed by the Naden Harbour carcase). Of course, an inherent problem when attempting to utilise photographs alone of the Naden Harbour carcase in assessing the morphology of Caddy (quite apart from not knowing for certain whether it actually *was* a specimen of Caddy anyway) is determining accurately how much of its visible morphology was not an artefact created by decomposition of its tissues and also by distortion effected upon its tissues by the gastric juices of the sperm whale that swallowed it.

What a tragedy that this tantalising specimen was not preserved and sent to a museum or some other scientific institution for formal examination. But what of the alleged living baby Caddy that was captured and then released in 1968 by William Hagelund? Might that have divulged this marine cryptid's secrets if scientifically assessed? Again, the fundamental issue is whether it actually *was* a Caddy, or merely some already-known species but one that was not familiar to Hagelund. In their paper, Bousfield and LeBlond cited it as a Caddy, but this identification has been disputed by Woodley, Naish, and fellow researcher Cameron A. McCormick in a *Journal of Scientific Exploration* paper of 2011 and subsequent interchanges with Bousfield and LeBlond. Woodley *et al.* reinterpreted the creature as a pipefish *Syngnathus* sp. (closely related to seahorses), and noted that Hagelund's description and drawing of it did not actually contain those specific morphological details that so readily characterise Caddy. To be honest, I struggle to see much resemblance between Hagelund's creature and a pipefish; then again,

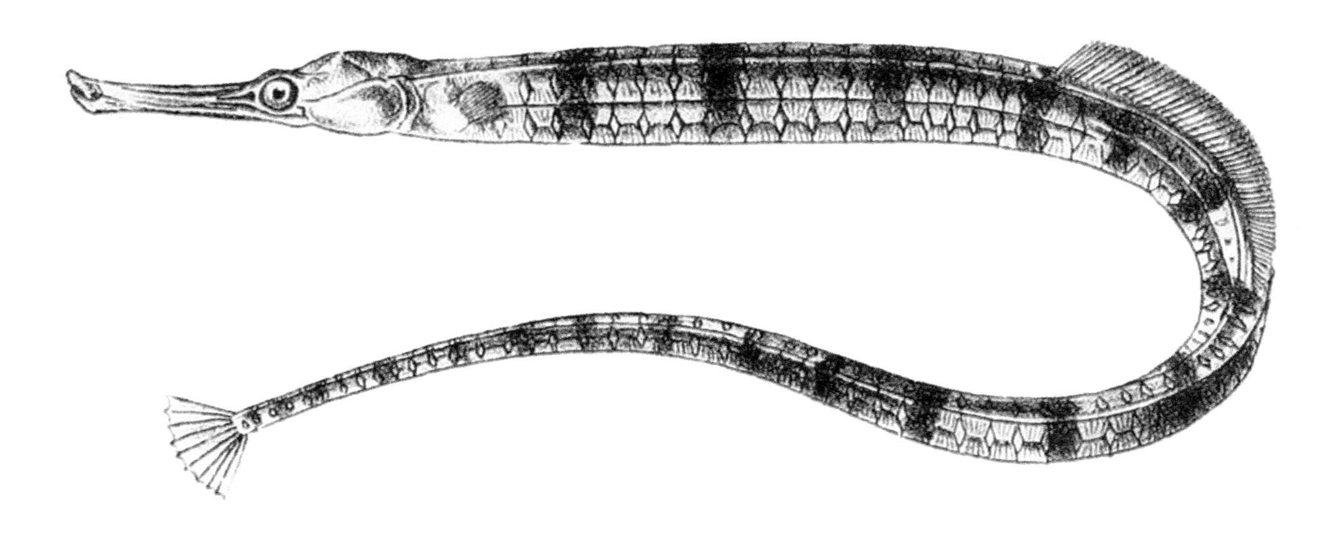

A *Syngnathus* pipefish

we have no idea how accurately Hagelund's sketch and verbal description portrayed the creature—and, as already noted, we have no idea whether it *was* a Caddy anyway.

So if it does truly exist, what exactly *is* the taxonomic identity of Caddy—a basilosaurine, a long-necked seal, a plesiosaur, a fish, or something else entirely? And does such speculation really matter in any case? As I have stated many times in many places: however interesting it might be to speculate what a given cryptid may or may not be, ultimately this is all that such an exercise can ever be—speculation, the expression of an opinion, nothing more. Facts only emerge when physical, tangible evidence is obtained. In the case of Caddy, two potential examples of such evidence were both lost, but all hope in solving the Caddy conundrum may not be. If Bousfield's researches concerning the reproductive behaviour of this mystery beast are correct, i.e. that Saanich Inlet is indeed used by Caddys from 15 June to 31 July each year as a breeding location, and that they move to shallow water with a sandy bottom or to deserted beaches to deliver their live young at night, then surely it would not be difficult to monitor this location during that period, using remote infrared cameras or possibly even some computer application such as Google Earth (in view of this species' large size). It's certainly well worth considering.

SEA CROCODILES, SEA LIZARDS, AND THE KIN OF THE KRONOSAURS

Cryptozoology's water monster files contain a number of reports that feature formidable creatures more than a little reminiscent of certain lineages of prehistoric reptiles *other* than the familiar long-necked plesiosaurs. In his sea serpent classification system, Heuvelmans lumped these reports together into a single category that he termed the marine saurian, and it is represented here by the following selection of sightings.

From *Sacramento* to Submarine

On 30 July 1877 while travelling in the mid-Atlantic from New York to Melbourne aboard a ship called the *Sacramento*, a helmsman and his commander, Captain W.H. Nelson, spied a 50-60-ft-long monster lying at the surface of the sea, with its head raised about 3 ft above the water. Its head was very like an alligator's, and about 10 ft behind this was a pair of flippers. Its body was reddish-brown, and resembled that of a very large snake. As the two men watched, the creature began to move away, and when it was 30-40 ft astern it lowered its head.

19th-Century representation of the *Sacramento* sea serpent

During a journey from New York to Belém, near the mouth of the Amazon River in Brazil, the steamer *Grangense* apparently encountered a similar beast, when, one morning in May 1901, passenger Charles Seibert and others aboard spied an unidentified animal frolicking in the sea nearby. Its head was crocodilian in shape, its neck short, its body greyish-brown, its jaws contained rows of 4-6-in-long teeth, and it twisted around in horizontal circles at the surface in seemingly playful mode before diving out of sight.

(NB—Although the latter two cases are traditionally housed within the marine saurian category of sea serpent, the eyewitnesses' description of the *Sacramento* beast's body as resembling a very large snake and the playful behaviour of the *Grangense* beast make me wonder whether these creatures were mammalian rather than reptilian; the *Sacramento* cryptid in particular sounds more like a serpentiform basilosaurine-like sea serpent, but see also later for an alternative interpretation of its body's snake-like description by its observers.)

One of the most dramatic sightings featuring a water monster from the marine saurian category occurred on 30 July 1915, directly after a German submarine called the *U-28* had torpedoed the British steamer *Iberian* off Fastnet Rock, Ireland. Less than a minute after the steamer had sunk, a violent explosion took place beneath the surface, blasting up out of the sea some pieces of wreckage—and a writhing sea monster! According to the *U-28*'s captain, Georg Günther Freiherr von Forstner, the creature was quite immense, measuring around 60 ft long and shaped like a crocodile, with a long tail tapering to a point, and four limbs with powerful webbed feet. Referring to it as "the underwater crocodile", he and the crew gazed at it in wonder before it fell back into the sea and vanished 10-15 seconds later.

As the creature was seen (however briefly) in its entirety, this is potentially a very important case in the water monster files—which makes it all the more frustrating that Forstner's account is surrounded by controversy. To begin with, he didn't even mention the sighting to anyone until 1933—which just happens to be the year in which the Loch Ness monster began to feature extensively in the news worldwide. Moreover, he estimated that the *Iberian* must have sunk to a depth of about 500 fathoms before the explosion took place—but as this supposedly occurred a mere 25 seconds or so after the ship had been hit, the speed of submergence involved would have approached a highly unlikely 90 miles per hour!

Forstner was even vague concerning the location of this incident, merely giving 'the North Atlantic' at first. Only in 1942, within his own, German translation of Rupert Gould's book *The Case For the Sea Serpent*, did he finally divulge that it had taken place off Fastnet Rock.

Also of interest here, as discovered by German cryptozoological investigator Ulrich Magin, is that although the local Irish newspapers had reported the sinking of the *Iberian*, they had not included any mention of a borne-aloft sea monster. Nor was it that these newspapers had any aversion to sea monster stories—only a week later, one of them published an account of a much less spectacular sighting. Consequently, Magin has speculated that such inconsistencies imply a hoax.

Interestingly, however, there is a second, much less familiar alleged sea serpent sighting of the marine saurian type on record that also features a German U-boat in 1915. As documented by Harold T. Wilkins directly below his coverage of the *U-28* incident in his book *Strange Cities of Old South America* (1952), a Captain Werner-Löwisch claimed that when he was officer of the first watch in the *U-20* one clear, moonlit night in 1915 he spied a huge marine monster, which he reported with great confidence in his diary as follows:

Contemporary depiction of the *U-28* sea serpent blasted out of the sea after the *Iberian*'s explosion

> Saw a sea serpent at 10 p.m., without possibility of doubt. The creature had a longish head, scales like a crocodile's, and legs with proper feet. The mate saw him, but when the captain came up from below, the monster had vanished. The monster was about 90 feet long.

In view of the correspondence in both year and type of craft between this report from Werner-Löwisch on the U-20 and the increasingly contentious, long-delayed report from Freiherr von Forstner on the U-28, could it be that the latter not only was fictitious but also had been directly inspired by the former, which in turn may have actually been genuine but was less dramatic and much less publicised?

On 15 April 1969, a 65-ft-long Alaskan shrimp boat called the *M.V. Mylark*, equipped with what was then state-of-the-art sonar surveillance equipment (a Simrad EH2A), was seeking shrimp off Raspberry Island, near Alaska's Kodiak Island, when it switched on its sonar to test that it was functioning properly. To the amazement of its operator, it suddenly revealed and duly recorded the presence of a gigantic sea creature, measuring 150-180 ft, swimming underneath the boat at a depth of around 55 fathoms. Determined to provide proof of the monster's reality, the crew managed to take a photograph of it too, and later described it as possessing two pairs of flippers, a long tapering tail, muscular neck, and snub-nosed head. Needless to say, the photograph has since been lost, but a printout of the sonar trace survived, and when American cryptozoologist Ivan T. Sanderson learnt of this case (via a friend, Captain Stanley Lee) he submitted the trace to no fewer than 14 leading academics variously specialising in biology, engineering, geography, law, naval operations, and oceanography for their opinions as to whether or not the trace was genuine. All 14 affirmed that in their view it had not been tampered with in any way, and that the sonar equipment itself was not faulty. The boat's crew stated that the creature was not a whale or anything else known to them from the area where it was seen. Had the photo survived, we might have conclusive proof for the reality of at least one species of marine saurian sea serpent.

As noted in my book *Mirabilis* (2013), for over a century reports have emerged from New Zealand of huge water monsters resembling flipper-limbed lizards but with crocodile-like heads encountered by terrified sea-goers in the waters around this dual-island country—as well as in at least one instance in a freshwater lagoon. One such animal, with noticeably big eyes, was spied about 18.5 miles off Lyttleton, on the eastern coast of New Zealand's South Island, in April 1971 by the crew of the *Kompira Maru*, who noticed its flippers when it dived underwater.

Another sighting, this time in freshwater and featuring what its terrified eyewitnesses termed "a giant lizard", 13-16.5 ft long and green in colour, occurred in 1990. This was when two young women sunbathing by a lagoon close to Taupo and Lake Taupo on North Island saw the creature in question swimming in the shallows. It even attempted (unsuccessfully) to catch a bird in its jaws, the front portion of its body emerging from the water during this attempt, but then it submerged again and swam into the depths. Although none is known to attain the claimed size of the creature reported at Taupo, amphibious species of monitor lizard are known from various regions of the world—but New Zealand is not one of them.

In 1993, after viewing it through a telescope, Earl Rigney from Canterbury, South Island, realised that what he had initially assumed to be a whale far out to sea was actually a colossal crocodile, roughly 30 ft long (i.e. almost a third longer than the largest estuarine crocodiles *Crocodylus porosus*), breaching at the water surface. And during September of that same year, while fishing off the Cook Islands (lying approximately 2,000 miles northeast of New Zealand), a vicar and his son allegedly spied a creature resembling a huge lizard, bigger than a whale, surface near to their vessel. This so alarmed them that they abandoned their fishing and sailed away with all speed!

And in 2006, while he was enjoying the sun on its deck, New Zealand sailor Ivan Levy's boat was allegedly attacked by two monstrous sea creatures described by him as resembling lizards with fins and measuring approximately 20 ft long. The terrifying onslaught continued for over an hour before they finally swam off, leaving a stressed-out Levy to bring his wrecked boat ashore. Subsequent suggestions that Levy had deliberately scuppered his vessel to claim insurance money were themselves scuppered when he revealed that the boat wasn't insured, thus gaining nothing from this ordeal other than brief local fame and news headlines.

The following South American report probably owes more to native legend and imagination than to fact, but I am presenting it here because whereas most of this section's water monsters are marine, it features a second potential freshwater version—and thus, if genuine, is of particular cryptozoological significance. Traveller Alfred G. Hales heard of it from local Indians while far up the Brazilian Amazon, and he included it in *Barney O'Hea: Trapper* (1932). It was reiterated as follows by

Harold T. Wilkins in his own book *Secret Cities of Old South America* (1952):

> This monster inhabits recesses of lush and steamy marshes and lagoons. One night, an Indian, fearing to sleep ashore, anchored his canoe in mid-stream. In the middle of the night he woke up and gazed ashore to the woods. There, his eyes were caught and fixed by two queer moons that seemed gently to sway in the tree-tops. They had a hypnotic force that made him take up his paddles and row ashore. The moons continued gently and rhythmically swaying, and turned golden green. The Indian, as one mesmerized, heard unearthly music and drums beat. Time seemed no more. He was nearing the shore and death, when, on a sudden, came a great wave of water under the bow of his canoe, and the moon, emerging from a break in the dense trees of the forest, lit up a scene from the Mesozoic Age. An immense reptile of species unknown rose from the waters and clashed his immense, tusked jaws on the neck of a giant anaconda. A frightful struggle followed and the tremendous coils of the snake were dragged from the tree down into the waters of the river. Mr. Hales makes the significant comment that snakes with these mesmeric powers are seen near ancient temples in the forests of Brazil, luring parrots, with hypnotic eyes that rapidly change their lights from green to crimson.

Even if we take this fanciful yarn seriously, the giant anaconda's equally enormous assassin might simply have been a very big caiman—those evil-tempered crocodilians that bedevil the larger rivers of Central and South America. Yet such creatures are only too familiar to the natives who live in constant dread of their lethal jaws. So why should Hales's particular reptile have been deemed to be an unknown species—unless, of course, it really was?

So what could the marine saurian be? The remit for this book is to concentrate upon identities of the potentially prehistoric kind, but one modern-day species worthy of brief mention here as a popular mainstream identity is the estuarine crocodile *Crocodylus porosus*. This impressive species, the largest known terrestrial predator in the world today, is certainly famous for forays far out to sea, well beyond its normal mangrove swampland, lagoon, riparian, and estuarine terrain. And on account of its size, gigantothermy may enable it to survive temperature below those normally acceptable to its species. However, even the largest confirmed specimens have not exceeded 22 ft (with more typical specimens measuring 14-17 ft), hence are smaller than the marine saurian (if eyewitness estimates of this cryptid have been accurate), and their limbs are normal, clawed limbs, not flippers as sometimes specified by marine saurian observers.

Pliosaurs, Mosasaurs, and Thalattosuchians

There are three totally separate taxonomic groups of extinct aquatic reptile known from the fossil record that provide much food for thought when reflecting upon the possible identity of the marine saurian category of water monsters documented here.

The pliosaurs were a group of plesiosaurs whose morphological evolution followed a very different path from that of their more familiar long-necked relatives. Like the latter reptiles they did have two pairs of long paddle-like flippers and a sturdy body, and they apparently

The author alongside a pliosaur painting (© Dr Karl Shuker)

swam in the same manner too—but instead of a long slender neck, short tail, small head, and short jaws, the pliosaurs were characterised by a very short, thick neck, a long powerful tail, and an extremely long, broad head and jaws—superficially like those of the crocodilians.

The most formidable and ferocious marine reptiles of all time, the pliosaurs include among their number the famously awesome 30-40-ft-long monster from Australia called *Kronosaurus* (estimates of its total length differing between researchers), and an as-yet-unnamed gargantuan specimen from Dorset, England, with a skull length marginally under 8 ft and an estimated total length of up to 52 ft (which if accurate would make it the largest pliosaur currently documented). They lived alongside (and actively preyed upon) their elasmosaurian (i.e. long-necked) plesiosaur relatives right up to the late Cretaceous Period, after which both groups officially died out (the pliosaurs slightly before the elasmosaurs).

Sharing the oceans with the pliosaurs and plesiosaurs during the late Cretaceous were the

Hainosaurus reconstruction (© Markus Bühler)

mosasaurs—a group of extremely large lizards related to the monitors (which include the famous 10-ft-long Komodo dragon), but highly specialised for an exclusively aquatic existence. Although usually dubbed sea lizards, in December 2012 the first recorded freshwater mosasaur species was formally described, and was named *Pannoniasaurus inexpectatus*. Its fossils were recovered from an open-pit mine in Hungary's Bakony Hills, and dated from the late Cretaceous.

The body of mosasaurs was long and streamlined, their limbs had become flippers, and their lengthy tail was flattened laterally. Their back and tail have traditionally been reconstructed with a prominent vertical fin (and often still are today), but this is erroneous. It resulted from some calcified tracheal rings' misidentification as a dorsal (nuchal) crest by S.W. Williston in a brief editorial note of 1898 (followed by a short 1899 article) documenting a beautifully-preserved specimen of *Platecarpus ictericus* (now *tympaniticus*) that had been uncovered in Logan County, Kansas, U.S.A.

Like the pliosaurs, mosasaurs had a very large head with long jaws that contained many conical teeth for breaking open ammonites and other shelled marine life. Some forms also attained a considerable size—North America's *Tylosaurus* was 25-30 ft long, Europe's *Hainosaurus* was at least 40 ft (some researchers have estimated as much as 50-56 ft for it), and Hoffmann's mosasaur *Mosasaurus hoffmannii* from the Netherlands may have attained a total length of up to 59 ft.

Pliosaurs seemingly had a smooth, scaleless skin, but the mosasaurs are known from some well-preserved fossilised remains to have been scaly, just like other lizards. And whereas the body of pliosaurs and plesiosaurs was relatively rigid, the mosasaurs swam by powerful horizontal undulations of their body, rather like snakes. Could it therefore have been its swimming mode rather than its morphology that drew comparisons by its observers between the *Sacramento* sea serpent's body and that of a large snake?

Less well-known and generally smaller than those reptiles mentioned here so far, but no less specialised for a marine lifestyle, the third group constituted the sea-crocodiles or thalattosuchians. Despite their name, they were not true crocodiles (although related to them), and most were up to about 12 ft long. In 2016, however, the discovery of a giant species, *Machimosaurus rex*, measuring up to 32 ft long, was announced. According to the fossil record, these were a relatively short-lived group that died out long before the pliosaurs and mosasaurs, having vanished by the early Cretaceous Period, around 130-120 million years ago, in the case of *M. rex,* but already gone by the end of the Jurassic for others.

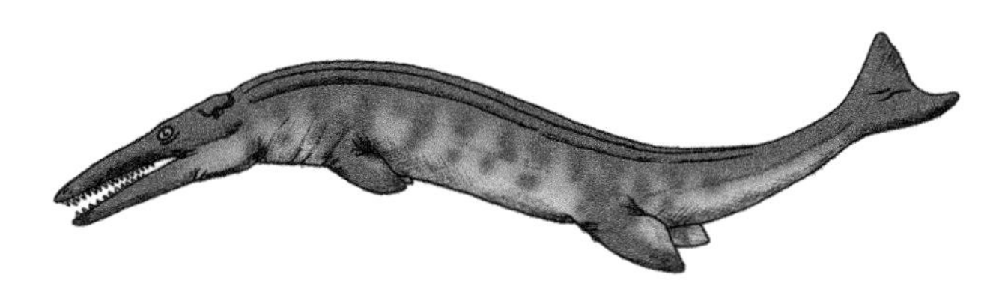

Thalattosuchian reconstruction (© Tim Morris)

Unlike today's crocodiles and alligators, the thalattosuchians had paddle-like limbs instead of fully-formed legs and clawed feet, an elongate, scaleless body, and a long slender tail that curved sharply downwards but bore a prominent dorsal fin. Moreover, by virtue of their unusually-shaped vertebrae, which bore a surprising resemblance to those of whales, they possessed the unique ability among reptiles to flex their backbone vertically, just like mammals.

This highly unexpected characteristic is of great cryptozoological significance—because, as French palaeontologist and thalattosuchian specialist Dr Eric Buffetaut has pointed out, it means that a reptile identity cannot be automatically dismissed in favour of a mammalian one when analysing reports of water monsters that flex their bodies vertically, especially if their head is reptilian in form.

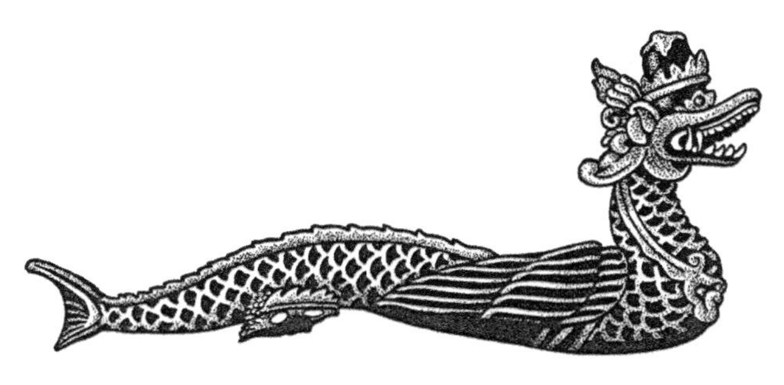

Thalattosuchian-reminiscent sea serpent depicted on a traditional Javan gong frame

In view of this, I was greatly intrigued when I uncovered a depiction of a sculptured sea serpent on a traditional Javan gong frame. Although highly stylised, it was still of cryptozoological interest—sporting a crocodilian head, a slender flippered body gently undulating vertically, and a vertical tail fin with a prominent dorsal fin. Just a coincidence?

Any of these three groups of marine reptile could be responsible for any or all of the water monster reports documented in this section, because the reports do not provide sufficient morphological details to enable an extensive taxonomic analysis to be attempted (although the giant flipper-limbed, crocodile-headed 'lizards' reported from the waters around New Zealand are certainly reminiscent of mosasaurs). All that can be stated with a degree of confidence is that such monsters are certainly more similar to these three reptilian identities than they are to any other well-known identity available.

In a 2001 *Fortean Studies* article on sea monsters, Dr Darren Naish made the valid point that cryptozoologists sometimes display a restricted viewpoint when selecting possible identities of the prehistoric kind for various cryptids, due to certain potential candidates of this type not being widely known outside palaeontological circles, and he mentioned such examples as various shastasaurs, hupehsuchians, choristoderes, and remingtonocetids. This is undeniably true, but for those cryptozoologists like myself who recognise that any proposed identification of a cryptid based only upon anecdotal, non-physical evidence can never be more than an opinion (not a fact), it is not a major concern, because opinions are just that—opinions (hence of relatively little scientific importance, despite the sad fact that some cryptozoologists waste inordinate amounts of time and energy fruitlessly arguing back and forth with one another as to the relative merits or otherwise of a particular identity for a particular cryptid). What is (or should be) a major concern for cryptozoologists, conversely, is collecting and documenting reports of cryptids, and whenever possible seeking tangible evidence that will categorically establish a cryptid's reality and thence its identity.

The Gambian sea serpent, based upon original sketches by Owen Burnham (© Dr Karl Shuker)

In other words, an actual specimen needs to be obtained—or at least a notable portion of one (such as the skull, some vertebrae, or a limb). Not a likely occurrence, you may think—yet as far as water monsters of the marine saurian category are concerned, it is possible that in 1983 this very nearly happened!

The Gambian Sea Serpent—A Stranger on the Shore

On 12 June 1983, wildlife enthusiast Owen Burnham and three family members encountered the carcase of a huge sea creature, washed up onto a beach in Gambia, western Africa. Most sea monster remains are discovered in an advanced state of decomposition, greatly distorting their appearance and making positive identification very difficult, but the carcase found by Burnham was exceptional, as apparently it was largely intact, with no external decomposition.

Now residing in England but having lived most of his childhood and teenage years in Senegal, Burnham was very familiar with all of that region's major land and sea creatures, but he had never seen anything like this before.

Realising its potential zoological significance, he made meticulous sketches and observations of its outward morphology, and noted all of its principal measurements.

In May 1986, *BBC Wildlife*, a British magazine, published a short account by Burnham describing his discovery, and including versions of his original sketches. Greatly interested, I wrote to him, requesting further details, in order to attempt to identify this remarkable creature. During our correspondence, Burnham kindly gave me a comprehensive description (plus his sketches) of its appearance. The following is an edited transcript of Burnham's first-hand account of his discovery, prepared from his letters to me of May, June, and July 1986:

> I grew up in Senegal (West Africa) and am an honorary member of the Mandinka tribe. I speak the language fluently and this greatly helped me in getting around. I'm very interested in all forms of life and make copious observations on anything unusual.
>
> In the neighbouring country of Gambia we often went on holiday and it was on one such event that I found this remarkable animal.
>
> June 1983. An enormous animal was washed up on the beach during the night and this morning [June 12] at 8.30 am I, my brother and sister and father discovered two Africans trying to sever its head so as to sell the skull to tourists. The site of the discovery was on the beach below Bungalow Beach Hotel. The only river of any significance in the area is the Gambia river. We measured the animal by first drawing a line in the sand alongside the creature then measuring with a tape measure. The flippers and head were measured individually and I counted the teeth. [In the sketches accompanying his description, Burnham provided the following measurements: Total Length = 15-16 ft; Head+Body Length = 10 ft; Tail Length = 4.5-5 ft; Snout Length = 1.5 ft; Flipper Length = 1.5 ft.]
>
> The creature was brown above and white below (to midway down the tail).
>
> The jaws were long and thin with eighty teeth evenly distributed. They were similar in shape to a barracuda's but whiter and thicker (also very sharp). All the teeth were uniform. The animal's jaws were very tightly closed and it was a job to prise them apart.
>
> The jaws were longer than a dolphin's. There was no sign of any blowhole but there were what appeared to be two nostrils at the end of the snout. The creature can't have been dead for long because its eyes were clearly visible and brown although I don't know if this was due to death. (They weren't protruding). The forehead was domed though not excessively. (No ears).
>
> The animal was foul smelling but not falling apart. I've seen dolphins in a similar state after five days (after death) so I estimate it had been dead that long.
>
> The skin surface was smooth, the only area of damage was where one of the flippers (hind) had been ripped off. A large piece of skin was loose. There were no mammary glands present and any male organs were too damaged to be recognizable. The other flipper (hind) was damaged but not too badly. I couldn't see any bones.
>
> I must mention clearly that the animal wasn't falling apart and the *only*

damage was in the area (above) I just mentioned. The only organs I saw were some intestines from the damaged area.

The paddles were round and solid. There were no toes, claws or nails. The body of the creature was distended by gas so I would imagine it to be more streamlined in life. It *wasn't* noticeably flattened. The tail was rounded [in cross-section], not quite triangular.

I didn't (unfortunately) have a camera with me at the time so I made the most detailed observations I could. It was a real shock. I couldn't believe this creature was laying in front of me. I didn't have a chance to collect the head because some Africans came and took the head (to keep skull) to sell to tourists at an exorbitant price. I almost bought it but didn't know how I'd get it to England. The vertebrae were *very* thick and the flesh dark red (like beef). It took the men twenty minutes of hacking with a machete to sever it.

I asked the men on the scene what the name of this animal was. They were from a fishing community and gave me the Mandinka name *kunthum belein*. I asked around in many villages along the coast, notably Kap Skirring in Senegal where I once saw a dolphin's head for sale. The name means 'cutting jaws' and is the term for dolphin everywhere. Although I gave good descriptions to native fishermen they said they had never seen it. The name *kunthum belein* always gave [elicited] a dolphin for reply and drawings they made were clearly that. I also asked at Kouniara, a fishing village further up the Casamance river but with no success. I can only assume that the butchers called it by that name due to its superficial similarities. In Mandinka, similar or unknown animals are given the name of a well known one. For example a serval is called a little leopard. So it obviously wasn't common. I've been on the coast many times and have never seen anything like it again.

I wrote to various authorities. [One] said it was probably a dolphin whose flukes had worn off in the water. This doesn't explain the *long pointed tail* or lack of dorsal fin (or damage).

[Another] decided it could be the rare *Tasmacetus shepherdi* [Shepherd's beaked whale] whose tail flukes had worn off. This man mentioned that the blow hole could have closed after death. Again the tail and narrow jaws seem to conflict with this. *Tasmacetus*'s jaws aren't too long and the head itself seems to be smaller than my animal's. *Tasmacetus* has two fore flippers and none in the pelvic region. The two flippers are quite small in relation to body size and pointed rather than round. *Tasmacetus* has a dorsal fin and 'my' animal didn't seem to have one or any signs of one having once been there. *Tasmacetus* even without tail flukes wouldn't have a tail long enough or pointed enough. The tail of the animal I saw was very long. It had a definite point and didn't look suited for a pair of flukes. Apparently, *Tasmacetus* is brown above and white below and this seems to be the only link between the two animals. I've been to many remote and also popular fishing areas in Senegal and I have seen the decomposing remains of sharks and also dead dolphins and this was so different.

[A third] said it must have been a manatee. I've seen them and believe me it wasn't that. The skin thickness was the same but the resemblance ended there.

Other authorities have suggested crocodiles and such things but as you see from the description it just can't have been.

After I think of the coelacanth I don't like to think what could be at the bottom of the sea. What about the shark (*Megachasma*) [megamouth shark] which was fished up on an anchor in 1976?

I looked through encyclopedias and every book I could lay hands on and eventually I found a photo of the skull of *Kronosaurus queenslandicus* which is the nearest thing so far. Unfortunately the skull of that beast is apparently ten feet long and clearly not of my find.

The skeleton of *Ichthyosaurus* (not head) is quite similar if you imagine the fleshed animal with a pointed tail instead of flukes. I spend hours at the Natural History Museum [in London, England] looking at their small plesiosaurs, many of which are similar.

I'm not looking to find a prehistoric animal, only to try and identify what was the strangest thing I'll ever see. Even now I can remember every minute detail of it. To see such a thing was awesome.

Presented with such an amount of morphological detail, quite a few identities can be examined and discounted straight away—beginning with *Tasmacetus shepherdi*. Although somewhat dolphin-like in shape, this is a primitive species of beaked whale, described by science as recently as 1937, and known from only a handful of specimens, mainly recorded in New Zealand and Australian waters, but also reported from South Africa. Whereas all other beaked whales possess no more than four teeth (some only have two), *Tasmacetus* has 80, and its jaws are fairly long and slender.

However, the Gambian beast's two pairs of well-developed limbs effectively rule out *all* modern-day cetaceans as plausible contenders, because these species lack hind limbs. They also eliminate the archaeocetes—even *Ambulocetus*. For although this 'walking whale' did have two well-formed pairs of limbs, unlike the Gambian sea serpent its teeth were only half as many in number, yet of more than one type. The Gambian beast's long tail and dentition effectively ruled out pinnipeds and sirenians from contention too.

Many 'sea monster' carcases have proved, upon close inspection, to be nothing more exciting than badly-decomposed sharks, but as the Gambian beast apparently displayed no notable degree of external decomposition, this 'pseudoplesiosaur' identity was another non-starter.

Indeed, after studying his detailed letters and sketches, it became clear that, incredibly, the only beasts bearing any close similarity to Burnham's Gambian sea serpent were two groups of marine reptilians that officially became extinct 66 million years (or more) ago.

One of these groups consisted of the pliosaurs—thus including among their number the mighty Australian *Kronosaurus* that Burnham himself had mentioned. Yet whereas their nostrils' external openings had migrated back to a position just in front of their eyes, those of the Gambian sea serpent were at the tip of its snout.

The other group constituted the thalattosuchians—always in contention here on

Reconstruction of Owen Burnham's discovery of the Gambian sea serpent carcase (© William M. Rebsamen)

account of their slender, non-scaly bodies, paddle-like limbs, and terminally-sited external nostrils. True, their tails possessed a dorsal fin, but a thalattosuchian whose fin had somehow been torn off or scuffed away would bear an amazingly close resemblance to the beast depicted in Burnham's sketches. Alternatively, assuming that a thalattosuchian lineage has indeed persisted into the present day, its members may no longer possess such a fin anyway.

Without any physical remains of the beast available for direct examination, however, its identity can never be categorically confirmed. In 2006, a team from the Centre for Fortean Zoology (CFZ) visited the site where, 23 years earlier, the headless carcase had apparently been buried shortly after Burnham had viewed it, but despite doing some digging there they did not uncover any remains. Nevertheless, this compelling incident provides a thought-provoking reason for wondering if not all of the giant reptiles of prehistory perished alongside the dinosaurs by the end of the Cretaceous.

Bring Me the Head of the Sea Serpent!

Equally interesting, and even more frustrating, than the Gambian sea serpent saga is the following little-known report, published by the *Jonesboro Evening News* on 11 May 1905. For not only might this have featured a specimen of the marine saurian category , but its head was actually on public display for a time, yet apparently without attracting any scientific interest:

> Honolulu is the possessor of the head of a real sea serpent. The intact bones of the curious head are on exhibition in the window of a store in that city, and hundreds of curious people crowd around the place waiting to get a glimpse of the strange object, says a recent report to the *New York Sun*. William Herbert Melton Ayres brought to Hawaii what is probably the first head of a serpent to be placed on exhibition. He came to Hawaii seven years ago, later went to Shanghai and a short time ago he returned to Honolulu quite unexpectedly.
>
> He walked into the office of the Bulletin and asked if that paper cared to have a story about a sea serpent. He was asked not to slam the door in going out. Nothing daunted, Ayres again descended on the office, bearing a large carpet bag. Depositing the package on the sporting editor's desk he insisted upon opening it, and revealed to the surprised gaze of all the bones of the head of a huge sea monster. The jawbones measured about four feet, the head being a couple of feet wide. There were 160 teeth, 80 upper and as many lower. The specimen was utterly unlike anything seen as far as the records go. Ayres stated that he had purchased the head from a Chinese fisherman who had found the body of the serpent washed upon the shore, the body measuring 78 feet, apparently having been killed by some passing steamer.
>
> Ayres believes that the serpent thus discovered is a descendant of the monster which inspired the dragon upon the [pre-Communist] flag of China.

I wonder what became of it—simply discarded when public interest waned? Of course, it is possible that the specimen was merely a toothed whale (odontocete) skull, but as no-one with zoological knowledge could even be bothered enough to examine it (had they done so, it would surely have become a familiar case in the annals of sea serpent history), we shall never know. Sadly, abject scientific apathy has been—and remains—an all-too-frequent obstacle encountered in cryptozoology.

A Prehistoric Gharial Alive in Australia?

Today, only two species of crocodile are known from Australia—the monstrous estuarine or saltwater crocodile *Crocodylus porosus*, and the smaller Johnston's or freshwater crocodile *C. johnstoni*. However, when Australian herpetologist Richard Wells lived in the Northern Territory, he received several consistent reports from a variety of informants, including fish poachers, old ex-crocodile hunters, and aboriginals, indicating that a third, very large but also very different and seemingly scientifically-unknown crocodilian existed Down Under, inhabiting at least the tidal parts of the Northern Territory's Mary River system. According to Wells's data, shared with me:

> [it] would appear to be totally aquatic with an elongated jaw with numerous exposed teeth more in keeping with that of some kind of gharial (*Gavialis*)—but it appears to reach a larger size, has paddle-like limbs, and is of nocturnal behaviour. Most reports of the creature have been dismissed as representing

> sawfish or crocodiles, but all the people who reported it were very familiar with sawfish and crocs, but were adamant that it is some sort of crocodile-like reptile and it scared the hell [out] of them.

Gavialis gangeticus, the only known modern-day species of *Gavialis* gharial, is not native to modern-day Australia. However, crocodiles feeding on a low-protein diet (as sometimes in zoos) are prone to developing exposed teeth that point laterally in the jaws. Could this explain the dental peculiarity of the Mary River mystery croc—is it based merely upon sightings of some abnormal specimens of a known species? (Johnston's crocodile does possess a noticeably long, thin, superficially gharial-like snout.)

19th-Century engraving of the gharial *Gavialis gangeticus*

Possibly—were it not for the claim that this cryptid's limbs are paddle-like. As stated to me by now-retired Queensland Museum fossil reptile expert Dr Ralph Molnar, this feature immediately calls to mind the prehistoric thalattosuchians. In contrast, although its limbs were not paddled (which may be just an exaggeration on the part of the eyewitnesses—perhaps the Mary River beast's limbs are simply more rounded than those of Australia's two known crocodiles, creating the illusion of being paddle-shaped?), less than two million years ago the Solomon Sea just above south-eastern New Guinea was hunted in by a very gharial-like species, known as the Murua crocodile and dubbed *Gavialis papuensis*.

Some authorities have challenged this creature's gharial classification, favouring a thoracosaur affinity, but Molnar, although initially sceptical, subsequently informed me that it may indeed have been a gharial. Its elongated snout is certainly reminiscent of one, and it is believed to have been a fish-eater, again like modern-day gharials. A round-limbed present-day representative of this species would correspond very closely to the Mary River cryptid.

TITANIC TURTLES

A final category of water monster that may involve a reptilian prehistoric survivor—and which Heuvelmans dubbed the 'father-of-all-the-turtles' in his sea serpent classification system—features those of a characteristically chelonian nature.

According to the fossil record, the largest sea turtle of all time was *Archelon ischyros*, from North America's late Cretaceous, measuring up to 13 ft long and 16 ft wide (from flipper to flipper), with a 25-ft circumference. Unlike most modern-day turtles, however, it lacked a thick, horny shell. Instead, it had merely a framework of transverse struts, constructed from the bony ribs sprouting from its backbone and probably encased in a rubbery skin like that of today's leathery turtle *Dermochelys coriacea*, which also happens to be *Archelon*'s closest living relative.

Verified as up to 7 ft long (but possibly even longer, judging at least from various scientifically-unconfirmed reports), the leathery turtle is the largest marine chelonian officially in existence today. However, there are reports on file of turtle-like beasts so enormous that some cryptozoologists consider it possible that the oceans conceal a modern-day descendant of *Archelon*—one, moreover, that has re-evolved a shell and has become even bigger than *Archelon* itself!

Archelon skeleton

Certainly, stories of gigantic sea turtles are common in many mythologies around the world, but mythology can hardly explain accounts such as these. On 30 March 1883, for example, while in the North Atlantic on Newfoundland's Grand Bank, Captain Augustus G. Hall and the crew of the schooner *Annie L. Hall* reportedly encountered a turtle so immense that at first they mistook it for a capsized ship, floating upside-down. When the schooner approached to within 25 ft, however, enabling them to obtain a closer look at it and also to gauge its size against their vessel's length, they found it to be a living turtle—at least 40 ft long, 30 ft wide, and 30 ft from the apex of its back to the bottom of its undershell (plastron), with 20-ft flippers.

During June 1956, in the vicinity of Nova Scotia (not far from Newfoundland), some of the crew aboard the cargo steamer *Rhapsody* claimed to have seen a gigantic turtle roughly 45 ft long, with 15-ft flippers—and made even more distinctive by its totally white shell. According to the crew, this remarkable creature could raise its head 8 ft out of the water.

Reports of turtle-like water monsters are not confined to Canada either. Some have emanated from the Indian Ocean, including the vicinity of Sri Lanka. In the 3rd Century AD, Roman scholar Claudius Aelianus (popularly known merely as Aelian), writing in his 17-volume treatise *De Natura Animalium*, mentioned the presence of immense turtles in the Indian Ocean whose shells were so large (up to 24 ft long) that they were used by the local people as roofing material!

Writing in his own magnum opus, *Geography*, which he completed in 1154 AD, Muhammad al-Idrisi, a notable Moroccan Islamic traveller, cartographer, and archaeologist, referred to comparably immense turtles, up to 20 cubits (33 ft) long, living in the Sea of Herkend, off the west coast of Sri Lanka.

Further examples of giant marine cryptoturtles are contained in a comprehensive chapter devoted entirely to mega-sized mystery chelonians in my book *Mirabilis* (2013).

Could they be unrecognised descendants of the mighty *Archelon*? If some of that novel roofing material is still in existence, we may yet find out.

Champ envisaged as a giant form of long-necked freshwater turtle (© Thomas Finley)

Presenting the Plesioturtle

In June 2003, while conducting sonar research in Lake Champlain's Button Bay area for a television show on Champ being produced by the Discovery Channel, a team from Fauna Communications Research led by Elizabeth von Muggenthaler detected a series of biosonar readings similar to those emitted by a beluga (white) whale *Delphinapterus leucas* or dolphin when seeking food underwater. Ten times louder than any known fish species in the lake, too irregular to be produced by a mechanical device or fish finder, and recorded on multiple instruments on three separate days (3, 4, and 10 June), some readings lasted up to 10 minutes, and von Muggenthaler believed that on one occasion the unidentified creature approached to within 30 ft of their boat. Currently, she has claimed to have recorded some 200 examples of 15 individual unknown animals belonging to the same species—but what might that species be?

Although cetaceans are normally the obvious choice when considering the origin of biosonar signals, not everyone is convinced by it in relation to these ones. For instance, in 2013 after viewing their echolocation charts and listening to the recorded sounds, Dr Lance

B. Lennard, a whale acoustics expert from the Vancouver Aquarium Marine Science Center, highlighted various signals that were not consistent with any that he had previously encountered with whales. Consequently, he concluded that the patterns were not of whales, were probably not even mammalian in origin, and ultimately were unknown to him. So if not whales, what other animals could be responsible for them?

In September 2005, a remarkable discovery was revealed that offered not only a possible non-cetacean explanation for these anomalous sounds but also a very thought-provoking identity of particular relevance to Champ. That was when Australian turtle researcher Jacqueline Giles submitted as her PhD thesis at Murdoch University in Perth, Australia, a very comprehensive study (accepted in 2006 and now accessible to read online) disclosing that a long-necked species of Australian freshwater chelonian called the northern or oblong snake-necked turtle *Chelodina oblonga* emitted a complex series of high-frequency sounds resembling echolocation while underwater. This was the very first scientific documentation of an underwater acoustic repertoire from an aquatic chelonian. Furthermore, according to the Acoustical Society of America, which closely examined and confirmed Dr Giles's findings, publishing in 2009 a paper on this subject authored by a research team headed by her, these sounds indicated the expression of a complex language that showed signs of high intelligence and social order.

Since then, recalling comparable chelonian-inspired thoughts relative to the Congolese mokele-mbembe (see Chapter 1), there has been cryptozoological speculation by American lake monster researcher Charles Pogan and others that perhaps an extra-large species of snake-necked (or similar long-necked) turtle exists in Lake Champlain. This might not only explain the strange sounds recorded here for more than a decade now, but would also offer as an identity contender for Champ a creature greatly resembling the plesiosaurian descriptions reported by so many Champ eyewitnesses. Accordingly, Pogan refers to this hypothetical reptile as the plesioturtle. In addition, such a species of turtle could survive in a temperate coldwater lake, especially one so plentifully supplied with suitable food species as Champlain, and is far less extreme a candidate to consider than a surviving, evolved plesiosaur.

Having said that, Pogan has not ruled out prehistoric survivorship entirely, suggesting that if his postulated plesioturtle does indeed exist, it might be a modern-day descendant of a truly gargantuan fossil freshwater turtle aptly dubbed *Stupendemys*. Estimated to have exceeded 13 ft in total length (its carapace alone was 11 ft) and currently known from two northern South American prehistoric species, this was the largest chelonian of all time. Moreover, it was related to the snake-necked turtles, and officially survived until as recently as the Pliocene's onset, 5 million years ago. An undiscovered modern-day *Stupendemys* representative might go some way to explaining a fair few freshwater long-necks, not just Champ.

GOING DIPPY OVER *DIPLOCAULUS*

Up to 3 ft long, superficially salamander-like in basic form, but belonging to the long-extinct nectridean taxonomic order within the equally erstwhile subclass of archaic amphibians known as lepospondyls, one of my favourite creatures from pre-dinosaurian prehistory has always been *Diplocaulus*—famous for its huge boomerang-shaped head, as exhibited by several species (plus a very close relative, the lesser-known but near-identical *Diploceratus*

Traditional arrow-headed reconstruction of *Diplocaulus* (CC0 Diagram Lajard)

burkei). The reason for its head's bizarre shape was its skull's pair of enormous but dorsoventrally flat, lateral bony projections known as tabular horns. These remarkable structures may have enabled its head to serve as a hydrofoil when this amphibian was swimming, or may even have prevented it from being swallowed by predators, by increasing its head's width beyond the gape of any carnivorous creature alive at that time that shared its distribution range.

Arising in North America, this extraordinary creature lived during the Permian Period (approximately 300-250 million years ago), although only *D. minimus*, currently the single known non-American representative (native to Morocco), occurred during the late Permian. Consequently, this makes various photographs and videos of alleged living *Diplocaulus* specimens that have surfaced online and elsewhere in the media during recent years nothing if not intriguing from the standpoint of prospective prehistoric survival.

The Maltese 'Hammerhead Lizard'

As I first documented in my book *Extraordinary Animals Revisited* (2007), in September 2004 the British magazine *Fortean Times* forwarded to me as their cryptozoological correspondent a short note from reader Stuart Pike enquiring about a photograph of unknown origin that had been circulating online and which depicted a bizarre-looking mystery beast labelled in accompanying internet reports as a hammerhead lizard.

Not long afterwards, Maltese journalist Tonio Galea independently contacted me to request details about this same photo, because according to local Maltese rumour its creature had lately been discovered alive on a rocky beach at Il-Maghluq, Marsascala, in the south of the island. I subsequently received several

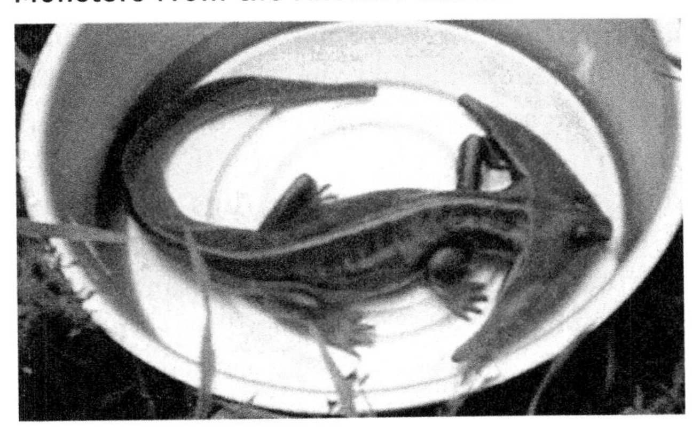

The 'hammerhead lizard' (USC Title 17 § 107)

more enquiries from other correspondents, and so too, it transpired, did various other scientists, including Malta University biologist Prof. Patrick J. Schembri, who wrote about it in a letter published on 21 November 2004 in Malta's *Sunday Times* newspaper.

In reality, however, what this intriguing photo depicts is a gypsum/non-urethane foam-based model of *Diplocaulus*. Investigations of mine eventually revealed via a Japanese model-making website (http://www.interq.or.jp/sun/mm-kas/tenji.htm) that it had been manufactured back in 1992 by an amateur Japanese model-maker in response to a magazine competition. I have not succeeded in discovering the model maker's name or the photographer, but I did manage to uncover a second, very similar photograph of the model itself.

Diplocaulus on Video

YouTube contains several videos purporting to show living specimens of *Diplocaulus* (with at least two totally different specimens featured—one pink, one green), but all of them have been filmed and uploaded by the same person—which means that either he/she is unaccountably successful at locating living specimens of an amphibian deemed extinct for at least 250 million years by palaeontologists, or all is not as it seems.

The person in question has the YouTube username SouldierTVSP, and has uploaded three separate videos of what is claimed to be the same specimen, which can be viewed in the following sequence of their filming at:

www.youtube.com/watch?v=F5GLOdwyhyM
www.youtube.com/watch?v=AaT8C3EWQKk
www.youtube.com/watch?v=5LOPk6zjasI

plus a much shorter video montage at:

www.youtube.com/watch?v=hGbJt1ys8VU

and a video of a visibly different specimen at:

www.youtube.com/watch?v=zd5783cnkoE

The first of the three videos allegedly showing the same specimen was filmed on 22 July 2011 (according to an on-screen caption). The second video consisted of three separate segments, filmed respectively (according to on-screen captions) on 26 July 2011, 1 August 2011, and 8 August 2011. No on-screen date was given for the third video, but as all three videos were uploaded by SouldierTVSP on 8 August 2011, this third video was clearly a continuation of the second one's ending section. Each of the three videos was accompanied by the following interestingly worded request: "Someone, please teach me this strange creature's true colors", beneath which for the third video was this additional, rather more forthright statement: "Diplocaulus Still Alive!! Paranormal Creature".

As will be seen when viewed, these three videos show what looks like a pink toy *Diplocaulus* amidst some vegetation debris floating in a current of water, seemingly a stream or river. There is nothing in the videos that can be used as an effective close-up scale to provide an estimate of size for the object (but unless the surrounding debris is very substantial, it would seem to be small). The object is moving entirely passively, drifting and buffeted by the current, with just an occasional slight movement of its tail or head, as might be expected, for instance, of a toy with a jointed tail and head, but no sign of any independent animate movements. In contrast, palaeontologists believe that in life, *Diplocaulus* would have probably swum by vertical undulations of its body, as its small weak legs and relatively short tail would not have been of much locomotory assistance. Of course, the object in the videos just might be a recently dead animal as opposed to a living one, but to my eyes the object looks as if it has never lived. Also, these three videos were filmed over an 18-day period, and yet the object's appearance does not change at all (in 18 days, a dead animal, conversely, would have shown considerable signs of decomposition, assuming that it hadn't already been devoured by a predator).

On 16 November 2012, SouldierTVSP uploaded a very short video montage entitled 'Diplocaulus Still Alive! Cryptid Exist [sic]'. It begins with a brief clip of a still photograph depicting someone holding a large creature to the camera, its somewhat salamander-like head pointing forward as it looks directly into the camera. The head bears a pair of fleshy flap-like lateral projections, but these are nowhere near as large or boomerang-shaped as those of *Diplocaulus*. The rest of the creature cannot be readily seen; indeed, based upon that photo alone, I'm not entirely convinced that the creature is an amphibian (not even a deformed one), rather than some unusual wide-mouthed fish. Nevertheless, an on-screen caption states in English and Japanese that as soon as its identity as *Diplocaulus* is confirmed, the full footage will be released (but as far as I'm aware, no such release has occurred to date). The remainder of the video shows what seems to be the earlier pink, apparent toy *Diplocaulus* resting on some vegetation (with a live wild duck of similar size close by, thus providing a useful size scale), then ending with a four-second clip of what looks like this same object floating in the water with its tail swishing from side to side, but very plausibly caused simply by the water current moving a jointed tail on a toy.

The final alleged *Diplocaulus* video by SouldierTVSP, uploaded on 11 January 2012, has the somewhat unexpected title of 'Kinky Cryptid Sightings', and showcases an iridescent green *Diplocaulus* with an enormous head moving across a pond (located near a waterfall) at the water surface in a seemingly active manner before its huge boomerang-shaped head becomes entangled in what looks worryingly like an item of female undergarment floating there (which presumably explains the title of the video!). For these reasons alone (not to mention the wide variety of anything-but-serious videos on other cryptozoological subjects that this person has also uploaded onto YouTube), I personally find it difficult to take this particular video seriously. In my view, some form of self-propelled model has been filmed here—but I would love to be proved wrong!

Oh, *Diplocaulus*, What Can the Matamata Be?

On 10 October 2015, a very striking photograph was tweeted to me by The Anomalist @anomalistnews, stating: "Is this a better diplocaulus hoax, or a previously unknown animal?" and "Only details on this photo is it was taken in 'Asia' and was posted on Facebook".

The photograph depicted what certainly looked like a *Diplocaulus*, resting half-in and half-out of some shallow water, and the degree of morphological detail visible was extensive.

19th-Century engraving of the matamata

But was it a living animal, or was it either a very realistic model or a very skilfully photo-manipulated image?

Not surprisingly, this mystery photograph attracted considerable interest and comments on Twitter, including this thought-provoking suggestion by Facebook friend Paul Willison: “IMO, a photoshop of a hellbender or giant salamander and baby mata mata turtle”. He also attached some photographs of these species in support of his opinion.

As already discussed in Chapter 1 of this book, the hellbender *Cryptobranchus alleganiensis* is North America’s giant salamander, beaten in size only by the giant salamanders of China and Japan. The mata mata or matamata *Chelus fimbriata* is a very bizarre-looking species of South American freshwater turtle. And there was no doubt that the ostensible *Diplocaulus* in the mystery photo did embody features from both of these species, so could that be the answer—a photographic montage or composite created by some ingenious morphing of hellbender and matamata images by person(s) unknown?

Spurred on by Paul’s suggestion, I spent some time Google-imaging matamata turtles, hellbenders, and giant salamanders in general, in search of corresponding photos or portions of photos, as well as *Diplocaulus*, in search of matching photos of models and restorations. Unfortunately, nothing turned up . . . until, that is, after scrolling down to the very bottom of the umpteenth Google-Image search page using the above and similar animal names as search words, suddenly the mystery photograph itself appeared!

It proved to be a photo of an exceedingly life-like *Diplocaulus* model created by expert Japanese model maker Goro Furuta. What’s more, it was just one of several photos (all copyrighted to him) of this wonderful model that were present in a publicly-viewable album on his Facebook page (this album can be viewed at: https://www.facebook.com.goro.furuta.39/media_set?set=a.507467349430212.1073741845.100005008544234&type=3).

Furuta has prepared many additional, equally spectacular animal models, and as I swiftly became a massive fan of his work after browsing pictures of them in his several albums on Facebook depicting his work, I am delighted that he is now a Facebook friend of mine.

Incidentally, I’d like to stress here that at no point has he ever claimed or sought to suggest that his *Diplocaulus* model was anything other than a model—the online confusion as to whether or not the photograph of it currently doing the internet rounds portrayed a living *Diplocaulus* is due entirely to misinformed speculation by people seeing the photo (copied from his FB album and circulated online by person/s unknown) but not knowing its origin and incorrectly assuming the model to be a real animal. (Having said that, I suppose it can be viewed as a backhanded compliment to his model-making expertise that his *Diplocaulus* model is so realistic that people have assumed it was alive!)

The mystery of the most life-like non-living *Diplocaulus* reported online so far was a mystery no longer. In a tweet of 10 October

2015 replying to the original one by The Anomalist and to those of Paul Willison, I stated: "It's a Diplocaulus model, by Japanese model-maker Goro Furuta: [and then I included the link to his relevant Facebook album]".

Sadly, however, it means that this boomerang-headed amphibian remains interred within the long-vanished Permian Period, but even back there it has offered up a startling surprise. Trace fossils have been found showing a pair of flaps or membranes linking the tips of its head's tabular horns to its body—in other words, *Diplocaulus* may not have been outwardly boomerang-headed as traditionally assumed, but might well instead have resembled in life the restoration of it with its horns enclosed inside membranes on display at the University of Michigan's Natural History Museum.

If so, then any videos or future photos of purported living *Diplocaulus* specimens that possess a boomerang-shaped head can swiftly be discounted—always assuming, of course, that an evolved modern-day *Diplocaulus* did not develop one during its 252-million-year continued evolution since the Permian??

For now, however, all of this is academic, because *Diplocaulus* is still defunct, but it remains one of my favourite prehistoric creatures too—even if it has lost its boomerang!

SILVER COELACANTHS AND SOUVENIR SCALES

The respective discoveries of two living species of coelacanth—*Latimeria chalumnae* in 1938 off South Africa and later in the sea around the Comoro Islands and Madagascar, and *L. menadoensis* in 1997 off the Indonesian island of Sulawesi (Celebes)—are among the greatest zoological events of the 20th Century. Moreover, there is some intriguing evidence to hand to suggest that history may at a later date repeat itself again, this time somewhere in the seas around North America, particularly the southern U.S.A. and/or Mexico.

One day in 1949, ichthyologist Dr Isaac Ginsburg at the U.S. National Museum received a short note from a souvenir seller in Tampa, Florida, enclosing with it a single fish scale for him to identify. She regularly purchased barrels of scales from fishermen, and used them in the manufacture of her souvenirs, but the scales in one of the barrels that she had recently obtained were very strange and unfamiliar in appearance—unusually thick and hard like plates of armour. Hence she was curious to learn more about the species from which they had originated. So too was Ginsburg once he had examined the scale that she had sent to him—because he could see that it greatly resembled those of the modern-day coelacanth genus *Latimeria*!

Yet *Latimeria* had never (and has still never) been recorded from the New World. Consequently, Ginsburg wrote back to the souvenir seller without delay, requesting more of the scales and further details regarding their origin—but he never received a reply. Equally tragically, the single scale was somehow lost, and has never been traced.

During the early 1970s, American naturalist Sterling Lanier was selling some of his famous brass figurines of fossil animals at an art show near Florida's Gulf of Mexico coast, when he noticed a very interesting necklace among the exhibits of a fellow artist. The necklace was composed of fish scales, and what had caught his attention was the fact that they looked very like those of *Latimeria*. According to its owner, he had extracted the scales from a pile of marine miscellany—seaweed, shrimps, etc—aboard a Mexican Gulf shrimp boat, and although he permitted Lanier to examine and even sketch the scales (sadly, Lanier has since

The Comoros coelacanth *Latimera chalumnae* (© Markus Bühler)

lost his sketches) he would not sell the necklace.

What may (or may not) be a noteworthy sequel to that incident occurred in 1992, when French naturalist Ronald Heu purchased three unusual fish scales in a souvenir shop at Biloxi, Mississippi. He felt that they were very like those of *Latimeria*, and sought the opinion of an ichthyologist colleague, who agreed that they were certainly similar in appearance. In the opinion, however, of a semi-anonymous correspondent from Chateaudun, as published in the French journal *Science et Vie* (October 1993), the scales were actually from a very large, known species of South American fish called the arapaima or pirarucu *Arapaima gigas*.

19th-Century engraving of an arapaima

It is a pity that these have not been examined further, to obtain a definitive identification.

Yet another morsel of mystery regarding controversial coelacanths derives from an Indian miniature painting dating from the 18th Century, which depicts a Muslim holy man standing beside a large fish. In a *Naturwissenschaftliche Rundschau* article from 1972, Berlin scientist Prof. Burchard Brentjes contemplated the possibility that the fish—with large fins and armour-like scales—was a specimen of *Latimeria*. After examining the miniature as reproduced in his article, however, I can see very little similarity between the two fishes—and when I showed the reproduced picture to French ichthyologist Dr François de Sarre, he discounted a coelacanth identity in favour of the Indian climbing perch *Anabas testudineus*.

However, the greatest controversy of all concerning a prospective link between the New World and still-undiscovered modern-day coelacanths comes from two small but very enigmatic artefacts that have become known as the silver coelacanths. In recent years, a major reappraisal of what they may truly represent and how old they actually are has occurred via an important scientific paper, which I documented in my recent book *A Manifestation of Monsters* (2015). Here is what I wrote:

In 1964, Argentinian chemist Dr Ladislao Reti purchased a small but very beautiful and extremely unusual silver ornament hanging in a village church near Bilbao, on Spain's northern Atlantic coast. Measuring 4 in long, it was a meticulously sculpted figurine of a fish—but of no ordinary species.

Its scales were thick and armour-like, and its first dorsal fin was borne directly upon its back—but its second dorsal fin, pectoral fins, pelvic fins, and anal fin were all borne upon fleshy lobes, resembling stumpy legs. As for its tail, it bore three fins, the third of which (whose

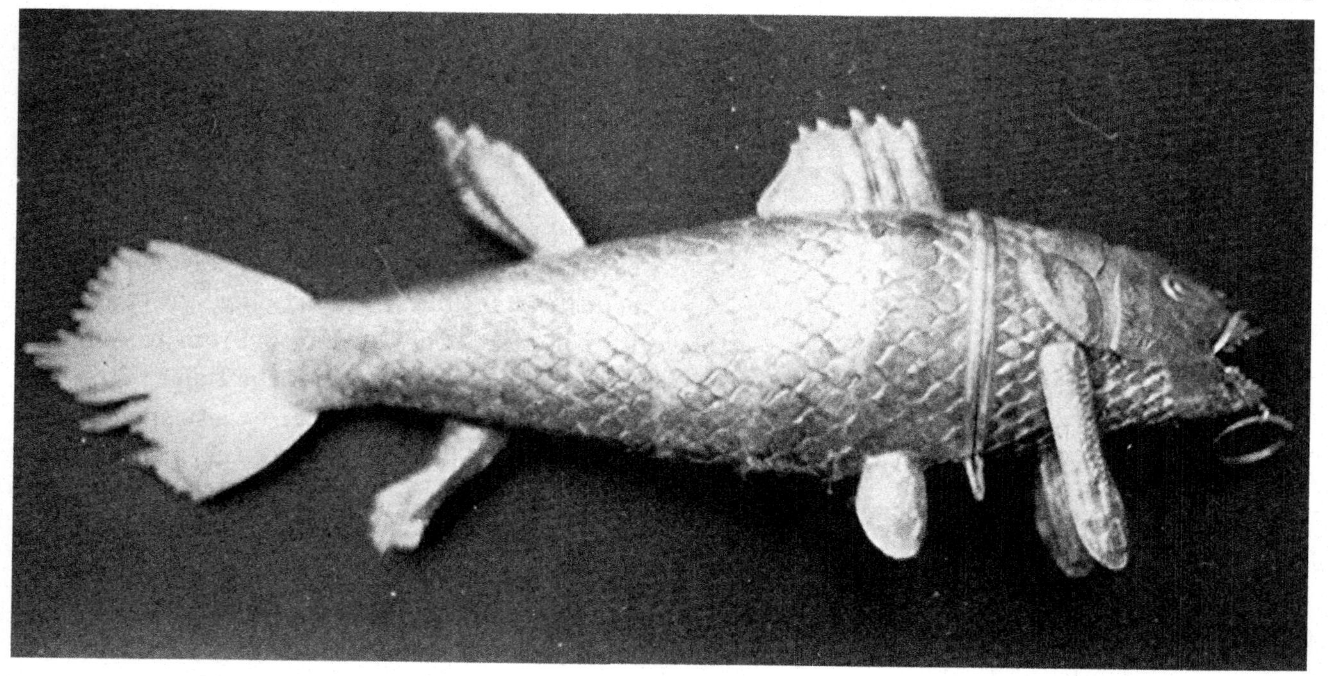

Reti's silver coelacanth ornament (courtesy ISC/J. Richard Greenwell)

base is just visible in the only known photograph of this ornament) was a tiny lobe sandwiched between the larger upper and lower fins. Back in 1964 (i.e. 33 years before the discovery of *L. menadoensis*), there was only one known species of living fish that possessed all of these features—in fact, the lobed fins and tripartite tail were unique to it—and that was *L. chalumnae*. Moreover, the silver objet d'art was instantly identifiable as an accurate three-dimensional image of that modern-day coelacanth species. Yet how could its presence in a Spanish church be explained?

Tragically, this remarkable figurine apparently has since been lost (although no details concerning such an event seem to have been made public). However, following examinations of the still-existing photograph of it (reproduced here) by a number of antique silverware experts, the consensus was that it was probably a religious votive that may well have been manufactured during the 17th or 18th Century. If correct, this is an astonishing situation—because *L. chalumnae* was not discovered by science until 1938! So how could so accurate a figurine of a coelacanth have been created two centuries or more earlier?

To make matters even more mystifying, in 1965 a second silver ornament in the form of a coelacanth was found, again in Spain. This one was spotted and purchased by Prof. Maurice Steinert, who was then a molecular biology student, while browsing through an antiques shop in Toledo, and is believed to be of similar age to the first.

Since the discovery of these two silver coelacanths, various solutions have been proposed to explain their anomalous—and ostensibly anachronistic—similarity to *L. chalumnae*. Perhaps they were modelled not upon the living *Latimeria*, but upon comparable fossil coelacanths from the ancient past. Yet if this were true, how can we explain these ornaments' uncannily accurate three-dimensional form? It is hardly likely that their sculptor(s) could have achieved such precise results using

The holotype of *Latimeria chalumnae*

only flat fossils as models. Alternatively, their existence may indicate that specimens of *L. chalumnae* had been brought to the Atlantic region prior to the species' scientific discovery in 1938—or even that there is an undiscovered population inhabiting the Atlantic.

However, coelacanth researcher Prof. Hans Fricke, from Germany's Max Planck Institute for Animal Behaviour, who was most intrigued by the silver coelacanths, offered a very different answer. While investigating their possible origin, he consulted the world's leading expert on Spanish silver art, Prof. José Manuel Cruz-Valdovinos from Madrid University's Department of Contemporary Art—who opined after examining photos of the two ornaments that they had probably been manufactured in Mexico, but while it was still under Spanish colonial rule, thereby explaining their eventual presence in Spain and their Spanish style.

In 2001, Fricke co-authored a scientific paper with Dr Raphael Plante from the Centre of Oceanography in Marseilles, France. In their paper, published by the journal *Environmental Biology of Fishes*, they revealed that they had conducted a close morphological comparison between the existing photograph of the lost Reti silver coelacanth with images of the holotype of *L. chalumnae*. In addition, during May 2000 they had made Steinert's silver coelacanth available for direct examination this time by Prof. Cruz-Valdovinos (rather than via mere photos of it as before).

Fricke and Plante concluded that in terms of body shape as well as fin form, position, and orientation, the Reti silver coelacanth appeared so similar to the holotype of *L. chalumnae* that it seemed reasonable to assume that it had been directly modelled upon the latter specimen. As for Steinert's silver coelacanth, Cruz-Valdovinos noted that because, to quote the Fricke-Plante paper, "this artefact lacks style marks, such as peculiar fantasy engravings, typical for baroque silver art pieces of the 16th or 17th century"—and also because of the sharp edges of its body scaling and fins (these edges being of a modern style and contrasting with the smoother edges present on older pieces that have resulted from frequent handling wearing down the edges)—this ornament was probably of recent manufacture. Consequently, he was now discounting his earlier opinion, based only upon photographic evidence, that it was centuries old. And when during May 2000 this specimen was examined independently of Cruz-Valdovinos by A. Jiminez, an expert on silver artefacts and the creation of silver fishes, he opined that it was a typical Spanish folklore art piece and that it was indeed of recent origin, dating from only the past 50-30 years, i.e. c.1950-1970.

As a result of the Fricke-Plante paper's conclusions, it has generally been assumed ever since that the mystery of the silver coelacanths has been solved. However, a careful reading of their paper reveals a couple of glaring contradictions, but which do not appear to have been brought to public attention anywhere—until now.

Although most of the the Reti silver coelacanth's fins are indeed comparable in shape and orientation to those of the *L. chalumnae* holotype, no mention is made by the authors in their paper that the first dorsal fin of this silver coelacanth looks nothing like that of the *L. chalumnae* holotype. For whereas the latter's first dorsal fin radiates outward from a narrow base and sports a convex upper edge (as is clearly visible in all of the most famous photos of this historic specimen, including the one reproduced here), the equivalent fin on the Reti silver coelacanth is almost triangular in shape, having a very wide base that tapers markedly upwards, with its uppermost edge much less wide and almost straight, bearing a horizontal line of extremely pointed rays running along it. Why such a noticeable discrepancy if this ornament were indeed based wholly upon the *L. chalumnae* holotype, and why no mention of this discrepancy by the authors in their paper?

Equally bemusing is the following outright contradiction. According to the paper's main text, Cruz-Valdovinos referred to a lack of fantasy engravings upon the Steinert silver coelacanth, which is a feature that in turn provided evidence that this ornament was of recent manufacture. Conversely, the caption to Fig. 2 in this same paper (Fig. 2 consisting of two photographs comparing the throat of a *Latimeria* specimen with that of this ornament) stated that the ornament's throat had fantasy engravings. Moreover, this statement was confirmed by these engravings' readily visible presence in the photograph of the ornament's throat. So how can this contradiction be reconciled?

Following on from that: because the Steinert silver coelacanth does indeed bear fantasy engravings (Cruz-Valdovinos's contrary statement notwithstanding), does this therefore mean that it is not of recent manufacture after all, but actually dates from a much earlier period? If so, this would dramatically pre-date the discovery of *L. chalumnae*, thus re-inciting all of the fevered speculation concerning whether the ornament is evidence for the existence of an undiscovered coelacanth population (or even species) that had been quelled since the publication in 2001 of the Fricke-Plante paper.

Somehow, I think that the seal on a final, unequivocal solution to the riddle of the silver coelacanths has yet to be set.

THE JAWS OF MEGALODON—SHARK OF NIGHTMARE . . . AND REALITY?

The following lines were penned by Victorian naturalist Philip H. Gosse, and appeared in his book *The Romance of Natural History* (1860):

A great white shark off Guadalupe Island (CC Elias Levy)

> Half concealed beneath the bony brow, the little green eye gleams with so peculiar an expression of hatred, such a concentration of fiendish malice, of quiet, calm, settled villany, that no other countenance that I have ever seen at all resembles. Though I have seen many a shark, I could never look at that eye without feeling my flesh creep, as it were, on my bones.

This graphic description vividly expresses the feelings of many people when confronted with sharks, especially the most feared species of all—*Carcharodon carcharias*, the great white shark. The world's largest living species of carnivorous fish (excluding plankton-eaters), it is known to attain a total length of up to 21 ft, but unconfirmed sightings of far bigger specimens have occasionally been recorded, mostly in tropical or sub-tropical waters. Could such sharks really exist—and, if they do, could they prove to be something even more terrifying than oversized great whites?

In his book *Sharks and Rays of Australian Seas* (1964), Antipodean ichthyologist Dr David G. Stead documented an astonishing account that had been narrated to him back in 1918 by some fishermen at Port Stephens, New South Wales. They claimed that their heavily-weighted crayfish pots, each measuring 3.5 ft long and containing several crayfishes (each weighing several pounds), had been effortlessly towed away by a ghostly white shark of enormous size. Estimates given by the fishermen ranged from the length of the wharf on which they had been standing, which measured 115 ft, to, in the opinion of one of the men, "300 ft long at least"! Even though Stead discounted these gargantuan estimates as the product of fear, he was clearly impressed by their claim, stating in his book:

> In company with the local Fisheries Inspector I questioned many of the men very closely, and they all agreed as to the gigantic stature of the beast. . . . And bear in mind that these were men who were used to the sea and all sorts of weather, and all sorts of sharks as well. . . . They affirmed that the water 'boiled' over a large space when the fish swam past. They were all familiar with whales, which they had often seen passing at sea, but this was a vast shark . . . these were prosaic and rather stolid men, not given to 'fish stories' nor even to talking at all about their catches. Further, they knew that the person they were talking to (myself) had heard all the fish stories years before! . . . The local Fisheries Inspector of the time, Mr Paton, agreed with me that it must have been something really gigantic to put these experienced men into such a state of fear and panic.

Surprise and shock at unexpectedly encountering an awesome Moby Dick of the shark world may well have helped to distort their assessment. Yet even if we accordingly allow a very generous margin of exaggeration, the result is still a creature of far greater size than one would expect for the great white shark. Perhaps the most telling aspect of this episode, however, is that the men were so shaken, after seeing whatever it was they saw, that they weighed anchors straight away, fled back to port, and refused to go out to sea again for several days. This is hardly the behaviour that one would expect from people who know that they will not earn any money if they do not go out to sea—unless their story is true, and they really were frightened by a monstrous shark.

An immense shark, sporting a square head, huge pectoral fins, a green-yellow body speckled with a few white spots (encrusted barnacles?), and measuring considerably more in total length than his 35-40-ft boat was spied in 1927 or 1928 by Zane Grey, while sailing off the French Polynesian island of Rangiroa (about 220 miles northeast of Tahiti) in the South Pacific's Tuamotu Archipelago. Grey was a famous writer of Western novels, but he was also a passionate angler and the author of eight angling books, including *Tales of Tahitian Waters* (1931), containing his account of his shark sighting.

Yet despite his experience in handling fishes of record-breaking size, Grey was unable to identify this immense specimen. A square head is certainly not reminiscent of a great white shark, of any size, but rather a whale shark *Rhincodon typus*. This harmless planktivorous species constitutes the world's largest fish of any type, with a maximum confirmed length of 41.5 ft (but likely to attain up to 50 ft), and it does have a very broad, massive head. Then again, read the next report . . .

In 1933, when about 100 miles northwest of Rangiroa aboard the *S.S. Manganui*, Grey's son, Loren, also caught sight of a gigantic shark, once again yellowish in colour but flecked with white, which revealed a great brown tail, plus a massive head that seemed to be at least 10-12 ft across, and a total body length estimated by Grey Jnr to be not less than 40-50 ft. However, he was convinced that it was *not* a whale shark. So what *was* it?

According to traditional beliefs of the Polynesian fishermen who work along the

The planktivorous whale shark (Sam Farkas/NOAA)

coasts of New South Wales, these waters are frequented by a frightening type of sea creature that they respectfully refer to as the Lord of the Deep. They liken it to a gigantic white shark, measuring about 100 ft in length. Is this what the Greys spied, and could this be what carried away the pots of the lobster fishermen in 1918?

In his book *Shark!* (1961), Thomas Helm documented his own (undated) encounter with a giant mystery shark. He and some other people were on board his 60-ft trawler in the Caribbean Sea when they spied a huge shark that he claimed was "not an inch less than thirty feet". He was able to estimate this accurately by comparing its length to that of his trawler; and he also noted that when it swam underneath, its pectoral fins were clearly visible on either side of the boat. He and the other eyewitnesses were unable to identify its species, but he stated that it "most closely resembled the [great] white shark".

During the 1970s, a Mrs T. Brinks and her keen sailor husband Dave were sailing their 40-ft boat about 100 miles west of Monterey Bay, California, when they encountered what looked like a great white shark but of huge proportions. When it swam alongside their vessel, they could see that in total length it equalled that of the boat. After a few moments, it veered to the west, swimming underneath their boat before disappearing (they actually felt the boat rise as it swam beneath it). The Brinkses later recalled their encounter with one of Mrs Brinks's work colleagues, Jon Ziegler, from Idaho, who presented the details in a letter published online by *Strange Magazine* in 2005.

More recently, in Season 3, Episode 7 (entitled 'Mega Jaws'), first screened on 18

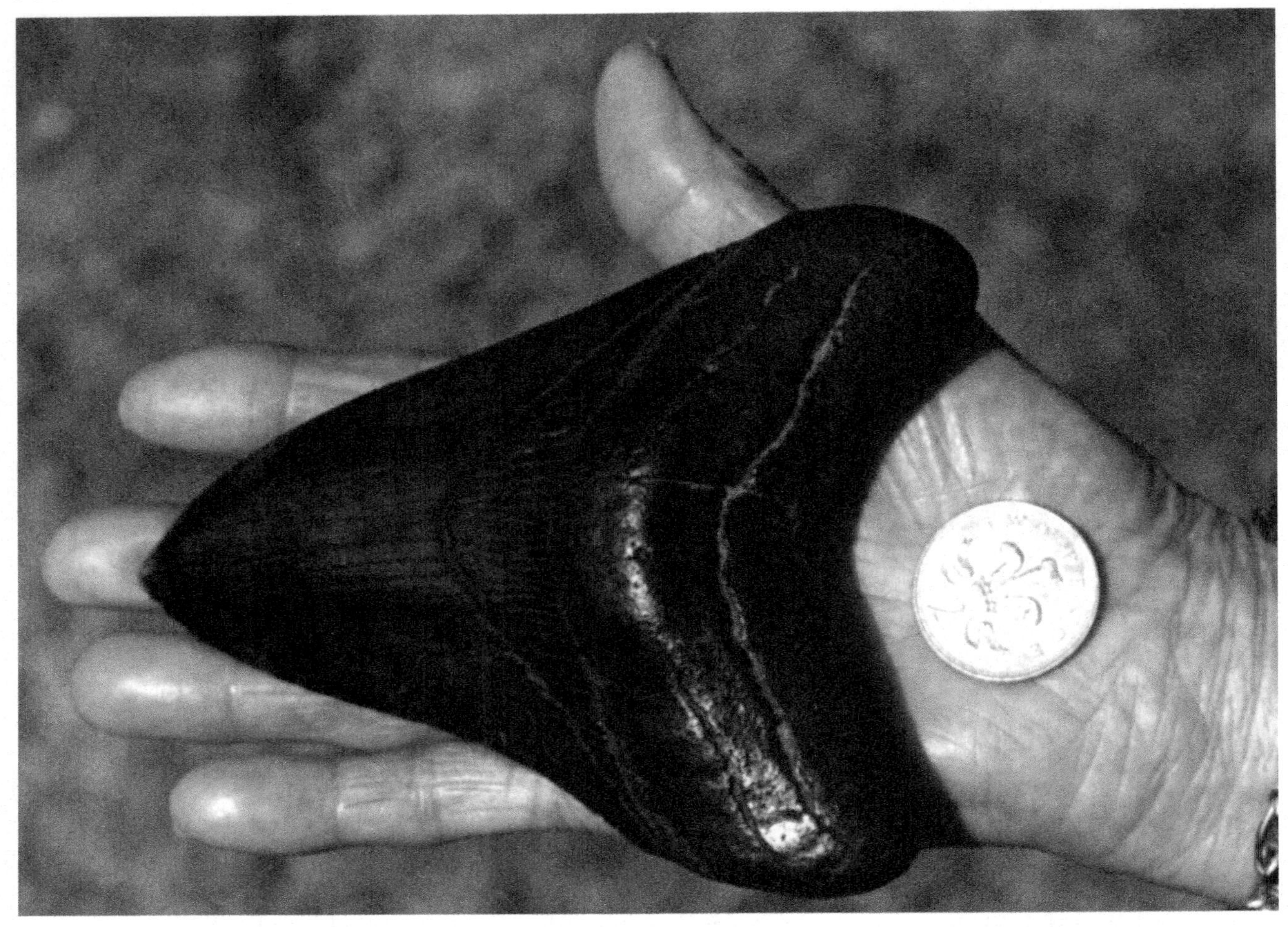

Fossil megalodon tooth (© Dr Karl Shuker)

March 2009, the cryptozoological TV show *MonsterQuest* unsuccessfully sought a giant black carnivorous shark occasionally sighted in the Sea of Cortez (Gulf of California), off Mexico's Baja California peninsula. Fisherman witnesses claim that it is 20-60 ft long, resembles a huge great white shark except for its dark colouration and massive tail, and have dubbed it El Demonio Negro ('the Black Demon'). Might it be a melanistic great white (a huge great white that was fairly dark dorsally and measured almost 20 ft long was hauled up out of the Sea of Cortez by commercial fishermen in April 2013, and parts of this sea are now known to serve as a great white shark nursery), or could it be something very different indeed?

Many ichthyologists are willing to consider the possibility that there are larger specimens of great white shark in existence than have so far been verified by science, but some cryptozoologists are far bolder. Their explanation for the Lord of the Deep is far more spectacular—a terrifying prehistoric resurrection, featuring a living leviathan from the ancient waters.

The great white shark once had an even bigger relative—the megalodon or megatooth shark *C. megalodon* ('big tooth'), sometimes placed in its own genus, *Carcharocles*. Named after its huge teeth, which were triangular in shape, up to 7.25 in high, and edged with sharp serrations, the megalodon was once believed to measure as much as 98 ft long, but this early estimate of its size was later shown to be incorrect, and was refined to a much more sedate yet still unnerving 43 ft. However, after various extra-large megalodon teeth, some almost 6 in long, were unearthed a while ago at the aptly-dubbed Sharktooth Hill near Bakersfield, California, ichthyologists conceded that certain specimens might have attained a total length of up to 55 ft.

The megalodon is presently known almost entirely from its huge teeth and some individual vertebrae. However, one notable exception is an associated vertebral column of approximately 150 individual centra (vertebra bodies) that range in state from fragmentary to nearly complete. In the major monograph *Great White Sharks: The Biology of Carcharodon carcharias* (1996), edited by Drs A. Peter Klimley and David G. Ainley, shark experts Drs Michael D. Gottfried, Leonard J.V. Compagno, and S. Curtis Bowman suggested on the basis of the previously-mentioned vertebral column's dimensions and other megalodon remains that in order to support its substantial dentition, the megalodon's jaws would have been "somewhat more robust, larger, and thicker, and with correspondingly more massive muscles to operate them" than those of the great white shark. In overall appearance, they proposed that the megalodon "would likely have had a streamlined, fusiform shape similar to, but more robust than, the [great] white shark and other lamnids, with more bulging jaws and a broader, blunter, and relatively more massive head".

If this reconstruction is accurate, might it explain the Greys' comments about the massive or square-shaped head of their respective giant mystery sharks? Moreover, it is believed that the fins of the megalodon were proportional to its larger size, and hence were bigger than those of the great white. Could this therefore explain the huge pectoral fins sported by the giant mystery shark sighted by Zane Grey?

The author alongside a life-sized recreation of megalodon jaws (© Dr Karl Shuker)

Once believed to be an exclusively near-surface, continental shelf dweller in tropical and subtropical seas, the megalodon is now thought to have been sufficiently adaptable to have inhabited a wide range of environments, from shallow coastal waters and swampy coastal lagoons to sandy littorals and offshore deepwater abodes, exhibiting a transient lifestyle, and of near-cosmopolitan geographical distribution. Adult specimens, however, were not common in shallow-water habitats (thus explaining the relative rarity of modern-day Lord of the Deep and other super-sized great white lookalike sightings?), and mostly lurked offshore, but may have moved between coastal and oceanic habitats during different stages of the life cycle.

The megalodon first appeared in the fossil record around 16 million years ago during the mid-Miocene, and was undoubtedly one of the most formidable marine predators of all time. So why, according to mainstream zoology, did it become extinct (if, indeed, it did!)? As yet, there is no definitive answer to this key question. However, the cooling of the oceans that occurred during the late Pliocene and Pleistocene in conjunction with the Ice Ages (an occurrence not conducive to the megalodon's survival, as it favoured warmer, tropical waters), coupled with the resulting migration towards colder, high-latitude regions by the larger whales that constituted its preferred prey (megalodon tooth marks on the fossil bones of such cetaceans are well documented), is the scenario most favoured as the cause of this giant shark's apparent extinction. Also, during the Ice Ages a substantial volume of seawater became locked inside continental ice sheets, thereby resulting in a significant worldwide fall in sea levels, which is something else that was not compatible with megalodon survival, restricting the number of nursery sites available for its juveniles' safe maturation.

Yet in view of how adaptable the megalodon was in terms of the variety of marine environments that it could inhabit, might it have once again been sufficiently adaptable to withstand these changes? True, the fossil record does not contain ample evidence of its survival in regions where water temperatures had significantly declined during the Pliocene. Then again, as pointed out by Gottfried *et al.*, this species may have existed in environments ". . . that have gone unrecognized due to preservational and/or collecting biases"—a significant but all-too-often ignored or neglected factor when making assumptions based upon the known fossil record.

Also, in view of the several exceedingly large whale species existing then, and still today (the whale hunting industry's depredations notwithstanding), the megalodon would not be short of suitable prey (big cetaceans, along with pinnipeds and fish too, are believed to have constituted its preferred diet). And what if, like the huge carnivorous sperm whale, it also sought out sizeable deepwater species such as giant squids, common in tropical as well as temperate seas, but for which, as is often true from deepwater habitats, there would be little if any readily available confirmation from the fossil record?

Even if faced with competition from today's largest carnivorous cetaceans, might there still be enough suitable prey out there in the vast oceans to sustain a viable megalodon population? After all, even large migratory whales like the blue whale and grey whale still spend part of their year in sub-tropical waters; and during those periods that these cetaceans spend in more polar zones, megalodons could subsist instead upon big fishes like the basking shark, whale shark, and abundant smaller species existing in sizeable shoals, plus giant squids.

Irrespective of the precise reason(s) *why* it died out, the findings of a 2014 study by American researchers Drs Catalina Pimiento and Christopher F. Clements (published by the journal *PLoS ONE*) suggest that the megalodon most likely did so approximately 2.6 million years ago, during the late Pliocene (a few have opined that it may have persisted into the early Pleistocene). However, these dates fail to take into account a dramatic, highly controversial revelation that occurred at the close of the 1950s. Back in 1875, the British oceanographic survey vessel *H.M.S. Challenger* had hauled up two megalodon teeth from the manganese dioxide-rich red clay deposit at a depth of 14,000 ft on the sea bed south of Tahiti in the Pacific Ocean. When, in 1959, these teeth were dated by Russian scientist Dr Wladimir Tschernezky, the scientific world received a considerable shock. Knowing the rate of formation of the manganese dioxide layer covering them, he had measured the thickness of the layer—and from the results that he had obtained, he announced in a paper published on 24 October 1959 in the prestigious scientific journal *Nature* that one of the teeth was only 24,000 years old, and the other was a mere 11,000 years old.

In short, *if* Tschernezky's results were accurate, the megalodon shark was still alive at the end of the Pleistocene epoch 11,700 years ago. And *if* this is true, it would again lend credibility to speculation among some cryptozoologists that this incredible species may still be alive today. It is nothing if not intriguing, incidentally, that these two enigmatic teeth were obtained in much the same (Tahitian) locality as that of the giant sharks respectively encountered by the Greys. Just a coincidence?

Having said that, there remains much contention among current ichthyologists and palaeontologists regarding Tschernezky's results. The main argument against them is that the teeth may have originally been reworked from older strata, as has been discussed earlier in this present book with respect to various alleged post-Mesozoic dinosaur and plesiosaur fossils. Also, there can be considerable variation in results obtained for the dating of manganese dioxide deposits, depending upon whether maximum or minimum deposition rates for them are being used, and such deposits also vary in relation to a number of fluctuating external factors such as the concentration in seawater of iron ions and photosynthesising plankton. Whether such variations can be so extreme as to yield a date as recent as only 11,000 years ago as opposed to one of at least 2.6 million years ago, conversely, has yet to be confirmed.

Also worthy of note here is the following statement from the earlier-cited paper by Pimiento and Clements:

> In a very small proportion of simulations (1.5%), the inferred date of extinction fell after 0.1 Ma. In six simulations (0.06%) the inferred date of extinction fell after the present day (and thus the species could not be considered as extinct). However, because in the vast majority of the 10,000 simulations (>99.9%) the extinction time was inferred to have occurred before the present day, we reject the null hypothesis (that the species is extant) and the popular claims of present day survival of *C. megalodon*.

In short, although too small in number to be considered statistically significant, from the vast array of fossil samples utilised in their simulations a few modern-day inferred extinction dates did occur, as well as some with an inferred extinction date of under 100,000 years. How can these be explained and which specific samples were responsible, I wonder?

All in all, if they still exist it would be very interesting to see those two teeth that were dated so contentiously by Tschernezky back in the late 1950s subjected now to modern-day dating techniques. The most common method for Quaternary (Pleistocene and Holocene) remains—which these teeth would be if Tschernezky's age estimates for them of 24,000 and 11,000 years respectively are correct—is radiocarbon (carbon-14) dating, but it generally cannot date specimens older than around 60,000 years. However, a more recent and potentially much more useful technique, which has already been proved to be effective with fossil teeth, is electron paramagnetic resonance (EPR).

As noted in a 14 February 2014 *Spectroscopy Europe* online paper authored by Dr Mathieu Duval from the Centro Nacional de Investigación sobre la Evolución Humana

(CENIEH) in Burgos, Spain, optimum time range application for EPR dating of tooth enamel lies between c.50,000 years and c.800,000 years. Moreover, in some specific conditions, the real time-range limits for EPR dating may be potentially pushed from present-day to around 2–3 million years. This means that EPR dating not only could demonstrate unequivocally whether Tschernezky's unexpectedly recent age estimates for these two very contentious megalodon teeth were correct, but also might still be able to provide an age for them even if they actually do date back to the time of the megalodon's official demise, i.e. approximately 2.6 million years ago—something that radiocarbon dating could not achieve.

One final comment regarding giant, ostensibly anachronistic shark teeth: in his authoritative work *The Fishes of Australia Part 1: The Sharks, Rays, Devil-Fish, and Other Primitive Fishes of Australia and New Zealand* (1940), Gilbert P. Whitley, then Curator of Fishes at the Natural History Museum in Sydney, Australia, stated:

> Fresh-looking [megalodon] teeth measuring 4 by 3 1/4 inches have been dredged from the sea floor, which indicates that if not actually still living, this gigantic species must have become extinct within a recent period.

Unfortunately, he didn't provide further details concerning these bold claims. Fossil megalodon teeth are generally black or grey, less commonly brown and even gold, but white specimens are also known—and although they too are fossilised, these latter ones can look deceptively recent in appearance, so Whitley may have been mistaken. As for the teeth noted by him, sadly I have no knowledge of where they currently reside.

In summary: Dr Stead considered that the shark responsible for towing away the fishermens' lobster pots could have been a living megalodon, but just how likely is this terrifying prospect? I am well aware that by virtue of its very nature, the megalodon must surely appear to be one of this book's least likely creatures to survive in the present day. Having said that: if, as noted here, this monstrous carnivorous shark dined upon large whales, pinnipeds, fishes, and (especially) giant squids, moving up and down through the sea depths in search of its varied prey, its huge food requirements could surely be met. And if, as predicted from palaeontological studies, it only occasionally entered the oceans' surface waters as an adult, this might explain how in spite of its size it has succeeded in eluding science, and why even fishermen in its general area of distribution only rarely catch sight of it.

Certainly, as someone who in 2008 flew from Santiago in Chile to Easter Island and, in so doing, spent no less than 4 hours travelling continuously across a seemingly limitless blue expanse of water with never so much as the tiniest speck of land in sight, yet knowing full well that this was in reality only a minute portion of the Pacific's full mid-oceanic extent, I feel qualified to offer the opinion that in such an unimaginably vast yet (for the greater part) only sparsely visited expanse of water relatively speaking, even creatures as huge as megalodons could surely exist just beneath the surface without ever being seen by humans for much if not all of their life. Here they could readily avoid the occasional cruiser or other sizeable sea vessel crossing the immense mid-ocean stretches of water upon which the various Pacific island groups are scattered like mere confetti, and only occasionally approach the shores of such islands where they may conceivably attract brief attention before travelling back out to the open seas once more.

Artistic impression of a megalodon encounter (© William M. Rebsamen)

We know that in Pliocene times megalodons occurred in coastal waters (albeit only rarely as adults), because the fossil record tells us so. But what if megalodons also lived in mid-oceanic stretches where any dead specimens either were consumed by other marine carnivores or became fossilised in locations where such remains can never be uncovered, such as the sea bottom—except, possibly, for a few anachronistic teeth dredged up by a research vessel?

And even if such a creature *is* spied once in a while when far out to sea, by some ocean-going tourists or bold fishermen venturing further out than usual from their coastal zone, what will they see? Just a triangular dorsal fin resembling a slightly larger-than-normal great white's, cutting silently through the water? Who would think to report that as anything special?

However, one could also argue that if the megalodon has indeed survived into the modern day, why was it not reported by whalers during the whaling age? Great white sharks were frequently attracted to harpooned, massively-bleeding whales, sometimes causing problems for whalers trying to land these huge, dying sea mammals or their carcases. How much greater a problem, therefore, would megalodons have posed? Yet I am not aware of any whaling records describing encounters with sharks that might have been megalodons. As for smaller, juvenile megalodons, surely these would be hooked or entangled in netting from time to time, just like similar-sized adult great whites are? Yet again, however, there do not appear to be records of this, unless any such juveniles that have been caught looked similar enough to adult great whites for anglers not to have considered them worthy of being brought to zoological attention?

Also, if the megalodon still exists there would surely be big whales out there that have survived a megalodon attack yet carry the scars created by such a monster's huge teeth, but again I am unaware of any records of this. Then again, any whale surviving a megalodon attack would need to be very big indeed, and such individuals probably remain far out of sight in the open oceans, and those not surviving such an attack would be devoured by the victorious megalodon, with any remains simply sinking to the ocean floor. Yet another anomaly if the megalodon is indeed still alive today is why no modern-day megalodon teeth have ever been found, bearing in mind that sharks shed numerous teeth every year, and that assemblages of shark teeth from other species have been procured from the sea floor. Then again, perhaps some modern megalodon teeth *have* been obtained, but, in view of how sought-after their fossil equivalents are by collectors (and expensive too!), have simply not been publicly revealed.

Having said all of this, there is a notable modern-day precedent for large sharks remaining hidden from science. In November 1976, a major new species of very large shark was accidentally captured by a research vessel anchored off the Hawaiian island of Oahu. Attempting to swallow one of the ship's anchors, it had choked to death, despite its enormous mouth, which swiftly earned its species a very appropriate name—the megamouth shark *Megachasma pelagios*. Measuring up to 18 ft long, this very distinctive species has since been recorded from waters all around the world, and observations of living specimens fitted with tracking devices have revealed that it undergoes vertical migration—staying in the depths of the sea during the day, and rising to the surface only at night. This explains how such a large and widely-distributed shark species had successfully managed to evade scientific detection until as late a date as 1976.

In fact, a megamouth—or some other very large, formally undescribed species of deep-water shark—may actually have been seen and photographed by a scientific team a full 10 years earlier. On 15 August 1966, the *San Mateo Times*, a Californian newspaper, carried the following very intriguing report:

> Undersea cameras of the Scripps Institute of Oceanography have photographed a colossal shark-like fish that is unfamiliar and may prefer living in the darkest depths of the Pacific.
>
> Scripps' Dr. John D. Isaacs, speaking at a weekend conference, estimated the fish at 15-20 feet in length and three to six feet thick at its widest.
>
> The species could not be determined, he said, because of the unmanned camera's limited field which only allowed picturing the fish's gills and pectoral fin.
>
> "It is probably a shark, but a shark the likes of which we have never seen before," he said. The fish was photographed at a depth of 6,000 feet off San Clemente Island, which is about 75 miles south of Los Angeles.

Since the first megamouth was caught off Oahu in 1976, several have been washed ashore or documented in waters off California, lending further support to the possibility that the Scripps's mystery shark was a specimen of this very big species—always assuming, of course, that it wasn't a juvenile megalodon. . .?

Incidentally, crypto-sceptics have suggested that the megalodon could not exist as a deepwater species because it would require all manner of morphological specialisations, but in view of the fact that the only physical remains that we have of it are teeth and vertebrae, how can anyone say with certainty that it didn't—or doesn't—possess any such specialisations?

Bearing in mind, therefore, that a mere 40 years ago the megamouth was still unseen and undiscovered by science, the prospect for prehistoric persistence of the megalodon cannot be entirely denied out of hand—however much we may wish to banish from our minds the disturbing image of a rapacious, flesh-eating shark at least twice the size of the current record-holder for the great white, cruising anonymously beneath the surface of the Pacific in the 21st Century.

Last—and least—of all but requiring a mention here if only because of how much confusion it caused (and still causes) among viewers not well-versed in cryptozoology is the infamous 'mockumentary'/'docufiction' *Megalodon: The Monster Shark Lives*, which was first aired on the American TV network Discovery Channel in 2013, concerning the alleged modern-day survival of this giant shark species. The programme has an entry on Wikipedia that summarises its history very succinctly:

> The story, with only short disclaimers at the beginning and ending indicating that it is fictional, revolves around the loss of a pleasure boat and crew off the coast of South Africa and an ensuing investigation that points to an attack by a member of the species *megalodon*, a prehistoric shark thought to be long extinct. Its format is that of a documentary that includes accounts of "professionals" in various fields related to *Megalodon*. It follows a similar format to another docufiction aired by Discovery Channel, *Mermaids: The Body Found*.
>
> The show, like *Mermaids*, came under equal criticism and scrutiny by both scientists and ordinary viewers due to the attempt to present fiction as a nonfiction documentary. Despite the disclaimers, some people actually believed they were watching a real documentary while others were offended that a docufiction show would be aired on a channel that had been known for true science shows. It should also be noted [that], unlike *Mermaids*, the disclaimers were barely even present, in addition to the talk show that was strongly saying and asking if people believed what was presented in *Megalodon* showed that the species was still alive. This misinformation likely caused the mass misconception that the shark species was still alive.

No it didn't—speculation on this subject was rife long before the programme was produced. As for whether the modern-day existence of the megalodon actually is—or is not—a misconception, this has already been discussed soberly and at length in the present section of this book. In my opinion, however, any attempt to do so in an equally rational, objective manner elsewhere is always likely to be overshadowed nowadays by the Discovery mockumentary's unhelpful contribution to the subject, which is a tragedy for those seeking to bestow gravitas and credibility upon serious cryptozoological debate.

A PAIR OF PLACODERMS DOWN UNDER?

Among the most bizarre of all fishes—indeed, constituting a taxonomic class all to themselves

The Hook Island sea serpent (© Robert le Serrec/Fortean Picture Library)

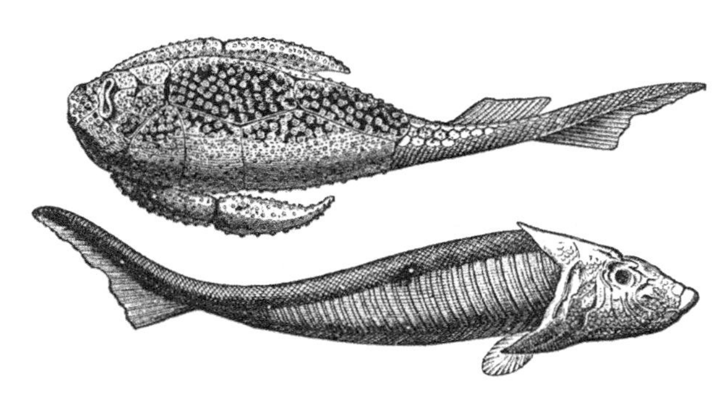

Two fossil placoderms—*Pterichthyodes* (above) and *Cephalaspis* (below)

—were the plated fishes or placoderms. Reaching their peak of evolutionary development during the Devonian Period, but extinct by the close of the Carboniferous (just under 300 million years ago), they were characterised by bony plates encasing their head and the anterior portion of their body. In 1986, renowned wildlife writer David Alderton cautiously investigated (but ultimately rejected) a living placoderm as an identity for two particularly curious and controversial sea monsters.

One of these was the 70-80-ft-long, horizontally-sinuous, tadpole-shaped creature that French photographer-traveller Robert le Serrec allegedly spied lying in shallow water on 12 December 1964, near Hook Island in Australia's Great Barrier Reef. According to le Serrec, it had a deep wound in one flank, and its quiescent nature enabled him to take some photos before it suddenly raised its head upwards. Not surprisingly, le Serrec fled in his boat, but when he returned later that day the animal had gone.

Most authorities have dismissed his monster and photos as a hoax (including Heuvelmans, even though one of his hypothesised sea serpent categories, the yellow belly, does resemble a giant tadpole). However, Ivan T.

The placoderm *Dunkleosteus* (CC-SA Kumiko)

Sanderson speculated that it might be an undiscovered, giant species of synbranchid or swamp eel. Alderton noted that its tapering body corresponded quite closely in shape with the placoderm *Cephalaspis*, but the photos contain no indication of body plates. Also, even the mightiest of placoderms, the 11-ft-long *Titanichthys*, falls far short of le Serrec's 70-ft monster.

The other putative placoderm was also reported from the Great Barrier Reef, surfacing on 3 January 1891 in a *Land and Water* letter penned by schoolteacher Selina Lovell, who claimed to have sighted it recently at Sandy Cape. Terming it the moha-moha (supposedly its aboriginal name), she likened it to "a monster turtle fish"—in a sketch, she depicted an incongruous composite that sported a snake's head and neck, a turtle's carapace, and a fish's tail fin.

After examining and discounting placoderms, modern fishes, and turtles as plausible identities, Heuvelmans dismissed the moha-moha as a zoological impossibility. Alderton also examined a placoderm identity, commenting that in *Pterichthyodes*, the forebody plates yielded a turtle-like shelled appearance, and its naked hind body resembled a typical fish's. However, it did not have an elongate snake-like neck (no known fish does).

As Alderton also noted, giant modern-day placoderms appearing at the surface are unlikely to escape scientific detection in an area so frequented by divers and fishermen as the Great Barrier Reef.

Representation of Lovell's sighting of the moha-moha, inspired by her description and original sketch of it (© William M. Rebsamen)

IN SEARCH OF SEA SCORPIONS AND SEA MILLIPEDES

On 11 March 1959, diver Bob Wall was 35 ft below the surface of the sea off Florida's Miami Beach when he noticed a large underwater cave and peeked inside to see if it contained anything of interest. What it did contain was an alarming creature with a cylindrical 5.5-ft-long body, raised 3 ft off the ground upon eight hairy legs, and a pointed head whose stalked, dollar-sized eyes were looking directly at him. Indeed, as the diver gazed at it, this startling apparition began to move towards him—at which point Wall decided that he had seen more than enough, and swam swiftly away. Nicknamed 'Specs' by the media, the next day it was sought by five divers from the Miami seaquarium, but declined to reappear, and has not been reported since.

Just over a year later, an even more unexpected water monster made its media debut—from the bottom of Dan Craig's 12-ft-deep well on his farm, situated 4 miles south of Lynn, in Randolph County, Indiana, U.S.A. In an interview published by the *Indianapolis News* for 8 June 1960, Craig revealed that he had known about his well's bizarre inhabitant for about a year, and described it as "an eerie beast with a dome-shaped head, two bulbous eyes, and eight flailing tentacles as long as a man's arm." According to his wife, who had also seen it, the creature resembled a mushroom as large as a plate, with long legs and feet.

Out of continuing curiosity, Craig decided to drain the well, and left behind only about 4 in of water at the bottom. Following this, an intrepid 12-year-old youth called Craig Lee, risking life and limb on a ladder, climbed down into its shadowy depths in search of its mysterious occupant. When he clambered back out, Lee claimed that he had spied something amid the debris on the well's gloomy bottom that was indeed the size of a plate, and looked like a greyish-yellow sponge mushroom with eyes on the top of its head, and with eight or ten legs as long as his own arm but with pincers on the ends.

It might well be mere coincidence, but a few days later a Randolph County conservation officer called Kenneth Yost hauled up out of the well a flesh-coloured ball of sponge rubber and a segment of garden hose, and the beast itself was never seen again.

On 30 January 1959, at Covington, in Kentucky, U.S.A., a grey octopus-like creature with ugly tentacles, rolls of fat surrounding a bald head, and a lopsided chest allegedly surfaced in the Licking River, and emerged onto the bank. Further back, on Christmas Eve 1933, a reputed 'octopus' (according to media accounts) was actually captured alive by two fishermen—Robert Trice and R.M. Saunders—after its tentacles appeared over the side of their rowing boat in the Kanawha River near Charleston, in West Virginia, U.S.A. When the body was brought ashore, it was found to measure 3 ft long from its head to the end of the longest tentacle.

Two very odd types of water monster have also been reported from certain mountain lakes in Argentina and Chile. One type is generally likened to a featureless expanse of cowhide. The other is serpentine with a claw-pointed tail—some descriptions claim that it has tentacles ending in hooves. Both forms are said to kill humans, dragging them beneath the surface of their watery domains and squeezing them to death.

Unlike the water monsters documented in this chapter's earlier sections, all of those reported here seem more likely to be invertebrates than vertebrates. In the revised edition of his absorbing book *Natural Mysteries* (1991), however, mystery beast chronicler Mark

A. Hall from Minnesota offered a more specific identity. He proposed that they are unrecognised modern-day descendants of one of the most extraordinary groups of prehistoric creatures known from the fossil record—the eurypterids or sea scorpions.

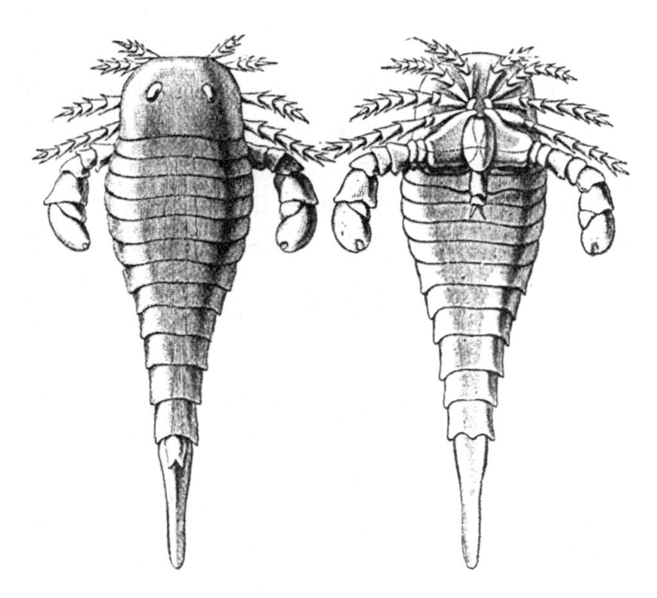

Fossil sea scorpions or eurypterids

They constitute a long-extinct group of arthropods—that incomparably diverse assemblage of invertebrates united morphologically by their jointed limbs. Nowadays split into several separate phyla after having been traditionally united within a single enormous phylum, arthropods include such familiar modern-day animals as the insects, crustaceans, arachnids, centipedes, millipedes, and horseshoe crabs, as well as rather more obscure creatures like the sea spiders, symphylans, and pauropods, plus some significant fossil groups, including the trilobites, and the sea scorpions themselves.

Including among their number the largest and most spectacular of all arthropods, attaining a total length of up to 8 ft in one species (*Jaekelopterus rhenaniae*), the sea scorpions were only distantly related to modern-day scorpions, as they were much more closely allied to the horseshoe crabs. They had reached their zenith of evolutionary of development by the late Silurian Period (approximately 420 million years ago) and (along with the trilobites) had disappeared entirely from the fossil record by the close of the Permian (around 170 million years later), victims of the mass extinction that occurred at that time.

Most sea scorpions were generally elongate—commencing with a broad prosoma (the combined head and thorax) whose rounded head sported a distinctive pair of laterally-sited compound eyes and elongated mouth, and whose short thorax bore six pairs of limbs. One pair, the chelicerae, were jaw-parts (which attained an exceptional degree of development in *Pterygotus* species, resembling pincer-like grasping organs). The others were true legs (which in some species were so stout that they may have enabled these particular forms to walk on land—see later).

Of those true legs, the most posterior pair was much larger and flatter than the rest, terminating in oval plates or paddles—undoubtedly constituting the principal swimming organs, but possibly serving as anchors too, or as scoops for digging up mud on the sea bottom where these creatures may have rested or hidden themselves. These wider legs earned the sea scorpions their scientific name—'eurypterid' translates as 'broad wing', alluding to the wide paddles.

The abdomen or opisthosoma was divided into two morphologically distinct sections. The forepart, known as the mesosoma, consisted of seven squat segments. The posterior part or metasoma consisted of five tapering cylindrical segments, the last of which bifurcated into two lobes—between which, concluding the sea scorpion's morphological roll-call, arose a long pointed spine (or, in some species, a short flat plate) called the telson or tail.

During their evolution, the sea scorpions adapted to living in brackish waters too, and even entered freshwater—a boon to mystery beast investigators seeking to link them to water monsters from a wide range of habitats. Indeed, based upon some exceptional eurypterid fossils unearthed during the 1980s at East Kirkton in West Lothian, Scotland, palaeontologists believe that certain sea scorpions might even have been terrestrial. And if their lineage has persisted into the present day, 245 million years of intervening evolution could have engineered many other changes too, yielding creatures quite different in form and lifestyle from their Permian predecessors.

Having said that, however, I am unable to share Hall's opinion—that all of the water monsters reported here are evolved eurypterids—for a number of fundamental reasons. To begin with: even allowing for over 200 million years of evolution, these monsters are far too diverse in form to belong to the same taxonomic group—especially an arthropod group. Even the most specialised eurypterid would surely be hard-pressed to resemble a piece of cow-hide, or an octopus with flailing tentacles (far-removed indeed from the jointed limbs that constitute the diagnostic feature of all arthropods).

Chile's traditions of hide-like freshwater monsters also extend to its coastal waters. According to the revised edition of Jorge Luis Borges's classic volume *The Book of Imaginary Beings* (1974), the marine counterpart is a mysterious type of octopus that resembles an outstretched ox hide (and is actually known locally as the hide), with countless eyes all around its perimeter and four larger ones in the centre of the body. However, as I pointed out in my own book *From Flying Toads To Snakes With Wings* (1997), this description scarcely recalls an octopus, but it is very reminiscent of some scyphozoans (true jellyfishes). For their bell is ringed by numerous eye-like sensory organs called rhopalia, and they also have four deceptively eye-like organs visible in the centre of their bell (these are actually portions of the gut, called gastric pouches).

However, scyphozoans are exclusively marine, so even if that identity explains the latter beast it is unlikely to be involved with the lake version. This might simply be a myth—inspired by reports of the real, marine creature, and utilised as a convenient 'bug-bear' by mothers to scare their children away from deep lakes where they could easily drown.

Certainly, non-cryptozoological explanations should not be denied a priori when examining cases such as these—as emphasised by zoologist Dr Aaron Bauer when reviewing Hall's *Natural Mysteries* in 1992 for the ISC's journal *Cryptozoology*.

The second type of Chilean lake monster, with slender body, tentacles, and claw-bearing tail, sounds a little more realistic—but little like a eurypterid. True, its tail recalls the pointed telson of some sea scorpions, but these animals never possessed limbs that could be remotely described as tentacles with hooves, and although their body was slender it was never used to squeeze their prey to death. Whether non-existent myth or corporeal monster, we need not dwell further upon its candidature as a eurypterid (evolved or otherwise).

As for the 'freshwater octopuses' reported earlier in this section, once again I doubt very much that these are in any way related to eurypterids. They seem much more like genuine octopuses—despite the fact that such animals are not supposed to be freshwater fauna. Octopuses belong to a taxonomic class of molluscs whose members are collectively known as cephalopods, and include the squids, cuttlefishes, and nautiluses too. Like the scyphozoans, however, cephalopods are traditionally thought of as being exclusively marine—but there are a handful of well-attested exceptions to this rule. The most famous of these is from 1902, and features the squids of Lake Onondaga, New York.

As reported in the journal *Science* (12 December 1902), a Syracuse fisherman called Mr Terry had lately netted a strange creature in the lake; when it was shown to science teacher Prof. John D. Wilson, he identified it as a squid. A second specimen was caught shortly afterwards in the same portion of the lake by a local restaurant owner called Lang. Fellow scientific colleagues shown the creatures by Wilson confirmed that they were squids.

Interestingly, however, the area of the lake in which the squids were found is just where the first salt springs were discovered and the first salt made in the Syracuse region by the early settlers long before salt wells were bored. So perhaps the water is sufficiently saline after all to sustain ostensibly marine beasts of this nature—which could have first reached the lake during the period in North America's post-glacial history when the lake was connected to the sea via the St Lawrence Valley. Whatever the explanation, cephalopods can occasionally be found in freshwater—so perhaps the 'living octopuses' should be treated more seriously. Even the way in which the Kanawha River specimen was measured—from its head to the end of a tentacle—is precisely the method one would use to ascertain the size of an octopus. Other supposed freshwater octopuses have been reported elsewhere across the U.S.A. too in more recent times.

The animated 'sponge mushroom'—even if it really was more than a misidentified blob of sponge rubber—hardly brings to mind images of armoured eurypterids either. *Pterygotus* had pincers, but these were limited to a single pair, utilised for grasping prey. Pincers would have served no useful purpose on the remaining limbs, which functioned as true legs, not as jaw-parts or prey-capturing instruments. If it really did exist, it sounds more like a flattened, crab-like crustacean than a eurypterid—perhaps one of the freshwater crayfishes.

Crustaceans also come readily to mind when considering Specs. The compound eyes of eurypterids, although large and noticeable, were never elevated on stalks. Yet this is a common feature among crustaceans. Moreover, eurypterids' heads were round, not pointed like those of some large crustaceans look when their long antennae project forwards; their bodies were flattened, not cylindrical like many large crustaceans again; and when standing or walking they may have held their elongate metasoma and telson curved upwards and forwards above their mesosoma and prosoma, like true scorpions—surely a very striking, memorable sight, yet one not mentioned by Wall.

Of course, an evolved eurypterid might display some (or all) of those features—and may no longer hold its metasoma in scorpionesque pose. However, it would also need to have lost its paddled swimming limbs, if it were to match the eight-legged body format of Specs. This is hardly likely, as these limbs were the eurypterids' principal organs of propulsion.

In my opinion, Specs is more than adequately reminiscent of a large crustacean for any serious consideration of other identities to be superfluous. An extra-large specimen of one of the American spiny lobsters *Panulirus*, which are fond of inhabiting large holes and crevices underwater, would yield a convincingly close match. Moreover, some such species of spiny lobster are noted for a pair of very large and conspicuous white markings, each with a big black spot at its centre, that certainly resemble spectacles and are sited just below its real but relatively inconspicuous stalked eyes.

Close-up of an American spiny lobster showing its very conspicuous and deceptively eye-like 'spectacle' markings (NOAA)

Might these markings have been mistaken for eyes by Wall when startled by his unexpected close encounter with Specs?

It does seem unwise to discount this very plausible identity in favour of one that requires the dramatic resurrection and notable morphological modification of a group of animals known only from fossils at least 245 million years old.

Speaking of which: a photograph that has been doing the rounds online for a number of years and has confused quite a few cryptozoology enthusiasts during that time portrays someone holding what appears on first sight to be a large eurypterid only recently dead. I have seen the following report accompanying this photo on several Facebook group and individual pages, but its earliest online appearance seems to be on the *Paranormal Geeks Radio* website on 13 April 2013. Here is the report:

> In 1971 farmer Ted Litton caught this weird animal alive in his artificial pond in Lilac, TX, & got his pic in the paper. 8 hours later his farm was besieged by Army soldiers wearing decontamination suits. They drained the pond, leaving an odd, spheroid cavity in the bottom. Litton says the Army dismissed his beast as a freak of nature yet they confiscated it, promising him 5 grand (which never materialized).

Needless to say, however, as I discovered when subjecting this photograph to a Google Image search, the reality soon proved to be very different. In fact, the eurypterid was a prop, an animatronic model to be precise, produced for the BBC television series *Sea Monsters* (2003) by the award-winning special-effects design company Crawley Creatures, based in the UK (a link to their website showing this exact-same photograph can be accessed at: http://www.crawley-creatures.com/Gallery.aspx?Job=76). The eurypterid model portrays the Ordovician genus *Megalograptus*, one of the earliest eurypterids on record.

Finally: during early May 1986, Channel 4, a British television station, broadcast a little-known true-life film that had been made in 1972 and was entitled *The Moon and the Sledgehammer*. Its subject was the Pate family, whose home was situated deep in some woods near Newhaven in East Sussex, southern England, and whose patriarch was Old Man Pate—who recollected seeing a marine creature that he referred to as a 'sea scopium'. According to his description, it "looks like an old ship's sailing cloth in the water. . . . They're black . . . . It's got a head like a frog, but it's all goldy colour round, and its eyes stick out a bit, bulge fashion. . . . It was about 7-8 ft under the water".

Pate had considered pulling it out of the sea to observe it more closely, but when he saw its large mouth, and realised that it was looking at him, he changed his mind. Afterwards, however, he regretted this, and in the film he announced plans to build a semi-submersible boat in which to look for the creature again.

Dorsal view of a sea scorpion of the shorthorn sculpin kind (CC-SA Genet)

Although the term 'sea scopium' naturally evokes images of eurypterids, there is a much simpler explanation available here. Pate's account recalls a *fish* called the sea scorpion—*Myoxocephalus scorpius*. Up to 2 ft long and black when fully mature, also referred to as the shorthorn sculpin, and a member of the taxonomic family of fishes known as cottid sculpins, this is a large-mouthed, frog-faced lurker among stones and seaweed on rocky seafloor with mud or sand. It is found in the English Channel, and the Atlantic north of Biscay, as well as the Arctic basin including the Siberian and Alaskan coasts. True, the head of this species is not golden, but in the film Pate stated that the sun was shining down through the water, so this no doubt gave the creature's head a golden sheen. Exit the 'sea scopium'.

Many years ago, in a report describing a new 3-ft-long species of *Mixopterus* sea scorpion, Norwegian palaeontologist Prof. Johan Kiaer recalled the thrill of its discovery:

> I shall never forget the moment when the first excellently preserved specimen of the new giant eurypterid was found. My workmen had lifted up a large slab, and when they turned it over, we suddenly saw the huge animal, with its marvelously shaped feet, stretched out in natural position. There was something so lifelike about it, gleaming darkly in the stone, that we almost expected to see it slowly rise from the bed where it had rested in peace for millions of years and crawl down to the lake that glittered close below us.

No doubt cryptozoologists share a similarly dramatic dream—to haul up a living eurypterid from the depths of the oceans or even from the muddy bottom of a large freshwater lake. And somewhere out there, perhaps there really are some post-Permian, present-day sea scorpions, indolently lurking in scientific anonymity. Based upon the evidence offered up so far in support of such a claim, however, this prospect seems no more likely than the resurrection of Kiaer's fossilised specimen from its rocky bed of Silurian sandstone.

Dredging Up Some Living Trilobites?

One of the best-known groups of fossil animal are the trilobites ('three-lobed'), this name deriving from the distinctive three-lobed structure of their body, which consists of the cephalon (head shield), the thorax, and the pygidium (tail shield). Ranging in size from a dinner plate down to a pea, they are famed for their segmented body form, numerous pairs of limbs, and extremely well-developed compound eyes.

Global but exclusively marine in distribution, this taxonomic class of arthropods was one of the earliest, with the first-known representatives in the fossil record dating back approximately 540-520 million years to the early Cambrian Period (though it is suspected that there may well have been earlier forms as yet unrepresented by documented fossils dating as far back as 700 million years, to the pre-Cambrian). During the lengthy course of their evolution, moreover, the trilobites became exceedingly successful, yielding a vast diversity of species (some 17,000 are currently recognised) as well as body forms and lifestyles before decreasing markedly in the Devonian, and finally dying out completely around 252 million years ago (in the mass extinction that occurred at the end of the Permian)—or did they? There is no well-established reason why they should have done.

As a zoologist living in the West Midlands, England, I am very aware that one particular trilobite species, *Calymene blumenbachii* from the Silurian Period, is so abundant in the fossiliferous limestone quarries of Wren's Nest in the Midlands town of Dudley that it is popularly known as the Dudley bug, and even appears on the Dudley County Borough Council's official coat-of-arms. Naturally, then, I've been a fan of trilobites ever since childhood, and my fossil collection contains several specimens, but my interest in cryptozoology would subsequently yield an additional reason for my being fascinated by them.

In the mid-1980s, I purchased American cryptozoologist Prof. Roy P. Mackal's classic book *Searching For Hidden Animals* (which had originally been published in 1980 in the U.S.A., but not until 1983 in the U.K.), and was delighted to find that it documented a wide range of lesser-known cryptids. However, one chapter that obviously attracted my particular interest was tantalisingly entitled 'Living Trilobites?', and included a discussion as to whether any representatives of these archaic arthropods might have survived the Permian mega-death and persisted in benthic anonymity on the ocean floor into the present day.

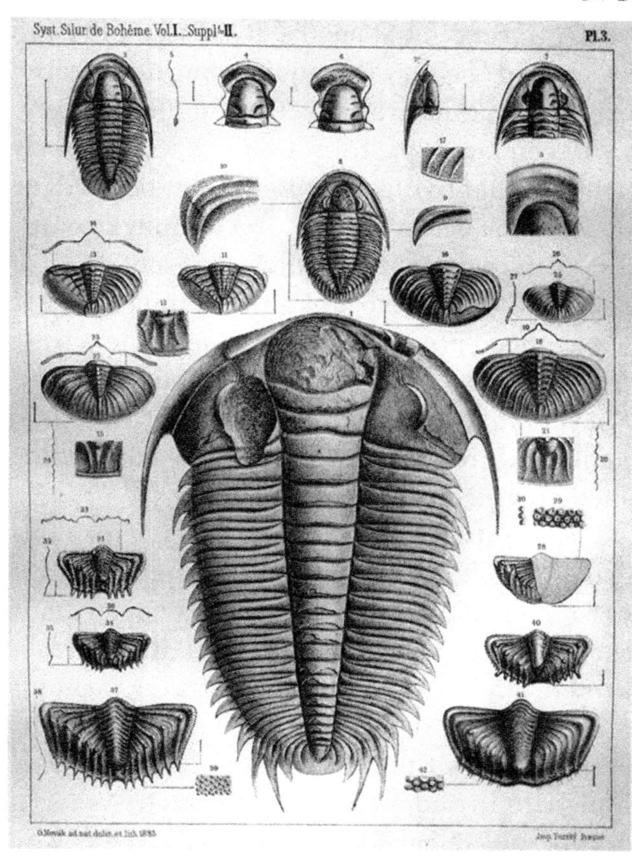

Beautiful 19th-Century illustration of trilobites, from Joachim Barrande's *Système Silurien du Centre de la Bohême*, published in 1852

As Mackal noted, many trilobites were shallow coastal dwellers (especially the later ones), yet no living trilobites from such localities have ever been discovered. Consequently, the only hope for modern-day survival is if "some forms adapted to a deeper, more obscure environment and there found refuge"—or if some that were already so adapted simply persisted. Should this scenario have indeed taken place, it could explain why no Cenozoic trilobite fossils have ever been found—because these would not be readily discovered or accessed on the ocean floor. But what about obtaining living specimens there?

Until reading Mackal's book, I hadn't been aware that the very first global marine research expedition, the voyage of HMS *Challenger* from 21 December 1872 to 26 May 1876, seriously believed that living trilobites might be dredged up from the ocean bottom. But although countless specimens that included representatives of over 4,000 hitherto-unknown animal species were indeed procured there, none of them were trilobites. Or, as worded in the authoritative *Encyclopaedia Britannica*'s eleventh edition, published in 1911, the "faint hope" of finding such creatures was not realised.

In reality, however, a few years *before* this expedition had even set out on its epic voyage of zoological discovery, a claim had been made that a living trilobite had already been obtained, and from a depth of 1,200 fathoms (7,200 ft). Moreover, this claim was actually believed for a time before the creature's true, non-trilobite identity was revealed. Veteran cryptozoological chronicler Willy Ley, who briefly reported the case in one of his many articles, didn't provide further details, but as noted by Mackal the timing and morphological similarities strongly suggests that the discovery in question was actually that of a certain Antarctic species of isopod crustacean (the taxonomic group which includes woodlice and sea slaters) that is astonishingly trilobite-like in outward appearance.

Brought to scientific attention in 1830, its first officially recorded specimen had actually been found inside the gut of a marine fish examined by American naturalist Dr James Eights while visiting the South Shetland Islands between Patagonia and Antarctica during the so-called 'Expedition of 1830'. Emphasising its remarkable morphological convergence, in 1833 Eights formally christened this memorable new species *Brongniartia* [now *Ceratoserolis*] *trilobitoides*. And it was indeed initially mistaken for one of these prehistoric arthropods by some observers, but it sports two pairs of antennae (a crustacean characteristic), whereas trilobites only had one.

Other modern-day creatures that have often been mistaken for living trilobites are chitons and water pennies. Chitons (or polyplacophorans, to give them their formal zoological name) constitute a taxonomic class of

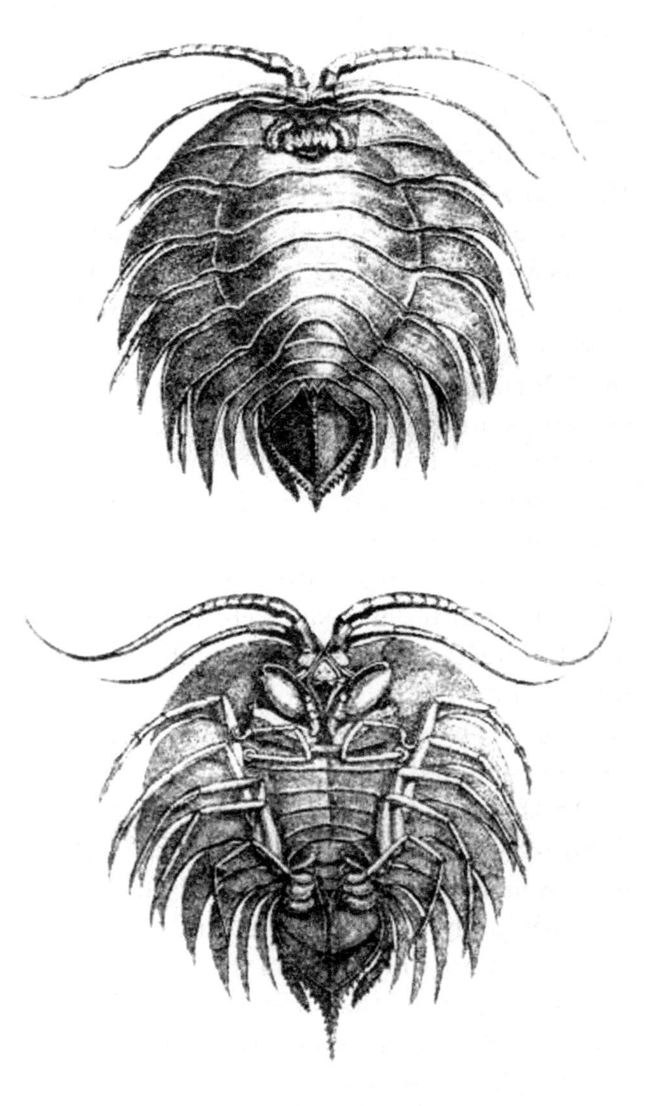

Dorsal (top) and ventral (bottom) view of *Ceratoserolis trilobitoides* as depicted in Eights's formal description of this trilobite-like crustacean species, 1833

molluscs characterised by their very distinctive shells, which are composed of eight separate but slightly overlapping plates, and afford these animals a superficially segmented appearance dorsally. If a chiton is turned over, however, its ventral body surface is seen to be non-segmented and only possessing a single, typically-molluscan foot, in contrast to the many limb pairs possessed by trilobites. As for water pennies, these trilobite imposters are the larval stage of certain aquatic freshwater psephenid beetles, belonging to the genus *Mataeopsephus*. Also superficially trilobite-like in outward dorsal appearance are both the larvae and the larval form-retaining adult females of lycid (net-winged) beetles belonging to the genus *Platerodrilus*, and which are therefore known colloquially as trilobite beetles. Native principally to tropical rainforests in India and southeastern Asia, some of these are brightly coloured. Finally, the juvenile stage of those famous 'living fossils' known as xiphosurans or horseshoe crabs is termed a trilobite larva, once again because of its superficial similarity to genuine trilobites.

Back in the 1980s, a bizarre story emanating from Australia briefly hit the news headlines, claiming that some trilobites had been found inhabiting Perth's storm drains. Not surprisingly, however, this was soon exposed as a hoax, featuring an old tyre that had been cut into the shape of a trilobite.

Numerous deepsea collecting expeditions have been launched since *Challenger*, but none has ever procured any living trilobites, and yet some tantalising indirect (or, to be precise, ichnological) evidence for such creatures *may* have been recorded, which Mackal described as follows:

> . . . in 1967, I was invited by Ralph Buchsbaum, professor of zoology at the University of Pittsburgh, to give a seminar on our researches at Loch Ness. During the social hour after the presentation one of his colleagues told me about experimental photography of the sea bottom that was in progress. He stated that photographs of fresh tracks identical to the *Cruciana* [sic—*Cruziana*], the fossilized trilobite tracks, had been obtained. He expressed the hope that traps could be lowered to catch whatever was making these highly suggestive tracks. As far as I know the nature of these tracks was never determined and nothing was ever trapped, because of a subsequent loss of funding for the project. The business of identifying sea-bottom trails and tracks is a tricky one and to infer living trilobites from a track is even more tricky. A marvelous collection of sea-bottom tracks and trails is presented in a book entitled *The Face of the Deep* by B. C. Heezen and C. D. Hollister. Only a tiny fraction of aquatic animal tracks have been identified, so that fertile ground for new discoveries is indeed abundant. . . . Underwater photography of the ocean floor . . . appears to be a promising tool for future cryptozoological expeditions.

Cruziana is a famous trace fossil taking the form of elongate, bilobed burrows that are roughly bilaterally symmetrical. As noted in a 2010 *Lethaia* paper by Dr Stephen Donovan, many examples are believed to be the tracks or trails yielded by trilobites while deposit-feeding, but certain others are deemed not to be, because they were present in freshwater environments (where no trilobite fossils have so far been found) and/or were of Triassic date, by which time all trilobites were supposed to have died out. But were these ostensibly anachronistic tracks actually made by surviving post-Permian trilobites for which direct fossil evidence has simply not been found as yet?

Incidentally, two other types of trace fossil believed to have been created by trilobites are *Rusophycus* and *Diplichnites*. The former fossils are excavations featuring little or no forward movements, and have therefore been interpreted as traces left by trilobites while resting or in defence/protection mode. In contrast, the latter fossils are believed to be traces left by trilobites walking upon the sediment surface.

Mackal ended his living trilobites chapter on a somewhat pessimistic note, concluding: "While not impossible, it is most improbable that living trilobites still exist". After that, this fascinating prospect appeared to have vanished from the modern world just as surely, it would seem, as the trilobites themselves—which is why I was so startled, but delighted, by a certain comment allegedly made by a well-respected current scientist.

On 24 June 2004, *Yahoo! News* released online a report concerning the receipt of a $600,000 start-up grant from the private Alfred Sloan foundation for a proposed 10-year international survey of the oceans' depths, at an estimated total cost of US $1 billion, to be funded by governments, companies and private donors, and officially dubbed the Census of Marine Life (CoML). As part of this grand-scale project, scientists led by researchers from the University of Alaska planned to use robot submarines and sonar to track down life forms in the Arctic Ocean's chilling deepwater domain, and expectations were that by the end of its decade-long course, the survey could easily have doubled the number of species known from this particular ocean.

All very worthy indeed, but what caught my eye amid all of these statements was one attributed in the news report to none other than Dr Ron O'Dor, chief scientist of the multi-nation CoML. According to the report, whose exact wording is quoted here as follows, Dr O'Dor "speculated that Arctic waters might hide creatures known only from fossils, such as trilobites that flourished 300 million years ago". It would seem, therefore, that the notion of finding living trilobites hasn't been entirely discounted by scientists after all.

Happily, the CoML did indeed take place, this very ambitious project ultimately featuring scientists from more than 80 different nations, and releasing the world's first-ever census in 2010—but no living trilobites were listed. Nevertheless, there is a notable precedent well worth mentioning here.

The monoplacophorans are a primitive taxonomic class of molluscs, whose youngest fossil species date from around 380 million years ago. On 6 May 1952, however, trawling off Mexico's western coast at a depth of almost 12,000 ft in dark, muddy clay, the Danish research ship *Galathea* hauled up 10 complete specimens and three empty shells of a small, seemingly unremarkable mollusc superficially resembling a limpet but which proved upon scientific examination to be a living monoplacophoran. This hitherto-unknown species was formally named *Neopilina galatheae*, since when further specimens of it, and of several additional modern-day species too, have been obtained (see my *Encyclopaedia of New and Rediscovered Animals*, 2012, for full details).

Structurally, these living monoplacophorans are very different internally from their archaic fossil ancestors, so if living trilobites do exist, these too are likely to be highly evolved species. Nevertheless, the discovery of *Neopilina* and kin readily demonstrates that it is by no means impossible for invertebrates deemed by their fossil record to have died out in very far-distant prehistoric ages to be represented, in fact, by living species that have simply evaded scientific detection.

Contemplating the Con Rit

When it debuted in his classic tome *In the Wake of the Sea-Serpents* (1968), Dr Bernard Heuvelmans's bold classification of sea serpents into no less than nine well-defined types was widely hailed within the cryptozoological community as a milestone in cryptid research, and it is still referred to today. However, the validity of certain of those sea serpent types has subsequently been challenged by various other researchers, due to revelations that cast doubt upon or totally discredit those types' proposed taxonomic identities.

Perhaps the most controversial of Heuvelmans's nine sea serpent types is his many-finned sea serpent *Cetioscolopendra aeliani*, for which he nominated a living species of armoured, scaly archaeocete as its identity. Unfortunately, however, as noted earlier in this present chapter, long before he had even categorised his types palaeontologists had already revealed that scales found in association with certain specimens of fossil archaeocete did not originate from them (as had initially been assumed following their discovery, and which had inspired Heuvelmans's identification of the many-finned as an armoured, scaly archaeocete). They belonged instead to various other creatures. In short, there are no verified specimens of armoured archaeocete in the fossil record, thereby greatly reducing the likelihood of any modern-day species existing.

But if many-finned sea serpents truly exist, what else could they be? In this present book's original 1995 edition, *In Search of Prehistoric Survivors*, I proposed a totally new identity, one that I still consider plausible, but regarding which, sadly, certain incorrect claims have been made in various subsequent online and hard-copy sources of cryptozoological data. Consequently, it is high time to refute in print these erroneous claims once and for all—by reminding readers precisely what I *did* propose in *Prehistoric Survivors* regarding the identity of the many-finned sea serpent. So here is the relevant section from that earlier book of mine, quoted in full as follows:

Another mystery beast that has been linked to the concept of surviving eurypterids is the so-called 'sea millipede'. In 1883, the headless, putrefying carcase of a remarkable, armour-plated sea monster was found washed ashore at Hongay in Vietnam's Along Bay. It was observed by several local Annamites, including an 18-year-old youth called Tran Van Con, who actually touched the body. Thirty-eight years later, he recalled this incident to Dr A. Krempf, Director of Indochina's Oceanographic and Fisheries Service.

The carcase was 60 ft long and 3 ft wide, and was composed of numerous identical segments—so hard in texture that they rang like sheet metal when one of the locals hit them with a stick. Each segment was dark-brown dorsally, light-yellow ventrally, measured 2 ft long and 3 ft wide, and bore a pair of 2-ft-4-in lateral spines. The terminal segment bore two additional spines, directed backwards like a pair of spiny tails. The stench from the decomposing carcase was so intense that the locals soon towed it out to sea where it sank, and they referred to the creature itself as con rit—'millipede'.

When contemplating this animal's possible identity in his book *In the Wake of the Sea-Serpents*, Heuvelmans briefly considered and rejected the sea scorpions as a likely candidate, together with crustaceans—favouring instead a hypothetical, highly-specialised form of

An imaginary scene featuring Heuvelmans examining the carcase of a con rit (© William M. Rebsamen)

evolved armoured archaeocete, which he dubbed *Cetioscolopendra aeliani* ('Aelian's centipede whale'), the many-finned sea serpent. A number of sightings are on file describing elongate sea monsters seemingly bearing numerous lateral fins or projections, which the ancient writer Aelian referred to as marine centipedes. I agree entirely with Heuvelmans that the con rit is unrelated to the sea scorpions, but I also have grave doubts that it is an archaeocete.

As already noted, the concept of armoured archaeocetes is no longer in favour; and in any case, even within his own selection of 'many-fins' Heuvelmans includes examples that simply cannot be mammalian. The most prominent of these is the 150-ft-long monster spied for about 30 minutes by a number of sailors on deck aboard HMS *Narcissus* on 21 May 1899, after the ship had rounded Algeria's Cape Falcon. In an interview concerning their sighting, a signal man made the following telling statement:

> *The monster seemed to be propelled by an immense number of fins. You could see the fins propelling it along at about the same rate as the ship was going. The fins were on both sides, and appeared to be turning over and over. There were fins right down to the tail. Another curious thing was that it spouted up water like a whale, only the spouts were very small and came from various parts of the body.*

Unless the numerous fins are in reality a pair of undulating lateral membranes extending the entire length of the creature's body—which does not seem likely from the above description—then the *Narcissus* sea serpent is neither a mammal nor any other form of vertebrate. Clearly, its fins were locomotory organs (creating via their propulsive movements the spouts of water noted by the signal man), not rigid spines like those reported from the carcase of the con rit. Consequently, my own feeling is that, in life, each pair of the con rit's spines had sheltered a pair of soft-bodied limbs beneath—but which, together with the remainder of this beast's soft tissues, had rotted away during decomposition, leaving behind only the hard dorsal cuticle. All of which is totally in accord with what one would expect from a crustacean—multiple locomotory limbs, hard dorsal armour that does not rot once the creature has died, and a soft body that very rapidly (and odiferously!) rots upon death.

The only major problem is the con rit's immense length—far beyond anything recorded so far by science from a known modern-day (or fossil) crustacean. It is well-known that the spiracular system of respiration utilised by insects (involving a vast internal ramification of minute breathing tubes) prevents them from attaining the gigantic proportions beloved by directors of science-fiction movies. However, crustaceans breathe via gills, and their bodies are buoyed by their all-encompassing watery medium. Hence the evolution of a giant aquatic crustacean is not wholly beyond the realms of possibility—and, to my mind, offers the only remotely feasible explanation to Vietnam's anomalous con rit or sea millipede.

It is perfectly clear from my extensive account quoted here that 'sea millipede' and 'sea centipede' are merely colloquial, non-taxonomic names for this cryptid, and that the identity for the con rit that I proposed in the original edition of this book back in 1995 was a crustacean—and *not* either a marine centipede (as wrongly claimed regarding my book in a number of websites), or a marine millipede (as wrongly claimed regarding my book in some other websites, as well as in an otherwise well-researched recent book—happily, its author has very kindly promised to include a correction in his book's second volume).

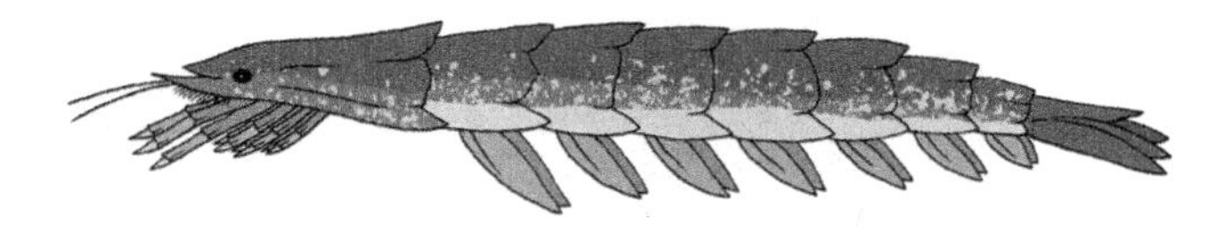

The con rit reconstructed as a crustacean in accord with my suggested identity for it (© Tim Morris)

To my mind, the con rit is one of the most fascinating if enigmatic marine cryptids on record, but with no modern-day sightings on file (at least not to my knowledge), whether it still does—or indeed ever did—exist remains as much a mystery today as the creature itself.

THE REAL LEVIATHAN?

How could a chapter on water monsters close without a consideration of the *real*, original leviathan? This colossal mystery beast of the Bible—"the piercing serpent . . . that crooked serpent . . . the dragon that is in the sea" (*Isaiah*, 27:1)—is, after all, the creature whose name has ultimately become an umbrella term for large, unidentified water beasts everywhere, as evinced by Tim Dinsdale's classic book on this subject, *The Leviathans* (1966, revised 1976).

The Old Testament contains four references to this monstrous sea creature, which provide several important morphological features—including its huge size, extensively scaled body, elongate shape, large plentiful teeth, shining eyes, powerful neck, smoking nostrils, and

'Destruction of Leviathan'—1865 engraving by Gustave Doré

Mosasaurs as Leviathan? (© William M. Rebsamen)

distinctive fins. Biblical scholars have nominated several different animals as the leviathan's identity, but whereas each possesses *some* of its characteristics, none has *all* of them.

The most popular identity is the Nile crocodile *Crocodylus niloticus*, which is indeed scaly and somewhat elongate, with an abundance of large teeth, a powerful neck, shining eyes, and a sizeable (albeit not enormous) body. However, it possesses neither fins nor smoking nostrils, and is not marine in habitat. Sharks are marine and finned, some are very large and fairly elongate, and many have plenty of large teeth, but not smoking nostrils, shining eyes, or scales.

Whales are also marine, finned, often very large and streamlined, and some have many large teeth. Moreover, the spray spouted upwards from around their blow-holes when they exhale could conceivably be distorted into smoke during the retelling of leviathan reports down through time. But whales are neither scaly nor shiny-eyed, and their necks are almost invisible.

And so it goes on—even identities as unlikely as the rock python *Python sebae* have been offered in a desperate attempt to reconcile this exceptional creature with a known type of animal.

Most probably, the leviathan is a non-existent composite, part-myth and part-reality. The latter component constitutes a hotchpotch of distinctive features drawn from all of the animals noted here, and possibly one other too—a bona fide sea serpent.

On account of its scaly skin, Heuvelmans considered the leviathan to be of the 'marine centipede' type. (Thus, according to his belief in what these creatures are, it is a modern-day armoured archaeocete; as we now know, however, such archaeocetes never existed.) Conversely, I consider that if the leviathan is either a sea serpent or a myth inspired in part by sightings of one, then it may conceivably be a living mosasaur.

Indeed, this identity uniquely combines *all* of the leviathan's features—its scaly body, elongate shape, shining eyes (typical of many large reptiles), powerful neck, fins, great size (as with the tylosaurs), large plentiful teeth, and smoking nostrils (as with the whale identity, no doubt a reference to the spouting of water displaced from around its nostrils when exhaling underwater as it surfaces). Yet only a complete specimen can conclusively test this hypothesis.

AND FINALLY. . .

"Canst thou draw out leviathan with an hook?" asked Job (*Job*, 41:1). Judging from science's singular lack of success in securing the carcase of any type of genuine water monster via any of the traditional means used in capturing aquatic animals, the answer to this question would appear, quite definitely, to be no! Clearly, then, it is time to develop a different means of obtaining such evidence—one commensurate with the sophisticated technology now available for scientific research—because until such evidence *is* obtained, all of the leviathans documented in this chapter, and many others besides, will remain an abiding mystery.

Nevertheless, if there is anywhere on our planet likely to be still concealing significant undiscovered species, it must surely be the great oceans that collectively constitute more than 70 per cent of its surface area. To quote naturalist Scott Wiedensaul, author of *The Ghost With Trembling Wings: Science, Wishful Thinking, and the Search for Lost Species* (2002):

> If cryptozoology is ever going to hit pay dirt, the jackpot is most likely to be marine. Even inshore waters are a mystery, and it is the height of hubris to think we've uncovered all the big surprises. It's certainly conceivable—perhaps not likely, but conceivable—that one or more large unknown species that fit the old "sea serpent" mold are hiding out there, too, ready to shock and delight us one of these days.

In fact, there is good scientific evidence for believing that, irrespective of their respective taxonomic identities, as many as 50 species of sea creature attaining relatively sizeable dimensions may indeed still await formal disclosure. How do we know this? By utilising statistics. To quote from a comment posted by Dr Darren Naish beneath his own sea monsters article that appeared in the online version of London's *Daily Mail* newspaper on 21 July 2011:

> Why say that there might be 50 undiscovered large marine animals out there? By plotting the discovery rate of large marine animals over time, we can generate a graph. One statistical trick you can do with graphs is work out whether the growth of any given curve has reached its 'end' (known as the asymptote), or if the curve is expected to continue on an upward trend. All the work done so far shows that discovery curves for big marine animals have *not* reached their asymptotes: it seems that between

> 10 and 50 new big marine animals await discovery (big = more than 2 m long along body axis). The lower figure is the more conservative one, of course.

The data cited above by Naish are from a *Journal of the Marine Biological Association U.K.* paper by zoologist Dr Charles Paxton, published in 1998. In it, he presented a cumulative species description curve for large open-water marine animals.

Back in 1886 (almost 80 years *before* Heuvelmans published his multi-species classification of sea serpents, incidentally), Charles Gould commented in his book *Mythical Monsters*:

> Let the relations of the sea-serpent be what they may; let it be serpent, saurian, or fish, or some form intermediate to them; and even granting that those relations may never be determined, or only at some very distant date; yet, nevertheless, the creature must now be removed from the regions of myth, and credited with having a real existence, and that its name includes not one only, but probably several very distinct gigantic species, allied more or less closely, and constructed to dwell in the depths of the ocean, and which only occasionally exhibit themselves to a fortune-favoured wonder-gazing crew.

It is a sad reflection of zoology's long-standing indifference to the subject of water monsters that those words are as relevant today as they were well over a century ago.

CHAPTER 4

Lions with Pouches and Horses with Claws—A Collection of Mammalian Methuselahs

The Queensland Tiger has always seemed to me to need further scrutiny: those reported fangs are suggestive of a specialized carnivorous marsupial, such as Thylacoleo [the officially-extinct pouched lion], as indeed Shuker points out. But as the Tablelands and Daintree Rainforests get tamed and explored, yield up their last mammalogical secrets, and get invaded by feral cats, we may lose the chance to find out. When the last rainforests have been logged, overrun by tourists, or dissected into tiny fragments, perhaps we will find a Thylacoleo skull on the mantelpiece of an old-timer, who will recount how his dogs killed the creature when he was a boy; and he will smile his enigmatic smile, and we will never know.

PROF. COLIN P. GROVES—*CRYPTOZOOLOGY* (VOL. 11, 1992)

Among a large collection of vertebrate fossils from the Lukeino Formation which I made recently I found the proximal phalanx of a large Chalicothere [an extinct horse-related ungulate with claws instead of hooves]. . . . On discovery of the Lukeino fossil, I described a Chalicothere to my field assistant, Mr Kiptalam Chepboi, who assured me that I had accurately described a Chemosit. I repeated the description to several other local people, all of whom gave the same opinion. The Chemosit is an animal of Kalenjin myth, on which the 'Nandi Bear' is supposed to have been based.

MARTIN PICKFORD—'ANOTHER AFRICAN CHALICOTHERE' (*NATURE*, 10 JANUARY 1975)

At the beginning of the Pleistocene epoch, just under 2.6 million years ago, a spectacularly rich, diverse mammalian fauna existed throughout the world, some of whose members attained dimensions far exceeding those of any representatives alive today—in what is termed the Holocene or Recent epoch. By the close of the Pleistocene 11,700 years ago, conversely, the vast majority of this megafauna had vanished. According to one school of thought, this is deemed to be a consequence of natural extinction. Conversely, some researchers have offered a very different explanation—overkill, i.e. the decimation of these mammals of magnitude not by nature, but by humans. And a third contingent supports the prospect of natural extinction and human-induced overkill being jointly to blame for megafaunal disappearance.

Irrespective of the cause, however, the effect may not be as black as is officially painted—for as will be revealed in this chapter, some of the most famous and outstanding members of the Pleistocene's fur-bearing megafauna may have avoided Holocene obscurity after all.

AUSTRALIA'S EXTRA ECHIDNA? NOSING OUT A SPINY SURPRISE DOWN UNDER

Many remarkable cryptozoological discoveries have been made not in the field but instead within the collections of museums. The following fairly recent case is a prime example of this, and it also happens to be a particular favourite of mine as it features those wonderfully archaic egg-laying mammals known as the monotremes, which are represented today only by the echidnas or spiny anteaters and the duck-billed platypus.

New Guinea is home to three species of long-nosed echidna or proechidna (all belonging to the genus *Zaglossus*) as well as to the much smaller, single species of short-nosed echidna *Tachyglossus aculeatus*. In contrast, the last-mentioned species is the only echidna known to exist today in Australia. For according to fossil evidence and ancient cave paintings, the long-noses died out here 30,000-40,000 years ago—a time when Australia and New Guinea were still united as a single land mass.

Thanks to a unique but long-overlooked specimen uncovered in the collections of London's Natural History Museum (NHM), however, this traditionally-accepted scenario appears to have been sensationally disrupted, yielding a hitherto-undisclosed but bona fide prehistoric survivor.

During a visit to the NHM in 2009, Smithsonian Institution zoologist Dr Kristofer Helgen found a skinned specimen of the western or Bruijn's long-nosed echidna *Zaglossus bruijnii* whose original data-tag totally astonished him. The tag revealed that this animal had been shot by Australian naturalist John Tunney in 1901, but not in New Guinea—instead, on Mount Anderson, a mountain in the vast, arid, sparsely-populated West Kimberley region of northwestern Australia!

Western (Bruijn's) long-nosed echidna, depicted in a 1919 print

It had then been stuffed and sent to Lord Walter Rothschild's private natural history museum at Tring, in Hertfordshire, England, and thence to London's NHM in 1939 after Lord Rothschild's death two years earlier. But it was never studied at either museum, thus remaining in scientific obscurity for over a century in total, until its long-unrealised significance was belatedly recognised by Dr Helgen.

For this unremarkable-looking specimen is proof that a species of long-nosed echidna was still living in Australia until as recently as the early 20th Century, and had not died out here all those many millennia ago after all. Moreover, when Mount Anderson aboriginals were questioned recently, they confirmed that their parents had hunted such a creature, which they positively identified from pictures shown to them.

Tunney's long-overlooked specimen and its potential zoological significance were formally documented by Helgen and fellow researchers in a paper published by the journal *ZooKeys* on 28 December 2012, and Helgen hopes to launch an expedition to the area where Tunney had procured it in search of a possible population of long-noses. Having said that, the provenance of Tunney's specimen is both vast and inhospitable, meaning that even if funds can be raised for an expedition to seek living long-nosed echidnas there, the chances of finding any are slim.

As noted by biologist Dr Adrian Burton in the April 2016 issue of *Frontiers in Ecology and the Environment*, however, this might not be required. Simply finding a few samples of their faecal droppings would be sufficient, because DNA extracted from them could be compared with that of Australia's short-nosed echidna, of Tunney's Australia-originating long-nosed echidna, and of long-nosed echidnas from New Guinea—and if a match were obtained with DNA from the long-noses, this would verify that such echidnas were indeed still living in Australia.

IN SEARCH OF THE MISSING MARSUPIALS

Roughly 40 million years ago, Australia broke away from East Antarctica, and has remained isolated from all other continental land masses ever since. The first pouched mammals (marsupials) had already entered Australia prior to its separation, but the placental or eutherian ('true') mammals—comprising virtually all of the non-marsupial mammals alive today—had not penetrated it in time. Consequently, except for some enterprising rodents that reached its shores many millennia ago via rafting and a few monotremes that had entered it before its separation, Australia's native mammals were exclusively marsupials.

In the absence of competition from placental mammals, the pouched clan underwent a dramatic evolutionary radiation—yielding species to fill almost every available ecological niche open to mammals. Moreover, because they were occupying niches filled elsewhere in the world by placentals, they even evolved convergently, yielding morphologically parallel forms. For example, wolves and dogs were represented in Australia by an extraordinarily dog-like marsupial called the thylacine, shrews were mirrored by tiny carnivorous marsupials inappropriately termed marsupial mice, 'flying' squirrels were faithfully duplicated by gliding marsupials called flying phalangers, anteaters by the numbat, ground squirrels by wombats, weasels by dasyures, rats and tree squirrels by possums, and so on.

Nevertheless, today's spectrum of marsupials seems to lack equivalents for certain major placental mammals elsewhere. There are no feline marsupials, no principally aquatic versions, no primate counterparts, and no representatives for the more cumbersome, burlier hoofed beasts such as rhinoceroses and tapirs. Yet as recently (geologically-speaking) as 10,000 or so years ago, such creatures did exist—and that is not all. As will be revealed, it is conceivable that these 'missing marsupials' may not be missing after all.

The Yarri—
Queensland's Mysterious Marsupial Tiger

During the years 1880-1884, Norwegian explorer Carl Lumholtz spent his time in Australia, studying its unique fauna and flora, and for most of that period he lived with the aboriginals of northeastern Queensland. As he later recounted in his book *Among Cannibals* (1889), the Herbert Vale natives spoke of two large mammals inhabiting the summit of the Coast Mountains that he believed to be unknown to science. He succeeded in discovering one of these, the boongary, which proved to be a hitherto-undisclosed species of tree kangaroo—

a group of marsupials previously known only from New Guinea—and was fittingly christened *Dendrolagus lumholtzi*. The other, conversely, has remained undiscovered. It was known to the aboriginals as the yarri or yaddi, and Lumholtz documented it as follows:

> From their description I conceived it to be a marsupial tiger. It was said to be about the size of a dingo, though its legs were shorter and its tail long, and it was described by the blacks as being very savage. If pursued it climbed up the trees, where the natives did not dare follow it, and by gestures they explained to me how at such times it would growl and bite their hands. Rocky retreats were its most favourite habitat, and its principal food was said to be a little brown variety of wallaby common in Northern Queensland scrubs. Its flesh was not particularly appreciated by the blacks, and if they accidentally killed a yarri they gave it to their old women. In Western Queensland I heard much about an animal which seemed to me to be identical with the yarri here described, and a specimen was once nearly shot by an officer of the black police in the regions I was now visiting.

By the time that Lumholtz was learning of the yarri from the Herbert Vale aboriginals, however, European settlers in Queensland had already begun to gain rather more direct proof of this beast's reality. In 1871, Brinsley G. Sheridan, police magistrate of Cardwell in Rockingham Bay, revealed in a letter published by the *Proceedings of the Zoological Society of London* on 7 November of that year that while walking with his teenage son along a path near to the shore, their terrier dog had picked up a scent from some scrub nearby that led them, half a mile further along, to a strange animal lying in long grass. According to Sheridan, it was:

> As big as a native Dog [dingo]; its face was round like that of a Cat, it had a long tail, and its body was striped from the ribs under the belly with yellow and black. My Dog flew at it, but it could throw him. When they were together I fired my pistol at its head; the blood came. The animal then ran up a leaning tree, and the Dog barked at it. It then got savage and rushed down the tree at the Dog and then at me. I got frightened and came home.

Sheridan's *PZSL* letter initiated a follow-up communication written by Walter T. Scott from Cardwell's Vale of Herbert, which was published by this same journal on 5 March 1872. In it, Scott reported that a Mr A.A. Hull, a licensed surveyor, had recently been working with a party of five men on the Murray and Mackay rivers, north of Cardwell, when, while in their tents one night between 8 pm and 9 pm, they heard a loud roar nearby. They swiftly emerged with guns at the ready, but although they failed to spot the creature responsible, they did note an unusual track precisely formed in the soft ground. In a 2012 *Journal of Cryptozoology* paper, Australian cryptozoologist Malcolm Smith evaluated this spoor and concluded that it did not appear to be referable to any known animal.

Moreover, on 5 November 1872, the *PZSL* published a second letter penned by Scott. In it, he revealed that shortly after Sheridan's sighting, a fawn-coloured creature with darker markings, a round head, long tail, and total size larger than that of a pointer dog had been reported by Robert Johnstone, an officer of the Native Police, who had spied it moving from tree to tree in scrub on the coast-range west of Cardwell.

Following famous Australian naturalist-author Charles Barrett's return to Brisbane in November 1936 after leading a four-man expedition up Mt Bellenden-Ker (at 5,000 ft, Queensland's second highest peak, and whose dense jungle was—and still is?—reputedly home to the yarri), the *Richmond River-Herald and Northern Districts Advertiser* newspaper carried a report on 24 November concerning his explorations. In this report, it was noted that although no yarri sightings had occurred on that occasion, Arnold Leumann, a noted North Queensland guide and bushman who had taken Barrett's expeditionary team up the mountain, had previously seen a yarri. According to his description, it was about the size of a dingo, but with a short, blunt head, rather like that of a tiger. Its body and tail were striped like a tiger's. It was perched on the branch of a tree, from where it snarled and spat at him. (Back in January 1889, fellow Australian explorer Archibald Meston had led an expedition to Mt Bellenden-Ker, expressing a desire to secure a specimen of the yarri, but none was forthcoming.)

Many similar reports followed during the early years of the 20th Century, which for the most part were sufficiently sober and consistent for science to look very favourably upon the Queensland tiger's status as a valid, undiscovered member of the Australian fauna. This was reflected in the decision by Taronga Park Zoo co-founder Albert S. Le Souef and Australian naturalist Harry Burrell to include it in their then-definitive guide to Antipodean fauna, *The Wild Animals of Australasia* (1926), which contained, under the heading 'Striped Marsupial Cat', the following field description:

> Hair short, rather coarse. General colour fawn or grey, with broad black stripes on flanks, not meeting over the back. Head like that of a cat; nose more produced. Ears sharp, pricked. Tail well haired, inclined to be tufted at end. Feet large; claws long, sharp. Total length about five feet; height at shoulder eighteen inches.

I would question their description of its stripes. In many yarri reports, its stripes are likened to unbroken hoops.

Queensland tiger restoration, with accurate coat patterning based upon eyewitness descriptions, but with enlarged canines (as in placental cats) rather than enlarged incisors—see later (© Dami Editore srl)

Equally optimistic in tone was the following short entry from the six-volume, partwork-issued *Purnell's Encyclopaedia of Animal Life* (1968-1970), edited by Drs Maurice and Robert Burton, which I originally read as a child during the late 1960s, and was the first information I'd ever encountered on the subject of this particular cryptid, sparking what has become for me a life-long interest in it:

> One animal known only from reports is the native tiger, reported at intervals in the rocky country and thick forests of northern Queensland, where it is said to prey on calves and kangaroos. It is apparently well-known to the aborigines, and the Europeans who have seen it describe it as being the size of a dingo, like a "cat just growing into a tiger". It is said to have transverse black stripes from the shoulders to the root of the long tail. The description also fits the thylacine, or Tasmanian wolf, presumed extinct on the mainland of Australia, but the native tiger is said to have a cat-like face whereas the head of the thylacine is dog-like. The identity of the native cat was nearly settled when one was shot. Unfortunately the body was eaten by wild pigs and the skin went bad.

The yarri's history has a most unusual aspect, flagrantly breaching the standard code of practice for all self-respecting cryptozoological creatures. For whereas such beasts are often seen and occasionally even photographed, but never captured or killed, during the 20th Century's early years yarri carcases apparently turned up with almost embarrassing frequency!

Around 1900, while walking through scrub on the Atherton Tableland, J. McGeehan heard a noise farther ahead that implied a confrontation between his dogs and a wild animal. When he reached the spot, he saw that the dogs had just killed some form of cat-like beast, seemingly a young specimen, but one that did not belong to any species known to him:

> It partly resembled a large domestic cat, excepting for the body, which was rather light in build. The most striking part of its appearance was the well defined hoops of colour which encircled its body. These hoops or bands appeared to be about 2½ inches in width, and the colours were white and dun alternating in perfectly marked circles. As far as I can remember the alternate colours did not extend to the head, legs or tail. I think that the colour of these parts was dun. . . . I noticed that, when the mouth was opened, the top and bottom jaws, at the front, contained long fine fangs, but regret that I cannot now recollect whether the number on each jaw was two or four. Whichever number it was the fangs were in sets of two and about a quarter of an inch apart. . . . From the point on the backbone immediately behind the shoulders to the butt of the tail, 14 inches. Length of front legs from the bottom of the paws to junction with the chest, 7½ inches. Height (from ground to back) 11½ inches to 12 inches.

In 1910, a Kuranda inhabitant evidently unfazed by these animals ("Most of the tiger cats which I have killed. . .") commented:

> Most of the tiger cats which I have killed were about four feet long and of fawn colour, with black stripes running across the body, which was fairly long, unlike an ordinary cat.

Sometime between 1900 and 1926, J.R. Cunningham and his dog encountered a yarri at Gootchie, near Maryborough. His dog had treed it, and when Cunningham rode up on his horse and saw it, he used his whip to lash the creature across its head, knocking it down onto the ground—where his dog promptly attacked and killed it. According to Cunningham's description, the creature had a long body and tail, with a total length of almost 6 ft, and a shoulder height of 20-24 in. Its pelage was fawn-brown, bore black vertical stripes, and was sleek rather than shaggy. Its legs were relatively short, and its teeth were long, as were its claws—but unlike those of true cats, they did not seem to be retractile.

Cunningham was convinced that it was neither a feral domestic cat nor any of those weasel-like marsupials known as dasyures or native cats. (Confusingly for the cryptozoologist, one of these species, *Dasyurus maculatus*, is widely referred to as the tiger cat—but, like all dasyures, it is spotted, not striped.) Moreover, in 1919 he witnessed the death of another yarri, this time at Eungella, near Mackay. It was cornered in a chicken run, where it had slaughtered virtually all of the poultry—but before it had chance to escape, their irate owner blew half of the luckless beast's head away with his shotgun.

In a similar incident, this time on a property at the head of the Mulgrave River adjoining an expanse of rainforest, one night sometime between 1928 and 1930 a Mr Woods heard a furious commotion in the hen house—and when he investigated he discovered a huge striped cat-like beast inside, surrounded by dead chickens. On that occasion, the creature made a swift, successful exit and fled up a nearby tree, but it was dislodged by a farmhand who dispatched it with an axe when it fell to the ground. To their great surprise, it proved to be the size of a kelpie dog—the large Australian sheepdog.

During the late 1920s, various Australian newspapers, including the *Brisbane Courier* (22 December 1928) and Mackay's *Daily Mercury* (17 June 1929), carried a short account concerning the yarri by a correspondent identified in the *Daily Mercury* report as a Mr M. O'Leary, who had lived in northern Queensland for 50 years, was familiar with the yarri, and claimed to have actually hit out at one. Here is his description of that incident:

> On the Tully highlands I once struck one of these animals in a large decayed tree, and was lucky enough to get it. It went 4ft. 7in. from nose to tall extreme, and was striped, with round head, not pointed, and croppy ears. There are still some of them left in those dense scrub lands.

It is not clear from his account whether O'Leary merely struck the yarri or killed it, but even if he did kill it the body was not made available in any way for scientific scrutiny. Back on 31 March 1923, journalist Reg Kendall and a cameraman had set out to seek the yarri in this same area, but in April, while attempting to cross the Tully River, Kendall was attacked by crocodiles, forcing him to abandon the raft that was carrying all of his equipment. Not surprisingly, therefore, their expedition proved unsuccessful in achieving its objective.

Two dead yarris were mentioned by Le Souef and Burrell in their tome. One, measuring about 5 ft long from nose-tip to tail-tip, had been shot by a settler in the Atherton district when it began stalking his goats, and was observed by naturalist George Sharp. Some time earlier, Sharp had seen a living yarri, moving through scrub at the head of the Tully River. The other example, about 18 in high and as long as a large cat (thus a young specimen?) had been trapped by a Mr Endres of Mundubbera. He declared that it had been very savage when caught, was striped ". . . but not right round", and possessed a very short head and neck.

On 12 December 1932, Brisbane's *Daily Mail* contained the following report, headlined 'Shot Near Tully—Marsupial Tiger':

Restoration of the yarri emphasising its enlarged front teeth or incisors, an unusual feature except in a certain officially-extinct feline marsupial discussed later here (© William M. Rebsamen)

> Tully, Sunday.—Mr. A. W. Blackman, of Upper Murray, and a party who made a tour of the Kirrima lands, about 30 miles from Tully, claim to have shot what is generally known as a marsupial tiger (called by the aborigines "yaddi"). The animal, which proved ferocious when captured, was half as big again as a domestic cat, and was striped like a tiger. It was captured on the fringe of extensive scrub on the Cardwell Range, and it is thought that with a careful hunt another of the species could be captured.

Merely an unusually large feral domestic cat, or a young Queensland tiger? And whatever happened to its carcase?

With such an array of specimens as the selection documented here, one might have expected the mystery of the Queensland tiger to have been resolved long ago. The tragedy, however, is that not one of those carcases was forwarded to a scientific establishment for identification and study. In the case of Woods's yarri, for example, the pelt was left unattended until the following morning—by which time it had been mauled to the point of total ruin by dogs. The head and body of the dead specimen seen by Sharp had been devoured by wild pigs, and as its pelt was not preserved it soon rotted away (this is the specimen referred to at the end of the Burtons' earlier-quoted account). All of the others were apparently discarded.

Incidentally, an enigmatic big-toothed carcase dubbed 'Jaws' was discovered and photographed by a group of unidentified beachcombers on a sandy beach in the Margaret River area of Western Australia in c.1975, but it was not retained, and some cryptozoologists later speculated that it may have been a young yarri. Recently, however, it was convincingly identified by Dr Darren Naish as merely a decomposed domestic cat, after studying the only known photo of it but which clearly reveals its major dental characteristics (*Journal of Cryptozoology*, 2012).

When scanning through the yarri reports on file, certain features occur time and again. The animals sport a distinctly striped pelage, appear inordinately aggressive, are very adept arboreally (shinning up trees at the first sign of danger), and have extremely prominent, protruding teeth at the front of their jaws.

The frequently-cited pre-1930 encounter of G. de Tournoeur and P.B. Scougall with a yarri is one that vividly captures the aura of danger and menace that this creature evokes. They had been riding from Munna Creek towards Tiaro when they unexpectedly came upon:

> . . . a large animal of the cat tribe, standing about 20 yards away, astride of a very dead calf, glaring defiance at us, and emitting what I can only describe as a growling whine. As far as the gathering darkness and torrential rain allowed us to judge he was nearly the size of a mastiff, of a dirty fawn colour, with a whitish belly, and broad blackish tiger stripes. The head was round, with rather prominent lynx-like ears, but unlike that feline there were a tail reaching to the ground and large pads. We threw a couple of stones at him, which only made him crouch low, with ears laid flat, and emit a raspy snarl, vividly reminiscent of the African leopard's nocturnal 'wood-sawing' cry. Beating an angry tattoo on the grass with his tail, he looked so ugly and ready for a spring that we felt a bit 'windy'; but on our making a rush and cracking our stock-whips he bounded away to the bend of the creek, where he turned back and growled at us.

An even more graphic eyewitness report of a yarri, encountered in Cape York Peninsula, was that of writer-traveller Ion L. 'Jack' Idriess, who documented it during the 1920s. His report, which also included a second, less startling yarri sighting made by him, reads as follows:

> Up here in the York Peninsular we have a "tiger cat" that stands as high as a hefty, medium-sized dog. His body is lithe and sleek and beautifully striped in black and grey. His pads are armed with lance-like claws of great tearing strength. His ears are sharp and pricked, and his head is shaped like that of a tiger. My introduction to this beauty was one day when I heard a series of snarls from the long buffalo-grass skirting a swamp. On peering through the grass I saw a full-grown kangaroo, backed up against a tree, the flesh of one leg torn clean from the bone. A streak of black and grey shot toward the "roo's" throat, then seemed to twist in the air, and the kangaroo slid to the earth with the entrails literally torn out. In my surprise I incautiously rustled the grass, and the

> great cat ceased the warm feast that he had promptly started upon, stood perfectly still over his victim, and for ten seconds turned me gaze for gaze. Then the skin wrinkled back from his nostrils, white fangs gleamed, and a low growl issued from his throat. I went backwards and lost no time getting out of the entangled grass. The next brute I saw was dead, and beside him was my much-prized staghound, also dead. This dog had been trained from puppyhood in tackling wild boars, and his strength and courage were known by all the prospectors over the country. The cat had come fossicking round my camp on the Alice River.

Idriess's extremely dramatic report has been frequently quoted in subsequent accounts of this mystery beast. More recently, however, it had been overshadowed by suspicion regarding its veracity, when yarri investigators pointed out the extreme similarity between it and what was then believed to be an earlier, fictional version that had appeared in a novel called *Kangaroo*. Written by none other than D.H. Lawrence (later to become infamous for his highly controversial novel *Lady Chatterley's Lover*), *Kangaroo* was first published in 1923, and had been researched by Lawrence the previous year. This led to the widespread albeit unconfirmed assumption that Idriess's report must be a fake, nothing more than a plagiarised version of the fictional one in *Kangaroo* but claimed by Idriess to be factual.

In fact, it turns out that the exact reverse situation is the true one. As detailed by Australian cryptozoologist Malcolm Smith in his blog *Malcolm's Musings* (8 November 2013), the report had originated with Idriess after all. Now known to have been written by him using the pen-name 'Gouger', it was first published in an Australian periodical entitled the *Bulletin* on 8 June 1922, which Lawrence had read while in Australia researching his novel and decided to incorporate into it, with only very minor word changes. Indeed, the character in *Kangaroo* who narrates this report actually then goes on to state that it had been published in the *Bulletin*, but the crucial significance of that latter statement had not previously been recognised!

Of particular interest is a sighting made in Ben Lomond National Park by Craig Black in 1961, because he not only claimed to have encountered while here an adult female yarri but also stated that he was "positive" he saw that "it was carrying a pouched cub". If true, this confirms that the yarri is indeed a marsupial.

During the 1960s and 1970s, the leading yarri investigator was naturalist Janeice Plunkett, who gathered many eyewitness reports. Peter Makeig featured several of them in an article within the *North Queensland Naturalist* for 1970, including, for example: ". . . an animal about as large as a medium-sized dog [that] rushed out and climbed a nearby tree. The animal was very savage. Its coat was beautiful and striped like a tiger".

That was how an eyewitness described a yarri spied in 1925 in the Bellenden Range, and a specimen confronted at Kuranda in 1945 had: ". . . a round face and four exposed 'tiger teeth' . . . the other salient point in my opinion was the fact that big savage pig dogs were terrified of it". According to a yarri eyewitness who saw one on Mount Molloy in 1953, its head: ". . . was a good deal larger than an old tomcat, with teeth a lot like the extinct sabre toothed tiger". As for the individual spotted on Mount Bartle Frere in 1968, its head: ". . . appeared round and broad, its nose shorter and broader than a dog's. Some of its teeth appeared to protrude out and upwards like tusks".

In later years, yarri reports have become far fewer, leading to speculation that the species was dying out. Indeed, within their book *Out of the Shadows* (1994), Tony Healy and Paul Cropper were concerned that the yarri may have already gone—a victim, they suggested, of strychnine baits meant for dingoes, and of the abundant, tasty-looking, but lethally poisonous cane toad *Rhinella marina* (introduced from South America by humans).

Australian cryptozoologist Rex Gilroy, conversely, believes that it does still survive, and has collected some compelling post-1970s accounts. In 1984, for instance, after investigating a lion-like roar emerging from trees on the opposite side of a creek where he was camped at Daintree, Bill Norman came upon a ". . . panther-sized striped cat-like animal" in a tree, consuming a dead sheep. Displaying, for a yarri, an unwonted degree of diffidence in the face of human proximity, the animal promptly descended the tree and fled into the bush.

During early 1987 near Hughenden, a hunter who had wounded a dingo was about to pursue it when a large hay-coloured beast with black body stripes appeared—and proceeded to rip the dingo to pieces, literally. Wisely, the hunter abandoned his quarry—and left, quickly.

In 1991, Don Moss, a trail biker, was exploring a scrubland track west of Townsville, a large city on Queensland's northeastern coast, in the dry tropics region of this Australian state, when his approach startled a large fawn-furred animal with black stripes, causing it to leap down from a tree just ahead of him and bound off into scrub. When he reached the tree, Moss noticed that the carcase of a dead goat was wedged in the fork of a branch about 15 ft above the ground. The animal, seemingly a yarri, had evidently carried the carcase up there, just as leopards do with their prey, and had been feeding upon it, recalling Bill Norman's 1984 sighting of a similar beast in a tree feeding upon a sheep.

During April 1995, the Discovery Channel aired an episode of the documentary series *Arthur C. Clarke's Mysterious Universe* that was entitled 'On the Trail of the Big Cats'. One segment of this episode presented reports of Australian mystery cats, two of which were very noteworthy as they may have featured the yarri.

One of them revealed how Ted Francis decided one Sunday to go to the river at Mount Mulligan, Queensland, to do some panning for gold. While there, he looked down and saw a very sizeable feline creature that looked back up at him. The creature had dark fur marked with white stripes.

In the second report, Nigel and Charlie Tutt were cutting hoop pine on Mount Stanley at the headwaters of Brisbane River in Queensland. While they were walking along a track there, Charlie saw what looked like a tiger to him. This creature half-rolled twice on a big old pine stump, about 3 ft across, and it spanned the stump's entire diameter. Its most distinctive features were two curved horizontal white stripes, surrounding which was dark-grey fur that merged into ginger. It had the appearance of a cat that was ready to spring upon prey, and Charlie averred that it was definitely a tiger-like cat.

It is clear from the reports given here that attempts to identify the yarri as anything so conservative as an over-sized domestic cat, misidentified dog or dingo, or the occasional escapee placental tiger are ill-founded. Characteristics such as its distinctive hoop-like stripes, protruding tusk-like teeth, tree-climbing prowess, and longstanding aboriginal knowledge of its existence (to the extent that it has its own name) at once discount all of those identities and label it as a bona fide native species—but there is none known today that offers the slightest correspondence to it. So what else is on offer?

The Once and Future Thylacine—In Australia and New Guinea?

More as an act of desperation than diligence, some investigators have sought to reconcile the yarri with the Tasmanian wolf or thylacine *Thylacinus cynocephalus*—the world's largest modern-day species of carnivorous marsupial, officially denounced as extinct in Tasmania since 1936, but which probably still persists here in small numbers, judging from an immense file of often reliable, detailed eyewitness reports that grows larger every year. It also existed on mainland Australia until 2,300 or so years ago—when competition with the introduced dingo harried this canine marsupial into extinction there.

A 1919 print of the thylacine

However, a dog-headed creature with stripes only upon the upper portion of its back is hardly a plausible identity for a cat-headed beast with hoop-like bands sometimes encircling its entire body. Incidentally, although once assumed to have been exclusively terrestrial, the thylacine is now known from historical records to have been adept at climbing trees, so the yarri's arboreal prowess can no longer be used to discount the thylacine as this cryptid's potential identity. Having said that, the yarri's often-reported behavioural predilection for spitting down in rage from its tree-borne vantage point at would-be attackers sounds far more feline than thylacine.

As an interesting aside, reports of elusive, unequivocally dog-headed beasts with thylacine-like stripes emerge every so often from the mainland—a good example being the so-called Buderim Beast in the 1990s (which I documented in my book *A Manifestation of Monsters*, 2015). Moreover, in 1966 a hairy thylacine carcase of controversial age was found in a cave at Western Australia's Mundrabilla Station—lending support to the possibility that this species has not died out here after all. A handful of photos allegedly depicting mainland thylacines are also on file, but none of these is remotely convincing to me.

In their afore-mentioned book, Healy and Cropper explored claims that thylacines were released in parts of Australia in pre-1930s times, which, if one day confirmed, and assuming that they survived and bred, would provide a wholly different explanation for this species' existence here. Corroborating this scenario is the fact that with just one possible exception currently known to me, mainland aboriginals have no name or folklore appertaining to the thylacine—which is unusual if this animal has indeed lingered on here as a native species, but is precisely what one would expect if it has merely been re-introduced in modern times.

The one possible exception to this was documented by me as follows in my book *Dr Shuker's Casebook* (2008):

> Another dog of the Dreamtime is the marrukurii, which, according to aboriginal traditions prevalent in the vicinity of South Australia's Lake Callabonna [until the 1830s], resembled a dog in outline, but was brindled with many stripes. They were believed to be dangerous, especially to human children, carrying away any that they could find to their own special camp at night,

where they would savagely devour them. When questioned, the native Australians denied that the marrukurii were either domestic dogs or dingoes. Is it possible, therefore, particularly in view of their brindled appearance, that these Dreamtime beasts were actually based upon memories of the striped Tasmanian wolf or thylacine *Thylacinus cynocephalus*? After all, this famous dog-like marsupial did not die out on the Australian mainland until about 2,300 years ago.

Moreover, during the 1800s a Melbourne naturalist called Cambrian examined the head, feet, and skin of a thylacine allegedly killed in the Blue Mountains near Sydney, New South Wales, and also a partial skin of a female thylacine supposedly originating in South Australia's Lake Torrens/Flinders Ranges. I wonder what happened to these potentially highly-significant specimens? Back in the late 1980s, it was suggested by American zoologist Victor Albert that the yarri may be a short-faced (and hence more feline-faced) mainland thylacine, but there is no evidence that mainland thylacines belonging to the modern-day species *Thylacinus cynocephalus* were short-faced.

Having said that, Queensland was home during the early Miocene to a primitive thylacine called *Wabulacinus ridei*. What makes this form so interesting is that unlike the dog-like head of the modern-day species, its head was apparently cat-like in shape. Needless to say, a cat-headed thylacine would offer a very tantalising option as an identity for the yarri, but as *Wabulacinus* vanished from the known fossil record more than 10 million years ago, it would not be among the more likely ones.

Following the discovery by Dr Susan Bulmer during April 1960 of a Pleistocene-dated thylacine mandible in the Eastern Highlands District of Papua New Guinea (the country occupying the right-hand half of the island of New Guinea), this species is now known to have formerly existed in New Guinea too—and again, it may still do so. When grazier Ned Terry visited the remote, little-explored highlands of Irian Jaya (Indonesian New Guinea, occupying New Guinea's left-hand half) during the early 1990s, the natives spoke of a mysterious beast called the dobsegna. Its description—as documented by Greg Morgan in a *Garuda* article—instantly recalls the thylacine:

> . . . the locals rarely saw the animal in full daylight. Usually it was seen hunting at dawn or dusk for small marsupials and birds. The people said they were very afraid of it because they associated it with evil spirits and used its faeces to perform magic on enemies. They said its dens were usually in rocks or caves and described its head and shoulders as being like those of a dog except that it had a huge, strong mouth and a long, thin tail almost as long as its body. From the ribs to the hips, they said, there were no intestines (indicating that it was very thin there) and that in that area it had stripes.

Seldom have I encountered a more accurate verbal portrait of a thylacine. Even the reference to its mouth is significant. As dramatically captured in a short film depicting a yawning thylacine in Hobart's Beaumaris Zoo circa 1935, this species could open its long jaws far wider than any placental canid, yielding an astonishing gape that may have reached 120°.

In March 1997, news emerged from Jayawijaya, the central mountain range of Irian Jaya, of a supposed kind of 'tiger' living in caves in the regions of the Kurima Tableland, Oksibil, and Okbibab, but coming out at night to hunt the native peoples' livestock. The source of this claim was Jos Buce Wenas, the Regent Head of District Level II, Jayawijaya, whose information source was testimony from local missionaries, who also alleged that these 'tigers' were 3 ft tall and hunted in packs. In contrast, Tasmanian thylacines were not as tall as this and only hunted in small family groups, but perhaps New Guinea thylacines were/are bigger and did/do hunt in larger numbers than their Tassie counterparts? Media reports of the 'Jayawijaya tiger' induced a number of private individuals to visit the area and seek this cryptid, but none succeeded in sighting any.

On 11 January 2013, however, responding to a post concerning the Jayawijaya tiger by Malcolm Smith on his blog for 18 October 2011, an anonymous reader claimed to have seen one when stationed in Jayawijaya for three years, and also stated that it was grey, not the brown colouration typifying Tasmanian thylacines.

Smith's blog post also received an interesting response on 19 October 2011 from American cryptozoologist Chad Arment:

> Sometime in the mid- to late-1990s, I was contacted by an exotic animal importer from Florida, who had collected reports of a thylacine-like animal from New Guinea. (I don't unfortunately recall the exact location.) He was primarily over there purchasing snakes and other reptiles that had been captured, for export to the U.S. While over there, he was told about a strange doglike animal that was consistent with a thylacine. He came back, hoping to find someone willing to invest in an expedition to look for it, but I never heard whether he was able to do so.

In 2003 veteran Irian Jaya explorer Ralf Kiesel confirmed to me that since 1995 there have been persistent rumours of thylacines existing in at least two sections of Irian Jaya's Baliem Valley—the Yali area in the valley's northeast region, and the NP Carstenz in its southwest. The latter area is of particular significance because in the early 1970s Jan Sarakang, a Papuan friend of Kiesel, had a most startling experience while working with a colleague in the mountains just west of NP Carstenz.

They had built a camp for some geologists near Puncac Jaya at an altitude of roughly 1.5 miles and were sitting by their tents that evening, eating their meal, when two unfamiliar dog-like animals emerged from the bush. One was an adult, the other a cub, and both appeared pale in colour, but most striking of all was their stiff, inflexible tails, and the incredible gape of their jaws when they yawned spasmodically. Clearly drawn by the smell of the food, the two animals walked nervously from side to side, eyeing the men and their food supplies, and approaching to within 20 yards. Eventually the cub became bold enough to walk up to the men, who tried to feed it, but when one of them also tried to catch it, the cub bit his hand and both animals then ran back into the bush and were not seen again.

Except for their seemingly unstriped form, which may well have been a trick of the moonlight, once again these animals recalled thylacines, especially with respect to their stiff tails (a thylacine characteristic) and huge gapes.

Incidentally, although nowadays exceedingly rare there, New Guinea was until fairly recently frequented by a dingo-like placental canid, the singing dog, famed and named for its yodelling cries, and now bred in numbers as a pet elsewhere in the world, but notably in the U.S.A. Some sceptics of the surviving New Guinea thylacine scenario have suggested that perhaps the singing dog was responsible for such reports. However, this latter form is not striped, its tail is flexible like that of other true canids, and it cannot match the thylacine's exceptional gape when it yawns. Also it is—or was—well known to native tribes, who even mimicked its yodels as a means of communicating with one another across considerable distances. Hence it seems highly unlikely that they would fail to recognise it if they should encounter it in the wild here.

If the thylacine's existence in Irian Jaya is eventually confirmed, this would constitute an exceedingly significant example of prehistoric survival. Bearing in mind that as recently as 1994, these self-same highlands revealed a hitherto scientifically-unknown species of large, black-and-white tree kangaroo, referred to the natives as the dingiso or bondegezou (and subsequently dubbed *Dendrolagus mbaiso* by science), it would clearly be rash to deny the likelihood that there are other zoological surprises still in store here.

When Lions Had Pouches

Not so long ago, geologically-speaking, there would have been no problem in identifying the yarri—thanks to the existence of a creature called *Thylacoleo carnifex*, the marsupial or pouched lion.

Evolving not from the dasyurid lineage that gave rise to Australia's most famous carnivorous marsupials, *Thylacoleo* claimed descent from the same family tree (that of the diprotodontids) that spawned the herbivorous phalangers, possums, koala, and wombats. Through a somewhat incongruous quirk of evolutionary adaptation, however, *Thylacoleo* occupied the hitherto-vacant ecological niche Down Under for a feline meat-eater. Morphologically, therefore, it was a counterfeit cat—taxonomically, it is aptly defined as a 'killer possum'.

Judging from reconstructions based upon its fossilised remains, *Thylacoleo* ('pouched lion') was a cat-shaped beast up to the size of a lioness or tigress, with a short, round head, powerful jaws, a strong tail, fairly long limbs (of which the forelimbs were especially well-developed), and, suggestive of arboreal ability if needed, a pseudo-opposable thumb on each of its very large forepaws, which would have been of great value in grasping branches, as well as prey, bearing at its tip an enlarged claw. Each of its other toes bore a sizeable claw too. (As announced in February 2016, incidentally, a study of ancient scratch marks, found by a Flinders University palaeontological team at high elevations inside a Western Australian cave and conclusively identified by it as having been made long ago by *Thylacoleo*, has confirmed that this marsupial lion was indeed capable of climbing, an issue that had previously been contentious.)

Moreover, its muscle mass distribution and limb proportions indicate that this formidable beast was an ambush predator, sneaking up on its prey on foot or leaping down upon it from an overhanging tree branch. Consequently, its pelage may well have been striped or brindled for camouflage.

And formidable *Thylacoleo* certainly was, as revealed in a comparative bite analysis study by Dr Stephen Wroe and co-workers (*Proceedings of the Royal Society, Series B*, March 2005). Its findings regarding *Thylacoleo* are summarised on Wikipedia as follows:

> The jaw muscle of the marsupial lion was exceptionally large for its size, giving it an extremely powerful bite. Biometric calculations show, considering size, it had the strongest bite of any known mammal, living or extinct; a 101-kg [223-lb] individual would have had

Fossil skeleton of *Thylacoleo carnifex* at Naracoorte Caves, South Australia

a bite comparable to that of a 250-kg [550-lb] African lion. Using 3D modeling based on X-ray computed tomography scans, marsupial lions were found to be unable to use the prolonged, suffocating bite typical of living big cats. They instead had an extremely efficient and unique bite; the incisors would have been used to stab at and pierce the flesh of their prey while the more specialised carnassials [their huge, blade-like premolars] crushed the windpipe, severed the spinal cord, and lacerated the major blood vessels such as the carotid artery and jugular vein. Compared to an African lion which may take 15 minutes to kill a large catch, the marsupial lion could kill a large animal in less than a minute. The skull was so specialized for big game [such as the now-extinct sthenurines and other giant kangaroos, some of whose remains in Victoria'a Lancefield Swamp have been found bearing *Thylacoleo* bite marks], it was very inefficient at catching smaller animals . . .

There was, however, one feature that instantly separated *Thylacoleo* from all true cats—its teeth, most especially its fangs. Whereas those of all placental carnivores consist of greatly enlarged canines, the pouched lion's had evolved from its 'front teeth' or incisors—becoming huge tusk-like structures projecting outwards like the beak of a parrot (in stark contrast, its canines were reduced to tiny pegs). Also, relative to its putative appearance in life, certain researchers have attributed to it a somewhat phalanger-like head (referencing its ancestry) rather than an entirely cat-like one.

Nevertheless, its overall outward form would have been sufficiently feline for one to assume quite reasonably that non-zoological observers of a living specimen might well compare what they had seen to some form of large cat, such as a tiger if it were striped. In addition, the reconstruction of *Thylacoleo*'s likely appearance in life presented here does recall eyewitness descriptions of the yarri. And if the yarri is indeed a living *Thylacoleo*, this would explain why its teeth have attracted such attention from eyewitnesses and have been likened to tusks. Not only is this an accurate description of them, but by being incisors they would be present at the front of the animal's jaws, and would therefore be more readily visible than even the large but laterally-sited canine fangs of placental cats.

One modern-day yarri report that, if genuine and accurately recalled, would seem to confirm beyond any doubt that this feline Australian cryptid is indeed a marsupial was documented by Rex Gilroy in his extensive book on Australian cryptozoology, *Out of the Dreamtime* (2006). During November 1990, fishermen Jim Spriggs and Tony Banks were on a fishing holiday in the Moreton area, on the Wenlock River, in Queensland's Cape York Peninsula. After camping there one evening, they were woken early the following morning by loud clattering sounds that they soon discovered were the result of a strange animal rummaging through their cooking utensils that had been left outside their tent. It had grey fur with black body stripes, but of great significance was the alleged observation by both men that it possessed a reversed pouch containing a youngster inside—hence it was clearly an adult female marsupial of some kind. Once the

Thylacoleo carnifex model owned by Australian cryptozoologist Rebecca Lang, created by Sean Young, and constructed/painted by Jeff Johnson (© Rebecca Lang/Sean Young/Jeff Johnson)

men appeared, the animal ran off into the trees. It seemed to them to be somewhat dog-like but with feline aspects too, especially its head and its pricked ears. It was roughly 5 ft long, about 1.5 ft tall, rather thickset in build, and its tail looked a little bushy.

As recently as 2008, the answer to the long-standing mystery of whether the yarri and *Thylacoleo* were indeed one and the same entity may have finally been forthcoming, but unfortunately it was not to be. Paul, one of the two eyewitnesses present (publicly identified only as Paul and Jennifer van H. . .), reported a most intriguing discovery to Australian cryptozoologist Paul Clacher, who posted his statement as follows in the Big Cats section of his website *The Fossickers Network* (at http://uqconnect.net/~zzpclach/index.htm):

> 20100807—Pearson's Lookout (NSW) Sighting—December 2008 around 4:00 pm Reported 07/08/2010

Hi Paul,
In December 2008, I was travelling south on the Castlereagh Hwy near Pearson's Lookout between the towns of Capertee and Ilford on the way back to Sydney after an overnight delivery trip through Orange, Dubbo and Mudgee in an 8 tonne rigid truck. No bonnet so I have unobscured vision. This is on the edge of the Capertee Valley and is about 900—1000 metres asl. The dropoff to the LHS (east) is steep to the valley. Late in the afternoon, say around 4.00 pm and about 200m south past the lookout, I noticed 2 X roadkills on the southbound side just on the shoulder of the road.

Doing about 80km/h, I had a 5 or 6 second look at a small dead roo and something else. The markings on the torso of the other animal were dark brown / black and the main colour was tan. The markings made me look closer and the carcass was intact. The ears were rounded, the head was stout and like a lion cub and the front paws were huge in comparison to its body size. The back paws and tail were obscured because of the position it landed in after being run over (probably feeding on the small roo). My first thoughts were of a small lion, but the dark marking's threw me. It was a thick set animal about 500—600mm long. For the rest of the trip to Sydney (2.75 hrs) I couldn't stop wondering what this thing was, and having told the story to several people, I still couldn't come up with a logical explanation.

The story in today's *Daily Telegraph* regarding the Blue Mountains Panther made me check out for more info on the net regarding Australian big cats / ferals, and my wife and I came across your page and pics of the Thylacoleo model at the Mt Isa Riversleigh Fossil Centre. This is what I believe to be the same animal, however it may have been juvenile due to its proportions. If this was an Australian Marsupial Lion, believed to be extinct for how many years, how many similar species are out there that are believed to be mythical? The abundance of roos and wallabies in this area is phenomenal, so a foodsource for a carnivorous predator to thrive in the Capertee Valley is probably salubrious.

As I was driving on a log book, I couldn't stop as I would have run over my driving hours, and didn't really consider the possible importance of the sighting. In hindsight, I wish I had turned around and stopped at the lookout and walked down to have a better look and at least taken photos or thrown the carcass on the truck for an expert to identify.

Personally, I am sufficiently convinced by the correspondence between *Thylacoleo* and the yarri to consider that if the latter does exist, it is indeed most likely a living species of pouched lion. Perhaps, for instance, it is a surviving representative of *T. carnifex* itself (although conceivably a dwarf form, as the yarri seems smaller); or it might be a related modern-day species not presently known from fossils, and possibly sporting an even more feline, rounded face than that of *T. carnifex*. (Back in the 1950s, incidentally, Australian philanthropist and businessman Sir Edward Hallstrom, patron of Taronga Zoo in Sydney, actually offered a £1,000 reward for anyone who captured a Queensland tiger that proved to be a living *Thylacoleo*.)

So many reports of yarris emanate from localities close to expanses of rainforest that it seems reasonable to assume that this remote, rarely-visited habitat is their natural terrain, from which they periodically emerge in search of prey if conditions necessitate. Most fossils of *T. carnifex* date between around 1.6 million and 50,000 years old, and some palaeontologists believe that it had died out before humans reached Australia.

Conversely, various other palaeontologists favour a different scenario—that the arrival in Australia of humans who changed great swathes of this continent's once-grassy landscape to one of desert scrub, thus depriving many large native mammalian herbivores of their diet and, as a result, starving carnivores like *Thylacoleo* that preyed upon those herbivores, brought about its extinction. However, this fails to take into account that humans also introduced large non-native mammalian herbivores, such as sheep, goats, and cattle, that could survive in this changed environment and would therefore provide very suitable replacement prey for *Thylacoleo*. In addition, the rainforests in the heart of yarri territory have remained virtually unchanged for the past 30,000 years, and are well-stocked with wallabies and other large native herbivores, so there is no good reason why they could not have supported a self-perpetuating *Thylacoleo* population into modern times.

Moreover, there is some fascinating inconographical evidence indicating that *Thylacoleo* may actually have survived until as least as recently as 6,500 years ago. In 1978, anthropologist Dr John Clegg documented an ancient aboriginal cave painting in the Northern Territory's Arnhem Land Plateau portraying an animal that could possibly be a striped cat. And in 1984, Drs Peter Murray and George Chaloupka conceded that this example and also one other painting here may indeed depict *Thylacoleo*.

In any event, a niche for a large feline mammal is still present within the Antipodean ecosystem. Judging from numerous reports of mysterious tawny or black 'big cats' reported from many parts of Australia during the past century, however, even if the yarri does exist it is having to share this niche with a motley assortment of escapee puma and black panthers (melanistic leopards), as well as some remarkably sizeable feral domestic cats—which begs a somewhat disturbing question. Could the real reason for the yarri's apparent decline in recent times be competition with out-of-place placental cats?

It is said that history never repeats itself. Let us hope that this is true. How terrible it would be if the decline and fall of the mainland thylacine at the hand (or paw) of the dingo is being—or has already been—reprised with a living, undiscovered species of marsupial lion, but this time by way of feral mega-moggies, pumas, panthers, and other exotic feline pets abandoned by their owners.

Mystery Lions of the Blue Mountains

Speaking of out-of-place felids, one or more African lions on the loose is the conservative explanation generally offered by naturalists when faced with the enigma of the Blue Mountains' maned mystery beasts. West of Sydney, New South Wales, the Blue Mountains have long been associated with rumours and reports of huge cat-like beasts of ferocious temperament. Interestingly, they are not confined to modern reports but were well known to the aboriginals who once inhabited this range. They called them warrigals ('rock dogs')—a name sometimes applied nowadays to the dingo, but it is clear from their descriptions that their warrigals were something very different from a dingo.

Rex Gilroy has made a detailed study of the Blue Mountain lions. According to his accounts of these animals included in one of his articles (*Nexus*, June-July 1992) and in his cryptozoology book *Out of the Dreamtime* (2006), the aboriginals described them as 6-7 ft long, around 3 ft high, with a large cat-like head, big shearing teeth that protruded from their jaws, brown fur (sometimes light, sometimes dark—sexual, or age, differences?), and a long shaggy mane. Testifying to these animals' continuing presence here, this is also an accurate portrait of the terrifying leonine beast that approached three young shooters in the Mulgoa district south of Penrith, close to the Blue Mountains' eastern escarpment, one day in 1977—fleeing into nearby scrub only when the alarmed trio fired at it. A similar creature had been reported from this same region in 1972, where it had allegedly been killing sheep.

Back in April 1945, a bushwalking party clambering down Mount Solitary's Korrowal Buttress made good use of their binoculars to watch four warrigals moving across Cedar Valley. And in 1988, some campers near Hampton, west of Katoomba, saw one for themselves—this area had been experiencing some severe cases of cattle mutilation, a feature that crops up time and again when charting sightings of warrigals.

Based upon the longstanding history of these animals, I find it difficult to believe that they could be escapee lions or suchlike. An undiscovered native species would seem to be a more tenable explanation—echoed by Gilroy, who proposes that the warrigal is a surviving species of pouched lion. As it differs markedly in appearance from the yarri, however, we can only assume that if Gilroy's hypothesis is correct and if the eyewitness accounts of such animals are accurate, there are two separate species of pouched lion currently prowling various portions of Australia's wildernesses. This is a remarkable concept, but nonetheless it would appear to be the only one that offers a satisfactory conclusion to this extraordinary saga.

Devil-Pigs and Diprotodontids

One of the most unlikely cryptozoological creatures on file is the New Guinea 'rhinoceros'. On 28 January 1875, *Nature* published the following letter from Alfred O. Walker:

> Lieut. Sidney Smith, late of H.M.S. *Basilisk*, reports that while engaged in surveying on the north coast of Papua, between Huon Bay and Cape Basilisk, being on shore with a party cutting firewood, he observed in the forest the "droppings" (excrement) of a rhinoceros in more than one place, the bushes in the neighbourhood being also broken and trampled as if by a large animal. The presence of so large an animal belonging to the Asiatic fauna in Papua is an important fact.

However, scepticism on the part of the *Nature* editors concerning this alleged fact was sufficient for Walker to submit an expanded account, published in the journal's next issue:

> The heap of dung first seen, which was quite fresh (not having apparently been dropped more than half an hour), was so large that it excited Mr. Smith's curiosity, and he called Captain [John] Moresby to see it. Neither of them knew to what animal to assign it. Quantities of dry dung were afterwards seen. Shortly afterwards, the *Basilisk* being at or near Singapore, Capt. Moresby and Mr. Smith paid a visit to the Rajah of Johore, who had a rhinoceros in confinement. Mr. Smith at once observed

> and pointed out to Capt. Moresby (who agreed with him) the strong resemblance between the dung of this animal and that they had seen in Papua.

Moresby was sufficiently intrigued to include his own account of this incident in his book *Discoveries and Surveys of New Guinea and the d'Entrecasteaux Islands* (1876), and the droppings do seem to have been much more substantial than those produced by a feral domestic pig—the only *known* ungulate (hoofed mammal) of note on New Guinea (droppings from carnivorous mammals are very different). Even so, the concept of an undiscovered form of rhinoceros existing there continued to be rejected, and by such acknowledged authorities as naturalist Luigi d'Albertis and former Papuan governor Sir William MacGregor. More recently, the reputed occurrence on New Guinea of the Javan rhino *Rhinoceros sondaicus* was similarly dismissed in a terse 1960 paper by Russian researcher Prof. Wladimir G. Heptner—since when this entire subject has faded into obscurity.

Yet it cannot be denied that *something* was responsible for the conspicuously large animal droppings spied by Smith and Moresby. So if, as seems certain (for basic zoogeographical reasons, let alone any others), there is no justification for believing that New Guinea boasts any species of rhinoceros (unknown or known), some other explanation must be sought. Today, the mainstream suggestion is that the mystery droppings must have been from Bennett's cassowary *Casuarius bennetti*, a modest-sized species of ratite bird native to New Guinea. However, there is also a very intriguing cryptozoological alternative as the identity of their source—namely, a mysterious creature graphically referred to by one very startled eyewitness as a "devil-pig"!

In a letter to *Nature* of 4 February 1875, and published alongside Walker's second communication, German zoologist Dr Adolf Meyer discounted any likelihood of a rhinoceros in New Guinea, but documented a personally-received account of another unexplained mammal there, one that even today still awaits scientific investigation. Meyer wrote:

> I beg leave to mention a report of a *very large quadruped* in New Guinea, which I got from the Papuans of the south coast of the Geelvinks Bay . . . when hunting wild pigs along with the Papuans, they told me, without my questioning them, of a *very large pig*, as they called it, fixing its height on the stem of a tree at more than six feet. I could not get any other information from them, except that the beast was very rare, but they were quite precise in their assertion. I promised heaps of glass pearls and knives to him who would bring me something of that large animal, but none did.

Meyer never did succeed in ascertaining this mysterious mammal's identity. Nor did Rev. Samuel Macfarlane, who claimed to have encountered buffalo-like tracks in the Papuan jungle during a 150-mile voyage with Luigi d'Albertis along the Fly River there in 1875. Just over 30 years later, however, a dramatically close encounter took place, with what would seem to have been *two* such beasts, during a major expedition led by Papuan pioneer Captain Charles A.W. Monckton to Mount Albert Edward in spring 1906.

On the fateful day in question, 10 May, Monckton sent on ahead two of his party's native members, Village Constable Oina and Private Ogi, to determine whether a track spotted during the previous day was negotiable. When they did not return, the party became anxious, and some more of its members were dispatched beyond the camp's perimeter to fire shots, in the hope that they would be answered by the two missing men. After a time, the shots were returned, and later Oina and Ogi were brought back, whereupon Ogi was found to be suffering from exhaustion and shock brought on by an astonishing experience.

Ogi asserted that during their investigation of the track, he and Oina had somehow become separated from one another, and while attempting to locate Oina again, Ogi had come upon two extraordinary beasts feeding upon a patch of grass. They were so bizarre in appearance that the petrified Ogi felt sure that they were pig demons or "devil-pigs"! According to his description, each of these 'devil-pigs' was approximately 5 ft long, and 3.5 ft high, had a horse-like tail, cloven feet, a very dark skin bearing a pattern of stripe markings, and a long snout. This description does not match that of any species currently known to inhabit New Guinea. Ogi tried to shoot one of these curious creatures, but whether through cold, fright, or both, his hand shook so much that he missed. What happened after that is unclear—whether Ogi chased the animals or whether they chased him is unknown—but eventually Oina discovered him in a pitiful state of shock, and took him back to camp.

Monckton was very perplexed concerning this incident, but when he documented it in his book *Last Days in New Guinea* (1922) he affirmed that the tracks of a very large cloven-footed animal had indeed been found on Mount Albert Edward. He also noted that Sir William MacGregor had mentioned a sighting of an unidentified long-snouted animal in his report of the Mount Scratchley Expedition. Even so, not everyone was convinced about the existence of the devil-pig, and the more cynical sceptics later nicknamed it 'Monckton's gazeka' (see later for this name's derivation).

Yet in view of the corroboration of Ogi's report by the Mount Scratchley sighting, by Meyer's information, and by Monckton's first-hand knowledge of cloven-footed tracks in the area (not to mention, of course, the extra-large excrement observed by Smith and Moresby), it would evidently be unwise to dismiss the devil-pig dossier out of hand. Moreover, in 1910 Walter Goodfellow learnt of a similar beast from the natives in the vicinity of the Mimiko River while participating in a British Ornithologists Union expedition in what was then Dutch New Guinea (nowadays Irian Jaya—Indonesian New Guinea). This in turn was the

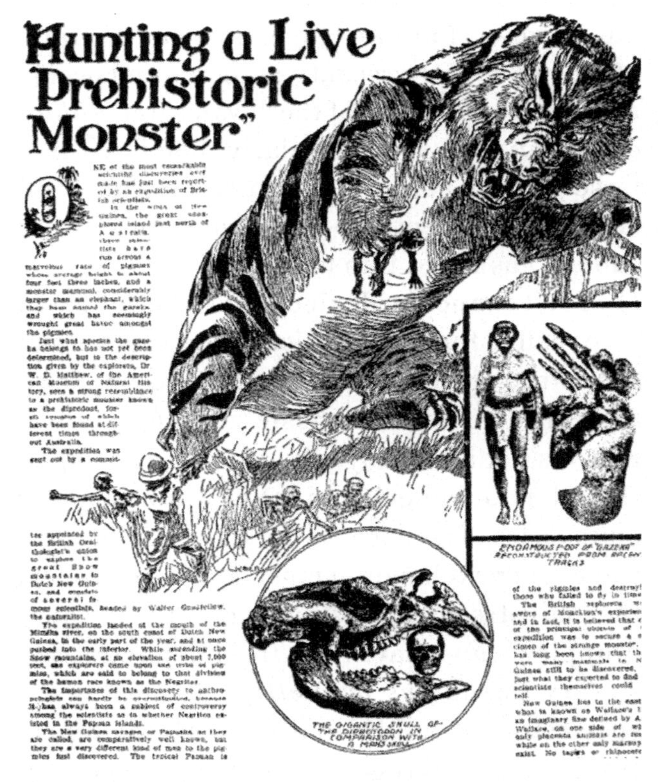

"Hunting a Live Prehistoric Monster"

Lurid depiction of the New Guinea devil-pig as an implausibly-rapacious, elephant-sized carnivore—from a highly-imaginative report in the *Stevens Point, Wisconsin, Gazette* in 1910

inspiration for a hilariously over-blown and exceedingly exaggerated account published later that same year in an American newspaper, the *Stevens Point, Wisconsin, Gazette*.

Interestingly, there is even a native New Guinea myth regarding a monstrous devil-pig, as documented by Monckton in a later book, *New Guinea Recollections* (1934). And it hardly need be said that the chronicles of cryptozoology are packed with folktales of unlikely creatures that have nonetheless eventually revealed themselves to be real after all.

So if the devil-pig does exist in reality as well as in reverie, what could it be? It is certainly not a feral domestic pig—for although there are many all over New Guinea, none attains the size of Ogi's creatures. Overall, it bears an intriguing similarity to the tapirs—those odd-looking ungulates that resemble the bewildered outcome of an unlikely liaison between a large pig and a small elephant, yet are most closely related to horses and rhinos. Of the five modern-day species, four are confined to the Americas, and the fifth exists only in Asia (see also later in this chapter).

The devil-pig's general size compares with the smaller species of New World tapir, as does its dark colouration (though its patterning of stripes is reminiscent of juvenile specimens), and in particular its long snout. Conversely, its horse-like (perhaps bushy?) tail is not a tapir feature; moreover, as tapirs are perissodactyls (odd-toed ungulates), the devil-pig's cloven (even-toed) feet pose something of a problem too. A much greater difficulty to be faced when attempting to ally it with tapirs, however, is its actual presence on New Guinea—because the fauna of this vast island is predominantly Australian, not Asian, in nature. That is, virtually all of New Guinea's native mammals are marsupials, thus severely limiting any possibility that the devil-pig is a tapir. This same argument also applies to comparisons that Monckton sought to draw between Ogi's beasts and those unusual wild pigs known as babirusas, native to the southeast Asian island of Sulawesi (Celebes) and outlying isles.

During their long period of evolutionary isolation, however, when Australasia's marsupials radiated to yield morphological and ecological counterparts to many mammalian types found elsewhere in the world, the niches occupied outside Australasia by the more robust, lumbering types of placental hoofed mammal (such as rhinos, oxen, and pigs) were for a time filled here by a taxonomic family of sizeable herbivorous marsupials called diprotodontids.

There were a number of different types of diprotodontid—ranging from the porcine *Zygomaturus* and bovine *Diprotodon* to the horned 'marsupial rhinoceros' *Nototherium* and, of especial significance to the devil-pig case, the forest-dwelling *Palorchestes*. Some palorchestids were of comparable size and build to true tapirs, and were equipped with the tapir trademark, an elongated trunk-like snout and upper lip; indeed some researchers have even dubbed them 'marsupial tapirs'.

Diprotodontids are generally assumed to have been contemporary with early humans in Australia. An enigmatic cave painting, believed to be about 10,000 years old, which may be a primitive depiction of *Palorchestes*, was found in 1986 by Percy Trezise on the ceiling of a cave in Queensland's Quinkin Reserve; and a *Nototherium*-like beast is portrayed on a rock face west of Alice Springs. The latter petroglyph, which is particularly distinctive—and intriguing—was described as follows by Australian explorer/gold-prospector Michael Terry in his book *War of the Warramullas* (1974):

> Etched in a smooth cliff face thirty feet above the ground, is the outline of an

Reconstruction of *Palorchestes* (© Tim Morris)

> animal rather like a rhinoceros. It has a long, thick body, stumpy legs, an upswept tail, and a horn curving backwards from the head.

Even so, based solely upon fossil evidence, the last diprotodontids of any kind officially died out at least 11,000 years ago (with the most geologically-recent palorchestid fossils currently obtained being around 13,000 years old). For quite a long time, it was thought that despite the occurrence of land-bridges linking Australia with New Guinea during the Pleistocene, diprotodontids never migrated from the Australian mainland to this latter island. However, fossil zygomaturids have now been found in New Guinea (and in association with early human remains), as well as the weird *Hulitherium* (see later).

In a letter written to me on 13 November 1990, University of Papua New Guinea mammalogist James I. Menzies kindly provided me with the following additional information on this subject:

> We have fossil, or subfossil, diprotodonts from various places in New Guinea but few have been dated. There are also diprotodont bones in a midden deposit dated (I recall) to some 14,000 years b.p. so that diprotodonts *were* contemporary with early human settlers. These are all small (pig-sized) zygomaturine diprotodonts.

Moreover, although no fossilised palorchestid has been uncovered here so far, there is a tantalising piece of iconographical evidence lending support to the possibility that these tapir-like beasts did exist on this island—and until a relatively late date.

A series of stone carvings collected in the Ambun Valley of New Guinea's Enga Province from 1962 onwards, and no older than a few millennia at most, depict a very distinctive type of mammal. Its body is basically spherical, with

Ambun mystery beast as represented by one of the Ambun stone carvings, and depicted here on a postage stamp issued by Papua New Guinea (NB—the stamp mis-spells 'Ambun' as 'Ambum') (Fortean Picture Library)

its forelimbs and hindlimbs clasping its belly, and its well-delineated neck bears a narrow head with large ears and eyes, which tapers directly into a short trunk-like snout curving downwards and sporting at its tip a pair of flaring nostrils. This long-nosed beast has traditionally been assumed to be one of New Guinea's three species of short-spined, long-nosed, egg-laying echidna or spiny anteater (*Zaglossus* spp.). In 1987, however, Menzies offered a very thought-provoking alternative identity for the Ambun beast—a palorchestid.

As Menzies pointed out, the shape of the Ambun beast's head, with its short tapir-like trunk, prominent external ears, and large eyes, corresponds much closer to reconstructions of a palorchestid's head than to that of *Zaglossus*—which only has very small eyes, ears, and nostrils, an almost non-existent neck, globular head, and a particularly long tubular snout. So perhaps palorchestids did exist in New Guinea after all—and in view of the devil-pig dossier, perhaps they still survive here today, amid its dense, little-explored forests.

Incidentally, a longstanding source of much confusion and misinformation regarding the Papuan devil-pig is the application to it by person(s) unknown of the nickname 'Monckton's gazeka'. It is often erroneously claimed by online cryptozoological websites to be a genuine, official name for this cryptid, supposedly applied to it by English-speakers in New Guinea to honour Monckton. In reality, however, it is actually a somewhat sarcastic put-down of the latter explorer.

The name derives from an entirely fictitious mystery beast called the gazeka, which was created by English comic actor George Graves, who introduced it in the stage musical *The Little Michus* at Daly's Theatre, London, in 1905. Although all but forgotten today, so popular did Graves's creation become during the early 1900s that it even inspired competitions to produce the best illustration of what his imaginary beast may look like.

According to Graves, the gazeka had been discovered by an explorer accompanied on his travels by a case of whiskey, and who thought he may have seen it once before, in some form of dream. Clearly, therefore, whoever shortly afterwards dubbed the Papuan devil-pig 'Monckton's gazeka' was utilising Graves's then-famous invented creature to make a sly, topical dig at Monckton's expense, implying that he had dreamt up the devil-pig, possibly while under the influence of alcohol!

In reality, conversely, Monckton may have actually obtained tangible evidence for the cryptozoological gazeka's existence. In his book *Some Experiences of a New Guinea Resident Magistrate* (1920), Monckton revealed that a native from Goodenough Island (the westernmost of the three largest D'Entrecasteaux Islands in Papua New Guinea's Milne Bay Province) gave him as a present a very big, extremely old, almost circular tusk. The tusk was mounted in native money (small circular discs formed from the hinges of a rare shell), and was hung upon a sling to be worn round the neck. According to local native folklore, it was a fang from a giant snake that lived at the island's summit.

Monckton, however, felt that it resembled a wild boar's tusk, albeit of unusual shape and size, and when he later showed it to Scottish-born naturalist Sir James Hector, who was by then a major scientific figure resident in New Zealand, Hector deemed it to be from a babirusa. But as noted earlier, these strange-looking wild pigs with famously huge, jaw-piercing tusks are not native to New Guinea, and Monckton came to believe that Hector's identification of it was incorrect and that it had probably originated instead from New Guinea's

The original gazeka—a depiction of George Graves's version from 1905 as featured in a Perrier's Water advertisement

mystery devil-pig. Monckton gave this intriguing tusk to a friend, Richard Burton, from Longner Hall in Shrewsbury, England, but whether it is there today, somewhere amid the present-day Burton family's massive collection of curios (dating back to medieval times), is unknown. This is a great pity, as a knowledgeable mammalogist could surely accurately identify the creature from which it had derived.

Red 'Elephants' in New Guinea

On 4 September 2007, Loren Coleman posted on the cryptozoological website *Cryptomundo* a very unusual but highly intriguing communication that he'd received earlier that day from a Japanese reader, Shunsuke Yokota, and which in my view may have bearing upon the devil-pig. Yokota had written:

> I have read a Japanese book about cryptozoology by Mr. Tatsuo Saneyoshi. In the book the author mentioned red elephants that were sighted in the Nassau Mountains, Indonesia by two American Navy pilots on June 15, 1952.
>
> Mr. Tatsuo Saneyoshi is a well-known cryptozoologist in Japan.
>
> In his book he also states that those pilots who sighted the red elephants discovered a huge canyon, much like Grand Canyon in the U.S., at the same time they saw the elephants. It has been confirmed in 1955 that the canyon existed, but no one has seen the elephants since.
>
> The author mentions that Dr. Lawrence at University of Massachusetts commented regarding the event stating the pilots probably saw other mammals that resembled elephants, and the red color is probably either dust or the reflection of the sun light. But Mr. Saneyoshi argues that there are no mammals that resemble elephants in New Guinea in Indonesia.
>
> The author says that there are no known fossils of ancient elephants in New Guinea, but there are many found in nearby Java, including Stegodon. He theorizes the possibility of ancient elephants crossing the ocean long ago and surviving to this day, unseen by humans.
>
> I was trying to find any information about [the U.S. Navy pilots sighting] in English but [have had] no luck yet. Do you know anything about it?

Hardly surprisingly, a prospect as ostensibly bizarre as red elephants existing in New Guinea was not one that Coleman had encountered before (nor had I—certainly no such creatures are known to exist anywhere in New Guinea, the Nassau Mountains being in Irian Jaya and including some of this island's highest peaks). So he asked if any *Cryptomundo* readers were aware of any sources of information concerning these mystery beasts—a query that yielded some very interesting responses.

Some readers speculated whether the creatures might truly have been bona fide elephants, especially as one would expect American Navy pilots, trained in conducting accurate observations, to be reliable eyewitnesses. If so, however, for fundamental zoogeographical reasons it is highly unlikely that they would constitute a native elephant species. Perhaps, therefore, they could be Asian elephants *Elephas maximus* brought to New Guinea at some point in the past by Indo-Chinese colonisers, because they would be useful domestic beasts of burden, but transporting such massive creatures here from mainland Asia would not be an easy task by any means. Alternatively, might the SeeBees or similar Australian units have carried out such introductions, construction units in World War II utilising Indian, Burmese, or Sumatran elephants for moving heavy objects? Again, however, the difficulty in carrying out such introductions onto New Guinea remains a major problem to resolve with regard to such a proposal. In addition, the exceedingly mountainous, lofty terrain present in the Nassau range is hardly an ideal habitat for elephants.

In contrast, if these 'elephants' were actually some native form of large long-snouted marsupial, such as a palorchestid species, none of the difficulties posed above would apply. Moreover, they might look superficially elephantine, sufficiently so at least to confuse Navy pilots not likely to be extensively versed in detailed zoological identification (especially regarding officially-extinct forms such as palorchestids) into thinking that they had seen true elephants—particularly if their sighting was only brief and/or from some considerable distance.

Having said that, those sighted beasts' red colour seemingly distinguishes them from the long-nosed putative palorchestid already reported from New Guinea—the afore-mentioned devil-pig—but this discrepancy may be readily resolved. Elephants and other large mammals are often fond of giving themselves mud baths, rolling and wallowing in mud that is frequently red; ditto for dust baths and dust. Indeed to quote multi-award-winning Kenyan-born artist Guy Combes regarding Kenya's famous 'red elephants of Tsavo' (one of which was the subject of his spectacular oil painting 'The Rainmaker'):

> Iron oxide . . . is also one of the primary compounds of the rich volcanic soil of Tsavo and many other geological areas of East Africa. Elephants in these areas, as with any other places they inhabit, adorn themselves with the colour of the soil by continually rolling in mud and dusting themselves with dry dirt, until they become so entirely covered and 'dyed', that this appears to be their natural hue.

So perhaps the creatures seen by the pilots were not elephants but devil-pigs (and thence palorchestids?), and not genuinely red in colour but merely covered in red mud and/or red dust.

Checking online, I have learnt that Tatsuo Saneyoshi has authored several Japanese-written animal-related books, but I have been unable to determine which (if any) of the ones noted by me there contain his account of the supposed red elephants of New Guinea. As it would be interesting to identify and trace the book in question, then translate the relevant section from it into English, however, if anyone reading this present book of mine could assist in doing so, I would be extremely grateful.

The Bewildering Bunyip—An Aquatic Anomaly

Diprotodontids have been implicated in the identity of at least three other Australasian mystery mammals too—the bunyip, the gyedarra or kadimakara, and the yahoo—of which the most famous is the bunyip. Nowadays, this term is popularly used Down Under as a colloquial noun for anything puzzling or unusual, but in cryptozoological parlance it is the specific name of an unidentified aquatic mammal widely reported for centuries from Australian lakes and rivers, particularly in the southeast.

Having said that, some examples are clearly based more upon legend or vivid imagination than upon any bona fide beast. These include such marvels as the two-headed wonder of

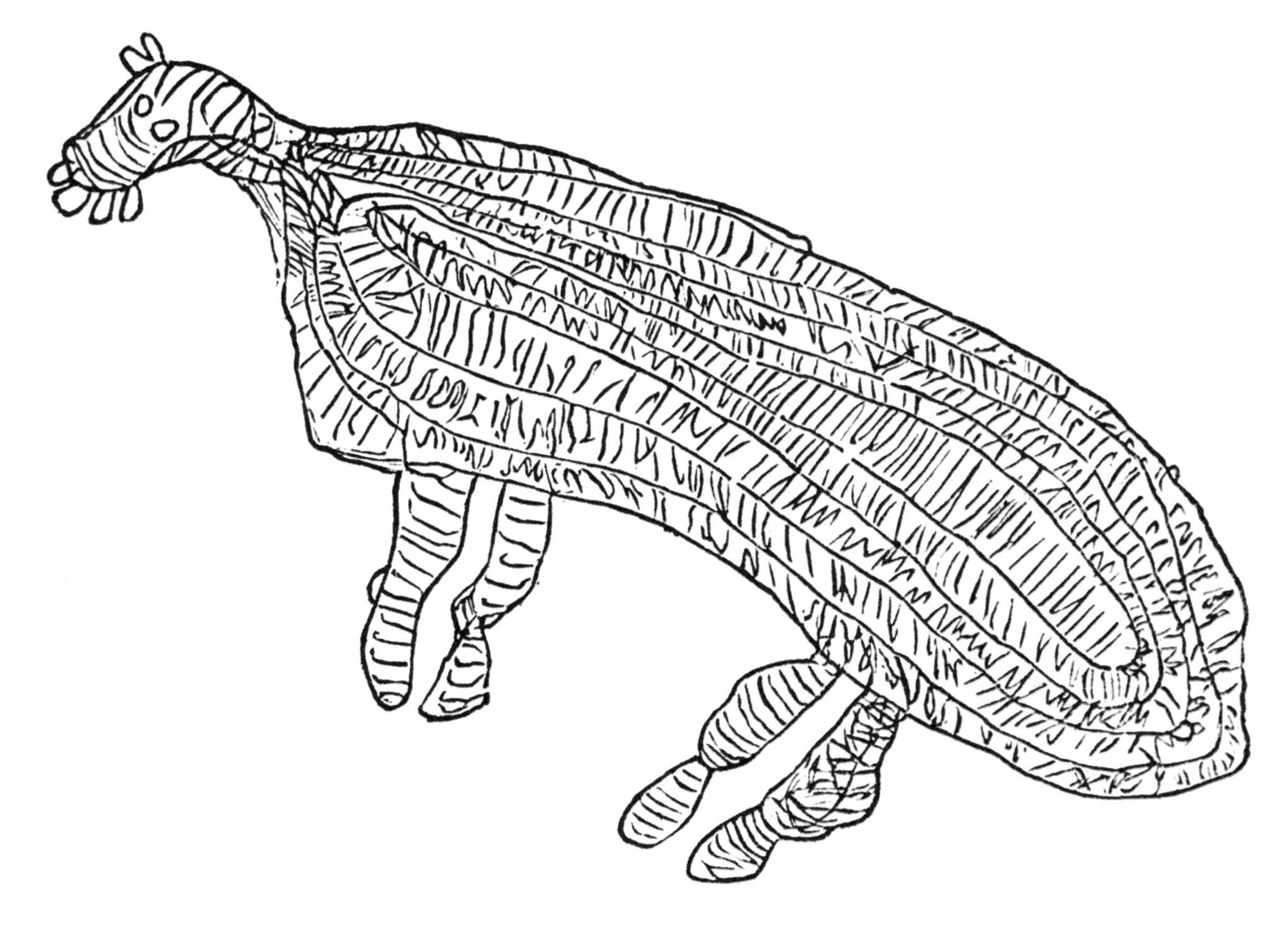

Aboriginal drawing of 1848, depicting a bunyip from the Murray River vicinity

Tuckerbil Swamp near Leeton, New South Wales, which could allegedly swim in both directions without changing gear, and was spasmodically reported until summer 1929/30; and the even more amazing three-headed bunyip, said to be 60 ft long and covered in shiny scales, that was reputedly spied in July 1972 by fishermen near the mouth of Gudgerama Creek, 200 miles east of Darwin, in the Arnhem Land Aboriginal reserve.

Fortunately for the bunyip's credibility, there are also many much more sober reports on file. One morning in November 1821, a jet-black water beast with a bulldog-like head was sighted 100 yards away in the marsh running into Lake Bathurst South, New South Wales, by farmer E.S. Hall, who also claimed that it made a noise like a porpoise. Another example with a head like that of a bulldog was a black shaggy specimen 4-4.5 ft long, spied in Tasmania's Lake Tiberias during autumn 1852; a comparable bunyip was seen in January 1863 by Charles Headham in Tasmania's Great Lake. And in 1872, the men on board a punt floating upon Victoria's Lake Corangamite were so startled by a close encounter with an inquisitive surfacing bunyip—likened by them to a big retriever dog with a round head and tiny ears—that men and punt abruptly parted company!

Sightings such as these are somewhat reminiscent of seals. Even more so was the shark-tailed creature whose head was specifically described by its eyewitness as seal-like, and which was spotted surfacing near Dalby, in Queensland, in 1873. There is little doubt that certain reports can indeed be explained as simple misidentifications of seals, plus other known animals, such as dugongs and crocodiles. Occasionally, a more exotic identity is exposed. In 1950, police unmasked the elusive booming bunyip frequenting a swamp near Wee Waa, New South Wales, as a heron-related bird called the bittern. Similarly, in 1960, the mysterious kitten-headed bunyip of Sydney's Centennial Park, with a neck like a tortoise and a tail like a porcupine's quills as described by eyewitnesses, proved to be a musk duck.

But what of the rest? How can we explain reports of emu-necked bunyips, for instance, or the very sizeable bunyip spied during 1848 in the Eumeralla River near Port Fairy, Victoria, by a shocked stockman—who described it as being as heavy as a very large bullock and brown in colour, with a kangaroo-like head, long maned neck, and a huge mouth brimming with teeth? After all, these seem more akin to the long-neck category of water monsters documented in Chapter 3 of this book than to the shorter-necked bulldog-headed bunyips recorded here.

To-scale rendition of the long-necked bunyip category (© Connor Lachmanec)

A detailed article on long-necked bunyips by Australian cryptozoologist Gary Opit appeared in the December 2001/January 2002 issue of *Nexus*, in which he referred to this bunyip category as the true bunyip (dismissing the dog-headed forms as seals), and noted that it has a variety of native names, including the katenpai, kinepratia, and tanatbah reported from New South Wales's Murrumbidgee River, and the tunatpan from southern Victoria's Port Phillip district. As for the possible identity of long-necked bunyips, Opit cautiously suggested that a late-surviving species of palorchestid may be responsible.

Also in need of an identity is the pale-coloured specimen "with the face of a child" and a body the size of a large dog that horsemen showered with stones while riding across Canberra's Molonglo River in 1886.

If only scientists could receive some physical remains of bunyips for study. In fact, at least one bunyip skull may have been made available—only to be wrongly identified?

In 1846, naturalist William S. Macleay identified a fragment from a supposed katenpai bunyip skull—lately found by Atholl Fletcher on the banks of the Murrumbidgee River and independently claimed to be from a bunyip by several aboriginal groups—as possibly that of a young camel or a deformed colt. And in a sight-unseen pronouncement, London's Prof. Richard Owen breezily dismissed it as that of a calf. As there is an appreciable degree of difference between a calf's skull, a camel's, and a colt's (deformed or otherwise), it is odd that three different identities were proffered for the same skull—thus suggesting that it was by no means readily identifiable with a known species. This intriguing specimen was deposited in the Colonial Museum of Sydney (now the Australian Museum), and attracted great crowds of viewers when first displayed there, but in best bunyip tradition it later vanished.

As far back as 1924, Dr Charles W. Anderson of the Australian Museum had suggested that stories of the bunyip could derive from aboriginal legends of the extinct diprotodontids. This is a view reiterated more recently by Australian zoologists Drs Tim Flannery and Michael Archer, in the multi-contributor volume entitled *Kadimakara* (1985), nominating the palorchestids as plausible candidates.

However, these latter creatures' tapir-like snouts conflict with the dog-like heads frequently described by bunyip eyewitnesses—but comparisons with *Diprotodon* yield a greater correspondence. Nevertheless, until a specimen becomes available for study (and preferably a complete one, to avoid the type of ambiguous identification meted out to the Murrumbidgee skull), reconciling the bunyip with *any* type of diprotodontid necessarily remains an exercise in speculation.

Gyedarra/Kadimakara—Diprotodontids in the Desert?

According to the aboriginals of Central Australia, this vast region's harsh, inhospitable desertlands were once home to a huge grass-eating quadrupedal beast known variously as the gyedarra or kadimakara. No such beast is known to modern-day zoology, but back when this land was less arid, its wetlands were inhabited by great herds of the rhinoceros-sized *Diprotodon*. As their lakes and swamps dried out, however, huge numbers died of starvation and dehydration, so that today their fossilised remains are abundant here.

The most famous example is that of Lake Callabonna, in northeastern South Australia. Once encircled by lush vegetation, it was a favourite feeding haunt of *Diprotodon*, which grazed there in great numbers. About 17,000 years ago, however, the climate became warmer, and the lake began to shrink. Hundreds of these placid but cumbersome beasts died of thirst and hunger in the parched terrain, or became so weak that they were trapped by the thick sticky mud that had once been cool water. During two expeditions to this locality, in 1893 and 1953, as many as a thousand complete skeletons of *Diprotodon* were unearthed.

When an aboriginal from the Gowrie water holes was shown diprotodontid bones by palaeontologist Dr George Bennett from Sydney's Australian Museum in the early 1870s, he claimed that they were from the gyedarra, a beast as large as a heavy draught horse, and stated that his forefathers had recalled seeing them in deep water-filled holes in this area's riverbanks, from which they would emerge only to feed. As orally-preserved traditions rarely survive longer than a few centuries without becoming greatly distorted, and as the description given here compares well with the anticipated appearance and lifestyle of *Diprotodon*, this animal may have persisted into much more recent times than palaeontologists currently acknowledge.

Hamilton Hume (1797-1873) was one of Australia's most eminent, successful explorers. During his several major expeditions, he made such notable, famous geographical discoveries in New South Wales as Lake Bathurst and the Goulburn Plains during an 1817-18 expedition with deputy surveyor-general James Meehan, and the Darling River (the Murray River's longest tributary) in early 1829. No less notable (at least from a cryptozoological standpoint) but far less famous, conversely, was a claim made by him in 1821 that back in 1818 he and Meehan had discovered at Lake Bathurst some very

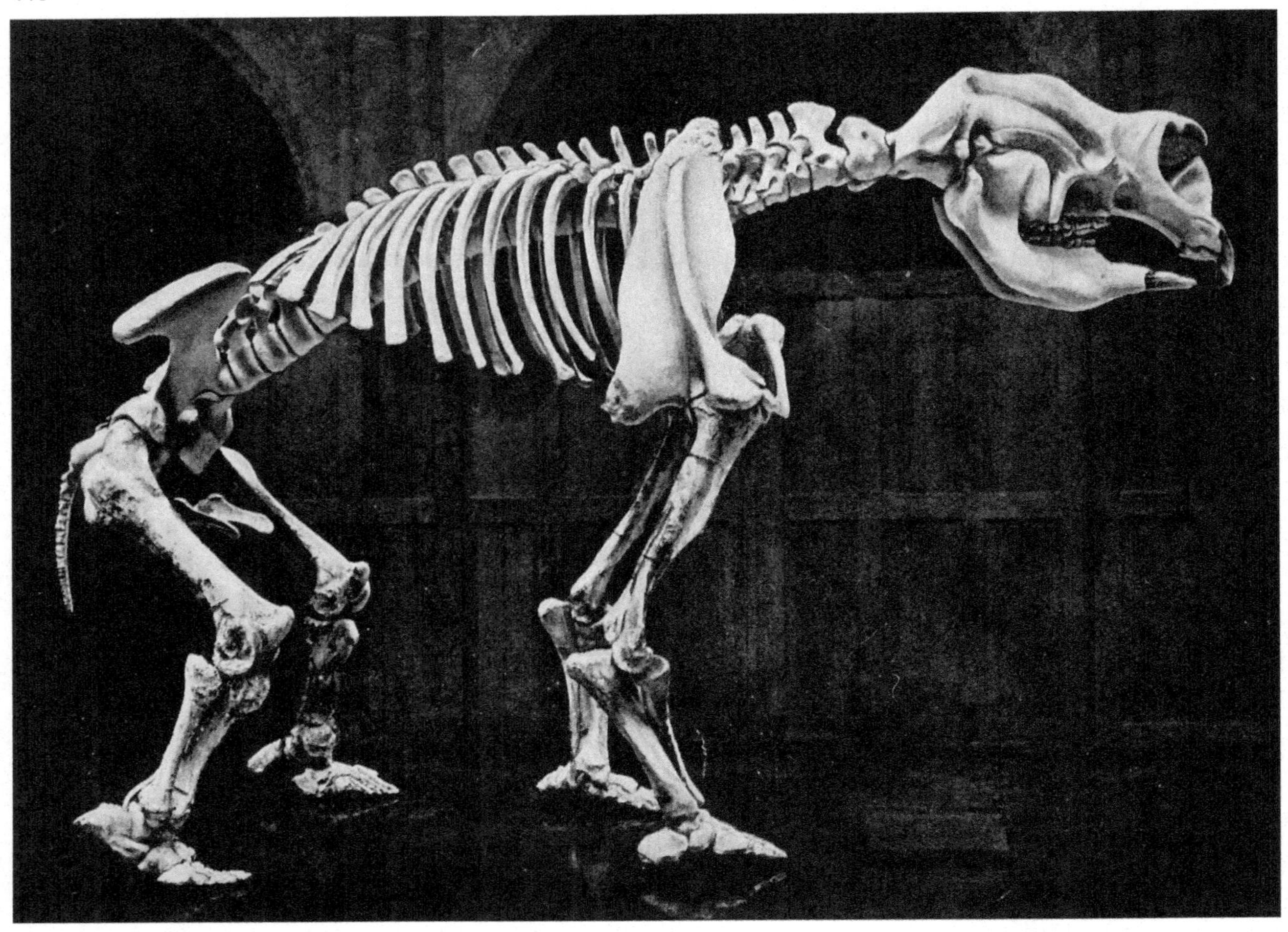

Diprotodon skeleton depicted in 1910 print

large bones, which in his opinion looked as though they had originated from "some sort of manatee or hippopotamus". However, he did not bring any of these bones back with him, and there is no clue as to whether they were of recent age or were fossilised.

Nevertheless, the recently-established Philosophical Society of Australasia was sufficiently intrigued by his claim not only to chronicle it in its minutes for 19 December 1821 but also to offer to reimburse any expenses that Hume may incur if he would return to the lake and procure any tangible evidence of this creature's reality, such as some of the bones that he had previously found there, or even capture a living specimen. Sadly, however, he never did go back. Could this "manatee or hippopotamus" have been a recently-deceased, modern-day diprotodontid? To a non-zoologist like Hume, sturdy diprotodontid bones may well have reminded him of a hippo's. Interestingly, as already noted earlier in this section, during November 1821 a different observer spied in a marsh running into this same lake a jet-black, bulldog-headed water beast. This was popularly deemed to be a bunyip, but whatever it was, its description does not recall a manatee or a hippopotamus—or indeed a diprotodontid for that matter.

During the 1840s, German explorer Dr Ludwig Leichhardt (1813-?1848) investigated rumours of *Diprotodon*-like beasts existing amid

Western Australia's arid interior after he discovered some fossil teeth from these giant mammals along with bones from other megafauna at Isaac's Station during his scientific explorations of the Darling Downs in 1844. In a letter that he sent to his friend John Archer, a settler, on 14 March of that same year, Leichhardt wrote:

> It seems there was a time when gigantic Kangaroos of the size of a bullock or of a rhinoseos [sic] gambolled over the downs and who knows whether such strange things do not still exist in the tropical interior.

But after setting off in 1848 from the Condamine River on what was planned to be a 2-3-year exploratory expedition to reach the Swan River, during which he intended to pursue such matters further, Leichhardt's expedition vanished without trace. The last recorded sighting of him was on 3 April 1848 at McPherson's Station, Coogoon, on the Darling Downs. Almost 170 years later, the reason for Leichhardt's disappearance and that of the other six members of his party (together with all of their horses, mules, and meat-providing cattle) remains a mystery, but recent investigations indicate that they perished somewhere in the Great Sandy Desert of this vast island continent's parched, forbidding interior.

Certainly, Australia's dry heartland is among the world's most environmentally-hostile (and therefore least-explored) areas, but who can say whether some diprotodontids gradually adapted to the changing habitat here, acquiring the capability to withstand the harsh new regime imposed by the deserts' formation? Indeed, as pointed out in *Kadimakara* by Victoria Museum palaeontologist Dr Tom Rich, *Diprotodon* may actually have owed its successful existence in the late Pleistocene to the spread of arid conditions in Central Australia during this period.

If they did adapt, there seems no reason why they could not have survived into the present day, rarely spied even by the aboriginals and undisturbed by Western settlers.

The Yahoo—a Marsupial Ape?

Australia's longstanding isolation from other continental land masses must surely make it one of the last places on Earth where we might expect to find reports of undiscovered primates. Yet at least two different types appear in the cryptozoological chronicles, and are often confused with one another. One is an alleged ape-man or man-beast called the yowie. The other seems to be more ape than ape-man, and is known as the yahoo.

The yahoo's leading investigator is Canberra scholar Graham C. Joyner, who believes that its name is derived from the orang-utan, which was also termed the yahoo in England when the first specimens to reach that country arrived during the early 1800s. This in turn suggests that the yahoo is indeed ape-like—but is it actually an ape? Excluding the highly unlikely possibility that as long ago as the 19th Century's onset a sizeable number of escapee apes from captivity were roaming the Australian countryside, and recalling Australia's virtual monopoly of marsupials among its contingent of native mammals, it is far more likely that the yahoo is a marsupial that resembles an ape—but today's array of pouched mammals are noticeably bereft of anthropoid counterparts, so what could it be?

One of the most striking yahoo reports was published in 1912 by the *Sydney Sun* newspaper, documenting surveyor Charles Harper's observation, made via campfire light, of a large and very bizarre beast in the New South Wales bush. According to Harper:

> Its body, legs, and arms were covered with long, brownish-red hair . . . but what struck me as most extraordinary was the apparently human shape, but still so very different. . . . The body frame was enormous, indicating immense strength. . . . The arms and forepaws were extremely long and large, and very muscular. . . . All this observation occupied a few minutes while the creature stood erect, as if the firelight had paralysed him. After a few more growls, and thumping his breast, he made off, the first few yards erect, then at a faster gait on all fours through the low scrub. Nothing would induce my companions to continue the trip, at which I was rather pleased than otherwise, and returned as quickly as possible out of reach of Australian gorillas.

Although the chest-beating activity is certainly reminiscent of a male gorilla's warning behaviour, the rest of the description offers scant similarity to this species, or indeed to any other mammal currently known to science. Other reports, generally consistent with this description, furnish additional details—collectively portraying a nocturnal beast principally quadrupedal but capable of standing erect on its hind legs and thereby attaining a height of 5-6 ft; covered in dark fur upon much of its torso but lighter on its neck, limbs, and belly; equipped with very long, muscular arms and elongated toes; but lacking a tail.

From this reconstruction, Joyner initially sought to identify the yahoo as an undiscovered, surviving species of giant wombat. One pertinent example is the mighty *Phascolonus gigas*, standing 3 ft at the shoulder and probably resembling a small bear, whose fossils have indeed been found in southeastern Australia but which officially died out during the Pleistocene's last glacial period. In addition, eyewitness reports of non-yahoo cryptids that also apparently resemble giant wombats have emerged from several regions of Australia, according to Rex Gilroy, who has documented a selection in his book *Out of the Dreamtime* (2006). As revealed by American cryptozoologist J. Richard Greenwell, however, thoughts regarding the yahoo subsequently turned toward a far more intriguing candidate.

A wombat—could the yahoo be a giant form? (CC Yvonne81)

In 1986, a radically new species of Pleistocene diprotodontid with a highly domed head and a markedly short muzzle was formally described by Drs Tim Flannery and M.D. Plane from some 38,000-year-old fossils discovered in New Guinea. What made this particular species, the mountain diprotodontid *Hulitherium tomasettii*, so interesting was that it eschewed the lumbering, weighty body form characterising most diprotodontids in favour of a smaller and much more agile skeletal construction, with extremely mobile limbs.

The result of this dramatic deviation from the typical diprotodontid blueprint was a creature that would have borne more than a passing external resemblance to a bear—or an ape (two taxonomically dissimilar creatures that nonetheless are frequently confused by eyewitnesses—a recurring motif in bigfoot and yeti history, for instance). Indeed, as suggested by Flannery in *The Antipodean Ark* (1987), in terms of its ecological niche *Hulitherium* can be looked upon as New Guinea's answer to Africa's gorilla, South America's spectacled bear, and Asia's giant panda.

Hulitherium would have been perfectly capable of rising up, and even walking for a short time, on its hind legs, just like a bear or an ape—and just like the creature spied by Charles Harper. Perhaps the geographical range of *Hulitherium* stretched beyond New Guinea, to include the Australian mainland. If so, a population surviving largely undetected by Western settlers could very satisfactorily explain the otherwise bewildering file of yahoo reports. And on New Guinea itself, perhaps a late-surviving, mountain-confined relict population might even offer an alternative identity for the sizeable but still-mystifying devil-pig.

'Giant Rabbits' Down Under

In *On the Track of Unknown Animals* (1958), Dr Bernard Heuvelmans referred to reports by gold prospectors in the deserts of Western Australia telling of encounters with 'giant rabbits' that could vanish at extraordinary speed, and which were sometimes specifically referred to by them as "kangaroos 12 ft high". Heuvelmans suggested that these latter could indeed be giant kangaroos, of which several different species existed until the end of the Pleistocene.

Brown University mammalogist Prof. Christine Janis later nominated the now-extinct short-faced kangaroos or sthenurines, including *Sthenurus* spp. (up to 8 ft tall) and particularly their largest member, the goliath kangaroo *Procoptodon goliah* (up to 9 ft tall), as suitable Pleistocene-dated identities for these 'giant rabbits'. In 2014, however, a *PLoS ONE* paper co-authored by Janis revealed that due to their immense weight (over 500 lb), their upright stance, plus their extremely rigid spine and tail, the sthenurines were probably only able to support themselves on one foot at a time. This in turn indicates that rather than hopping on their two hind feet like modern-day kangaroos, these huge short-faced relatives probably walked (or lumbered) on their hind legs instead, and were therefore much slower. However, such activity—or lack of it—does not correspond with the inordinately speedy 'giant rabbits' reported by the desert-exploring gold prospectors.

Having said that, in a Facebook conversation with Prof. Janis on 10 June 2015 I learnt from her that although the biggest sthenurines (like *Procoptodon*) probably couldn't hop, "the smaller ones probably used hopping for faster speeds or rapid escape) but walking for slower speeds". When I asked her what would have been the largest sthenurine that might have been able to leap as well as walk, she replied: "Something like *Sthenurus stirlingi*, which would still be twice the mass of a red kangaroo [*Macropus rufus*, the largest kangaroo species known to exist today] (around 150 kg [330 lb])". So perhaps there is hope for reconciling the gold prospectors' 'giant rabbits' with sthenurines after all?

In 1978, excavations by Sydney University anthropologists at a swamp in Lancefield near Melbourne, Victoria, unearthed fossils confirming that humans had coexisted with giant kangaroos, such as the 6-ft-tall titan kangaroo *Macropus titan*, for several millennia after arriving here around 40,000 years ago. However, it is likely that these huge yet inoffensive creatures were looked upon as an important source of meat and thus were gradually extirpated.

Yet as with the quadrupedal *Diprotodon* in relation to the gyedarra and kadimakara, if relict populations belonging to these browsing bipedal examples of Australia's Pleistocene megafauna found sanctuary from human persecution in the more inaccessible and uncongenial reaches of the Australian deserts, they could still be there today. Whether they could explain the 'giant rabbit' reports, conversely, is another matter entirely now, in light of the study by Janis *et al.* referred to here. However, the titan kangaroo was a 'true' kangaroo (i.e. of the genus *Macropus*), and although very large was still much lighter than the sthenurines, so it may indeed have bounded rapidly upon its hind legs, just like its more familiar, smaller relatives.

Restoration of *Megatherium* (GDB)

GIANT GROUND SLOTHS, AND A BEAST WITH THE BREATH OF HELL

When we think of sloths, we generally picture those famously sluggish, dog-sized, tree-dwelling beasts that spend much of their time hanging upside-down from branches in modern-day Central and South America. Millions of years ago, however, there were several additional, very different morphological types—of which the most famous and dramatic were the ground sloths.

Most of these were primarily terrestrial, some were rather bovine in appearance but with shaggy fur, and many were considerably larger than their arboreal relatives. Although principally quadrupedal, ground sloths were capable of squatting erect on their hind legs, supported by their long sturdy tails, to browse upon high-level foliage, and their distribution range included not only tropical mainland Latin America, but also North America as well as various of the Caribbean islands.

There were four separate taxonomic families containing ground sloths. The largest species were the megatheriids, typified by *Megatherium* ('big beast') from the Pleistocene of Patagonia, which attained the size of an elephant (recently split from the megatheriids into their own taxonomic family are the nothrotheriids). At the other extreme were the megalonychids, some being no bigger than a large cat, but also including the ox-sized *Megalonyx* ('big claw'), which earned its name from the huge claw on the third toe of each of its hind feet. This latter family also contains today's two-toed tree sloths.

Intermediate in size between those groups of ground sloth were the mylodontids—which are of particular cryptozoological interest. For although the last representatives of all types of ground sloth officially died out several millennia ago, reports of mysterious creatures resembling these supposedly bygone beasts have emerged from several different Neotropical locations in modern times. These include in particular some compelling evidence to suggest that Brazil may harbour a species of living mylodontid, still eluding scientific discovery yet well known to the native people sharing its secluded jungle domain, who refer to this formidable cryptid as the mapinguary. But let's begin at the beginning.

Monkey-Faced Monsters and Hairy Pangolins

Perhaps the first indication that ground sloths might still survive came from a French traveller-priest, Father André Thévet, who noted in 1558 the alleged existence in southern Patagonia and southern Chile bordering the Straits of Magellan of a strange beast called the su or succurath. The engraving accompanying his account depicted a very bizarre quadruped. Even allowing for distortion due to the prospect that this drawing was based not upon direct observation of the su by the artist but rather upon a second-hand description, the beast cannot be identified with any modern-day South American species.

Engraving of the su, from Edward Topsell's *The History of Four-Footed Beasts and Serpents*, 1658

Conversely, the mylodontids were short-faced species with a notably long tail, powerful chest narrowing to slimmer abdomen, and paws equipped with powerful claws—thus yielding a tolerably close comparison with the depicted su. Just a coincidence?

During the 1870s, a no-less eminent eyewitness than Argentina's then Secretary of State, Ramon Lista, encountered a compelling fusion of ambiguous su and anachronistic *Mylodon* while exploring Santa Cruz Province in southern Patagonia. According to his account, the unidentified beast that he and his companions briefly spied emerging from and re-entering the thick foliage through which they were travelling resembled in general shape those quasi-reptilian, insectivorous mammals the pangolins, native to Africa and Asia. Instead of possessing their unique pinecone-like scales, however, Lista's beast was covered in long reddish-grey fur—and, seemingly, an invisible suit of armour, because although they fired at it several times, their bullets did not harm it at all.

This last-mentioned characteristic is highly significant because it unifies other aspects of the ground sloth saga. Early explorers of Patagonia had been informed by the Tehuelche Indians that this great region of Argentina is reputedly home to a frightening ox-like animal with short dense fur, huge claws, and the remarkable capability to withstand arrows or bullets—but until Lista's sighting, such stories had been dismissed by science as unfounded native folklore. In reality, however, during the close of the 19th Century a hitherto-obscure cave at South America's southernmost tip unveiled what may well be the key to this mysterious invulnerability.

An engraving depicting the preserved *Mylodon* skin found in a cave at Last Hope Inlet, Patagonia, in 1895, showing the bony ossicles

The Last Hope that Almost Provided the First Hope—For a Living *Mylodon*

So far, proof of a living ground sloth had been wholly anecdotal, but then a fascinating discovery was made that offered scientists physical evidence—and, with it, their first glimmer of real hope—for the continuing existence of such a creature. Or so it seemed.

In January 1895, former German soldier Herman Eberhardt decided to explore an enormous cave called Cueva Ultima Esperanza (Last Hope Cave) situated at the back of his sheep-farm settlement at southern Patagonia's Last Hope Inlet, near the Straits of Magellan. Inside, he found a human skeleton, some non-human mammal bones, plus a dried furry skin, 5 ft long and about 2.5 ft wide, that he afterwards draped from a tree—enabling successive periods of rain to wash away the great quantity of salt heavily encrusting its upper surface.

This bore many very distinctive coarse reddish-grey hairs, but the under surface was even more distinctive, because it was embedded with numerous tiny white bones (dermal ossicles)—as if the animal from which it had originated had been encased in a coat-of-mail.

This peculiar pelt was the subject of much local discussion, giving rise to speculation that it may be a cowhide encrusted with pebbles, but Eberhardt discounted this, favouring instead the possibility that it was from the skin of an undiscovered species of seal. In the event, neither of these identities was correct. A year later, Eberhardt was visited by Swedish explorer Dr Otto Nordenskjöld, who cut off a portion of the skin and sent it to Dr Einar Lönnberg of Uppsala Museum for formal examination. In November 1897, Eberhardt received another interested scientist, Dr Francisco Moreno from the La Plata Museum, and gave him what remained of the skin (once it had become famous, several chunks had been removed by souvenir hunters). Moreno deposited it at the La Plata Museum, and identified its owner as a mylodontid ground sloth.

Similar pelts have since been found elsewhere in South America, and it is now known that dermal ossicles—located in the shoulders, thighs, and back—were only possessed by members of this particular ground sloth family. Interestingly, the protection afforded by them would have been enhanced by another mylodontid characteristic—their shield-like ribcage, constituting a series of ribs so closely aligned alongside one another that they were almost physically united.

On account of its seemingly fresh appearance, the Eberhardt skin was widely considered by scientists—most especially Prof. Florentino Ameghino from the Buenos Aires Museum—to constitute tangible proof that the mylodontids had not died out after all, that they were instead represented by a bona fide modern-day form. Furthermore, on account of its impregnable 'armour' of dermal ossicles, this elusive species was deemed to be one and the same as the mysterious bullet-proof beast spied by Lista and reported for centuries by the Tehuelches. Ameghino even went so far as to christen it *Neomylodon listai*.

Moreno, conversely, was far from convinced by this line of thought—and in a lecture to London's Zoological Society on 21 February 1899 concerning the skin, he announced that in his opinion it was not of recent age at all. On the contrary, he firmly believed it to be several millennia old, and had been mummified by freak, hermetic conditions present in the cave. (More recently, carbon-14 dating confirmed that it was around 5,000 years old.)

Even so, Moreno did concede that in 1875 an old Patagonian cacique called Sinchel had showed him a cave near the Rio Negro reputedly inhabited by an unidentified beast locally termed the ellengassen—said to resemble a hairy ox with a human head, and an armoured body whose only vulnerable spot was an unprotected region on its belly.

Moreno's lecture inspired two other excavations in Cueva Ultima Esperanza during 1899, but far from providing a clearer insight into the possibility of contemporary ground sloths, they succeeded only in inciting further controversy. The first excavation was by Dr Nordenskjöld's nephew, Erland, who uncovered a layer of *Mylodon* dung *beneath* a deposit containing the bones of llama-related creatures called guanacos that had been killed by early humans. From this, Nordenskjöld Jnr deduced that humans and *Mylodon* had both occurred within this cave, but had not coexisted here—that the ground sloths had become extinct long before humans had first entered it.

Conversely, the second excavation, conducted by La Plata Museum geologist Dr Rudolph

Hauthal and investigating a different region of the cave floor, found ground sloth dung in a layer *above* some man-made hunting implements. To him, this implied that humans had existed here *before* the arrival of ground sloths. Moreover, after finding what seemed to him to be preserved samples of chopped hay, he then boldly proposed that humans and *Mylodon* had *coexisted* here, and even that the sloths had been maintained by humans in a corral, like domestic cattle.

In short, the excavations had yielded contradictory finds that had in turn given rise to diametrically opposing theories. How could this paradox be resolved? It took many years, but in 1976 it *was* resolved—revealing a classic 'blind men and the elephant' scenario.

During that year, an English researcher called Dr Earl Saxon excavated a previously-unexamined portion of the cave floor, and discovered that it contained not one layer of sloth dung but two, sandwiched between which was a stratum of human remains. Suddenly, everything became clear. Both Nordenskjöld Jnr and Hauthal had found the human stratum, but the former researcher had only uncovered the lower sloth layer, whereas the latter researcher had only uncovered the upper sloth layer—thus explaining why their finds had seemed to contradict one another and had thereby led to the formulation of mutually exclusive theories concerning the nature of *Mylodon*-human presence within the cave.

In fact, sloths and humans had never coexisted inside it; instead, their presence here had alternated. From around 13,000 to 12,000 years ago, it had been inhabited by ground sloths; from about 5,700 to 5,600 years ago, humans had lived here; and subsequent to this, perhaps up to as recently as 2,500 years ago, ground sloths had occupied it once more. Another *Mylodon* muddle had been sorted out—and, of cryptozoological importance, in a manner providing proof of this animal's survival into more recent times than hitherto recognised.

Tragically, however, this had not been the only sloth-related source of confusion to arise at the close of the 19th Century. During that same period, Moreno's rival Ameghino had further muddied the *Mylodon* waters by claiming this animal to be the identity of another South American cryptid—an extremely vicious, aquatic beast called the iemisch or jemisch. In reality, there is little doubt that descriptions of this creature actually refer to some type of very large otter—perhaps unknown populations of the saro or giant otter *Pteronura brasiliensis*, or possibly even a still-undiscovered, related species.

During the 20th Century, there were several searches for living ground sloths. One of the most famous of these took place in 1900-1901 and featured British explorer-adventurer Major Hesketh V. Hesketh-Prichard. It was financed by London's *Daily Express* newspaper and partly inspired Sir Arthur Conan Doyle's subsequent famous novel *The Lost World* (1912). Sadly, however, it failed to emulate the success of the latter's fictional Professor Challenger, despite Hesketh-Prichard travelling 10,000 miles during his quest—as documented in his book *Through the Heart of Patagonia* (1902).

During his search, Hesketh-Prichard had been contacted by fellow explorer Edward Chace, an American who went on to spend 30 years (1898-1928) in Patagonia. Chace had informed Hesketh-Prichard that in the Andes on the western border of southern Argentina's Santa Cruz Province (where Ramon Lista had lately seen his own hirsute mystery beast), an Indian had recently followed a trail of very strange prints, as if made by a wooden shoe with two cleats across the sole, that had led him to an extraordinary creature resembling a hairy pig but as big as a bull.

In 1932, while taking part in a British Museum expedition through Yucatan and Honduras, Mayanist archaeologist Dr Thomas W.F. Gann briefly spied a strange animal reminiscent of a ground sloth amid some marshy territory close to the Rio Hondo border of British Honduras (now Belize) before it disappeared into the nearby forest, seemingly untroubled by the close proximity of the expedition's members and their horses. According to Gann's description as recounted by Harold T. Wilkins in his book *Secret Cities of Old South America* (1952), the beast trotted like a large ape, had a sizeable body covered in black shaggy fur, and a head with a white mane covering its face. According to American cryptozoologist Ivan T. Sanderson, such creatures are referred to by Belize natives as cave cows.

Moreover, in his own book *Discoveries and Adventures in Central America* (1928), Gann recalled how, many years earlier, while exploring the then-uninhabited and very wild bush, swamp, and savannah terrain near the headwaters of the Mopan River near the Belize-Guatemala border, a friend, naturalist Frank Blaucaneaux, had a very frightening experience that may have featured a ground sloth. He had sent his servant-companion, a powerful youth named Joe, armed with a rifle, into some scrub to investigate what was causing a sturdy, 20-ft-tall cuhoon palm tree to sway back and forth despite there being no breeze. Suddenly, from the direction that Joe had taken, a terrifying succession of agonised shrieks pierced the silence, and when Blaucaneaux pushed his way through the dense undergrowth to reach the tree, noticing that the bush around it had been trampled down, as if by some heavy animal, he found Joe lying fatally injured in this cleared space. His chest and abdomen had been "gouged in a series of great parallel furrows", and the left side of his face was extensively mutilated too.

Blaucaneaux deduced that these hideous wounds had been made by the claws of some very large, powerful mammal. Just before he died, Joe was able to whisper that he had found what he called the old devil himself, under the palm tree, and that this entity had ripped him up and had then run into the bush. Moreover, Blaucaneaux soon found in the vicinity of the tree what Gann, in his book's retelling of this incident, described as "a pretty obvious trail, along which some heavy and powerful animal, in its passage through the scrub, had bent and broken branches and torn off leaves and twigs". After scooping out a shallow grave with his machete beneath the tree and burying Joe there, Blaucaneaux cautiously followed this trail for some considerable distance, armed with his machete and also a Winchester rifle, as it led through the scrub and then through savannah before entering virgin forest, through which it progressed for a couple of miles, the unseen beast then turning into the dry bed of a stream, with its trail finally leading him to the mouth of a large cave in an almost vertical limestone cliff at the foot of a high hill.

Very warily entering the cave, Blaucaneaux finally saw the mystery beast's footprints, perfectly preserved in the soft brown earth on the cave floor. He described these to Gann as "almost exactly like the thumb and two first fingers of a gigantic human hand, each digit armed with a great claw". By now, however, dusk had fallen, so Blaucaneaux had no option but to retrace the trail to the palm tree while there was still some light to see by. Unfortunately, however, he became lost en route back, and had to make camp in the bush for the night, sleeping in the blanket that he had been carrying rolled-up on his back, before successfully finding a village two days later. Consequently, Blaucaneaux never did discover the identity of the creature that had killed Joe, and which he had tracked for so long afterwards. Nevertheless, the trail left by the creature, and the wounds that it had inflicted upon Joe, are consistent with what a decent-sized ground sloth's bulk, strength, and fearsome claws could accomplish.

Mário Pereira de Souza, a mine worker, claims that in 1975 he came face to face with a very large ground sloth near Brazil's Jamauchim River. The creature was emitting a foul stench and gave voice to a nightmarish scream as it lurched towards him on its hind legs, so de Souza lost no time in making his escape. And during the late 1980s, I learned from the then ISC Secretary J. Richard Greenwell that one of his colleagues had once seen a beast greatly recalling a ground sloth at a cave in Ecuador.

In 2001, Hesketh-Prichard's great-grandson, Charlie Jacoby, led a month-long, two-man expedition to Patagonia, during which he retraced his ancestor's epic search for a living ground sloth here exactly a century earlier, in the hope that this time one would indeed be found. Sadly, the quest was unsuccessful. Undeterred, however, in 2005 Jacoby returned to Patagonia, this time in his capacity as principal expert for the filming of 'Giants of Patagonia', an episode in the History Channel's documentary series *Digging For The Truth*. This particular episode, first aired in the U.S.A. in April 2006, examined the history of the ground sloths, and concluded that a species may indeed still exist today.

The Malodorous Mapinguary

The most extensive recent researches concerning surviving ground sloths are those of Dr David C. Oren, from Brazil's Goeldi Natural History Museum in Belém, who has spent many years investigating a monstrous beast from the Mato Grosso called the mapinguary (also spelt 'mapinguari').

Traditionally, this unidentified creature has been labelled by cryptozoologists as an undiscovered bipedal ape-like primate—comparable to the elusive didi of Guyana and certain other man-beasts reported elsewhere in the Neotropics.

In stark contrast, however, Oren's researches have led him to the conclusion that the mapinguary is actually a modern-day *Mylodon*—with a remarkable defence mechanism so devastatingly effective that it inspired one writer to dub this hidden creature 'the beast with the breath of Hell'! During his many field trips in Brazil since 1977, Oren has learnt much from the native peoples concerning the mapinguary, and he believes that it exists in small numbers within westernmost Brazilian Amazonia, in the states of Amazonas and Acre.

Drawing upon around 100 interviews conducted with alleged eyewitnesses among the local Indians and rubber tappers (seven of whom claim to have killed a mapinguary), Oren has compiled a detailed description of this cryptid. And in a *Goeldiana Zoologia* paper from August 1993 (followed several years later by an update), he painstakingly demonstrated how each of its major features is characteristic of a mylodontid ground sloth.

The mapinguary has reddish fur, feet that turn backwards, a miraculous invincibility to wounding from weapons (except around its navel), a monkey-like face, the height of a 6-ft-tall human when standing on its hind legs but a weight of about 500 lb, tracks that either are like those of a human but pointing backwards or are as round as a pestle or bottle bottom, a diet of bacaba palm hearts and berries, horse-like faeces, and loud vocalisations

Confrontation with mapinguaries depicted as living ground sloths (© William M. Rebsamen)

sometimes low and resembling thunder but sometimes higher and similar to a human shouting.

All fur samples of mylodontids obtained from mummified specimens have been red in colour; and based upon reconstructions from fossil evidence it would seem that ground sloths walked with their long claws curved towards the centre of the body—thereby yielding tracks that look as if they are pointing backwards. Mylodontids are the only ground sloths to have possessed dermal ossicles, and, as already noted, when coupled with their shield-like ribcage a near-impenetrable body armour would be the inevitable result. Only in the vicinity of the navel, i.e. below the bottom edge of the ribcage, where there are no dermal ossicles for protection either, would a mylodontid be vulnerable to wounding.

Although the faces of ground sloths, as reconstructed from fossil remains, do not appear very monkey-like, those of tree sloths are certainly somewhat simian, so a mylodontid species possessing such characteristics is not impossible. As for the typical image of ground sloths as huge, hulking beasts, some species were actually much smaller—indeed, the dwarf megalonychids that may still have existed upon certain Caribbean islands when the first Europeans arrived there were no bigger than large domestic cats. Hence a human-sized mylodontid is perfectly plausible, and a relatively small species would obviously have more chance of remaining undetected by science amid Amazonia's vast rainforests than would a larger one.

The backward-pointing tracks are consistent with those of mylodontids, and the strange round tracks could well be the impression of a mylodontid's long heavy tail tip, especially if using it for support when rearing up on its hind legs. Oren owns a clay mould of an alleged mapinguary footprint, about 1 inch deep, with three large toes pointing backwards, showing that whatever made the print walks on its knuckles.

The mapinguary's diet also agrees with that of ground sloths, as do descriptions by locals of the way in which it twists palm trees to the ground to obtain their heart and berry-like fruits. And preserved mylodontid faeces do resemble horse droppings.

Yet when DNA extracted from alleged mapinguary droppings given to Oren by some natives was compared with DNA extracted from a ground sloth skin obtained back in 1898 in Chile, they did not match. Instead, the droppings' DNA was found to be horse-related, suggesting that they were from a tapir.

Conversely, according to a short *Cincinnati Enquirer* report of 31 May 2001, and also documented in a Facebook post by American cryptozoological researcher Randy Merrill on 13 May 2012, geneticist John Lewis from the Society for the Search for Cryptozoological Organisms and Physical Evidence (SCOPE), was leading a six-person SCOPE expedition in Brazil's Amazon Basin throughout much of spring 2001 seeking the mapinguary when on 22 March he stepped in a pile of faeces from a very large animal, which he duly collected. When DNA extracted from the droppings was analysed, it was apparently an exact match with that of dung from extinct ground sloths. As yet, however, I have not seen any scientific confirmation of this exciting claim, but some additional background information is present on SCOPE's website (at: skunkape.veryweird.com), which although no longer directly online is still accessible (via: web.archive.org).

Finally, after studying the extremely well-developed hyoid (throat) bones of fossil ground sloths at Florida State University, Oren felt that they were probably capable of producing loud vocalisations—thereby supporting yet again the native claims for the mapinguary. Moreover, during a search for this cryptid conducted by British wildlife scientist Pat Spain for an episode of the American TV show *Beast Hunter* entitled 'Nightmare of the Amazon' and first screened on 4 March 2011, Spain blasted a loud call consisting of a slowed-down tree sloth call (thereby making it sound deeper, like the version claimed by natives for the mapinguary)—and he actually received a response, from an unseen creature but voicing a very similar call to Spain's slowed-down tree sloth call. Could the creature have been a mapinguary?

So far, Oren's identification of the mapinguary as a living *Mylodon* seems well-founded —but there are two further details disclosed by the natives that initially posed quite a problem when seeking to reconcile the former beast with the latter one. According to a bizarre but widespread claim by local hunters, the mapinguary has two mouths—one in the normal location, the other in the centre of its belly! And as if that were not grotesque enough, when threatened by capture this creature releases a foul-smelling gas redolent of faeces and rotting flesh that asphyxiates anyone approaching too closely. Apparently, three mapinguaries have actually been captured by various brave hunters in quite recent times, but all three escaped because their captors were unable to withstand the hideous stench emitted by these animals. This was also the reason given by all seven of the hunters who claimed to Oren that they had killed a mapinguary for their discarding its remains.

Seeking to explain these anomalies, Oren suggests that the mapinguary's 'belly mouth' is a specialised gas-secreting gland, whose mephitic discharge acts as an effective defence mechanism, warding off anything—or anyone—not sufficiently disheartened by its body armour to have abandoned hope of killing it.

The mapinguary may be the largest native land mammal in South America today, as well as a major prehistoric survivor, so its formal discovery and identification as a living ground sloth would be a sensational zoological event. It certainly seems to be one of the most likely creatures in the cryptozoological annals to be officially unveiled one day by science, especially if sought during the autumn, when it reputedly descends from the Andes—and provided, of course, that its would-be discoverer remembers to take along a very effective gas-mask!

Having said that: I have learnt from Dr Andrew Johns that in July 1953, a hunter confronted by an enraged mapinguary and forced to hide from it inside a fallen hollow tree had apparently succeeded in killing it when, bellowing loudly, it had reared up onto its hind legs—thus exposing its vulnerable navel, at which the hunter had taken aim with his 16-calibre shotgun after loading it with heavy shot that he normally set aside for hunting tapirs. The hunter had encountered the mapinguary deep in the jungle about 5 hours from his camp along the Rio Açaituba, in a sparsely-populated region of northern Brazil's Pará State.

After returning to his camp and telling everyone there about his dramatic experience, the hunter and some friends journeyed back to where he had killed the mapinguary and carried its carcase to the river. There, a passing boat took word of the killing to a local priest, who came and not only observed but also took photographs of it. Moreover, one of these photos, depicting the beast's head and shoulders, supposedly appeared in an issue of *Diario de Commercio*, a Manaus newspaper. But what happened to the carcase and the photos afterwards is apparently unknown. Also, I have been unable to trace this potentially highly-significant newspaper report containing the photo mentioned here, so if any Brazilian readers could do so, and make it available to me, I would be extremely grateful.

The Saytoechin or Yukon Beaver Eater

In September 1989, the then recently-formed British Columbia Scientific Cryptozoology Club (BCSCC) was contacted by a Canadian First Nation member named Dawn Charlie concerning a mysterious beast featuring in their oral traditions relating to Yukon's wildlife. The beast in question was referred to as the saytoechin (which translates as 'beaver eater'), and was described as being bigger than even the biggest grizzly bear, and feeding principally upon beavers, which it apparently captured by flipping up their lodges and then seizing the exposed beavers inside. When Native Americans living in the area were shown a book of extinct mammals, they selected an illustration of a ground sloth as the saytoechin, and the most recent reported sighting of one dates from the mid-1980s. As documented in 1990 by BCSCC co-founder Prof. Paul LeBlond in #4 of the Club's newsletter after interviewing Dawn Charlie, the details given by her concerning this sighting are as follows:

> The latest report was from Violet Johny, my husband's sister, who was fishing with her husband and her mother at the head of Tatchun Lake 4 or 5 years ago. An animal came out of the woods, 8 or

9 feet high, bigger than a grizzly bear. It was a "saytoechin" and it was coming towards them. They panicked, fired a few shots over its head and finally managed to get the motor going and took off. There are other reports. There is also a report that a white man shot one in a small lake in that area. Beaver eaters are supposed to live in the mountainous area east of Frenchman Lake.

Although ground sloths are generally thought of as tropical Latin (particularly South) American creatures, before their official extinction at the end of the Pleistocene some species had migrated northwards and had indeed established themselves in parts of North America. At least five genera are currently represented by fossils discovered in various locations here, including a single species, *Megalonyx jeffersonii*, in Yukon. So in terms of zoogeography alone, a Yukon ground sloth is already known, but obviously a living one is another matter entirely—as is the saytoechin's apparent dietary proclivity for beavers.

For according to traditional palaeontological belief, all forms of terrestrial non-aquatic ground sloth were exclusively herbivorous. Having said that: in 1996, Drs Richard A. Fariña and R. Ernesto Blanco from the Universidad de la República in Montevideo, Uruguay, published a thought-provoking if controversial paper in the *Proceedings of the Royal Society*, in which they proposed that *Megatherium* could have used its fearsome claws to overturn, stab, and kill glyptodonts as prey.

From analysing a *Megatherium* skeleton, Fariña and Blanco discovered that its olecranon (the elbow portion to which the triceps muscle attaches) was very short. This adaptation is found in carnivores, and optimises speed rather than strength. These researchers opined that this would have enabled *Megatherium* to use its claws like daggers, and they suggested that it may have commandeered kills made by the sabre-tooth *Smilodon* (such behaviour is known as kleptoparasitic) in order to add nutrients to its diet. Moreover, based upon the estimated strength and mechanical advantage of its biceps, they proposed that *Megatherium* could have overturned adult glyptodonts as a means of scavenging or hunting them.

However, this proposal has not gained widespread acceptance. In particular, palaeontologist Dr Paul S. Martin considers it "fanciful", noting that in terms of their dentition, ground sloths lack the carnassials that characterise predators, and that to suggest even that they were scavengers (let alone predators) is a reach. In addition, ground sloth dung deposits studied by him in Arizona's Grand Canyon and also in caves in Nevada, New Mexico, and western Texas contained no traces of bone. So far, therefore, at least as far as the palaeontological world is concerned, the case for carnivorous ground sloths in the past (not to mention in the present) has yet to be convincingly made.

Having said that: Indian tribes in parts of Brazil speak of a large red-haired mystery beast seemingly distinct from the mapinguary that they refer to as the xolchixe, and which cryptozoologists call the tiger sloth. This is because, based upon native descriptions, this little-known cryptid apparently resembles a lion or tiger-sized ground sloth but moves much faster than such beasts are believed to have done, is partly carnivorous as opposed to being entirely herbivorous, and is also partly arboreal rather than being exclusively terrestrial.

As for the saytoechin: as discussed by Canadian cryptozoologist Sebastian Wang in a *BCSCC Newsletter* article (winter 2006) documenting this little-known cryptid, although the

Native Americans selected a ground sloth from a book of extinct mammals as resembling it there is little else that actually links the two creatures directly. Other, less dramatic identities for it include an unusually large grizzly bear or black bear, plus some cryptozoological ones, such as a bigfoot, or even a surviving short-faced bear *Arctodus* or giant beaver *Castoroides* (both discussed in their own right later in this chapter), although the idea of a giant beaver habitually preying upon normal beavers does not seem very likely. As far as I am aware, no specific search has ever been made for the mystifying Yukon beaver eater, so it is surely time for someone to rectify this oversight.

Grounds Sloths in New Zealand and England?

It was cryptozoological archivist Richard Muirhead who kindly brought to my attention what must surely be the most unexpected claim ever made regarding alleged living ground sloths, which can be found in British retired submarine officer Gavin Menzies's book *1421: The Year China Discovered the World* (2002). In it, he claims that from 1421 to 1423, during China's Ming dynasty under the Yongle Emperor, the fleets of Admiral Zheng He, commanded by the captains Zhou Wen, Zhou Man, Yang Qing, and Hong Bao, discovered the Northeast Passage, the Americas, Australia, New Zealand, and Antarctica; circumnavigated Greenland; attempted to reach the North and South Poles; and circumnavigated the world a century before Ferdinand Magellan carried out the first officially-recognised circumnavigation. Mainstream historians do not agree with his claims, but such matters lie outside the scope of this present book of mine.

What does lie within its scope, however, is Menzies's suggestion in his own book that on one of their ships the Chinese took aboard some mylodontids captured in Patagonia but that upon reaching New Zealand in c.1421 a pair escaped when the ship was wrecked in Dusky Sound in Fjordland at the southwestern tip of South Island. Moreover, in 1831 the *Sydney Packet*, a ship from Sydney, Australia, visited Dusky Sound, where two sailors from the ship saw an animal that according to Menzies fitted the description of a ground sloth.

If so, this would indicate that the escaped mylodontids from the 1400s had not only survived in New Zealand but must also have established a population that was still in existence there four centuries later—always assuming of course that the beast seen was indeed a ground sloth, which is a massive assumption to say the least, and even more so when an independent source of information concerning this latter cryptid is examined (see below). Also, the wrecked ship was not Chinese, but an English vessel called the *Endeavour*, and was wrecked in 1795, not 1421.

Further information concerning this very strange state of affairs was presented in Robyn Jenkin's fascinating book *New Zealand Mysteries* (1970), which contained the following detailed account of the sailors' mystery beast sighting:

> Even more bizarre was a story, also reported to the Collector of Customs in Sydney when the *Sydney Packet* returned home in 1831. One of the ship's gangs which had been stationed at Dusky Sound told of the discovery of an enormous animal of the kangaroo species.
>
> The men had been boating in a cove in some quiet part of the inlet where the rocks shelved from the water's edge up to the bushline. Looking up they saw a strange animal perching at the edge of

the bush nibbling the foliage. It stood on its hind legs, the lower part of its body curving into a thick pointed tail, and when they took note of the height it reached against the trees, allowing five feet for the tail, they estimated it stood nearly thirty feet in height!

The men were to windward of the animal and were able to watch it feeding for some time before it spotted them. They watched it pull down a heavy branch with comparative ease, turn it over and tilt it up to reach the leaves it wanted. When it finally saw them, the animal stood watching the men for a short time, then made one almighty leap from the edge of the bush towards the water's edge. There it landed on all fours but immediately stood erect before making another great leap into the water. The men were able to measure the first jump and found it covered twenty yards. They watched the animal plough its way down the Sound at tremendous speed, its wake extending from one side of the Sound to the other.

Here again one is tempted to think the rum was talking, and for an Australian going away from home for months on end, what other animal would stir the imagination but a kangaroo? But how much more romantic to think that perhaps they really had seen some prehistoric animal living out its days in the remote fastnesses of the West Coast Sounds.

Romantic it may be, but the mundane reality is that no ground sloth is suspected to have behaved in the highly dramatic manner ascribed to the creature described here, or to have attained its colossal dimensions, which even dwarf those of the mighty *Megatherium*. In any case, as no comparable accounts appear to have been filed in this dual-island country since that one, it is surely safe to say that if a living ground sloth is indeed discovered one day, it will not be anywhere in New Zealand!

Finally: British author and wildlife educator Clinton Keeling was a renowned historian of zoological collections, including zoos, private menageries, and travelling sideshows featuring exotic animal exhibits. In a short but fascinating *Animals and Men* article (July 1995), he proposed that two of these latter shows travelling through England during the 1700s and 1800s respectively may have exhibited living specimens of a notorious African cryptid known as the Nandi bear (which is extensively documented later in this present chapter). However, as I fully discussed in my book *A Manifestation of Monsters* (2015), after writing to Keeling for more information and studying the very detailed letter that he kindly wrote to me, I consider it more likely that those exhibited by the later of these two shows were in fact living ground sloths, as now summarised here.

In November 1869, a huge travelling animal-based show called Mander's Menagerie included in its list of exhibits a pair of creatures referred to as 'Indian prairie fiends'—a name that Keeling believed Mander coined personally, as he was wont to do for creatures unfamiliar to him. Having said that, Mander was extremely knowledgeable regarding most animals, so these 'Indian prairie fiends' must have been extremely unusual in order to have received such a moniker from him.

According to Mander's description of them as included within an extremely lengthy advert, placed in a York newspaper during November 1869 and which was effectively a stocklist of the show at that time, they were:

> Most wonderful creatures. Head like the Hippopotamus. Body like a Bear. Claws similar to the Tiger, and ears similar to a Horse.

In his letter to me, Keeling discounted their 'Indian prairie fiend' name by accurately stating that nothing resembling them is known from North America. But what if they had come from South America instead? The 'Indian' reference could simply have been to whichever native Indian tribe(s) shared their specific distribution in South America. And could it be that 'prairie' was nothing more than an alternative name for 'pampas', perhaps substituted deliberately by Mander as he knew that 'prairie' would be a more familiar term than 'pampas' to his exhibition's visitors?

But does the South American pampas harbour a creature resembling those cryptids from York? Until at least as recently as the close of the Pleistocene epoch, this vast region (encompassing southernmost Brazil, much of Uruguay, and part of Argentina) did indeed harbour large shaggy bear-like beasts with huge claws, noticeable ears, plus sizeable nostrils and mouth. I refer of course to the ground sloths—and as already noted here, this same region has hosted several modern-day sightings of cryptids bearing more than a passing resemblance to ground sloths, and thence to the York mystery beasts.

Specimens of many other South American beasts were commonly transported from their sultry homelands and exhibited in Europe back in the days of travelling menageries there. Could these have included a couple of ground sloths? In addition, armed with such huge claws a cornered ground sloth might well be more than sufficiently belligerent if threatened or attacked to warrant being dubbed a fiend. Last, but certainly not least: once exotic specimens in travelling shows died, their corpses were often sold by the show owners to museums—so could the vaults of one such institution contain two presently-overlooked ground sloth cadavers less than 150 years old? It certainly wouldn't be the first time that surprising and highly significant zoological discoveries have been made not in the field but within hitherto unstudied or overlooked collections of museum specimens.

Life-sized model of glyptodont (© Miroslav Fišmeister)

THE MINHOCÃO—AN ANOMALY IN ARMOUR

Until the close of the Pleistocene, the armadillos in South America and southern North America shared their world with a group of distant relatives called the glyptodonts. These resembled armadillos to a certain extent—but on a gigantic scale, measuring up to 13 ft long. Their body armour was also of colossal proportions, constituting as much as 20 per cent of these animals' entire weight. Consisting of a huge domed shell of fused polygonal bony plates on their back, with a bony covering on top of their head too, it undoubtedly conferred upon these beasts a distinct similarity to an armoured tank. As for their tail, this was positively medieval—long and armour-encircled, with a mace-like, spike-bearing ball of solid bone at the tip, and probably used in the same way as a mace, flailing it at potential attackers.

According to the fossil record, the glyptodonts had died out by the end of the Pleistocene, around 11,700 years ago, but if they had lingered into the present day, we might expect their unique appearance to be sufficiently memorable for anyone spying these animals to provide a readily recognisable description. In reality, although there is evidence on file that has at one time or another offered hope to cryptozoologists that the glyptodonts are indeed still with us, the morphological comparability between the beasts seen and bona fide glyptodonts is of very varying quality.

The inhabitants of southern Brazil and Uruguay have often spoken of anomalous, furrow-like trenches of great depth that have suddenly appeared in the ground for no apparent reason (but often near to some sizeable lake or river), and which they claim to be the work of a mysterious serpentine creature called the minhocão. It was a paper by zoologist Prof. Auguste de Saint Hilaire, published by the *American Journal of Science* in 1847, that first brought the minhocão to Western attention. Revealing that its name is derived from 'minhoca'—Portuguese for 'earthworm'—he stated:

> . . . the monster in question absolutely resembles these worms, with this difference, that it has a visible mouth; they also add, that it is black, short, and of enormous size; that it does not rise to the surface of the water, but that it causes animals to disappear by seizing them by the belly.

In his paper, de Saint Hilaire gave several instances in which horses, cattle, and other livestock had been supposedly pulled beneath the water to their doom when fording the Rio dos Piloes and Lakes Padre Aranda and Feia in Goyaz, Brazil. He believed that the minhocão was probably a giant version of *Lepidosiren paradoxa*, the eel-like lungfish of South America.

19th-Century engraving of the South American lungfish

Thirty years after de Saint Hilaire, German zoologist Dr Fritz Müller, residing in Itajahy, southern Brazil, provided further details regarding the minhocão, when in 1877 his account of its activities appeared in the journal *Zoologische Garten*—later reiterated and recycled in other European publications, including the eminent English journal *Nature*. Here it was revealed that the channels excavated by the minhocão are so deep that the courses of entire rivers have been altered, roads and hillsides have collapsed, and orchards have fallen to the ground—and it also offered some new insights into this enigmatic excavator's morphology.

In the region of the Rio dos Papagaios, in the Brazilian province of Paraná, circa 1840s:

> A black woman going to draw water from a pool near a house one morning, according to her usual practice, found the whole pool destroyed, and saw a short distance off an animal which she described as being as big as a house

> moving off along the ground. The people whom she summoned to see the monster were too late, and found only traces of the animal, which had apparently plunged over a neighbouring cliff into deep water. In the same district a young man saw a huge pine suddenly overturned, when there was no wind and no one to cut it. On hastening up to discover the cause, he found the surrounding earth in movement, and an enormous worm-like black animal in the middle of it, about twenty-five metres [75 ft] long, and with two horns on its head.

In 1849, Lebino José dos Santos, a wealthy proprietor, was travelling near Arapehy in Uruguay when he learnt about a dead minhocão to be seen a few miles off, which had become wedged into a narrow cleft of rock and consequently died. Its skin was said to be ". . . as thick as the bark of a pine-tree, and formed of hard scales like those of an armadillo".

And in or around 1870, one of these beasts visited the environs of Lages, Brazil:

> Francisco de Amaral Varella, when about ten kilometres [6 miles] distant from that town, saw lying on the bank of the Rio das Caveiras a strange animal of gigantic size, nearly one metre [3 ft] in thickness, not very long, and with a snout like a pig, but whether it had legs or not he could not tell. He did not dare to seize it alone, and whilst calling his neighbours to his assistance, it vanished, not without leaving palpable marks behind it in the shape of a trench as it disappeared under the earth.

In a detailed letter, published by the *Gaceta de Nicaragua* (10 March 1866), Paulino Montenegro included accounts of a similar creature from Nicaragua. Said to be covered with a skin clad in scales or plates, it ". . . is described in general as a large snake, and called 'sierpe,' on account of its extraordinary size, and living in chaquites [pools or ponds]".

In his summary of this fascinating cryptozoological case, the editor of *Nature* offered two possible identities for the minhocão. One, echoing de Saint Hilaire, was a giant lungfish. The other, which no doubt by virtue of its more sensational potential has attracted much more attention during subsequent years, was a living glyptodont.

I have always viewed this latter theory with more than a little scepticism, for several reasons. First and foremost, I cannot believe that anyone would liken a beast as bulky and tank-like as a glyptodont, with a domed carapace on its back, to a giant worm or snake. Anything less serpentine than a glyptodont would be hard to imagine! In contrast, *Lepidosiren* is a notably elongate, anguinine (and anguilline) beast.

Although the description of horns and upturned nose could refer to the ears and snout of a glyptodont, it could equally apply to the slender, anteriorly-positioned pelvic fins of *Lepidosiren*, which also has a somewhat pig-like snout.

Another problem with the glyptodont identity arises when contemplating the exceedingly fossorial (burrowing) nature of the minhocão. It seems extremely unlikely that anything bearing such an immense amount of body armour—evidently for protection from attack by predators—would have either the need or the inclination for a fossorial mode of existence.

Creatures sharing this lifestyle, such as earthworms and moles, are conspicuously devoid of body armour—because they are not likely to encounter predators as frequently as if they spent their lives on the surface, and also because such armour would greatly impede burrowing activity. Fossil remains of glyptodonts provide no evidence at all for any extensive degree of underground activity. On the contrary, they appear to have been fully-terrestrial grazing herbivores, and the excessive development of their carapace and the mace-like construction of their tail clearly imply an expectation of frequent confrontation by large surface-dwelling predators equipped with fangs and claws.

True, the comparison by some eyewitnesses of the minhocão's scales to those of armadillos might seem to favour a glyptodont identity—but in reality, the reverse is true. Whereas the armadillos' armour is composed of a series of rings, in the glyptodonts their characteristic domed carapace consists of an elaborate mosaic of plates that bears no resemblance to armadillo armour. As recently as the late Pleistocene, there was a group of creatures somewhat midway in form between armadillos and glyptodonts, called pampatheres. Native to South America and also the southern U.S.A., some attained glyptodont dimensions, but their armour was of the ringed, armadillo form. Once again, however, they did not seem to be principally fossorial.

In contrast, although *Lepidosiren* lacks external scales this scalelessness is a modern development, as the more primitive Australian lungfish *Neoceratodus* is profusely scaled, like ancestral lungfishes. Hence its scales do not exclude the minhocão from a lungfish identity if we postulate that it may be an unknown, scaled species.

Even its burrowing activity is consistent with a lungfish. Like some African lungfishes (*Protopterus* spp.), during the dry season *Lepidosiren* aestivates—i.e. it secretes a protective cocoon around itself, and remains buried in the mud at the bottom of ponds or river beds in a self-induced state of suspended animation until the rainy season begins, whereupon it breaks out of its encapsulating cocoon and swims away.

According to the *Nature* report, the minhocão's deep trenches mostly appear after continued rain, and seem to start from marshes or river beds. This is just what one would expect of a giant lungfish—emerging from its subterranean seclusion at the onset of the rainy season. Conversely, although armadillos can swim, they only do so when required to—they are not normally aquatic; the same was probably true for the armour-laden glyptodonts.

In short, although the size estimates for the minhocão are certainly exaggerated, taken as a whole I feel that its description is more applicable to an extremely large lungfish than to a glyptodont. I do have misgivings concerning the minhocão's supposed propensity for hauling livestock down into its watery domain—this is hardly what one might expect from a lungfish, even a giant one. In reality, however, it may simply be an effect of turbulence or a type of localised vortex for which the minhocão is being wrongly held responsible.

There is a further identity, however, that offers an even closer correspondence to the minhocão, yet which had never been suggested prior to the publication of this present book's original edition, *In Search of Prehistoric Survivors*, in 1995. Namely, an enormous form of caecilian.

Native to the tropics of Africa, Asia, Mesoamerica, and South America, and spending virtually their entire lives burrowing underground, the caecilians are little-known limbless amphibians with outwardly segmented bodies that are extraordinarily similar in external appearance to earthworms—except for their readily visible mouth, and a pair of sensory tentacles on their head that resemble horns or ears when protruded. This description corresponds perfectly with that of the minhocão as penned by de Saint Hilaire.

In addition, although their skin feels smooth and slimy, many caecilians do possess scales (unlike other modern-day amphibians), embedded within the skin.

Engraving of *Siphonops* sp., a caecilian

The largest living caecilian known to science, Colombia's *Caecilia thompsoni*, is marginally under 5 ft long. However, a giant species with well-developed scales and a capacity for excavation matching its great size would make an extremely convincing trench-gouging minhocão. And if its scaling mirrored its body's external segmentation, it would resemble the ringed armour of armadillos—clarifying why eyewitnesses liken the minhocão's scaly skin to this armour.

South America's *Typhlonectes* caecilians inhabit rivers and lakes, so even the minhocão's aquatic inclinations are not incompatible with these apodous amphibians. Moreover, terrestrial caecilians often emerge above ground after heavy rainstorms—another minhocão correspondence. Also, caecilians are carnivorous, and grab their prey from below—a giant species with comparable behaviour might therefore resolve reports of livestock pulled under the water when crossing rivers and lakes reputedly frequented by minhocãos.

All in all, the identity of a giant caecilian for the minhocão provides so intimate a correspondence, not only morphologically but also behaviourally, that I personally see no reason for looking elsewhere for an explanation of this mysterious subterranean monster.

On 4 October 2010, I received the following communication from correspondent Samwell Rowan concerning a truly remarkable but hitherto-undocumented minhocão-reminiscent mystery beast encountered by his mother during the late 1980s or early 1990s in a Peruvian rainforest:

> She told me she was walking by herself in the jungle and saw what she initially thought was a large, black snake moving through the leaves on the forest floor. She then noticed it had armoured plates and may have had numerous small legs. Both my mother and I are aware that there are giant centipedes in that region, but the size of it does not match up. She described it as being well over one foot thick and never saw its head nor tail even though she observed it for several minutes. She guessed it must have been at least twenty feet long. She didn't mention her sighting to me in full detail until a couple of years later because she assumed it was [a] centipede and was not aware of the minhocão. I was not aware of the minhocão either until about a year ago when I first found your website.

If it were limbless, it does indeed recall the body form reported for the minhocão—such a pity that she didn't catch sight of its head. If, conversely, it had numerous small legs, then the latter mystery beast is instantly eliminated from further consideration. I also tend to discount a giant centipede, in favour of a giant millipede, specifically one of the armoured species. This is because centipedes tend not to be black and, although multi-limbed, their legs are relatively fewer and bigger than those of millipedes—some of whose larger species are indeed black and equipped with numerous small legs.

Even so, the world's biggest millipede specimen—an African black millipede *Archispirostreptus gigas* owned by Jim Klinger of Coppell, Texas, and sporting an impressive complement of 256 legs—only measures 15.2 inches long and 2.6 inches in circumference. This is a far cry indeed from the monstrous dimensions claimed by Rowan's mother for the unidentified vermiform creature encountered by her in Peru (even allowing for unintentional exaggeration or over-estimation of size). Moreover, for fundamental physiological reasons, no known species of terrestrial arthropod attains anything even remotely approaching those dimensions, so Peru's worm-like wonder currently remains an enigma.

The minhocão is not the only cryptozoological controversy to which the glyptodonts have been linked, albeit only temporarily in the following case, which began on 7 October 1967 with a short report in *Science News*. The world's largest species of modern-day armadillo is the giant armadillo *Priodontes giganteus* of eastern South America, up to 5 ft long and weighing as much as 130 lb. Although not a common species, it has never been deemed extinct. According to the *Science News* report, conversely, the giant armadillo had been written off as extinct by zoologists until the recent capture in Argentina of a male—one, moreover, that must have been of prodigious size. For the report gave its length as 6.5 ft, and its weight as a colossal 2,200 lb (1 ton)!

Faced with such dramatic dimensions, and the fact that the true giant armadillo does not occur in Argentina, mystery beast investigator Michael J. Shields speculated whether the *Science News* beast could actually be a living glyptodont. This in turn prompted Ivan T. Sanderson to pursue the matter with his usual zeal via his Society for the Investigation of The Unexplained (SITU), but the truth was a great disappointment. It proved to have been a 'ninth-hand' story derived from an unknown source in Mexico's Argentinian embassy—in other words, a classic case of the Chinese whispers syndrome operating on overdrive!

Finally, in summer 1991 David H. Hinson brought to cryptozoological attention a gold and jade pendant recovered from an archaeological dig site in Panama before World War II and now housed in the University of Pennsylvania's museum. It is officially believed to represent either a crocodilian reptile or a jaguar, but Hinson drew attention to features that he likened to those of a glyptodont. These include its claws, tail, head, and, of particular note, what seems to be a slightly domed carapace. Hinson pointed out that even if the glyptodonts

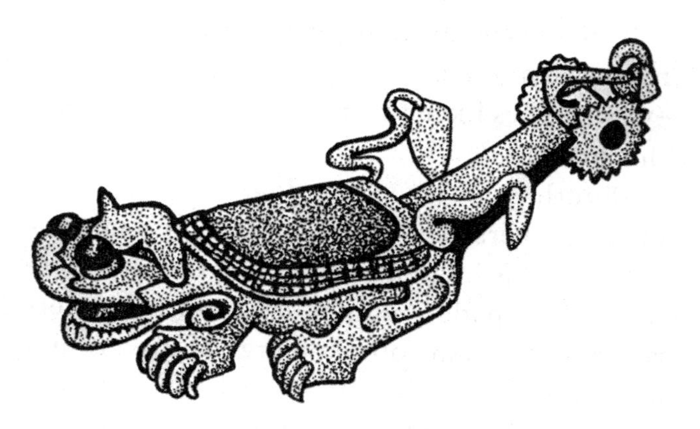

Sketch of the glyptodont-reminiscent pendant publicised by David H. Hinson

had survived the Pleistocene's close after all but were extinct at the time when the pendant was made—more than a millennium ago—perhaps they had survived beyond the Pleistocene sufficiently long to have become incorporated in Panamanian mythology.

Like so many other examples of tantalising iconography noted by me in this book, the answer will probably never be known. Certainly, the beast is much more reminiscent of a carapace-bearing creature than a crocodile or jaguar, but beyond that we just cannot be sure.

A SCALY TALE FROM RINTJA

Sandwiched between the large island of Flores to the east and the much smaller island of Komodo to the west, within Indonesia's Lesser Sundas, is the even smaller island of Rintja (Rindja). Komodo's fame in cryptozoological circles is due of course to its giant monitor lizard, the Komodo dragon *Varanus komodoensis*, which remained undescribed by science until 1912, despite the fact that it is the world's largest living species of lizard. However, Komodo's glory may yet be overshadowed by its smaller neighbour, Rintja, thanks to reports of an extraordinary cryptid said to frequent this diminutive dot of land.

In 1963, French traveller Pierre Pfeiffer's book *Bivouacs à Borneo* included details of a visit to Rintja. While hunting there one night with an old native hunter, he was informed about a very interesting mystery beast that the island's inhabitants greatly fear, and which they refer to as the veo. According to the hunter, the veo is as big as a horse, has a long head, and fur on its belly, but its flanks are covered in scales, and its feet are equipped with very large claws. During the day it stays in the mountains, but at night it descends to the mangrove coasts, where it feeds upon crabs and shellfish. Sometimes its distinctive cry can be heard during the evening, sounding like "hoo-hoo-hoo".

In light of this description, it should come as no surprise that the hunter vehemently dismissed Pfeiffer's bizarre suggestion that the veo might be a dugong! Moreover, he stated that he had actually seen a veo once, while hunting at night at a locality on Rintja called Loho Buaji, in the company of a native policeman from Labuanbadja on Flores. When they encountered the veo, they were so frightened that they fell to the ground at once, and lay there watching it but without moving until the creature disappeared. In reply, Pfeiffer claimed that if he were ever to meet a veo, he would simply shoot it, but the hunter assured him that this would make no difference, as the veo's scaly skin protected it from bullets.

Pfeiffer concluded that the veo tradition probably stemmed from distant, distorted memories of the armoured horses ridden by the first Portuguese explorers to visit this island, some five centuries earlier—but not everyone agrees with him.

During the mid-1990s, Dr J. Zahrádka, a physician from Teplice in the Czech Republic, came upon a Czech translation of Pfeiffer's book, and sent a copy of the chapter documenting the veo to Jaroslav Mareš, one of his country's leading cryptozoologists. Mareš was extremely interested in the information—so much so that he wrote to an Indonesian friend called Uning, from Jakarta, Java, requesting any additional details that he may be able to supply, because he knew that Uning had spent some time on Rintja.

Sure enough, while there Uning had indeed learnt of the veo, and after closely questioning the native people he had been able to prepare a very detailed, precise description of it, which is as follows. Measuring at least 10 ft in total length, the veo has a long head, and much of

Pangolin, 1830s illustration

its body is covered in very large scales that overlap one another like roofing tiles. On its head, throat, belly, the inner side of its limbs, and the end of its tail, however, it has hair. Its feet bear long claws, and if disturbed it can be very dangerous, sitting up vertically on its hind legs and slashing at its antagonist with the huge razor-sharp claws on its front paws. Its diet consists primarily of termites and ants, but it will also eat crustaceans and other small sea creatures left stranded upon the beach by the outgoing tide.

When Mareš first received this report from Uning, he assumed that the veo was one and the same as the Komodo dragon. Despite its name, this mighty lizard is not confined entirely to Komodo; it is known to exist on Rintja too, as well as on Flores and Padar, and may also exist on Sumbaya, though this has yet to be confirmed. However, when Mareš suggested this identity, Uning replied that the Rintja natives are familiar with the Komodo dragon, and readily distinguish it from the veo.

This is not surprising, because Uning's description is tantalisingly reminiscent of an extremely distinctive type of creature very different from any lizard. The veo's impressive body armour of large overlapping scales, its lengthy head, the large sharp claws on its feet, and the presence of hair rather than scales on its underparts and inner limbs—these are all morphological characteristics of the pangolins or scaly anteaters, native to Africa and Asia.

So too is the veo's dietary preference for ants and termites. As for its behaviour: some pangolins are indeed predominantly nocturnal, but they are also primarily timid beasts that prefer to roll up into a ball, rather than attack, when threatened. However, one could readily imagine that a pangolin as big as the veo might not be quite so timid, and hence may be more liable to confront an enemy in the daunting manner described by the Rintja natives. In addition, pangolins can run with their front limbs raised completely off the ground, and when walking they will often pause periodically and raise themselves up like scaly kangaroos, squatting vertically on their hind legs, and supported by their long tail—as described for the veo.

Nevertheless, considerations of size also expose a major inconsistency between the veo and a pangolin identity for it, because none of today's known species of pangolin is anywhere near as large as the veo. Even the biggest, the giant pangolin *Manis* (*Phataginus*) *gigantea* of Africa, does not usually exceed 5 ft in total length. Of course, estimates of the veo's size could be exaggerated, due to the fear that it engenders among the people of Rintja. However, there is also a second, much more dramatic, thought-provoking possibility.

In 1997, Mareš included a detailed account of the veo in one of his cryptozoological books, *Svet Tajemných Zvírat* ('*The World of Mysterious Animals*'), and noted that during the Pleistocene epoch (2.59 million to 11,700 years ago), a huge species of pangolin, *Manis palaeojavanicus*, measuring over 8 ft long, occurred in the Greater Sunda islands of Borneo and Java. Accordingly, Mareš deems it possible that either this species, or some other, even larger and still-unknown pangolin, survives today on little-explored Rintja, where it is called the veo by the natives, but remains undiscovered by science.

In 1912, Komodo startled naturalists with its giant monitor. Perhaps Rintja will one day emulate its famous neighbour, courtesy of a gigantic pangolin, and thereby add another cryptozoological success story to the history of the Lesser Sundas.

NORTH AMERICA'S GIANT BEAVER—MORE THAN A MYTH?

Until at least as recently as the end of the last Ice Age, around 12,000 years ago, during the late Pleistocene epoch, when the first human colonists had already arrived here, North America was home to two separate species of bona fide giant beaver.

The better known of these, and as big as the American black bear *Ursus americanus*, was the northern giant beaver *Castoroides ohioensis*. Fossil evidence reveals that it was up to 8 ft long, with 6-in-long incisors, and weighed 132-220 lb. Moreover, the second species, the southern giant beaver, *C. leiseyorum*, was slightly bigger, but its remains are currently known only from South Carolina and Florida, whereas *C. ohioensis* was widely distributed in the midwestern U.S.A. south of the Great Lakes and also in Canada north of them, as well as in Alaska. Currently, there is no official confirmation that *Castoroides* constructed lodges and dams, but some experts deem this to be likely.

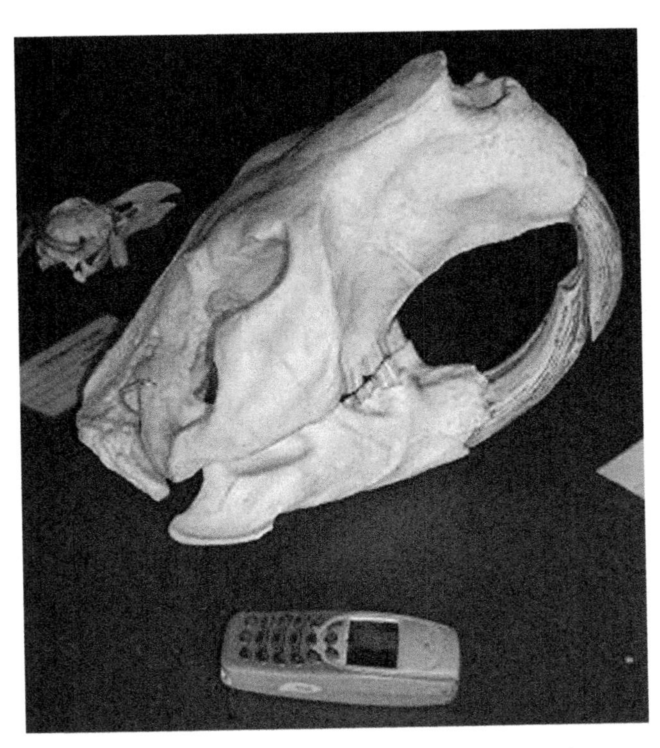

Castoroides skull (© Markus Bühler)

But what has *Castoroides* to do with prehistoric survival? This is the tantalising question that surfaced in my book *Mirabilis* (2013) —and here is how, and why, I answered it:

Approximately 3 ft long and generally weighing 33-77 lb, the American beaver *Castor canadensis* is second in size only to the capybara among present-day New World rodents.

According to Amerindian traditions, however, there was once a gigantic form of beaver existing in the U.S.A. and Canada that exceeded even the capybara in stature.

The Malecite are an Algonquian-speaking Native American people indigenous to the St John River valley, crossing the borders of New Brunswick and Quebec in southeastern Canada and Maine in the northeastern U.S.A. Their time-honoured orally-preserved lore contains a detailed legend of how Gluskap, their mythical culture hero and transformer, angrily pursued a giant beaver for failing to show due respect to his latest creation, man, and for building huge dams that blocked the river. However, the giant beaver escaped his clutches, fleeing far away.

Moreover, such stories are by no means confined to eastern North America's Malecite culture. In British Columbia, southwestern Canada, for instance, there is a longstanding Salishan tradition of a huge beaver-like beast known locally as the slal'i'kum inhabiting Cultus Lake. A legend from the culture of the Pocumtuck tribe inhabiting the region around Deerfield in Massachusetts, U.S.A., tells of how Lake Hitchcock—a large lake in the Connecticut River Valley and dating back to the Pleistocene epoch—harboured a giant beaver that sometimes came ashore and attacked people, until one of their bravest hunters killed it. They also claim that several lakes here were created as a result of dams constructed by gargantuan beavers.

The lore of the Tlingit people living around Sitka on southeastern Alaska's Baranof Island contains the story of how a gigantic beaver-like beast once devastated an entire village. And the Montagnais-Naskapi people, an Algonquin tribe inhabiting Labrador, allegedly named a major river there Mishtamishku-shipu ('Giant Beaver River') as a direct reference to a pair of enormous beavers, said to be larger than seals, that were killed there long ago to prevent them from breeding.

Moreover, claims concerning giant beavers have been made in much more recent times too. During the 19th Century, reports of a mystifying brown-furred water beast likened to a huge beaver or immense otter emerged from Fowler Lake in Saskatchewan, Canada, though, sadly, the beast itself did not do likewise. Accounts of what some investigators deem likely to have been an extra-large beaver-like creature are also on file from Utah's Bear Lake and Utah Lake during that same time period. And sightings of creatures recalling the slal'i'kum have been reported in British Columbia's Cultus Lake as recently as the 1990s. Moreover, some fascinating reports from the late 20th Century and early 21st Century were collected and published by American cryptozoological chronicler Michael Newton in his *Encyclopedia of Cryptozoology* (2005).

In March 1993, Tom Greene, horticultural superintendent for Moline, Illinois, claimed that a 5-ft-long beaver weighing more than 75 lb and extremely strong was gnawing down trees at the Marquis Harbor Yacht Club. It had already sprung and escaped from two traps set to snare it.

During 2000, a Ms J. Greenwald supposedly spied a horse-sized beaver close to Bullfrog Marina, on Lake Powell—a sizeable body of freshwater that spans the southern Utah-northern Arizona border in the southwestern U.S.A. Two years later, she finally reported her sighting formally to the National Institute of Discovery Science in Las Vegas, Nevada. In her report, she stated that the creature was brownish-black, weighed an estimated 700-800 lb, was unbelievably big, and could certainly kill a person if it so chose. Yet even assuming that she had undoubtedly over-estimated its size, if her claim is true then it still must have been a seriously big beaver!

Interestingly, in August 2002 an online account posted by someone identified only via the user name Staci told of how her family had encountered a beaver "the size of a horse" at Lake Powell. In view of the very same comparison of stature (horse-sized) and location (Lake Powell) appearing in both reports, could Staci have been one and the same as J. Greenwald, I wonder, reporting an anonymous version of her sighting before going public with it during that same year?

Even more recently, in 2007, I was contacted on several occasions by Canadian cryptozoologist John Warms who is actively investigating reports of alleged giant beavers, following his own sighting in 2006 of what he believes to have been one such creature. Based in Manitoba, he had first learnt of their apparent existence back in 2003, after speaking with residents of a Cree Nation reservation in northern Manitoba. Since then, he has been in touch with a number of other eyewitnesses, all describing similar creatures but this time without realising that extra-large beavers were unusual.

As for his own sighting: after hearing about bygone observations of huge beavers in Manitoba's Assiniboine River, Warms decided to spend a night in his Jeep beside this river during April 2006, when it was in flood stage—and here, in his own words, is what he saw:

> Just before dark, not expecting to see anything, I saw this large head and what I took to be its back and tail behind it, moving along with the current. As I bent down to have a better look through the branches, it immediately plunged and slapped the water like an ordinary beaver. I was stunned to think that I had perhaps seen the "extinct" giant beaver, and had to convince myself ever since that what else could it be? The head was the size of a basket ball, and the whole body was seven or eight feet long. I have seen hundreds of ordinary beaver, and there was just no comparison.

Warms informed me that he has also found some very large underwater tunnels, 3 ft in diameter, that he considers may have been constructed by giant beavers, particularly in southern Manitoba, and he believes that they occasionally build lodges and dams too, like America's (and Europe's) smaller, known species of beaver. Warms is hoping to elicit funding to conduct extensive research, and during late December 2009 his ongoing quest for evidence of these creatures' existence led him to Utah and Arizona, as featured in various newspaper articles.

The prospect of giant beavers may seem highly improbable, at least on first sight. In reality, however, nothing could be further from the truth, because thanks to the Pleistocene reality of *Castoroides*, there is a very significant, fully-confirmed precedent.

Could it be, therefore, that the northern giant beaver's dramatic form and presence had exerted a sufficiently profound impact upon North America's early human colonists for its memory to have been preserved down through countless generations and incorporated into their orally-preserved folklore, and which also travelled westward with them as they traversed the continent from east to west? Or is it even conceivable that this dramatic species actually lingered on beyond the last Ice Age into historical times, and perhaps, just perhaps, even into the present day in certain remote localities within its prehistoric range?

The last word on this thought-provoking subject comes from a paper published in spring 1972 within the highly reputable periodical *Ethnohistory*, and which first alerted me to the fascinating presence of giant beavers in cryptozoology. Written by Jane C. Beck, it was entitled 'The Giant Beaver: A Prehistoric Memory?', and evaluated the frequent presence (with only minor variations) in Native American tradition of the basic folktale concerning a deity pursuing a giant beaver. In her conclusion, Beck offered the following hypothesis to explain the unexpected abundance and persistence of what would ordinarily be a very local, obscure legend:

> Thus it seems that all evidence points to the giant beaver tale being a folk memory of a prehistoric creature. Actual proof is not yet possible, but it is important to look to the future and suggest a road that the archaeologist and historian might well follow.

This suggestion has already borne fruit. Just over 20 years after Beck's paper was published, the second, southern species of giant beaver was formally described and named, and continued findings of fossil remains has expanded the northern species' known prehistoric range very considerably. In short, new revelations concerning North America's one-time giant beavers are still occurring. So who knows—perhaps in the future, further tangible, physical discoveries will be made that will verify post-Pleistocene persistence of these mega-rodents.

MOUNTAIN TIGERS AND SABRE-TOOTHED TIGERS—ONE AND THE SAME?

While working as a hunting guide in the West African country of Chad during the 1960s, Christian Le Noël heard of a mysterious striped cat of great size that reputedly inhabits the mountains of Ennedi in the north of the country, and is consequently known to the French-speaking people of this area as the tigre de montagne—'mountain tiger'. When he questioned the Zagaoua tribe concerning it, they furnished him with several additional details. According to their description, the mountain tiger is larger than a lion and strong enough to carry off big antelopes, but lacks a tail, possesses red fur patterned with white stripes, has long hair on its feet, makes its home within the numerous caverns pitting this remote, mountainous locality, and has a pair of huge teeth that protrude from its mouth.

As with the mokele-mbembe and certain other African mystery beasts, reports of animals clearly equivalent to the mountain tiger are not confined to a single region of Africa. According to the Hadjeray tribe of southwestern Chad, it also occurs here, and is referred to by them as the hadjel. Similarly, Le Noël learnt that this formidable animal is said to frequent the Ouanda-Djailé district of the Central African Republic too—where, once again, it is termed the tigre de montagne by French-speakers, and is known to the Youlou tribe as the coq-djingé or coq-ninji. Other names given to it here include the gassingram and vassoko.

On one occasion in 1975, Le Noël himself had what would seem to have been an uncomfortably close encounter with just such a creature. While leading an eland hunt and accompanied by an old native game tracker, he was approaching a very large cavern in Ouanda-Djailé when suddenly they heard a terrible roar that Le Noël, despite his great experience as a hunter, could not identify. His tracker, conversely, experienced no such problem, readily identifying its unseen owner as a mountain tiger and refusing to go on any further. When pressed for details, the tracker gave him a verbal description that corresponded remarkably closely with *Machairodus*—the African sabre-toothed tiger! (Worth noting here, however, is

Reconstruction of mountain tiger's alleged appearance, based on eyewitness reports (© Tim Morris)

Artistic restoration of *Machairodus kabir* (© Hodari Nundu)

that sabre-toothed tigers were not closely related to true tigers, belonging instead to a separate taxonomic subfamily of felids, Machairodontinae.)

Greatly surprised by this, Le Noël resolved to return here the following year, and when he did he came armed with illustrations of numerous animals, living and extinct. Without any prompting, the picture that his tracker colleague positively identified as the mountain tiger was a reconstruction of *Machairodus*—a beast that supposedly died out during the Pleistocene epoch, more than 11,700 years ago. Interestingly, in 2005 a scientific team that included Stéphane Peigné revealed in a *Comptes Rendus* paper that fossils of a very large (weighing an estimated 770-1036 lb, around 8 ft long, and up to 4 ft tall) but hitherto-undescribed species of African sabre-tooth, formally dubbed *Machairodus kabir*, had been uncovered in the late Miocene hominid locality of TM 266, Toros-Menalla, in Chad. This also happens to be a country from which modern-day reports have been filed for the elusive but similar-sized, notably big-fanged mountain tiger.

In addition, Le Noel once saw a hippopotamus in southern Chad that had died of strange wounds which could only have been given by a cat armed with exceptionally well-developed upper canine teeth. Nor are big-fanged mystery cats unique to Africa.

According to a fascinating online article published in Chinese (http://tieba.baidu.com/p/806911199?lp=5027&is_bakan=0&mo_device=1), but which was kindly summarised in English for me by Canadian cryptozoologist Sebastian Wang after alerting me to its existence, an extremely mysterious striped felid may exist in China that is very different indeed from the known South China tiger *Panthera tigris amoyensis* (and which may well be extinct in the wild now anyway).

As recently as May 1994, in the Shennongjia region of China's northwestern Hubei Province, the article's author allegedly spied a giant cat measuring 12-15 ft long (and therefore much larger than any normal tiger), with white fur bearing vertical yellow stripes. An extreme white-furred variant of the tiger, known as the snow tiger (see my book *Cats of Magic, Mythology, and Mystery*, 2012, for a detailed account), does possess only very pale, yellowish stripes (indeed, some specimens bear no or virtually no markings at all), but it does not measure 12-15 ft long. Moreover, both the snow tiger variant and the more familiar white tiger variant with brown or dark grey/black stripes have only ever been recorded from the Bengal tiger *P. tigris tigris*, not from the South China subspecies or any other.

Possible appearance of the Chinese sabre-toothed mystery cat (© Dr Karl Shuker)

In any case, the Shennongjia mystery cat also reputedly sported a pair of huge canine teeth up to 9 in long and therefore reminiscent of a prehistoric sabre-toothed tiger's.

The sighting's precise location was on the tallest peak of the eastern Shennongjia region, at an altitude of just over 9,000 ft. The article's author subsequently learnt that a few such cats had previously been killed by local hunters. If only a pelt or skull had been preserved for scientific examination—but perhaps some hunter does possess such objects, as trophies displayed proudly in his home. If so, he may own specimens of immense cryptozoological significance.

The most famous sabre-tooth was *Smilodon* of the New World, a huge beast with massive canines and very short tail. With the appearance of the Panamanian isthmus 2 million years ago, this great cat's North American homeland became linked to the former island continent of South America, so *Smilodon* migrated there too, penetrating Brazil, Peru, and as far south as Argentina. By the Pleistocene's close, it had apparently died out in the northern continent, where it was represented by *S. fatalis*—but can we really say the same for the southern one, represented there by the enormous *S. populator* (see later)? Some tantalising reports on file lend faint hope to the chance that a smaller modern-day descendant of this exotic beast still survives in South America, undetected by science.

The author with life-sized *Smilodon fatalis* replica skull (© Dr Karl Shuker)

In his book *The Cloud Forest* (1966), traveller Peter Mathiessen noted that an itinerant seaman called Picquet described to him a rare cat frequenting the rainforests of Colombia and Ecuador that is said to be very shy, not quite as large as a jaguar, striped rather than spotted, and equipped with a pair of very large protruding teeth. Science knows of no striped species of cat native to anywhere in South America. True, Peruvian zoologist Dr Peter Hocking has collected reports of a similar beast reputedly inhabiting hilly and lowland forests within Peru's provinces of Ucayali and Pasco, and in August 1994 I learnt that he had obtained a cat skull purportedly from this unidentified felid, with an American cat specialist to whom he later showed a photo of it deeming the skull to be from a wholly new species. When it was examined more recently by a team of biological researchers that included Hocking and British palaeontologist Dr Darren Naish, however, it was identified in their 6 March 2014 *PeerJ* paper as merely the skull of a jaguar.

An extraordinary cat initially claimed to be a bona fide *Smilodon* but later reclassified as a 'mutant jaguar' by the authorities was supposedly shot in the forests of Paraguay in 1975 and was afterwards examined by a zoologist called Juan Acavar, who was amazed by its immense canines. No details as to the ultimate fate of its corpse are known, however, and I have not succeeded in tracing the whereabouts of Acavar—so the identity of this perplexing find remains unconfirmed, just like that of so many other 'lost' cryptozoological relics. However, in 1984 another living sabre-tooth was reputedly encountered in Paraguay, this time by François Piquet, a French sailor, but he did not shoot it (*Science Illustrée*, September 1998).

On 12 September 2001, I received a very interesting email from Spanish biologist Gustavo Sanchez Romero, based in Tenerife, one of Spain's Canary Islands. It reads as follows:

> The purpose of my mail is that one year and half ago I traveled to Venezuela (I have relatives there, since a lot of people from the Canaries migrated there back in the fifties, sixties, seventies and even eighties) including my parents.
>
> Well once there I took a trip to visit Salto el Angel (Angel Falls) the largest waterfall in the world, being almost 1000 m. high. It departures from the Auyan Tepui, a lofty flat top mountain (Tepui being the local name) only found in Venezuela. Once there I heard about Alexander Laime, the person you mention in your book (Prehistoric Survivors) who saw the prehistoric aquatic dinosaurs [sic—plesiosaur-like beasts] once bathing in a lagoon [as documented by me in Chapter 3 of this present book]. Also the guide in the zone told me about "El tigre dantero" meaning the "Danta eating tiger". Danta is local name for tapir, and he told me that it was the size of a cow, and the surprising characteristic about it is that it was supposed to have huge fangs (canines), just like the prehistoric saber tooth tiger does have! I thought about it a little and then I have read in some books about similar descriptions from Paraguay and Ecuador. I thought that maybe you would like to hear this little story, so I hope it is useful to you!

It certainly is, because although the guide's claim contained little in the way of morphological details other than the creature's body size and huge canines, it nonetheless extends considerably the geographical distribution of reports appertaining to mega-fanged mystery cats in South America.

In 1863, C.C. Blake provided the following cryptozoological appetiser:

> At Timana, in New Granada, sculptured stones have been figured by Mr Bollaert, representing a feline animal, the proportions of whose teeth slightly exceeded those of existing cats, and might possibly indicate a modified descendant of the extinct *Machairodus* [*Smilodon*] *neogaeus* of Brazil.

Smilodon populator (=*neogaeus*), native to Brazil and Argentina during the Pleistocene, was bigger than a lion, and presumably preyed upon various of the very large ungulates peculiar to South America. When they died out, so did *S. populator*. Today, this continent has few notably large mammals—hence a contingent of *S. populator* would have great difficulty in sustaining itself. A smaller version, conversely, could well succeed in carving out an adequate niche, especially in relatively inaccessible, undisturbed areas, such as remote, mountainous cloud forests, preying upon medium-sized ungulates such as deer and tapirs. Is it just a coincidence that this is precisely the type of terrain from which reports are emerging of a mysterious, medium-sized cat with notably large fangs? Its striped pelage would provide effective camouflage in this habitat, and its shy disposition would further enable it to avoid undue human scrutiny.

It is even possible that some sabre-tooths not only survived, but also sought to expand their ecological confines—by becoming modified for a secondarily amphibious lifestyle. Certainly, there are a number of unidentified

Smilodon populator (at back) and *S. fatalis* (at front) (© Hodari Nundu)

mystery beasts on record from South America that strongly recall sabre-tooths of a distinctly aquatic persuasion. The water-dwelling yaquaru or water tiger of Patagonia, for example, with strong tusks, shaggy yellow pelage, and savage disposition, is a creature whose lifestyle, appearance, and behaviour do not wholly correspond with either of the alternative identities previously offered for it—the jaguar, or a huge otter. Only a water-dwelling sabre-tooth can reconcile all of its characteristics.

The Guyanan aypa, a tiger-headed water beast with extremely large teeth, may well be of similar identity, and is probably one and the same as another Guyanan anomaly—the greatly-feared maipolina. This fawn-coloured aquatic creature is said to inhabit riverbank caves, to measure just under 10 ft long, and to sport a pair of enormous fangs whose appearance as described by one eyewitness apparently compared with those of a walrus.

If amphibious sabre-tooths have evolved in South America, one might expect to encounter reports of similar beasts in Africa too—and such reports are indeed on file. A number of rivers in the Central African Republic, for instance, are reputedly home to a strange beast variously referred to as the nze-ti-gou ('water panther'), mourou n'gou ('water leopard'), dilali ('water lion'), and mamaimé ('water lion'). DR Congo is supposedly home to the simba ya mai ('water lion'), Angola has the coje ya menia ('water lion'), and Kenya the dingonek.

Judging from these names, the creature is evidently feline in basic form, and is described as such by its native observers, but is of aquatic habitat, larger in size than terrestrial felids, and with very large, protruding fangs. Details vary a little as to its coat colour—from yellow or brown with stripes for the mourou n'gou (though see also below for an eyewitness claim regarding spots for this cryptid), to red with stripes or spots for the nze-ti-gou. Yet even if a genuine reflection of the species' morphological range (rather than erroneous documentation or translation by those recording the eyewitness accounts), such variations are no greater than those exhibited by familiar species such as the leopard and lion. As for its large size, this can be expected for water-dwelling creatures, whose surrounding medium's gravity-combating properties of buoyancy enable them to attain much greater sizes than land-dwelling counterparts (as with whales, various pinnipeds, etc).

One of the most detailed accounts of such a beast came from adventurer John Alfred Jordan, whose book *The Elephant Stone* (1959) contains his dingonek sighting on the River Maggori running into Lake Victoria. According to Jordan, the creature was 15-18 ft long, with a massive head, two large fangs as thick as a walrus's tusks projecting down from its upper jaw, a broad tail, a spotted back as wide as a male hippo's, and spoor as large as that animal's too, but with long claws.

Intriguingly, Jordan stated that the dingonek was scaly, like an armadillo—which ostensibly rules out any affinity with a sabre-tooth. However, these 'scales' might simply be clumps of matted fur, which would resemble the genuine articles when viewed from a distance. A more radical, less plausible option is the existence of some bizarre form of scaly reptile that otherwise duplicates the appearance of an aquatic felid.

Oddly, the precise year when Jordan allegedly spied a dingonek is unclear. It was given as 1905 in *The Elephant Stone*; as 1907 in a letter by him to London's *Daily Mail* newspaper that was published on 16 December 1919; and as 1908 in an article by Jean-Pierre Anselme (*VSD Nature*, December 1993).

Possible appearance of the dingonek? (© Dr Karl Shuker)

In another of his books, *Elephants and Ivory* (1956), Jordan noted that it was considered bad luck to kill a dingonek. This belief is also prevalent concerning DR Congo's equivalent mystery beast, the ntambo wa luy or simba ya mai, as recorded in Charles Mahauden's book *Kisongokimo* (1965). Such taboos have undoubtedly assisted in saving these potentially dangerous creatures from extirpation.

At much the same time as Jordan's sighting, another hunter spied a dingonek floating on a log down the Mara River (also running into Lake Victoria) while in high flood—but it quickly slid off and into the water. Near Kenya's Amala River, the Masai call it the ol-maima and sometimes see it lying in the sun on the sand by the riverside—but if disturbed, it slips into the water at once, submerging until only its head remains above the surface.

A fairly recent search for one of Africa's amphibious feline cryptids took place from

December 1994 to January 1995. This was when cryptozoologist Eric Joye led a two-man Belgian expedition, dubbed 'Operation Mourou N'gou', to the Central African Republic, seeking the eponymous sabre-fanged water monster said to inhabit this country's lakes and rivers. Although they failed to spy it themselves, Joye and his team-mate, hunting guide Willy Blomme, gathered some interesting anecdotal evidence.

The dingonek resembles a bipedal sabre-toothed dinosaur(!) in this somewhat bizarre illustration accompanying a November 1917 *Wild World Magazine* article by Jordan on unknown African animals

Claiming to have narrowly avoided being propelled into the Bamingui River by one of these animals as he sat fishing in February 1985, a native guide called Marcel disclosed to Joye that the mourou n'gou hunts in pairs—one waiting in the river to seize any prey chased into the water by the other one.

Marcel likened its shape and size to a leopard's, and stated that its pelage is ochre in colour, dappled with blue and white spots that are very distinct upon its back but less well-defined upon its flanks. He also claimed that its long tail is hairier than the leopard's, its head is like that of a civet (does this mean that, like the civet, it has a dark face mask?), but its teeth are very large, like those of a big cat such as the lion. Marcel followed the mourou n'gou's trail, which was like that of a leopard but bigger, and he said that when it runs it leaves behind the impression of claws (not usual for a leopard).

As I discussed in my books *Mystery Cats of the World* (1989) and *Cats of Magic, Mythology, and Mystery* (2012), the reality of water-dwelling cats is not the unlikely fusion of mutually-exclusive concepts that it may initially seem. And when beasts of such similarity to one another are reported over as great an area of Africa—with as diverse a range of human cultures and beliefs—as is true with the water panthers and their kin, there must surely be more than native myth and superstition at the core.

Indeed, in South Africa's Orange Free State there are even cave paintings depicting dingonek-like beasts, complete with walrus tusks and scaly patterning. The most famous one (portrayed here), which appeared in *Rock-Paintings in South Africa* (1930) by George W. Stow and Dorothea Bleek, can be found in a cave in Brakfontein Ridge, which was at one time (and perhaps still is?) contained within the grounds of a farm called 'La Belle France'.

Perhaps the most unexpected putative sabre-tooth survivors of all, however, are those that have been reported from North America,

Walrus-like cave painting in South Africa

a continent far less likely to house such creatures in modern times than either South America or Africa. Yet some tantalising snippets of information concerning this radical prospect are known to me. In December 2009, I learnt from British cryptozoological researcher Richard Muirhead that in or around 1913, two such cats had allegedly been shot dead by the U.S. Cavalry in Arizona. His source for this information was an American cryptozoologist, Andrew Ste. Marie, who in turn had heard about the incident from Joe Taylor, a museum curator. No-one apparently knew what had happened to the cats' carcases afterwards, unfortunately, so there seems no way of confirming their taxonomic identity.

Also in 2009, Richard mentioned to me that yet another of his American cryptozoological contacts, Jerry Padilla, claimed that one night in 1946 "a very close relative now deceased" saw a sabre-tooth in northern New Mexico on an old, remote mountain road near the Philmont Scout Ranch in the Sangre de Cristo Mountains. The cat's colour was that of a lion. (Further details can be found in *Varmints: Mystery Carnivores of North America*, by Chad Arment.) With its only eyewitness no longer alive, however, once again there is no direct way of investigating this cryptid further, except perhaps by interviewing local residents there in case others have seen such an animal.

In 1998, the magazines *Science Illustrée* and *Illustreret Videnskab* each contained a frustratingly brief snippet claiming that in 1994, a Roberto Guitierez had observed a creature resembling a living sabre-tooth in northern

Mexico. (*Science Illustrée* also contained the report mentioned earlier by me concerning a French sailor supposedly seeing one in Paraguay in 1984.) More details, anyone?

Most recently of all, in June 2016, Richard Muirhead sent me a series of newspaper articles concerning a mysterious, unidentified animal that he speculates may have been a living sabre-tooth from Mexico. The first of these is a fascinating article that had appeared in the 2 August 1928 issue of the *Morning Oregonian* newspaper. Reporting the appearance outside this newspaper's office of a Colonel E.R. [sic—should be R.B.] Pearson, who both looked like and had lived a life very like a latter-day Buffalo Bill or Wild Bill Hickock, it paid particular attention to the large covered-over cage that Pearson had brought with him on the back of his truck, because of the quite extraordinary creature that was supposedly concealed inside it, and which this newspaper had humorously dubbed a 'whatizit':

> . . . he declared he had an animal, which has yet received no name—either as a scientific identification or as a family name such as pets often have—hidden away in that cage. Tarpaulins, rags, sheets of tin, close screen netting and what-nots entirely covered the cage, and so the crowd was not allowed to see enough to satisfy its curiosity. Two lucky newspapermen were the only ones allowed to peek, in and they'll tell the world there was a Whatizis in there because it growled at both of them at once.
>
> Beast Comes From Mexico.
>
> The beast was captured in Mexico southwest of Mexico City about three months ago, the colonel explained, and he is now taking it east to Washington, D. C., where it will be given a scientific identification. So far the colonel does not know what it is that rides with him.
>
> Anyway, it weighs 800 pounds, looks like a combination of tiger, lion and cougar; growls like all of them together; has claws like a cat, has stripes like a tiger, a square snout, shaggy hair around its shoulders and tusks in its mouth.
>
> "What does it eat?" an onlooker queried.
>
> "One calf every 2½ days," the colonel replied. "When I can't get a calf I have to buy fresh meat from a butcher shop. That's why I look so disreputable myself."
>
> Going east from Portland the colonel expects to "percolate" over the Columbia highway to Pendleton and over the Old Oregon trail to Yellowstone Park and thence on eastward to Washington.

A second report appeared on 18 August 1928, in the *Idaho Statesman*, reporting that the Colonel and his captive mystery cat had now entered Idaho en route to Washington D.C. It also provided some additional details concerning the cat itself:

> Colonel Pearson is en route to Washington to present the beast to the National Zoological society. The animal is about twice as large as a mountain lion, with peculiar beard and markings similar to those of a tiger. Its ears resemble those of a gorilla. The colonel expects the animal, which he captured in southern Mexico, will be identified at the zoological gardens.
>
> Six men assisted Mr. Pearson in roping the beast in its native haunts, where

> it was believed to have killed several humans. One of the assistants was severely injured by the big cat, and a second victim, the colonel said, was a spectator who thought the animal a "show animal" and moved too close to the cage. His face was laid open by a sweep of the cat's paw.

The *Morning Oregonian* article had been accompanied by a close-up photograph of the Colonel and the covered cage on his truck. But this did not reveal anything of the animal, as the cage was entirely hidden beneath its covering. It is a pity that neither of the two newspaper reporters who, allegedly, were allowed to see the creature was named in that article, as they may have been traceable. But what about the central figure in this story, the Colonel himself—what do we know about him?

Richard Muirhead and fellow cryptozoological researcher Chad Arment have collected some other newspaper articles concerning him, which Richard has kindly made available to me. One of them (*Denver Post*, 11 January 1921) referred to Pearson by the nickname 'Idaho Bill' and revealed that he was born in Hastings, Nebraska, but drifted into Idaho and later served as a scout for Buffalo Bill himself, before becoming a bronco businessman, renting out broncos (wild, unbroken horses) to cowboy associations for rodeo shows and other cowboy-related events.

But did Pearson ever reach Washington D.C. and present his mystery beast to the National Zoological Society? Apparently not—because an article published on 28 November 1930 by the *Council Bluffs Nonpareil* newspaper revealed that three weeks earlier, his truck had been overturned in an accident while travelling through Iowa and his mystery cat had escaped. Curiously, however, and in stark contrast to the descriptions given in the previous articles quoted here, in a telegram sent to the *Council Bluffs Nonpareil* Pearson now claimed that "it resembled a gorilla and was not dangerous and he begged hunters not to harm it". He even referred to it as his pet—a far cry indeed from the indisputably feline and exceedingly dangerous creature that he had previously claimed it to be.

And it is here where the trail goes cold. Was Pearson's mystery pet ever recaptured, or, as suspected by Muirhead and Arment, was the entire episode just a hoax? What I can be sure of, however, is that the creature's description as given in the *Morning Oregonian* and *Idaho Statesman* articles does not correspond with any species presently known to exist in Mexico. But whether, Pearson's 'whatizit' really was a living sabre-tooth, or perhaps was merely some ordinary creature modified to look exotic (like the kangaroo with painted-on stripes and affixed fur-covered wings of wire that was once exhibited as a Jersey devil—see my book *A Manifestation of Monsters* for full details), or was nothing at all, just a figment of someone's over-blown imagination, presently remains—like so much else in cryptozoology—an unanswered riddle.

ATROX ASCENDING?—MANED MYSTERY CATS IN NORTH AMERICA

The possible present-day survival in North America of the American lion *Panthera (leo) atrox* as an explanation proposed by some cryptozoologists for maned mystery cats reported in modern times across this continent is a subject that I documented at length in my very first book, *Mystery Cats of the World* (1989). However, because that latter book not only has been out of print for many years now but also has become highly collectable (second-hand copies having been regularly sold for

Representation of the Elkhorn Falls/Wayne County mystery cats (© William M. Rebsamen)

prices of up to £200), its contents may no longer be as readily accessible to everyone as they were back in the late 1980s and 1990s. So I am reprinting below what I wrote therein on this subject, together with a few minor amendments and updates.

According to many documented eyewitness accounts from the eastern U.S.A. (plus an occasional report from the western states and Canada), a number of mystery cats resembling fully-maned African lions *Panthera leo*, not to mention lionesses and lion cubs too, are on the loose there. The principal sources of information concerning these felids are the writings and researches of veteran American cryptozoologist Loren Coleman, who has been instrumental in bringing these intriguing mystery cats to the general attention of cryptozoologists in various of his books and articles (particularly those in *Fortean Times*).

Possibly the most famous of the early U.S. lion episodes concerns a lioness-like creature nicknamed Nellie, making headlines in central Illinois during the early summer of 1917 by attacking a butler picking flowers. Despite huge searches being instigated straight away in the Sangamon River region's woodlands nearby, Nellie was never captured, but an interesting discovery was made: she appeared to have a maned mate! For a beast whose description closely matched that of a male African lion was sighted during the hunt and several times afterwards—a "large yellow, long-haired beast", according to eyewitness James Rutherford, who spotted it near a gravel pit on 31 July 1917. Like Nellie, however, this mystery cat was never caught.

A similarly shaggy felid pursued a party of four adults and two children at Indiana's Elkhorn Falls in the early evening of 5 August 1948—happily they escaped unscathed. The eyewitnesses claimed that it resembled a lion, with a long tail and bushy hair around its neck. Two days later, a comparably leonine beast was spied nearby and at close range by two farm boys—and in the company of a second strange cat that apparently resembled a black panther. One of the boys fired his rifle at them, whereupon they turned away, along a lane. The very next day two creatures identical to these were sighted by some farmers north-east of Abington, and the following morning in Wayne County, after which they were heard of no more. And in November 1950 an unseen creature that supposedly roared like an African lion was blamed for the killing and consumption of 42 pigs, 4 calves, 4 lambs, and 12 chickens in Peoria County—all in a single night!

Ceresco, Nebraska, was the scene of a prolonged but unsuccessful lion-hunt just 12 months later, which had been sparked off by a number of reports of a mysterious maned cat in the area. Over the next few years, further sightings were made at various locations close by—including Rising City and Surprise (the latter hosting an alleged sighting featuring a lion and lioness).

Lions are not exclusive to the U.S.A. either, as demonstrated by the observation of a 5-ft-long maned felid standing at least 3 ft tall and possessing a long tail that terminated in a tufted tip, reported by Leo Dallaire. He spied it on his farm near Kapuskasing in Ontario, Canada, during June 1960.

A beast sighted by an Ohio tourist in Oklahoma's Big Cabin country and described by him as being an unmistakable African lion initiated a major lion hunt in mid-March 1961. During the same period, an animal identified as a lion was also spotted by two nurses amidst shrubbery in the grounds of Oklahoma State Hospital. Local inhabitants affirmed that it had been in the area for the past two months, claimed to have heard it roar, and accused it of having eaten many of their chickens. Rogers County sheriff Amos Ward and Tulsa Zoo director Hugh Davis both felt the mystery beast did indeed exist, and a rumour (albeit unsubstantiated) of a circus truck having overturned in that vicinity some time earlier was resurrected as an explanation for its origin. Like its predecessors, however, it was never caught and reports petered out.

During the mid-1960s a young African lion was allegedly killed by two deer-hunters near Georgia's Blue Ridge. Its origin is unknown, but it was rumoured to have been a pet that had become unmanageable; a photo of this cat still exists.

In May 1970, Illinois re-emerged as a centre of lion sightings when yet another large-scale lion-hunt was launched. Its quarry was an 8-ft-long beast with a mane and long tail, observed at Parthenon Sod Farm near Roscoe by the farm owner, George Kapotas, together with Tom Terry and five of his fellow workers. The creature was not found. Nonetheless, a week later a number of livestock on Lyle Imig's farm close by hurtled through two barbed-wire fences as a result of an encounter with a mystery beast—which, although remaining unseen itself, left behind some formidable tracks. These were described as being 5 in long and 4.75 in wide, with an inter-print length of 40 in.

On or around 1 March the following year, a most peculiar cat was sighted skulking near the home of Howard Baldrige, near Centralia, Illinois. Baldrige described it as resembling a sort of shrunken lion. About twice the size of a large domestic cat and yellow in colour, it had a long tail, very short legs, and a face "like something on television". Make of that what you will!

In 1976, and at a distance of 50-100 ft, a much larger mystery cat with a mane around its neck was sighted and shot at unsuccessfully by farmer J.H. Holyoak in his pasture within Georgia's Berrien County. The same year saw yet another police lion-hunt too, its quarry having a "shaggy black mane, light brown body and a black tuft at the end of a long tail", reported by several people in Tacoma, Washington. The best that could be found, however, was a collie-Alsatian mongrel dog called Jake at the city dump.

In the late autumn of 1977, a two-month-old lion cub was discovered by Police Lieutenant Ronald King in Muscatine, Iowa, but despite his putting out a teletype report enquiring whether anyone had lost a lion, the cub's origin remained obscure. On 23 January 1978, a lady in Loxahatchee, Florida, was taken aback to see a lion just outside her window. Following the by-now-familiar pattern, a police search was duly instigated and returned home empty-handed. Moreover, the nearby Lion County Safari Park and two residents known to keep lions as pets all affirmed that none of theirs was missing. In August 1979, a lion cub appeared outside a back door in Ohio's West Chester. Although happy to be fed on cat food by the family's children, the unexpected visitor was eventually traced back to an amusement company from which it had escaped. And on 10 November 1979, a growling 300-400-lb lion prowling through Fremont, California, was sighted by several people and was encountered at close range in the Alameda County Flood Channel by Police Officer William Fontes, but it succeeded in eluding the large-scale search-party attempting to track it down.

The late 1970s also saw a supposed puma allegedly killed in the North Carolina mountains turn out to be an African lioness whose body had been rescued from a dumpster. And in 1982, a skeleton found in a ditch, again in North Carolina, proved to be that of a young African lion.

A sighting of a maned lion was made during July 1984 in a suburb of Cleveland, Ohio. Another was recorded near Texas's Fort Worth Zoo during late February 1985. Needless to say, the zoo itself was checked, but no lions were missing, and although it was later seen by two police patrolmen who identified it unhesitatingly as a lion, it eluded all attempts at capture and was not sighted again.

Maned mystery cats are still being sighted across North America. One notable case on file from West Virginia in 2007 was reported widely by the media, together with another one the following year from Virginia and one that same year from Colorado too. There were also some highly confusing reports of unidentified 'lions' from Georgia in 2009 that may—or may not—have been maned cats; and a suspiciously canine maned mystery 'lion' videoed in Los Angeles, California, during 2014.

There would appear to be three identities on offer for North America's maned mystery cats:

1) Large Dog?

This is a very popular 'official' explanation for reports of mystery cats. Yet although it is true that various dog breeds can appear surprisingly panther-like or puma-like under certain viewing conditions, very few dogs (regardless of such conditions) can be mistaken for a full-sized, fully-maned lion.

Probably the closest approximation to a leonine dog is the Chinese chow—and by chance there is at least one episode on record involving a maned mystery cat in which a specific chow was put forward as the felid in question. This was during the case of the Fremont lion of November 1979. After Fontes had made his sighting of the mystery beast in the flood channel, a local inhabitant came forward to claim that his own 40-lb chow puppy was the animal responsible—to which Fontes soon retorted: "The puppy in no way resembled the 300 to 400 pound animal observed in the flood control channel". Considering that not even the very largest of adult chows can attain such dimensions, it is hardly surprising that Fontes should so readily dismiss a chow puppy from contention.

Even more remote is the possibility that anyone could confuse a lion with a Brittany spaniel, yet this was a highly regarded 'official' solution to the lion spied near Fort Worth Zoo in 1985. Its police eyewitnesses, however, begged to differ. And who can blame them? Anything less like a lion than a white-and-orange/chestnut/black dog of maximum height 20 in and bearing a pair of fluffy pendulous ears would be hard to imagine—except, perhaps, for mongrel Jake! In any case, whereas many maned mystery cats have been reported roaring, I have yet to read of a case in which one of these animals barked! Sometimes, a dog suffering from a severe case of mange has appeared superficially lion-like, having lost much of its hair except for a fringe around its neck and at the tip of its tail, but obviously its dimensions were much less sizeable than those of a real lion.

2) African Lion Escapees?

Despite being anathema to some mystery animal aficionados, the exotic escapee theory is a far more reasonable solution to North America's maned feline mysteries. Lions are indisputably very popular animals—not only in public zoos and circuses but also in private animal collections and even as household pets. That escape (or release) of lions from captivity in North America does indeed occur has already been demonstrated by the discovery of the various dead specimens recorded here, and it seems highly unlikely that a lion or lioness would have much trouble in surviving in much of this continent's countryside. Moreover, we have already learnt that cubs are not averse to escaping.

Of great pertinence is the horrific reality that persons in Texas and elsewhere in the U.S.A. buy African lions from zoos, circuses, and pet owners for the express purpose of inviting would-be big-game hunters to come and shoot them, usually (although apparently not always) within fenced enclosures. Clearly the possibility of escapees must be a very real one. During April 1988, for instance, news emerged that two African lion carcases had been discovered on a reservoir site set aside at Texas's Wallisville by the U.S. Army Corps of Engineers and that lion hunts had indeed been occurring on that site. Yet as lion hunting was technically not illegal there, U.S. Federal officials feared that they could do little to prevent it. As such activity is not limited to Texas either, we surely need not look much further for explanations of American maned lions.

Nonetheless, a third explanation is on offer—one involving lions that in America are not out-of-*place* but out-of-*time*, i.e. involving a putative prehistoric survivor.

3) Surviving American Lions?

During the Pleistocene, leonine cats existed not only throughout much of the Old World but also, as a result of the former presence of a Bering Sea land bridge connecting northern Asia with Alaska, in the New World, thriving

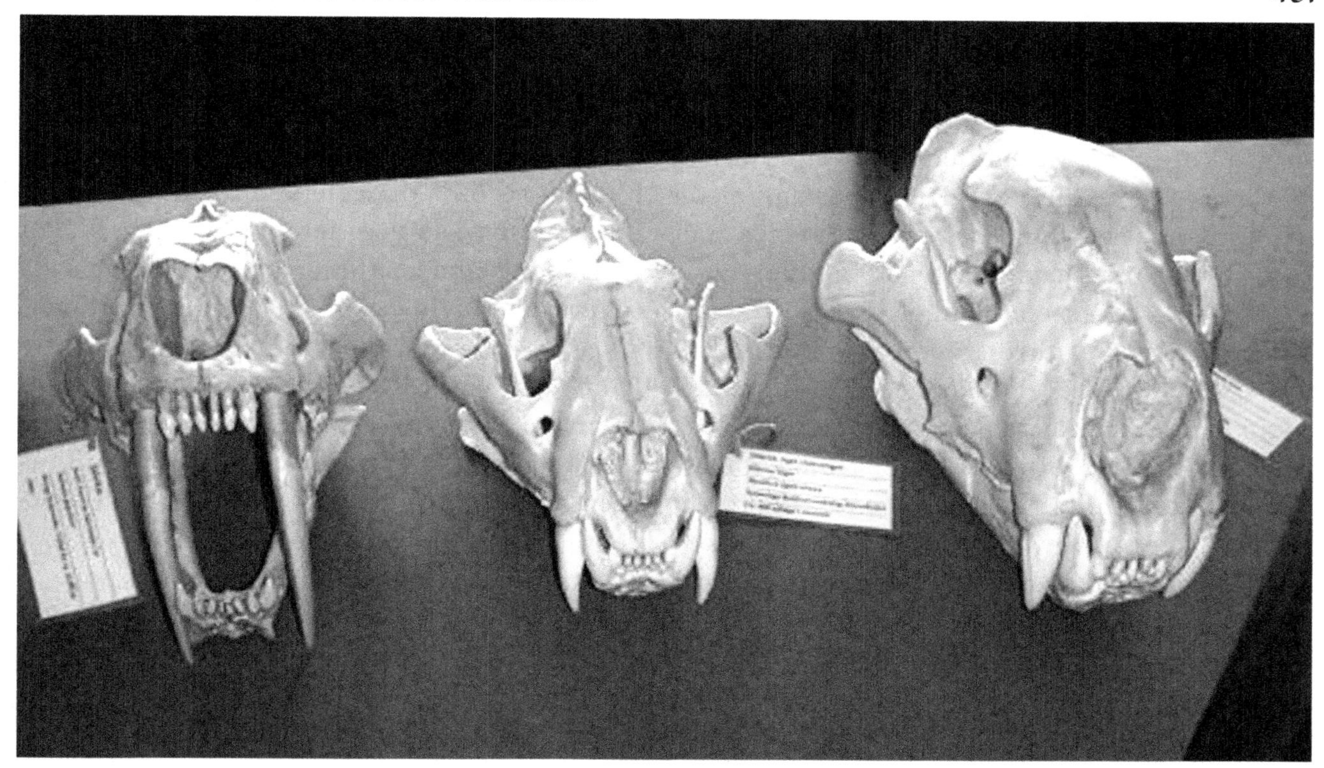

From left to right: skulls of the North American sabre-tooth *Smilodon fatalis*, the Siberian tiger *Panthera tigris altaica*, and the American lion *P. atrox*, showing the latter cat's huge size (© Markus Bühler)

in North America, Central America, and as far south as north-western South America. This New World felid, which, as revealed a little later, was truly enormous in size, is most commonly dubbed the American lion (but should not be confused with America's mountain lion or puma *Puma concolor*, which is a much smaller, very different species of felid).

Although originally categorised as a gigantic jaguar, until very recently the American lion had traditionally been classed as a leonine subspecies, but separate from all of its Old World counterparts. Deemed to be a sister lineage to the Eurasian cave lion *P. (l.) spelaea*, it has been named *P. leo atrox*. During the past few years, however, many (but not all) palaeontologists have considered it sufficiently distinct from all lions to be reclassified as a separate *Panthera* species in its own right, thus becoming *P. atrox*. It appeared to die out at the end of the Pleistocene—no fossil evidence dating from the Holocene is known. However, developing a suggestion first put forward by fellow American cryptozoologist Mark A. Hall, Loren Coleman has postulated that the unexpected maned felids being sighted across North America today may actually be surviving male individuals of *P. (l.) atrox*. Furthermore, he has proposed that the equally mystifying black panthers frequently sighted in many parts of North America but never formally identified could be *P. (l.) atrox* females.

In support of this theory, Coleman has commented in his book *Mysterious America* and elsewhere that the behaviour of these two types of American mystery cat compares closely with that of genuine lions elsewhere—with the maned beasts being proud but cautious, like typical male lions, while the pantheresque felids are less retiring and more aggressive, like typical lionesses. He has added that the classification of these two unidentified felid forms as the two sexes of a single species can also explain those very occasional sightings on record of maned cats and black panther-like cats having been seen together.

Altogether a quite fascinating theory, whose scientific development could constitute a stimulating intellectual exercise. Sadly, however, in practical terms it suffers from a number of fundamental problems, as brought to public attention by British researcher Mike Grayson in a *Fortean Times* communication (winter 1982).

First and foremost of these is the radical difference in pelage colouration between the maned and the black panther-like cats. As Grayson has pointed out, if we assume that these do indeed constitute the two sexes of the same species, the exhibition of normal leonine colouration by the male whereas the female displays very pronounced melanistic tendencies would constitute an example of extreme sexual dimorphism totally without parallel amongst other mammalian species. The male-only mane of the lion is proof that this species is indeed capable of evolving a marked degree of sexual dimorphism, but even allowing for the fact that some lions have dark belly and neck manes, it is still exceedingly improbable that such dimorphism could aspire to the exceptional level required by the Coleman-Hall theory.

Equally contentious is the total absence of sightings in modern-day North America of mystery lion prides. As Coleman himself has noted, lions are social, and, according to felid specialist Dr Helmut Hemmer in a *Carnivore* paper published in 1978, the high degree of cephalisation—brain development—displayed by the American lion makes it possible that this felid was too. So if Coleman's *P.* (*l.*) *atrox* theory is true, why no records of prides? And, indeed, why, out of the plentiful maned mystery cat records documented by him and others, are there so very few that involve one of these felids and a pantheresque cat being seen together? Having said all of that, however, since its redesignation as a species separate from Old World lions, the American lion has been looked upon by some researchers as more likely to have been solitary, which would thus seem to favour Coleman's theory after all—except for the following dilemma.

In the Old World lion, which is a polygamous social species, the darkness of a male's mane serves as a significant visual clue regarding that male's quality in relation to potential mates and rivals—the darker the mane, the more dominant the male. As stated in an *American Scientist* article (May-June 2005) by Peyton West, a member of the research team that uncovered the hitherto-obscure purpose of the lion's mane:

> Female lions live in prides consisting of related females and their dependent offspring. As the cubs grow, young females typically join their mother's pride, and young males form "coalitions" and disperse to look for their own pride. This creates a system in which a small group of males can monopolize many females, leading to severe reproductive competition. Predictably, males compete intensely for mates.

Yet although the mane's very visible demonstration of a male's fitness will therefore be very beneficial in attracting mates and repelling rivals, it comes at a high cost to the lion's health, inasmuch as a dark mane inhibits dissipation of heat from the lion's body in hot climates. Consequently, if the American lion were indeed solitary, lacking the complex social, polygamous lifestyle of its Old World relatives, we would expect it not to possess a mane at all, or only an insignificant one, because the thermal advantage in not having one would outweigh the dominance-signalling advantage of having one (which would be far less important in a solitary, monogamous species). Yet if it did lack a (notable) mane, it clearly cannot present itself as a plausible candidate for North America's maned mystery cats.

An important point highlighted by Grayson is the undeniable fact that animals identical to North America's black panthers are also being reported in regions of the world where survival of prehistoric lions is totally incongruous (such as Great Britain), as well as from regions where lions have never existed at any time (such as Australia). In comparison with the escapee theory, the *P.* (*l.*) *atrox* explanation is clearly inadequate here.

In addition to Grayson's comments regarding the *P.* (*l.*) *atrox* concept, another equally formidable and significant objection to this exists. Namely, the blunt fact that, judging from reconstructions of the American lion in life based upon fossil evidence, it simply did not resemble the maned cats being seen there today.

Whereas North America's maned mystery cats resemble modern lions both in overall size and in physical appearance, the American lion was very different indeed. To begin with, it was at least one quarter larger than the largest of modern lions—hence its earlier, alternative name, the great cat. Furthermore, relative to today's lions the American lion's face was shorter, with a broader nasal region, and its limbs were notably longer—so much so, in fact, that it is considered to have been a truly cursorial felid, far more than any modern lion. There is also the afore-mentioned issue of its putative manelessness to keep very firmly in mind.

Arguing in favour of post-Pleistocene survival of *P.* (*l.*) *atrox*, Coleman has noted that only one American lion fossil skeleton has been recovered for every thirty of its sabre-tooth contemporary *Smilodon fatalis* from the famed tar pits of Rancho La Brea, near Los Angeles, California. This is a startling ratio, one that has been put forward by many palaeontologists as evidence for the intellectual superiority of the American lion. Coupled with this is the fact that, relative to its body size, the brain of the American lion was larger than that of any Old World counterpart, either from the Pleistocene or from the present day. Consequently, it would be expected from this that *P.* (*l.*) *atrox* would stand a better chance of persisting undetected into modern North America than would *S. fatalis*.

Yet in spite of the American lion's intellectual level, we cannot ignore the seemingly irreconcilable morphological and behavioural problems noted here. Hence, although most certainly a captivating concept, the theory that America's maned and black pantheresque mystery cats constitute a relict population of *P.* (*l.*) *atrox* is ultimately untenable, at least in my opinion—with the exotic escapee theory reasserting itself as the most plausible solution to these felid forms. But what of those occasional episodes involving a maned lion-like and a black panther-like cat being seen together by various eyewitnesses? The most reasonable explanation is that in each of these exceedingly rare instances, they were a couple of escapees (probably from a circus or private collection) consisting of a lion and a melanistic leopard that had originally been reared and maintained together prior to their escape. Rearing cubs of different big cat species together is by no means uncommon in captivity, especially in smaller, private zoos, and is sometimes even done purposefully in order to facilitate interspecific matings in the hope of obtaining exotic-looking hybrids.

THE KAMCHATKAN IRKUIEM—IS IT LONG IN LEG AND SHORT IN FACE?

In spring 1987, amid the far northeastern Kamchatka peninsula region of what was then the Soviet Union but is now Russia, hunter Rodion Sivolobov obtained the skin of a giant white bear. To most eyes, it might simply look like the pelt of an oversized polar bear, but according to Sivolobov, and the area's local reindeer breeders, it is something very different—and very special. They believe it to be from a huge and extremely distinctive species of bear still awaiting formal scientific discovery—an imposing, ferocious creature called the irkuiem (aka irquiem).

For 10 years, Sivolobov had been collecting reports of this creature, much rarer and twice as big as Kamchatka's notably large brown bears, with a height at the withers of 4.5 ft and weighing as much as 1.5 tons. According to local testimony, the irkuiem has a relatively small head, short back legs, and a highly unusual running gait—throwing down its forepaws and heaving the back ones up to meet them, yielding an extraordinary 'looping' mode of locomotion almost like a caterpillar! As for its luxuriant, snow-white fur, when Sivolobov succeeded in obtaining a skin of one of these bizarre-sounding 'caterpillar bears' he promptly sent samples from it, together with a photograph of the entire pelt, to a number of zoologists in Moscow and St Petersburg for their opinions. Inevitably, however, the general consensus was that dental and cranial samples would also be needed in order to attempt a conclusive identification of its species, so Sivolobov hopes to procure these necessary specimens one day—but surely the very considerable advancements in DNA analyses that have taken place since then would yield some significant results with the pelt?

Meanwhile, he has good reason for remaining optimistic that the irkuiem's eventual discovery will prove to be a major cryptozoological triumph. For according to no less august an authority than internationally-esteemed Russian zoologist Prof. Nikolai K. Vereshchagin (d. 2008), this elusive creature could well prove to be a surviving representative of one of the Pleistocene's most impressive mammalian carnivores—the short-faced bear *Arctodus simus*.

Up to 6 ft high at the shoulder, up to 12 ft tall when standing erect on its hind legs, boasting a 14-ft vertical arm reach, and weighing as much as 2 tons, this monstrously huge bear, one of the largest of all mammalian land carnivores, was distributed from Alaska down as far as California (where it was particularly common) on the North American continent. (A second, less-famous, smaller species, *A. pristinus*, was confined to the southern U.S. states—especially Florida—and also Mexico.)

Also termed the bulldog bear, *Arctodus* was characterised not only by its squat-looking face (actually an optical illusion engendered by its short nasal regions and deep snout), but also by its relatively short body and very long legs. The result was an uncommonly gracile bear wholly unlike any species known today—so much so, in fact, that the true nature of its hunting mode remains a subject for much debate.

Its gracility argues against *Arctodus* being able to use sheer physical strength to overcome its prey, and yet its great bulk equally argues against it being able to use its lengthy limbs to chase after prey in a fleet-footed, flexible, cheetah-like manner. Consequently, a popular

The author (5′ 10″ tall) alongside a life-sized model of a short-faced bear (© Dr Karl Shuker)

theory is that this giant bear was a klepto-parasite—i.e. using its formidable size and undoubted aggression to frighten away smaller carnivores from their kill and then steal it from them.

Yet regardless of which modus operandi it employed in obtaining prey, *Arctodus* was undoubtedly successful at doing so for some considerable time, having originated around 800,000 years ago during the mid-Pleistocene epoch and persisting throughout the remainder of this geological time period. Nevertheless, the eventual extinction of the American mammoths, mastodonts, camels, horses, and other herbivorous megafauna upon which it preyed, plus changing climatic conditions, and encroaching competition from the smaller but highly resourceful brown bear *Ursus arctos*, all played a part in its own gradual demise, so that by the end of the Pleistocene, the short-faced bear had supposedly died out—but had it?

During the latter part of the Pleistocene until around 15,500 ago, Alaska was joined via the Bering land-bridge to Siberia. In 1988, Calgary University zoologist Prof. Valerius Geist suggested that the brutal belligerence of *Arctodus* might actually have impeded primitive humans' passage from the Old World into the New World via the land-bridge. However, that self-same continental connection might also have featured prominently in this bear's own movements. Could *Arctodus* have migrated across it from northern North America into eastern Asia, subsequently dying out in its original New World homeland, but persisting undetected by science amid Kamchatka's remote, harsh terrain?

If so, continued evolution may even have modified its limbs, reducing their length to yield a body shape more comparable to its chief competitor, the brown bear, but retaining its greater body size as a further means of combating the brown bear's ecological rivalry—thus yielding the irkuiem as described by the Kamchatkan reindeer breeders. Having said that, the 'caterpillar bear' locomotory aspect of the irkuiem remains a major enigma, but an explanation may present itself should supplementary information be forthcoming one day.

Interestingly, certain findings show that even in North America the short-faced bear survived to a more recent date than traditionally believed. In March 1992, Utah palaeontologists Drs David D. Gillette and David B. Madsen documented their excavation four years earlier at central Utah's Huntington Reservoir of a partial cranium and isolated rib belonging to a short-faced bear that dated to less than 11,400 BP (Before Present day)—i.e. over a thousand years more recent than the previous record for the youngest remains of this species. Moreover, they speculate that relict populations may have persisted until 10,000 years BP, or even later—i.e. beyond the Pleistocene, into historic times.

Whereas cryptozoological sceptics condemn attempts to reconcile the irkuiem with *Arctodus simus*, or a modified version of it, as little more than wishful thinking, Prof. Vereshchagin remained convinced that the prospect holds promise:

> I personally do not in any way exclude the possibility that there is an eighth species of bear in the world today. The theory that it could be a close relative of an extinct Ice Age bear does not seem so far-fetched either.

Perhaps a future expedition by Sivolobov or some other intrepid investigator will vindicate Vereshchagin's opinion?

Incidentally, 'irkuiem' is not the only name that has been applied to this particular cryptid.

It has been referred to as the god bear too, which is somewhat confusing, however, because this latter moniker has also been used in relation to a second type of huge (yet very different) ursine mystery beast of Kamchatka, one that is instantly distinguished from the irkuiem by virtue of its jet-black fur.

Long before the irkuiem became news, the forested peninsula of Kamchatka was already noted for very large bears, though these were long-haired brown bears, which in 1851 were dubbed *Ursus arctos beringianus*, the Kamchatka brown bear—the largest Eurasian subspecies of brown bear. Officially, the mighty Kodiak bear *U. a. middendorffi* of southwestern Alaska's Kodiak Archipelago, sporting an average total length of 8 ft and shoulder height of 4.33 ft in the male, is the largest subspecies of brown bear alive today anywhere. However, in 1936, Swedish scientist Dr Sten Bergman noted in a *Journal of Mammalogy* paper that Kamchatka may house a gigantic, short-furred, jet-black bear form that exceeds in size all other bears.

Dr Bergman had been shown the pelt of one of these mysterious out-sized beasts in autumn 1920 during a 1920-1922 Swedish expedition there, and he also recorded an equally colossal skull allegedly from one such bear, plus an enormous bear paw print measuring just under 15 in long and 10 in wide. Both the skull and the paw print had been observed (and, in the case of the paw print, photographed) by fellow Swedish scientist René Malaise, during his nine-year sojourn in Kamchatka.

The existence of such a bear form in this region has been supported to some extent by Russian sources, according to David Day, who noted in his book *The Doomsday Book of Animals* (1981) that weights of 2,296 lb, 2,227 lb, and 2,311 lb have been recorded by Russian hunters from specimens here. But as the most recent records concerning such huge bears date back to the early 1920s, it must be assumed that they have since disappeared. (Incidentally, some researchers have erroneously assigned the taxonomic name *U. a. piscator* specifically to these ursine giants, but in reality this name had already been coined long before such creatures had become known to scientists, having originally been applied, albeit synonymously, to the Kamchatka brown bear back in 1855.)

Having said that, rumours persist that some specimens do still exist in certain remote Siberian localities closed off by the Soviet military during the Cold War, so who knows? Perhaps it may be premature to write off Bergman's black-furred mega-bear just yet. Nevertheless, the morphological variability of *Ursus arctos* is notoriously, infamously immense—inciting the description and naming at one time or another of no less than 96 different taxa of brown bear in North America alone, plus another 271 in the Old World! Moreover, the researches of Russian bear biologist Dr Igor A. Revenko from the Kamchatka Ecology and Environmental Institute, who spent several months a year for some years from the mid-1980s to the early 1990s studying brown bears in Kamchatka, have revealed that not only is this region's brown bear population much bigger than hitherto suspected (estimated at 8,000-10,000 in 1994) but also that the genetic variation within this population is huge.

All of which means that even if it does still survive, Kamchatka's giant short-furred mystery black bear is more likely to represent a mere (albeit spectacular) non-taxonomic variant rather than a discrete taxonomic form in its own right. But until, if ever, some physical evidence can be made available for DNA analysis, its true zoological identity seems forever destined to bemuse and mystify in best cryptozoological fashion.

THE WAHEELA—GREAT WHITE WOLF(?) OF THE ARCTIC WASTES

As a zoologist with a passionate interest in cryptozoology, Ivan T. Sanderson received numerous reports of many different mystery beasts. Few, however, were more intriguing than those that he collected regarding the huge, highly elusive, ghostly-white 'wolves' occasionally encountered in the barren Arctic wastes of northern Alaska and Canada's Northwest Territories, and which are known to the native Indians as waheelas.

One source of waheela data was a decidedly hard-headed, no-nonsense truck mechanic referred to by Sanderson as Frank Graves, who had visited the Nahanni Valley in the Northwest Territories sometime during the early 1960s. It was here, while hunting for food one day with a local Native American companion, that he gained firsthand experience of a waheela. The Native American was attempting to flush out some likely targets from the dense forests encircling the base of a small plateau on which Graves was standing—ready to shoot anything that emerged. Eventually, something did emerge—but it was certainly not what Graves had expected.

No more than 20 paces from him was what he took at first to be an enormous snow-white wolf, roughly 3.5 ft tall at the shoulder, with an unusually wide head, and sporting a luxuriant pelt of very long shaggy hair. Later referring to it as "the grand-daddy of all wolves", Graves was so alarmed by the sheer size of this extraordinary creature that he immediately shot at it with both barrels of his 12-gauge, loaded with birdshot and heavy ball, and felt sure that he had hit it on the left flank. However, the wolf—or whatever it was—appeared both unharmed and unalarmed, merely gazing at him for a moment before turning around and casually ambling back into the forest. When Graves's Native American friend turned up shortly afterwards, and learnt what had happened, he became very quiet and at first would not speak of it at all. Later, however, he confessed that he was well aware of this animal, but affirmed that it was not a wolf, stating instead that it was an entirely separate type of beast.

He claimed that such beasts not only were much larger than wolves, but also could be readily differentiated by their much wider heads and thicker tails, smaller ears, rather short legs, and splayed feet. They were behaviourally distinct too—hunting alone, avoiding true wolves, and functioning as scavengers rather than active predators. In addition, they may exhibit one further, and particularly sinister, difference.

Nahanni Valley is often referred to as the Headless Valley—seemingly for good reason. According to Sanderson, the Canadian mounties have on their records a number of unsolved cases featuring prospectors and other travellers who either have never returned after visiting this locality or have been found lying dead in their cabins there with their heads bitten off. Moreover, in a *North American BioFortean Review* article from January 2006 surveying Headless Valley lore, North American cryptozoologists Gary A. Mangiacopra and Dwight G. Smith actually provided a chronological listing of persons who suffered one or other of these grisly fates there.

This mode of dispatching humans is not typical of true wolves, nor of bears—so what was responsible for their deaths? As far as his Native American companion was concerned, the mysterious species encountered by Graves is the most likely candidate, and confirmed that whereas in most areas they are rare, spending much of their time in the far north's tundra zone, they occur throughout the year in Nahanni Valley.

Artistic rendition of the waheela's possible appearance (© Tim Morris)

Similar stories, in which descriptions of the creatures in question tallied closely with Graves's, had already been mentioned to Sanderson by an old friend, professional cameraman Tex Zeigler, who had learnt of these animals during a trip to Alaska—which is why Sanderson had been so intrigued by Graves's wholly independent testimony. An early colonial account was later obtained by fellow cryptozoologist Loren Coleman from the historical archives of Indiana University, which featured a fatal encounter by three trappers with a waheela near a lake in northern Michigan; and other stories reached Sanderson's ears from elsewhere amid the forested portions of northern Canada.

What is so significant about this body of anecdotal evidence is that if such creatures genuinely exist, they may indeed prove to be something far removed from true wolves. Up to around 5.5 million years ago in North America (and even more recently in Pakistan), representatives of a very unusual group of carnivores still existed. They were known as amphicyonids or bear-dogs, because their massive forms seemed to unite the most characteristic morphological features of bears and dogs—partnering the burly body, short limbs, and plantigrade feet of bears with the dentition and face of wolves. In reality, however, they were neither, constituting instead a distinct taxonomic family of their own—but they would have compared very favourably in appearance with the elusive waheela.

The amphicyonids' eventual extinction is thought to have resulted from undue competition with other mammalian carnivores—but in the bleak, inhospitable terrain of the tundra, seldom visited by humans and offering far less rivalry for prey than the more temperate regions further south, the tantalising possibility of modern-day survival inevitably presents itself. For it is here, if anywhere on Earth, that a species of amphicyonid could have persisted—undetected by science, undisturbed by humans, and unchallenged by its fellow fauna.

Having said that, the common wolf *Canis lupus* is an exceedingly diverse species genetically and morphologically (so diverse, in fact, that recent researches have elevated certain subspecies to species in their own right). Hence it may well be that the waheela is merely an extra-large variety adapted for existence in an Arctic environment. Supporting this option are reports of a remarkably similar Arctic cryptid in Greenland, the amarok.

Since Sanderson's death in 1973, the waheela has become one of cryptozoology's forgotten mystery beasts. Judging from the case presented here, however, the wastelands of North America possess considerable potential for ultimately hosting a sensational zoological discovery—always assuming, of course, that they can attract the attention of someone bold enough to pit himself against a creature capable of decapitating a man with a single bite.

Incidentally, in a posthumously-published *Pursuit* article from 1974, Sanderson used the term 'dire wolf' when referring to bear-dogs as a possible explanation for the waheela; this incorrect terminological usage has caused some degree of confusion ever since, because the dire wolf *Canis dirus* was an extinct species of true wolf, i.e. a canid, not an amphicyonid. (Having said that, in his book *Bird From Hell and Other Mega Fauna* (2010), Gerald McIsaac discussed a mysterious canine entity known to the First Nations people of north-central British Columbia, Canada, as the wilderness wolf, which he considered may constitute a relict representative of the dire wolf. This latter New World species was larger than

Restoration of *Amphicyon ingens*, a North American bear-dog (Roman Uchytel)

the common wolf *C. lupus*, but it officially became extinct near the end of the Pleistocene, just under 12,000 years ago.)

Also, the description of the waheela seen by Graves that I have given in my coverage of this cryptid here is not Graves's original account but rather an expanded version of it compiled by Sanderson that he included in that same *Pursuit* article of his from 1974. Until recently, I had never seen Graves's original account, but I lately came upon a reprinting of it in the October 2004 issue of the *North American BioFortean Review*—so for the sake of completeness, here is the relevant section from it in which Graves documented his encounter with the waheela:

> But then an enormous white thing that I at first thought must be a Polar bear just sort of wandered out of the trees. It wasn't a bear; it looked more like a gigantic dog. It stood straight up on rather long legs, more like a dog or a wolf. I had seen plenty of wolves and some of them are enormous enough up there; but this thing was twenty times the size of any wolf I had ever heard of. By a sort of reflex action I fired at it — and it was less than twenty paces away

Dramatic representation of the ferocious Nandi bear (© Markus Bühler)

and only partly screened by little bushes. I hit it with two barrels of ball-shot. It didn't even jump, but turned away from me, and just walked back into the forest. I reloaded and fired again, and I know I hit it in the rear, but it just kept on walking. Shortly afterwards, my Indian friend bobbed up, asking what I had got. I didn't know what to say for a bit but, when I told him, we did another of our famous disappearing acts, and this time we loaded the boats and pushed off up river — real fast.

The original publication source of Graves's account, as told by him to Sanderson, is currently unknown.

THE NANDI BEAR—HORRIFIC HYENA, BLOOD-THIRSTY BABOON, UNEXPECTED URSID, OR CRYPTIC CHALICOTHERE?

'Nandi bear' is an umbrella term that has been indiscriminately applied to all manner of mystery beasts reported from eastern Africa (especially western Kenya's Nandi district). In *On the Track of Unknown Animals* (1958), Dr Bernard Heuvelmans sought to disentangle the many wholly separate creatures that had been erroneously lumped together by previous Nandi bear investigators—including all-black aged ratels (honey badgers) *Mellivora capensis*, aardvarks *Orycteropus afer*, unusually big baboons, and abnormally large and/or aberrantly-coloured hyenas (but no true bears!). Ratels and aardvarks do not constitute prehistoric survivors, so they fall outside this book's remit, but at least four other beasts that might be components of the Nandi bear composite (including two not pursued by Heuvelmans) may indeed involve such creatures.

A Short-Faced Shocker from the Pleistocene

Let us begin with the hyena member of this potentially anachronistic quartet, because I consider it possible that some of the Nandi bear reports attributed to hyenas did not simply involve freak individuals of modern-day forms. Instead, the true explanation may feature a certain species of awesome prehistoric hyena—a creature that would be formidable enough in appearance and ferocious enough in temperament to account for even the most terrifying reports of the infamous Nandi bear.

What Europeans call the Nandi bear has a vast range of local African names—including chemosit (chimiset), kerit, koddoelo, gadett (geteit), and khodumodumo. Of these, most reports concerning the first-listed seem to describe beasts belonging to the baboon category of Nandi bear (see later), but there is one notable exception—documented in 1927 by Captain William Hichens, a local magistrate and government official based in East Africa (see also Chapter 1).

He had been sent to a small Nandi village to investigate the death of a six-year-old girl, who had been killed and carried off by an unidentified animal that had battered its way through the mud-and-lattice wall of a native hut in order to reach her. The natives claimed that it was the work of a chimiset, and according to the village chief it inhabited a small forest-clad kopje (boulder hill) some 5 miles away. One night not long after this, during Hichens's unsuccessful search for it, the alleged chimiset snatched away his own hunting dog, tethered to one of his tent's poles. In so doing, it pulled the pole away, causing the tent to collapse and envelop Hichens, which prevented him from seeing it—but he certainly heard it:

> . . . the most awful howl I have ever heard split the night. The sheer demoniac horror of it froze me still. . . . I have heard half a dozen lions roaring in a stampede-chorus not twenty yards away; I have heard a maddened cow-elephant trumpeting; I have heard a trapped leopard make the silent night a rocking agony with screaming, snarling roars. But never have I heard, nor do I wish to hear again, such a howl as that of the chimiset.

A trail of blood on the sand showed where the beast had carried away Hichens's dog, and also revealed a clue as to the nature of its abductor:

> Beside that trail were huge footprints, four times as big as a man's, showing the imprint of three huge clawed toes, with trefoil marks like a lion's pad where the sole of the foot pressed down. But no lion, not even the giant nine feet four and a half inches long which fell to Getekonot, my hunter, at Ussure, ever boasted such a paw as that of the monster which had made that terrifying spoor.

These tracks led to the forested kopje, but although other examples were found there on more than one occasion during the several days Hichens and his team spent exploring its domain's gloomy, fearful interior, the creature itself remained resolutely hidden from view.

As Heuvelmans pointed out, Hichens's description of the tracks points towards a hyena rather than a large baboon or big cat, and there is no better candidate than a hyena for emitting the bloodcurdling howl described here—but a hyena of leonine proportions?

Yet such a beast could also explain certain mystery creatures documented by traveller Roger Courtney, in his book *Africa Calling* (1936). One of these animals was reputedly encountered by a white settler in the Trans-Nzoia district:

> This man, who was a teetotaller, a non-smoker, and ordinarily truthful, swore that he had been attacked one night, when he was alone in his hut, by something that could only have been a Nandi Bear. The creature broke down the door to get in at him, and was, he said, about eight feet high and like a grey polar bear. This terrible creature, its red eyes blazing and jaws slavering, went straight for the man, and a chase round the table followed. Luckily, however, the man managed to grab a revolver that was hanging on the wall and fire into the animal's chest, at which it turned and, growling horribly, made out through the doorway and off.

A second incident featured Courtney himself:

> One day, when hunting in the forest, I was shown by a palpitating game-scout a pair of very peculiar animal footmarks. They were two enormous pug marks, the size of dinner-plates, in a soft patch of ground. They were spade-shaped and turned inward. The claws must have been non-retractile, as I could distinctly see the small cuts where they had dug into the earth. The fact that they turned inward revealed a bear-like character; otherwise I would have said that they had been made by the grandfather of all hyenas. But, then, a hyena enormous enough to leave footprints as big as those would have himself been a fabulous beast.

With no physical evidence for true bears—ursids—existing in Africa today, the giant hyena identity once again arises (but see later for more about African ursids).

Many reports of Nandi bear depredations, whether upon livestock or upon village inhabitants, feature grisly details of how it tears open its victim's head—with such force that often the top of the skull is entirely detached—and scoops out the brain to eat. Circa 1919 in the Lumbwa area of Kenya, one such brain-eater, which was termed a gadett by the locals, was supposedly hunted down and killed—whereupon it was found to be an unusually large specimen of the spotted hyena *Crocuta crocuta*. There is no proof, however, that this beast was genuinely the gadett—and sceptics have suggested that hyenas, unlike baboons, would be unlikely to carry out the type of assault laid at the gadett's deviant door.

According to Pitman, however, such activity is far from unknown for hyenas:

> Such injuries to human victims as scalping and crushed skulls are often cited as resembling the work of a bear. It is true that bears often thus mutilate their victims, but to assign these injuries to some strange bear-like animal is to betray ignorance of hyena habit. Bold and hungry hyenas often maim horribly porters sleeping in the open or natives in flimsy huts, and it is curious that in the majority of cases the head is selected for attack—dozens of cases could be quoted

of scalping, skulls crushed, and faces literally torn off by spotted hyenas.

Around 1957-1958, Douglas Hutton, the manager of Chemomi Tea Estate in the Nandi district, shot two specimens of a strange, still-unidentifed species and sent their carcases to the tea factory, where several members of staff viewed them. During the early 1980s, Nairobi-based naturalist Dr G.R. Cunningham van Someren interviewed three of these eyewitnesses independently, and received detailed, largely consistent descriptions.

Standing almost 3 ft high at the shoulders, which bore a heavy mane of long hair, the creatures had broad, short heads with small ears, broad chests, rearward-sloping backs that emphasised their long heavy forelegs and shorter hind legs, and short tails. Only when enquiries were made as to the colour of their fur was there any noticeable disagreement between the trio of eyewitnesses—two claimed that it was grey-brown with light tips, the third stated that it bore black spots on a lighter background colour. None had ever seen such animals before.

The creatures' skeletons had been left in the bush to be cleared by ants, after which Hutton sent them to Nairobi Museum for examination. In the report that he later received, their species' identity had been given, somewhat enigmatically, as a 'giant forest hyena'.

In 1962, the father of Nandi-born white hunter Jamie McLeod shot a creature whose unresolved species McLeod himself has since seen, and which he too referred to as a giant forest hyena. According to his description, it is twice the size of the spotted hyena, with long shaggy brown hair that tends to be very dirty on its belly, a lion-sized head, large carnivorous teeth, and a sloping back (but not as pronounced as the spotted hyena's).

At the end of July 1981, Ken Archer, General Manager of Eastern Produce Company, was informed that farmers working on plots of land on the Nandi Escarpment road from Chemilil to Nandi had sighted a mysterious animal that they were unable to identify. On 12 August, he and van Someren visited the farmers to ascertain the animal's appearance—and discovered from the vivid description given by one of the observers that it matched that of the beasts shot by Hutton back in the late 1950s. He denied emphatically that it had been a baboon or a pig, and did not think that it was a spotted hyena either.

From such reports as these, I too consider it unlikely that 'giant forest hyenas' of the type documented here are spotted hyenas. Occurring throughout eastern Africa south of the Sahara, this latter creature is the largest of the three modern-day species of carnivorous hyena, weighing as much as 120 lb, with a total length of up to 7 ft in exceptional specimens, and a shoulder height sometimes reaching 3 ft. Yet these still fall well short of the dimensions required by any hyena seeking to assume the role of the mystery beasts reported here.

As for the long, shaggy, drab-coloured fur usually ascribed to them, this is much more reminiscent of another modern-day species, the brown hyena *Hyaena brunnea*. A little-known nocturnal species inhabiting the plains, savannahs, and even the seashores of southern Africa, it can come as something of a surprise even to the natives, let alone Westerners (though I was fortunate enough to see one in 2008 when on safari in South Africa)—and can be quite ferocious if threatened. In terms of size, however, the brown hyena is much smaller than the spotted hyena. But if a creature resembling a gigantic brown hyena existed, it could make an extremely convincing 'giant forest hyena'—which is why I am so intrigued by the fact that, not long ago, something very like that did indeed exist in the Dark Continent.

The species in question was *Pachycrocuta* [formerly *Hyaena*] *brevirostris*, the short-faced hyena—named after its abrupt, bear-like muzzle—which inhabited not only Africa but also Europe and Asia during the Pleistocene. Although basically hyaenid in outline, its dimensions were those of a lioness, making it the largest of all hyenas, and its canine teeth were enormous. Moreover, unlike modern-day hyenas, this massive species was a much more active predator, less dependent upon carrion-feeding and scavenging—and the sight of one of these creatures hunting must have been awe-inspiring in the extreme. As noted by renowned Finnish vertebrate palaeontologist Prof. Björn Kurtén in his book *Pleistocene Mammals of Europe* (1968):

> Anybody who has seen an angry, snarling brown hyena with its big shaggy mane on end will appreciate what a terrifying apparition its gigantic extinct ally must have been to the primitive men of its day.

Two brown hyena photographs—one vintage, one recent—showing this species' distinctive mane (bottom © Markus Bühler))

Moreover, research by anthropologists Drs Noel Boaz and Russell Ciochon on Peking man *Homo erectus* fossils unearthed alongside those of short-faced hyenas in China's famous Zhoukoudian cave system has attributed scoring and puncture patterns observed on some *H. erectus* long bones and skulls—marks originally thought to be signs of cannibalism—to predation by this huge hyena.

It was in a *Fortean Times* article (February 1993) that I introduced the hypothesis that some Nandi bear reports could be convincingly explained by the undiscovered contemporary survival of the short-faced hyena—and in both appearance and behaviour this ferocious, immensely powerful species of enormous size undeniably provides a much closer comparison to those animals grouped within the giant hyena category of Nandi bear than does any other creature from the present or the past.

Nor can there be any mystery why the Nandi bear is so greatly feared if the short-faced hyena does indeed still survive. The concept of a creature as large as a lioness, equipped with the brutal strength and crushing jaws of a hyena, and functioning not as a skulking scavenger but as an active predator is too hideous to contemplate even from the safety of an armchair in England, let alone amid the primeval darkness of the Nandi forests.

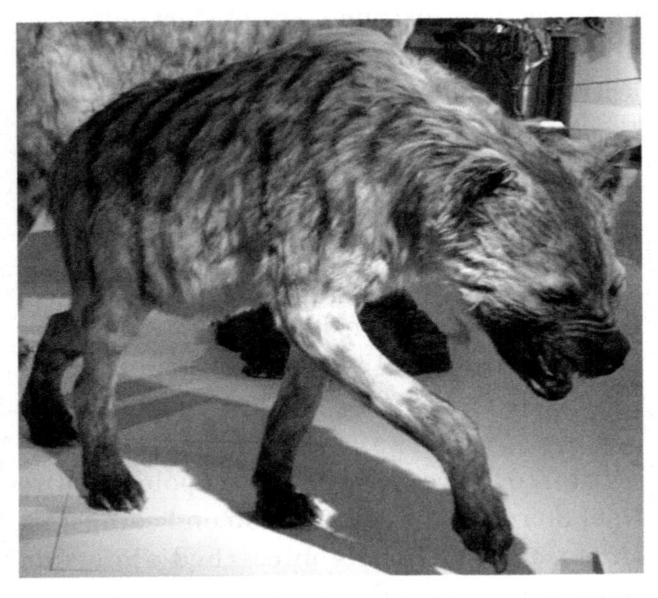

Reconstruction of the short-faced hyena (Tiberio)

Incidentally: in the earlier, ground sloth section of this present chapter, I noted that zoological collections historian Clinton Keeling had suggested that living Nandi bears may have been separately displayed by two travelling shows in England during the 1700s and 1800s respectively. As I discussed there, I consider it more likely that the two creatures exhibited by Mander's Menagerie while at York in 1869 could actually have been living ground sloths. Conversely, the creature exhibited at Halifax during the 1730s by the other show (name and proprietor unknown) holds more Nandi bear promise, at least on first sight. According to a preserved advertising bill from that venue, the show's collection of animals exhibited there included:

> A young HALF and HALF; the head of a Hyena, the hind part like a Frieseland Bear.

Keeling wondered whether the "Frieseland [sic] bear" that the Halifax 'half and half' was likened to was a polar bear *Ursus maritimus*. In reality, however, the only bears native to Friesland, which is part of the present-day Netherlands, are brown bears *Ursus arctos*. Consequently, this suggests that the animal's hind parts resembled a brown bear's, not a polar bear's (i.e. brown-furred, not white-furred). But could this hyena-headed, bear-bodied beast truly have been a subadult Nandi bear, as Keeling believed?

I documented Halifax's 'half and half' fully in my book *A Manifestation of Monsters* (2015), and, as I pointed out there, it sounds very reminiscent of a certain scientifically-recognised but publicly little-known species—one that has already been noted here in this present chapter, whose distinctive appearance would undoubtedly have made it a most eyecatching exhibit. Namely, the brown hyena, which just so happens to combine a hyena's head with a dark brown shaggy-furred body that is definitely ursine in superficial appearance, and which is seldom seen in captivity even today (let alone almost 300 years ago).

So might the Halifax mystery beast have been a subadult (or even an adult) brown hyena, captured alive alongside various more common African species and then transported to Britain with them, where it was destined to be displayed to a wide-eyed public that had never before seen this exotic-looking species? It is certainly not beyond the realms of possibility, and is, I feel, a more plausible identity than a Nandi bear.

Then again, if this latter cryptid is real but was more common back in the 1700s than it seems to be today, who can say for sure that a possibly still-young specimen wasn't captured alive by some brave, enterprising animal collector in Kenya and brought back to England for exhibition purposes there?

A final, previously-obscure Nandi bear report relevant to the hyena category appeared on 9 July 1923 in the *San Diego Union*, a Californian newspaper. It reads as follows:

> American Kills Mystery
> Beast in E. Africa
>
> It is thought that the strange man eating beast which for years past has terrorized the Kericho district of East Africa has been killed at last. It is certain that a weird animal hitherto unknown to science has fallen to the gun of J. Herman Burge, an American sportsman, who is at present on a shooting expedition to Kenya colony. The beast was "spotted" early one morning outside the camp, and was shot in the act of charging one of the party. The animal is described as a species of giant man-eating hyena of a strength and bulk unheard of, and with jaws as powerful as a lion's. The skin is stripped [sic—striped] like an ordinary hyena's, but unlike that animal, its hind and fore-quarters are of an equal height. It has a mane of long, stiff bristles, which are capable of being erected to form a terrifying fringe to the massive face. Not long ago this animal, which is known by the natives as the ketet [=geteit], or Nandi bear, dragged a live ox over a five-foot wall. A few days later it killed a native with one blow of its paw.

It is clear from the above account that those who observed this singular beast considered it to be a hyena, albeit one of massive proportions, which makes the claim that its hindquarters and forequarters were of equal length so puzzling. Also, although its stripes and mane recall the morphology of Africa's familiar striped hyena *Hyaena hyaena*, the latter species' mane does not fringe its face when erected. Most mystifying of all, however, is what happened to this potentially significant specimen. Was it (or at least its head) retained as a trophy by Burge? If so, it needs to be traced. (I have already checked online for Americans with the name J. Herman Burge, only to uncover a surprising number of matching persons and whose age in 1923 was also compatible with the hunter's likely age.) Certainly, it is not every hunter (or the descendants of one) that may have a major still-undescribed species among their collection of spoils.

Nightmarish Nandi Bears as Monstrous Mega-Baboons?

The short-faced hyena would assuredly be a petrifying creature to encounter, but a second potential prehistoric survivor that may feature in the complex composite that we know as the Nandi bear may be no less terrifying.

Could certain Nandi bear reports feature a persisting species of giant Pleistocene baboon exhibiting horrific, bloodthirsty behaviour, as suggested by Heuvelmans? I first examined this nightmarish concept as part of a chapter on baboons in my book *The Menagerie of Marvels* (2014)—here is what I wrote:

True baboons (i.e. belonging to the genus *Papio*) typically subsist upon such mundane sustenance as grass, fruit, leaves, roots, tubers, insects, birds' eggs, and small rodents, but as a number of wildlife researchers and observers will testify, these aggressive primates are more than willing to set their sights upon larger, fleshier food sources too, should the opportunity to do so arise. American zoologists Prof. Sherwood L. Washburn and Prof. Irvin DeVore have spent many years studying baboons in several different areas of East Africa. In one reservation, baboon troops often visited the same water holes as smaller monkeys called guenons. Normally, the two species coexisted with no conflict apparent between them, the guenons moving freely among the baboons. One evening, however, the two zoologists were startled to see a baboon suddenly seize hold of one of the guenons, kill it, then promptly devour it. On another occasion, in Kenya's famous Amboseli National Park, they saw two large male baboons boldly kill and eat two newborn Thomson's gazelles, despite the efforts of their respective mothers to protect their young.

At least, however, these unfortunate victims were dead before they were consumed. In a letter to a German wildlife magazine, a South African observer, L. MacWilliam, reported how he had once seen a baboon capture a hare, then sit down and gaze for a time at its terrified victim, gripped firmly in its paws, before casually biting off one of the hare's ears and eating it, then tearing a large chunk of flesh out of its struggling victim's body and eating that, and then another chunk, and another, before finally discarding the hideously mutilated yet still-living creature and rejoining its troop nearby.

Occasionally, baboons have even been daring enough to steal human babies. One such incident took place in 1965, at an immigration camp established in Brakpan, South Africa, when a particularly bold baboon seized a baby out of its carriage in the presence of its mother, then promptly killed it by biting the poor child's head repeatedly. Uganda Protectorate game warden Colonel Charles R.S. Pitman reported that five native children were attacked in separate incidents within a single year in Uganda's Bunyoro District, and that children guarding crops were sometimes seized by a baboon and disembowelled via powerful downward kicks of its muscular clawed feet.

The largest baboon species of any kind alive today is the gelada *Theropithecus gelada*. Native to the highlands of Ethiopia, its head-and-body length is 20-30 in, with a tail length of 12-20 in, but males average 41 lb in weight (females average 24 lb).

Unlike true baboons, however, the gelada is a vegetarian, existing as a graminivorous grazer. It is the only modern-day species in the genus *Theropithecus*, but in prehistoric times there were other, even larger members. Of these, the most spectacular species must surely have been *T. oswaldi*, dating from the early to mid-Pleistocene of South Africa, Kenya, Tanzania, Ethiopia, Morocco, Algeria, and Spain, because this enormous baboon was as big as a gorilla. Mercifully, however, just like the gelada, it was strictly herbivorous.

But not all giant prehistoric baboons were plant-eaters. Bearing in mind how savage and bloodthirsty today's *Papio* baboons can sometimes be, how much more so might an extra-large baboon of meat-eating persuasion have been back in Africa's far-distant past? Take, for instance, the very aptly-named *Dinopithecus* ('terrible monkey') *ingens*, which inhabited South Africa during the Pliocene epoch, 5.3 million to 2.58 million years ago. Adult males were up to 7 ft long, 5 ft high, and 200 lb in weight. This monstrous monkey shared its domain with our distant ancestor *Australopithecus africanus*, upon which it may well have preyed, because in a notable size-reversal, *A. africanus* was no bigger than a present-day baboon whereas *Dinopithecus* was the size of a full-grown present-day human male. How fortunate, then, that *Dinopithecus* is long-extinct—or is it?

Of particular relevance to the putative giant baboon identity for the Nandi bear is the koddoelo—the Nandi bear representative that allegedly frequents the dense forest of the lower and middle valley of the Tana River, which at just over 600 miles long is the longest river in Kenya. During the early years of the

20th Century, the District Officer for that Tana region was a Mr Cumberbatch, who provided the following details to British anthropologist C.W. Hobley, and which in 1912 Hobley published in the *Journal of the East Africa and Uganda Natural History Society*:

> . . . the German missionaries who have lived for many years at Ngao state that the Pokomo natives know of a forest beast called the 'Koddoelo,' and one is said to have been killed near Ngao some years back. On one occasion one of the missionaries found that the whole population of the biggest Pokomo settlement in Kina Kombe district had deserted their village and crossed the river because this animal was roaming about in the bush near the village.
>
> The animal was described to the District Officer by a Pokomo (who, however, admitted that he himself had not seen it) as being as large as a man, as sometimes going on four legs, sometimes on two, in general appearance like a huge baboon, and very fierce.

Although not common, bipedalism in baboons is by no means unknown. In 1976, for instance, Dr M.D. Rose published a detailed paper in the *American Journal of Physical Anthropology* documenting bipedal behaviour within a troop of olive baboons *P. anubis*, noting that it occurred in a wide variety of situations, of which feeding was by far the most common one. Moreover, in a *Folia Primatologica* paper from 2013, Drs F. Druelle and G. Berillon produced quantitative data revealing

Restoration of *Dinopithecus ingens* (© Hodari Nundu)

that in a captive situation, bipedal posture occurred more frequently in juvenile than in adult olive baboons.

In 1913, Hobley published further details regarding the koddoelo in the *Journal of the East Africa and Uganda Natural History Society*, deriving his information from a Mr Rule, who had enquired about it among the Wa-Pokomo people. Here is a detailed description of this cryptid as supplied by the Wa-Pokomo to Rule and thence to Hobley, together with an additional account that had been passed on to him by Tana's then-Assistant District Commissioner:

> Colour, reddish to yellow; length, about 6 feet; height, about 3 feet 6 inches at the withers; hair long, and all accounts agree on the point of a thick mane; tail short and very broad; claws very long; head, fairly long nose, teeth long but not so long as a lion; fore-legs said to be very thick.
>
> The Pokomo state that several have been killed, and one man says that he killed one himself a good many years ago. It is said to be very fierce, and to visit villages and carry off sheep. On these occasions the natives either cross the river until it leaves the neighbourhood or frighten it away by beating drums. The Waboni hunters know the beast well, but say that they prefer to leave it alone.
>
> The Assistant District Commissioner on the Tana also sends a further account of the animal, based on recent inquiries, and it was described to him by Pokomo, who said they had seen it, and their account was as follows:—Light in colour, long hair on neck and back, usually goes on fore-legs but can go on its hind-legs, not known to climb trees, rather smaller than a lion, tail about 18 inches long and some 4 inches broad, is nocturnal in its habits, fore-legs very thick; said to leave a track with one deep claw mark behind the marks of its four toes (this is rather obscure). They are agreed about its ferocity, and say it attacks a man on sight. One is said to have killed a rhino near Makere, but this is rather difficult to credit. One tried to raid a goat kraal last January, but was driven away by the noise made by the villagers when the alarm was given.

It is fully confirmed that baboons in South Africa will kill and devour sheep and goats. Rhinos, conversely, are another matter entirely—unless *Dinopithecus* or something like it has indeed persisted into the present day? Of course, the idea of such a large, distinctive, aggressive creature surviving undiscovered by science in modern times seems very remote. Nevertheless, there is little doubt that the descriptions given here do recall a baboon in overall morphology, but one of seemingly much bigger size than any species officially existing today.

Having said that, it has long been known that there are a few reports on file concerning exceptionally large specimens of known modern-day baboon species, including the aforementioned olive baboon. One such report was published by the earlier-mentioned Colonel Charles R.S. Pitman more than 70 years ago, in his second of two autobiographical books, *A Game Warden Takes Stock* (1942), and reads as follows:

> Several years ago I was informed that an outsize race of baboons frequented certain parts of the forest [Uganda's

Might a giant, highly aggressive form of baboon explain at least some Nandi bear reports? (© William M. Rebsamen)

Mabira Forest]. They occurred either singly or in small parties, were reputed to be nearly the size of a full-grown man, extremely wary and rarely seen. Further, it was stated that the attitude of these nasty creatures towards unarmed natives was extremely truculent and, naturally the local folk were very afraid of them. I was promised a specimen if one could be obtained. Eventually an enormous example was collected, but though interesting it proved to be nothing new, the scientific verdict at the British Museum (Natural History) identifying with *Papio anubis anubis*—the green baboon.

Similarly, when considering the Nandi bear in his first autobiographical book, *A Game Warden Among His Charges* (1931), Pitman had stated:

> Some of the male baboons I have seen on the Uasin Gishu Plateau [in Kenya] have been of colossal size, capable of killing children with ease; large dogs have been almost torn to pieces, the victim held in its arms. The ape [sic—baboons are of course monkeys] practically disembowels it with downward sweeps of its muscular nail-tipped legs. A great male baboon indistinctly seen in grass or amidst bushland might well be taken for an unknown species.

So if native descriptions of the koddoelo have been liberally seasoned with a generous helping of exaggeration coupled with imperfect observations, it may well be that this belligerent cryptid is simply based upon extra-large specimens of modern-day baboon species, without having to contemplate the formidable prospect of *Dinopithecus* resurrection.

Even so, a giant baboon has also been entertained as a plausible identity for two further Nandi bear incarnations—the chemosit and the kerit. In his own book's coverage of this subject, Heuvelmans postulated that perhaps surviving descendants of *Dinopithecus*:

> . . . have given rise to the widespread legends about the chemosit and koddoelo? At all events a reconstruction of a giant baboon is extraordinarily like most of the natives' and many of the settlers' descriptions of them.

Sadly, however, this entire issue may well remain forever within the realm of unsubstantiated speculation, because Nandi bear sightings seem to have entirely dried up. Indeed, I am unaware of any reports of such encounters from within the past few decades, leading to the inevitable conclusion that even if it were indeed a real creature, the Nandi bear may now be extinct.

Yet perhaps this may not be such a bad thing. After all, as judiciously noted by Heuvelmans:

> If such a beast survived it is easy to see why the chemosit arouses such terror. The mere thought that there may be a living animal as huge and strong as a gorilla and as brutally savage as a baboon is frightening enough.

Amen to that!

Not a Hyena but a Hyena Bear?

Heuvelmans did not include a true (ursid) bear as an identity for the Nandi bear, as it is known from the documented fossil record that with one prehistoric exception such creatures never penetrated sub-Saharan Africa. But that one exception is a very telling one, and in my view is well worth further consideration rather than dismissing it out of hand.

Known as the African hyena bear *Agriotherium africanum* (one of several species belonging to this now-extinct but once-widespread ursid genus) and dying out around 5 million years ago, its fossils have been found not only in South Africa (famously), but also in Ethiopia (situated directly north of Kenya). Moreover, as revealed in a 2005 *Revista de la Sociedad Geológica de España* paper, a related species, *A. aecuatorialis*, is represented by remains in upper Miocene/basal Pliocene-dated deposits at the Tugen Hills in Kenya itself. *Agriotherium* fossils of comparable age are also on record from the western rift in Uganda (situated directly west of Kenya). In Ethiopia, this ursine genus did not vanish until around 3.6 million years ago during the mid-Pliocene, and in France it persisted into the early Pleistocene, surviving until a mere 2.5 million years ago.

Up to 9 ft long and possessing dog-like crushing teeth, *Agriotherium* is believed to have been predominantly carnivorous (estimates of its bite force calculated by researchers and released in November 2011 suggest that it was the most powerful yielded by any terrestrial mammal—hence 'hyena bear'). And it would have been big enough to prey upon even the largest ungulate species inhabiting its domain, let alone smaller mammals, such as primates.

Heuvelmans sought to explain all remotely ursine Nandi bears as extra-large ratels. However, certain eyewitness accounts of Nandi bears certainly recall a bona fide—and notably

Artistic restoration of *Agriotherium* (© Hodari Nundu)

large—bear, as with the huge, previously-mentioned 'grey polar bear' encountered in the Trans-Nzoia district and documented by Courtenay in his 1930s book. If described accurately, this cryptid would have been far too big, and the wrong colour, to have been a ratel. Some bear-like footprints have also been observed, as previously noted here too.

If a true bear as monstrous as *Agriotherium* has somehow succeeded in lingering on in scientific anonymity amid the wilder, more remote regions of Kenya, it is little wonder indeed why it has incited such terror among those humans unfortunate enough to live within or venture forth into its territory.

The Chalicothere Connection

Then again, in 1923 a discovery was made in Central Africa that spawned what must surely be the most unusual but also the most intriguing of all Nandi bear identities currently on offer—an identity so fascinating in its own right that I sorely hope such a creature does indeed still linger on. This was when E.J. Wayland, director of the Geological Survey of Uganda, sent to the British Museum a small collection of fossils lately disinterred from the vicinity of Albert Nyanza. Among these were some fragments that British Museum palaeontologist Dr Charles W. Andrews at once recognised to be from a remarkable creature known as a chalicothere.

The chalicotheres were among the most curious of all fossil mammals. Taxonomically, they were perissodactyls—members of the odd-toed order of ungulates, which also contains the horses, rhinoceroses, tapirs, and the extinct titanotheres (see later). Much of their cranial and skeletal features were similar to those of the horses—with one major exception, their limbs. The forelimbs of chalicotheres were notably longer than their hind limbs, yielding a hyena-like, rearward-sloping back. Most extraordinary of all, however, were their feet—for despite being hoofed mammals in the strict taxonomic sense, chalicotheres sported large claws instead!

The purpose of such highly unexpected accoutrements has never been satisfactorily explained, but the most popular theory is that these herbivorous browsers used them for digging up roots and for pulling down tree branches to within their jaws' reach.

Chalicotheres were already known from fossils found in North America (including *Moropus*, the most familiar genus), Asia, and Europe, but Wayland's were the first recorded from Africa. They were also the most recent,

Restoration of an African chalicothere (© Hodari Nundu)

geologically—for whereas those from elsewhere were mostly pre-Pleistocene, these African remains dated from the late Pleistocene.

The okapi's sensational discovery in 1901 remained uppermost in the minds of zoologists (especially those associated with Africa) for many years afterwards. In the case of Andrews, its impact sparked off a highly thought-provoking line of speculation when considering in 1923 the novelty of the chalicothere's onetime, and relatively recent, existence in Africa:

> The finding of a Chalicothere in Central Africa is of especial interest because a species occurs [i.e. in fossil form] in Samos associated with *Samotherium*, which is very closely similar to the Okapi, the discovery of which a few years ago attracted so much attention. It seems just possible that a Chalicothere may still survive in the same region and may be the basis of the persistent rumours of the existence of a large bear- or hyena-like animal . . . the "Nandi Bear," stories of which are constantly cropping up. Whatever it may turn out to be, the beast seems to be nocturnal in its habits and to resemble a very large hyena, an animal in which the proportions of the fore and hind limbs are much as in some Chalicotheres.

Unlike so many other cryptozoological proposals, the chalicothere connection evidently has appeal for even the most traditionally-minded of zoologists, because it has since surfaced not only in cryptozoological works but also in several mainstream zoological publications.

For instance, the eminent anthropologist Prof. Louis S.B. Leakey referred to the subject

The formidably-clawed foot of a chalicothere skeleton (© Dr Karl Shuker)

in an *Illustrated London News* article for 2 November 1935 entitled 'Does the Chalicothere—Contemporary of the Okapi—Still Survive?' His palaeontological work at Olduvai had disproved earlier speculation from sceptics that Wayland's chalicothere remains were derived or reworked fossils (i.e. fossils from older deposits that had somehow become re-embedded in a younger fossil bed, as already discussed in previous chapters of this present book), because chalicothere fossils unequivocally contemporary with those of late Pleistocene humans had been unearthed here too. Accordingly, in his account Leakey reiterated the possible synonymity of the Nandi bear with a species of living chalicothere.

A few years earlier, in *A Game Warden Among His Charges* (1931), Charles Pitman had also examined the chalicothere's candidature as a Nandi bear—and provided what could well be a very perspicacious explanation for the paucity of modern-day Nandi bear reports:

> The fact of its being an ungulate has proved of the utmost importance in connection with certain evidence acquired

> in the course of investigations to ascertain whether there was any reasonable foundation for the stories of the Nandi bear; and if so what the chances were of its existence at the present day. If not, was it possible that the species became extinct only recently? Now, the Nandis and allied tribes . . . assert that the Nandi bear was never plentiful; but, prior to the devastating rinderpest epizootic which swept through the whole of Africa at the close of the nineteenth century, practically exterminating the buffaloes, it was not uncommon. The species is said to have suffered so terribly from the ravages of the malady that it was reduced to the verge of extinction, and has never recovered. This story has often been cited as proof positive of the absurdity of the Nandis' tales. How, the sceptics ask, could a carnivorous beast suffer from rinderpest? But, if the Nandi bear is connected with the presumably extinct *Chalicothere* I see no reason why it should not have been susceptible to the disease in common with the majority of the ungulates.

Having said that, rinderpest, a now-eradicated morbillivirus, affected only artiodactyls (even-toed ungulates, e.g. cattle, antelopes, pigs), not perissodactyls (which chalicotheres were). In 1995, however, a team of Australian scientists documented a distant, hitherto-unknown relative of rinderpest, the Hendra virus, that was comparably deleterious to horses (which, like chalicotheres, are perissodactyls). So could a morbillivirus (or a related virus) have indeed wiped out a chalicotherian Nandi bear? None of the other Nandi bear identities would be affected by such a disease, so if only those identities were components of the Nandi bear composite (i.e. with no ungulate component ever involved), we would expect Nandi bear reports to be still surfacing, whereas in reality none has emerged for many years.

In *On the Track of Unknown Animals* (1958), Heuvelmans pointed out that although a living chalicothere may well *resemble* the hyena-like creatures described by Nandi bear eyewitnesses, a herbivorous animal could not be responsible for their bloodthirsty onslaughts—which he blamed upon spotted hyenas. In short, genuine hyenas (or at least carnivorous beasts of hyena-like appearance) were carrying out the attacks attributed to the Nandi bear, but a chalicothere might also be sufficiently hyena-like in superficial, external form for eyewitnesses encountering one to believe that they had seen a Nandi bear.

As I revealed via one of this present chapter's opening quotes, the involvement of the chalicothere, however innocuously, in the Nandi bear saga gained further credibility in 1975. This was when, in a short *Nature* report of 10 January, palaeontologist Dr Martin Pickford from London University's Queen Mary College revealed that after finding some chalicothere fossils within the Lukeino Formation in Kenya's Baringo district and describing the chalicothere's likely appearance in life to his native field assistant Kiptalam Chepboi, he had been assured by Chepboi that this was an accurate description of the chemosit—an opinion reinforced by other natives to whom he gave the same description.

Unlike Heuvelmans, Leakey, and Andrews, however, Pickford doubted that the chalicothere was still alive today. But he conceded that it may have survived until the recent past, entering local mythology to yield tales of the chemosit before finally dying out.

In his authoritative palaeontological tome, *Mammal Evolution* (1986), Prof Robert J.G.

Savage offered a more optimistic opinion on the matter:

> Periodically come reports from the Kakamega forests in Kenya of sightings of the Nandi bear. The beast is described as having a gorilla-like stance with forelimbs longer than the hind, with clawed feet like a bear and with a horse-like face. Could the beast be a survivor of the chalicothere, thought to have become extinct in East Africa during the Pleistocene? The description above would fit with the skeletal remains of these extraordinary animals.

The same view was aired in the *Macmillan Illustrated Encyclopedia of Dinosaurs and Prehistoric Animals* (1988), a multi-authored volume with Prof. Savage as a consultant.

A second, entirely different source of evidence implying recent chalicothere survival was revealed by Brown University ungulate specialist Prof. Christine Janis. Not only does she support the likelihood of a chalicothere component within the multi-faceted Nandi bear identity, but also she has amassed some diverse and stimulating iconographical corroboration for the late survival elsewhere of several different types of ungulate officially believed to have died out much earlier.

These include two gold belt plaques from the frozen tombs of Siberia's Sakik culture (500-400 BC), in the collection of Peter the Great in the Hermitage, St Petersburg (formerly Leningrad). Each plaque (4.75 in long, 3 in wide, 6 oz in weight) depicts an odd beast referred to in the collection as a "fabulous wolf-like animal". As Janis recognised, however, its head and body are more equine than lupine (even sporting a horse-like mane), which, when combined with its clawed feet, yield a surprisingly chalicotherian appearance. True, they may simply be stylised wolves—but as she also noted, other plaques in this collection depict beasts readily identifiable as canids. So if these two creatures really are wolves, why should they alone have been portrayed in stylised form?

The author with a chalicothere model, showing its superficially hyena-like appearance (© Dr Karl Shuker)

(Eastern tombs, incidentally, appear to be good sources of cryptozoological material. In 1991, Dr Nikolai Spassov from Bulgaria's National Museum of Natural History documented two silver plaques recovered from some tombs in northern Mongolia dating from the 1st Century BC that portray the unmistakable form of the musk-ox *Ovibos moschatus*. This species still survives today in North America, but from the fossil record it was believed to have died out in the Old World long ago—since the early Holocene in Europe, and in Asia since becoming extinct in Siberia's Taymyr Peninsula approximately 2,000 years ago, but much earlier elsewhere on that continent.)

Finally, there is one aspect relating to the postulated connection between the Nandi bear and a surviving species of chalicothere that does not appear to have been considered before my discussing it in this present book's original edition. Just because chalicotheres were herbivorous, there is no reason to suppose that they were incapable of exhibiting belligerent behaviour. Anyone who has encountered a bull elephant on the rampage, a maddened hippopotamus, or a charging Cape buffalo or rhinoceros will readily confirm that an abstinence from meat consumption does not guarantee an absence of aggression! Perhaps the most dramatic examples are the South American peccaries. Anatomically, there is nothing to indicate a pugnacious personality, but the vitriolic savagery of these modest-sized pig-like ungulates is renowned—and greatly feared—throughout their distribution range.

What if chalicotheres were not as placid as palaeontology has painted them? A creature the size of a cow liberally armed with sharp, curving claws would make a formidable assailant if roused, or if habitually temperamental. Indeed, it might even constitute a veritable Nandi bear . . .

A HERD OF HOOFED MYSTERY BEASTS

In addition to chalicothere considerations, the cryptozoological literature contains numerous

The North American chalicothere *Moropus* utilising its mighty claws very effectively to ward off *Daphoenus*, an amphicyonid or bear-dog

unresolved cases regarding other unexpected ungulates. Among these cases are several dealing with controversial beasts that may be unrecognised prehistoric survivors—including the following selection.

Miniature Mammoths, and Memories of Mastodonts

Few cryptozoological subjects exude such an air of romantic wonder as the notion that a few woolly mammoths and hairy mastodonts may still be lurking unknown to science amid the remote backwaters of the modern world—which makes it all the more poignant that few such subjects are based upon so insubstantial and equivocal a body of evidence.

In the case of the woolly mammoth *Mammuthus primigenius* surviving in Siberia, there is little doubt at all that most such reports have actually originated from sightings of preserved, frozen mammoths by peasants so frightened by what they had seen that they never realised the truth—that the creatures were dead, and had been so for many millennia.

Then again, Chinese historian Sima Qian (c.145/136-86 BC) wrote that animals present in Siberia included: ". . . giant boars, northern elephants covered with bristle, and northern rhinoceroses". Was his mention of the bristly northern elephants based merely upon reports of long-dead ice-entombed woolly mammoths, or upon sightings of living specimens?

Also intriguing was a mystifying creature termed the wes, which in 1549 was name-checked as a member of Siberia's fauna in *Rerum Moscoviticarum Commentarii* ('*Notes on Muscovite Affairs*'). This was a major work on Russia penned by Carniolan diplomat, writer, and historian Baron Sigismund von Herberstein (1486-1566), who had twice visited that country as ambassador of the Holy Roman Emperor. According to Russian ethnologist P. Gorodkov (writing in a 1911 essay entitled 'A Travel to Salym Region of Siberia'), in the language of the Hant people inhabiting the area of Siberia where the unidentified creature in question supposedly lived the name 'wes' referred to a mysterious beast covered in thick hair and bearing tusks. A woolly mammoth, possibly, but if so, why no mention of its very visible trunk?

The one example that I do find convincing is the oft-quoted story that in 1920, M. Gallon, the French consul at Vladivostok, learnt from an old Russian hunter of how in 1918 he had spent several days following a trail of huge round footprints in Siberia's vast taiga forest, until at last he came upon the animals responsible—which he described as a pair of huge elephants covered in dark-chestnut hair, and with large, very curved, white tusks.

Did Gallon's hunter correspondent truly encounter a pair of living woolly mammoths in Siberia?

What is interesting about this particular account is that the hunter had described seeing them in a forest rather than on the barren ice plains traditionally deemed to be this species' true habitat, and also that he was unaware of mammoths as known from frozen examples. Soviet scientists ultimately proved that in reality the woolly mammoths had inhabited grassy steppes with trees rather than ice- and snow-carpeted tundra—but this was many years *after* the hunter's claim. So perhaps his story of encountering living mammoths in the taiga really was genuine. After all, a forest covering over 3 million square miles would have no problem sheltering a veritable multitude of mammoths from scientific scrutiny, let alone a single pair!

Also worthy of inclusion here is an all-too-brief report of 10 September 1903 from a Californian(?) newspaper called the *Evening Express* (a near-identical version also appeared in New Zealand's *Oamaru Mail* on 29 October 1903). It reads as follows:

> A Live Mammoth
>
> Dr. Fritzell, a prominent scientist frequently retained by the Government, says a San Francisco telegram, announces that he personally saw the fresh tracks of a mammoth on Unimake Island, in the Arctic regions, in his last journey. He gives minute details of the tracks, which sank four inches into the frozen earth. He hopes some time to search for the animal. Dr. Fritzell insists on the exactness of his observations, although he says he knows they may make him seem ridiculous.

Unimake (or Unimak) Island is the largest and easternmost member of the Aleutians, an island chain off southern Alaska, and is home to mammals as large as brown bears and caribou—but with no apparent sequel to Dr Fritzell's dramatic claim, mammoths do not appear to be among them.

One evening in early September 1998, a herd of mammoths was allegedly observed by gold prospectors in the Yakutia (Sakha) region of far-eastern Russia. Veteran Russian hominologist Prof. Dmitri Bayanov, who also has a longstanding interest in cryptozoology, received this information from crypto-colleague Georgy Sidorov, who was researching relict hominins in that area. Bayanov duly passed it on to French cryptozoologist Jean Roche, by whom, via Charles Perrier, it was posted in the cz@egroups.com cryptozoological chat group on 31 August 2000. It transpired that Sidorov had been visited at his home in Tomsk during May 2000 by a gold prospector named Oleg Rusin, who recounted the sighting made by himself and five fellow prospectors. Twelve mammoths were supposedly observed, at a distance of only around 80-90 yards, in a vast area completely devoid of human habitation. Once the mammoths saw the prospectors, however, they stampeded up a creek and disappeared within an expanse of larch trees.

For every ostensibly plausible report like those above, however, there are also plenty of far more dubious ones. In 1873, Edward Newman cautiously published a report in *The Zoologist* of how a Russian escaped convict called Cheriton Batchmatchnik had allegedly chanced upon a herd of living mammoths within a cave in the Aldan mountains of far-eastern Russia's Yakutia (Sakha) Republic. This would have been dramatic enough in itself, but the climax to his tale was an outrageously improbable confrontation between one of the mammoths and an equally implausible lake-dwelling reptile that Batchmatchnik described as a type of crocodile-serpent—30 ft long, scaly, and armed with huge fangs, which attempted to crush the mammoth within its powerful coils for over an hour before its victim succeeded in breaking free and escaping. As even the famed fecundity of reptilian evolution has yet to engender a constricting, serpentine crocodile (or a constricting, crocodilian serpent either, for that matter!), I see little point in examining this particular episode any further. And indeed, it was eventually exposed as a hoax, thought up by the *New York World*, the newspaper that first published it.

Cryptozoologically, the *Holy Bible* is famous for mentioning two very mystifying creatures—the leviathan and the behemoth, both of which have been documented earlier in this present book. Less well-known, however, is that another sacred work also contains a pair of highly mysterious animals. The following quote is from *Ether* 9:19 in the *Book of Mormon*: "And they also had horses, and asses, and there were elephants and cureloms and cumoms; all of which were useful unto man, and more especially the elephants and cureloms and cumoms". Mormon theologists have suggested various possibilities concerning the taxonomic identities of these animals. They include: some Central American species with which Joseph Smith was unfamiliar, such as the tapir or jaguar; a yet-undiscovered but probably now-extinct species; or, most intriguing of all, late-surviving mammoths.

The most famous proponent of the mammoth identity is probably the mathematician and Mormon religious leader Orson Pratt (1811-1881), who was an original member of the Quorum of Twelve Apostles of the Church of the Latter Day Saints. In a discourse delivered by him in the Old Tabernacle, Salt Lake City, Utah, on 27 December 1868 (and reported by David W. Evans), Elder Pratt related the history of ancient times as laid out in Mormon doctrine and referred to "the elephant and the curelom and the cumom, very huge animals that existed in those days", and also "fresh air for the benefit of the elephants, cureloms or mammoths and many other animals".

It seems clear from these two, separate phrases by Pratt that the curelom and the cumom were indeed very big animals, and, more specifically, that the curelom and the mammoth were one and the same animal. Hence his usage of the word 'or' in the second phrase; had he intended to signify that they were different animals, he would surely have used the word 'and' instead. As there do not appear to be any detailed descriptions of cureloms and cumoms within Mormon texts or elsewhere, however, it is unlikely that their taxonomic identities will ever be conclusively resolved.

Ironically, it is the fossil record, traditionally the scourge of cryptozoology, which provides this fledgeling science with its greatest hope for the reality of undiscovered modern-day woolly mammoths. This species was traditionally believed to have died out around 11,000 years ago—until 1993, that is, when Dr Sergey Vartanyan from Russia's Wrangel Island State Reserve and two fellow researchers made an astounding disclosure.

They had discovered fossilised remains revealing that as recently as c.2,500 BC (i.e. the time of the Great Pyramid of Giza's construction in ancient Egypt), woolly mammoths still existed on Wrangel Island—sited off the coast of northeastern Siberia in the Arctic Ocean. These, however, were no ordinary mammoths—like many island forms, they constituted a miniature, dwarf race, only around 6 ft high instead of almost 11 ft. This discovery was so revolutionary that some other researchers initially disputed the accuracy of the fossils' dating, but the Wrangel trio conducted further tests that verified their claim. These mini-mammoths had survived long after their full-sized relatives had vanished probably because Wrangel boasts a high diversity of plant life, and not until 3,000 years ago did human hunters first reach it. This in turn begs the question of whether in a forest as immense as the taiga, some normal-sized mammoths may indeed

Comparison of woolly mammoth on left with American mastodont on right (CC Dantheman9758)

have successfully eluded extermination by humans.

Speaking of mini-mammoths contemporary with the Egyptian pyramids: in a *Nature* letter for 2 June 1994, Israeli archaeologist Baruch Rosen brought to attention a decorative scene painted upon the tomb wall of Rekhmire, a governor of Thebes in ancient Egypt during the reigns of the pharaohs Tuthmose III and Amenhotep II (c.1479-1401 BC). The scene includes the depiction of a small hairy elephant (only waist-high to the men—Syrian traders?—portrayed alongside it) with a domed head, convex back, and disproportionately large tusks that looks closer in appearance to a diminutive mammoth than to either a young Asian or African elephant or to the taxonomically-controversial pygmy elephant of Central Africa.

Recent researches have revealed that dwarf mammoths survived into the Pleistocene on certain Mediterranean islands, such as *Mammuthus lamarmorai* on Sardinia and *M. creticus* on Crete. Moreover, some such islands were also home to the extinct dwarf elephants *Elephas* (*Palaeoloxodon*) *falconeri* (from Sicily and Malta) and *E. tiliensis* (from the Greek island of Tilos), this latter species now known from radiocarbon dating to have survived until as least as recently as around 4,300 years ago. Could the Mediterranean mini-mammoths have actually persisted *beyond* the Pleistocene, like the Wrangel version? If so, perhaps the depicted creature was a vertically-challenged Sardinian or Cretan mammoth, which had been captured and transported to Egypt. Or might it even have been a late-surviving Mediterranean dwarf elephant, younger than any fossil remains unearthed so far, and which was still alive at the time of Rekhmire?

Whereas mammoths were closely related to modern-day elephants (especially the Asian species), mastodonts constituted a totally separate taxonomic family—whose best-known species, the 10-ft-tall North American mastodont *Mammut americanum*, was characterised by a long head, lengthy upward-curving tusks, and a hairy coat. Officially, this species died out around 11,000 years ago, contemporary only with early humans, not modern-day ones. Over the years, however, various items of evidence ostensibly contradicting this view have been put forward.

There are some curious Amerindian legends that tell of huge beasts that *might* be mastodonts. For example, the legends of northern

Labrador's Naskapi Indians tell of the former existence of a daunting monster called the katcheetohuskw—said to be very large, with a big head and ears, big teeth, a very long nose with which it hit people, stiff rigid legs, and tracks in the snow that were deep and round. One Indian, shown pictures of an elephant, identified it as this beast. Some ethnologists, however, claim that its name is correctly translated as 'stiff-legged bear'—despite its long nasal weapon and big ears.

Another story, from the legends of Maine's Penobscot Indians, speaks of giant monsters with long teeth that slept at night by leaning against trees—just like elephants. And the Algonkian myths tell of an immense extinct creature with an extra arm, growing out of its shoulder—traditionalists identify this animal as a moose, despite its surprising accessory.

Yet even if these various tales were indeed based upon mastodonts (or mammoths), they provide no evidence for the latter beasts' modern-day survival, because the creatures in them are always spoken of as being long-vanished. As suggested in 1980 by George Lankford in a *Journal of American Folklore* article, they are most probably preserved folk memories dating back to late Pleistocene times, when early humans were contemporary with the last members of North America's mammalian megafauna.

Nevertheless, there are also on record a few claimed sightings of living mastodonts in modern times. The most famous of these featured British sailor David Ingram and his alleged trek with two companions from the Gulf of Mexico to the St Lawrence River in 1568—during this epic journey, the trio claim to have spied elephant-like beasts with huge tusks. Some 'hairy elephant' sightings surfaced during the late 1800s too, emanating from northern Canada and Alaska, but were never confirmed.

One such report, uncovered by American Fortean researcher and writer Jerome Clark, appeared in the *Winnipeg Daily Free Press* newspaper on 28 March 1893, and was reprinted on the *Cryptomundo* website by Loren Coleman on 24 April 2006. It reads as follows:

> The Stickeen Indians positively assert that within the last five years they have frequently seen animals which, from the descriptions given, must have been mastodons.
>
> Last spring, while out hunting, one of the Indians came across a series of large tracks, each the size of the bottom of a salt barrel, sunk deep in the moss. He followed the curious trail for some miles, finally coming out in full view of his game, says *The Philadelphia Ledger*.
>
> As a class these Indians are the bravest of hunters, but the proportions of this new spectacle of game filled the hunter with terror, and he took to swift and immediate flight. He described the creature as being as large as a post trader's store, with great, shining, yellowish white tusks, and a mouth large enough to swallow a man with one gulp. He further says that the animal was undoubtedly of the same species as those whose bones and tusks lie all over that section of the country.
>
> The fact that other hunters have told of seeing these monsters browsing on the herbs up along the river gives a certain probability to the story. Over on Forty Mile Creek bones of mastodons are quite plentiful. One ivory tusk, nine feet long, projects from one of the sand dunes on that creek, and single teeth have been found so large that they would be a good load for one man to

carry. I believe that the mule-footed hog still exists; also that live mastodons play tag with the aurora every night over on Forty Mile Creek in Alaska.

The combination of sparse specific details (e.g. precise date, name of hunter, precise location) and a light-hearted vein of writing style (especially at its close), however, does not bode well for the veracity of this report, especially as journalistic spoofs involving incredible animals were notoriously commonplace in American newspapers during the late 1800s.

Other claims for mastodont survival involve certain ceremonial mounds and their contents. There is, for instance, one such mound in Wisconsin whose shape is distinctly elephantine—based, perhaps, upon sightings of post-Pleistocene mastodonts? A more conservative theory holds that it was originally a bear, but flooding or some other natural phenomenon added a trunk-like extension to its anterior portion.

Even more deceptive is the highly controversial 'elephant pipe', found in one of Iowa's famous Davenport mounds during the 1800s. A beautifully-carved pipe bowl unequivocally depicting an elephant, it was originally identified as a mastodont portrayal, but was later dismissed as a fake.

According to Prof. H. Barry Fell in his book *America B.C.* (1982), however, it was actually an artefact inspired by the African elephant (in recent years split taxonomically into two closely-related species), and created by pre-Columbian settlers from Libya. A vast quantity of Libyan iconography and stone-engraved script is now known from the New World, including South America, where some elegant depictions of elephants were again formerly deemed by some to constitute proof that mastodonts survived into the recent past.

One such artefact, which is not only elegant but also enigmatic on account of how little I have been able to discover concerning it, is the subject of the following section.

One of the two photographs of the enigmatic Tiahuanacan elephant figurine (source currently unknown to me, despite considerable effort to try and trace it; USC Title 17 § 107)

Fire Beasts, Lightning Beasts, and the Elephant of Tiahuanaco

The artefact in question is a small black-ware pottery figurine of an instantly-recognisable elephant, complete with a pair of large round flaring ears, four sturdy legs, a downward-curving trunk, and a pair of short tusks. Yet it was unearthed at Tiahuanaco (aka Tiwanaku), a pre-Columbian archaeological site near Lake Titicaca in western Bolivia, where all manner of items have been found that originate from the once-mighty Tiahuanacan empire, which had existed there from 300 AD to 1,000 AD.

I first learnt of this figurine from an online Russian forum called *Laiforum*, one of whose members (with the user name Mehanoid) had posted there on 15 April 2016 an excellent side-view photograph of it with a link to a Russian video on YouTube. When I accessed this video (at: https://www.youtube.com/watch?v=CnUIXVvdUkQ&feature=youtu.be&t=27m15s), I discovered that it had been uploaded on 7 April 2016 by an association called ProtoHistory, and was a filmed lecture given by Russian historian Andrey Zhukov to publicise his new book, *Secrets of Ancient America* (2016).

Just a few seconds after 27 minutes of the video have played, two photos of the elephant figurine are briefly screened. One is the side-view picture posted on *Laiforum*, the other is a front-view shot. Moreover, the side-view picture is then re-screened but this time revealing that it is actually part of a page from what appears to be an article dealing with it, written in Spanish. Using a screen-shot of this, I have been able to confirm that it is one of at least two pages written in Spanish and illustrated with several other animal-themed Tiahuanacan black-ware pottery figurines. However, I have not succeeded so far in obtaining any additional details concerning the article itself—no author name, article title, periodical title, or date of publication. Nevertheless, the text present just on these two pages is nothing if not intriguing, because of the very unusual zoological identity that is being considered there in relation to the elephant figurine.

Prior to its connection to North America after the isthmus of Panama had formed around 3 million years ago, South America had been an island continent for millions of years, just like Australia. And, again just like Australia, during that vast time period its mammalian contingent had given rise to several entirely new, unique lineages, especially among its ungulates. But due to convergent evolution, outwardly they came to look very like the respective ungulates elsewhere in the world whose ecological niches they were occupying in South America. Chief among these Neotropical impersonators were the horse- and camel-like litopterns, plus the ox-like and hippopotamus-like notoungulates.

Less familiar but no less distinctive, however, were the pyrotheres ('fire beasts', so-named because the first documented fossil specimens had been excavated from an ancient

Modern restoration of the likely appearance in life of *Pyrotherium romeroi* (CC Dmitry Bogdanov)

Modern restoration of the likely appearance in life of *Astrapotherium* (CC Dmitry Bogdanov)

volcanic ashfall). In their most famous representative, *Pyrotherium romeroi* of Argentina, the nasal trunk, long head, general body shape, and overall size (up to 10 ft long and standing 5 ft at the shoulder) rendered it more than a little like a big tapir, or even a mastodont—both groups, incidentally, actually occurring first in North America, not reaching South America prior to the isthmus of Panama, thus enabling pyrotheres to occupy their niches here. Moreover, certain reconstructions of their likely appearance in life have given various large pyrotheres a somewhat pachydermesque form, complete with large ears, sizeable tusks, and a fairly long trunk. Could it be, therefore, as speculated in the Spanish article containing its two photos, that the Tiahuanacan elephant figurine was actually not an elephant at all but instead a late-surviving pyrothere?

Unfortunately, whereas certain litopterns and notoungulates are indeed known from documented fossil remains to have survived to at least as recently as the Pleistocene's close (as will be discussed in more detail later in this chapter), the youngest pyrothere fossils currently on record are over 20 million years old, dating back to the Oligocene. Moreover, even the most proboscidean-inspired pyrothere restorations are still far less elephant-like and much more mastodont-like than the Tiahuanacan figurine, so it seems prudent to assume that the latter does indeed represent an elephant rather than a pyrothere.

Worth noting here, incidentally, is that South America was once home to yet another taxonomic group of vaguely proboscidean ungulates—the astrapotheres ('lightning beasts'), which ultimately died out around 12 million years ago during the mid-Miocene epoch. Their most famous representative was *Astrapotherium*, which was slightly smaller than the pyrothere *Pyrotherium*, but, as with the latter beast, may have looked somewhat mastodont-like or even tapir-reminiscent in life, courtesy of its elongated head, nasal trunk, and tusks (very long in certain other astrapotheres, particularly Colombia's short-legged, long-bodied *Granastropotherium snorki*). However, its ears were only small, and it certainly would not have resembled the mystifying Tiahuanacan elephant figurine.

Of course, assuming that the latter figurine is indeed a depiction of an elephant (and it certainly looks like one, albeit a somewhat short-trunked individual), as opposed to a pyrothere, astropothere, or mastodont, the inevitable question requiring an answer is how can its origin at Tiahuanaco be explained?

Might it simply be a recently-created fake, therefore, as has already been postulated by many authors with regard to the depictions of comparably out-of-place and/or anachronistic animals upon the highly controversial Ica stones of Peru and to the equally contentious Acámbaro figurines of Mexico? Or is its existence further evidence of pre-Columbian trading and visitations by Libyans and other travellers from the Old World, as proposed by Fell? Clearly, this very perplexing figurine deserves a detailed study, and I would be very grateful to any readers who can offer me further information concerning it, including its current whereabouts plus the name and date of the periodical in which the Spanish article discussing it appeared.

In summary, it is cryptozoologically disappointing but archaeologically clear that there is currently no convincing evidence to suggest the modern-day survival of any proboscideans in the New World—only indications that if early humans had not been so successful at hunting, the endless plains and forests of the American West may have been home not only to bison, deer, and pronghorn, but to more than a few 'hairy elephants' as well.

The Beast of Båråboedaer

A mastodont-related oddity from the Old World features the magnificent stupa at Båråboedaer (aka Borobudur), which was built by migrants from India during the 9th Century AD, and is situated near Jakarta, on the Indonesian island of Java. In 1939, Michigan University geologist Dr E.C. Case brought to attention the intriguing fact that some of the water spouts at the tips of this spectacular terraced temple's conduits take the traditional form of a mythical monster called the makara, but, unique to this particular monument, possess carved teeth that are typically mastodont, in both form and number (hence very different from those of the Asian elephant *Elephas maximus*).

Makara spout, showing teeth, at Borobudur temple (CC Bryn Pinzgauer)

In explanation of this anomaly, Case suggested that the water spouts' model may have been a fossil mastodont skull, perhaps preserved as a temple exhibit. Tellingly, however, he also opined: "The possibility that the Mastodon survived in some remote locality until the ancestors of the builders of Båråboedaer saw it and preserved an account of it in legend is not too fantastic to be considered". It would not be the first time that depictions of an ostensibly fabulous beast were, in reality, based upon an animal undetected by science.

A Living Mastodont Captured in Borneo?

In March 2016, Dutch Facebook friend Loes Modderman drew to my attention a very intriguing report that had appeared in a Dutch newspaper, *De Indische Courant*, on 31 March 1926, and which she kindly translated for me. The report alleged that a young mastodont had been caught alive in Borneo, and was due to be shipped to Surabaya in eastern Java. It had been captured in the jungle near Persigan, at the foot of the Belajang mountains, by a Javanese zoologist called Sastrowidirdjo, and had been placed aboard the SS *De Weert* at Banjarmasin, which was due to leave for Surabaya on 28 March. Whether it did set sail, however, and, more to the point, whether the creature in question really was a mastodont, remains undetermined, as Modderman has been unable to locate any follow-up reports relating to this claim.

Could the supposed mastodont have simply been a hairy Asian elephant? There is a taxonomically-undetermined race of small-sized Asian elephant (sometimes dubbed the Bornean pygmy elephant) inhabiting northern and northeastern Borneo (specifically Sabah—one of two Malaysian states situated on Borneo; and northernmost Kalimantan—Indonesian Borneo). Alternatively, might it have been a Malayan tapir *Tapirus indicus*, officially extinct here for several millennia but which may still survive, judging from various modern-day sightings (see later in this present chapter).

Google searches by myself and others have failed to uncover information concerning any zoologist named Sastrowidirdjo, and the place names Persigan and Belajang have also drawn a blank, but Surabaya is a very famous Javanese city, and Banjarmasin is the capital of South Kalimantan on Borneo. Moreover, as fellow Facebook friends Karl J. Claridge and Carl Kelsall uncovered, the SS *De Weert* was also real. It was a Dutch-registered, steam-propelled cargo ship, based in India, and it did sail in that area, but it was sunk on 3 July 1942 following shelling by a Japanese submarine, 150 miles from Lourenço Marques (now Maputo), the capital of Mozambique in southeast Africa. Tragically, 69 of her crew were killed, and three survivors who were picked up by the British vessel SS *Mundra* were lost when that ship was itself sunk nine days later. No mention of any prior transportation of a mastodont by this vessel has been uncovered as yet, however, so that mystery currently remains unresolved.

A Jumbo-Sized Revelation from China?

As recently as 3,000 years ago, elephants were still living wild in northern China, which may come as something of a surprise to many people. But something even more surprising concerning them has lately been proposed.

It had long been assumed that these were Asian elephants, because this familiar modern-day species still exists today in southern China. However, research conducted by scientists from Shaanxi Normal University and Northwest University in Xi'an and from the Institute of Geographic Sciences and Natural Resources Research in Beijing, published on 19 December 2012 in a *Quaternary International* paper, sensationally claimed that the northern China elephants seemingly belonged to an entirely separate, ostensibly long-extinct genus, *Palaeoloxodon*—housing the straight-tusked elephants.

Until now, China's *Palaeoloxodon* species (as yet unnamed) was thought to have died out at the Pleistocene-Holocene boundary, approximately 11,700 years ago. If they are valid, however, the Chinese team's findings indicate that it was still alive at least 7,000 years longer, into historic times—i.e. a bona fide prehistoric survivor.

The team's revelations were based upon their re-examination of 3,000-year-old fossil teeth hitherto believed to have been from *Elephas* but now considered by them to belong to *Palaeoloxodon*; and their reinterpretation of 33 northern Chinese elephant-shaped bronze wares from the Xia, Shang, and Zhou dynasties (c.4,100-2,300 years ago). The trunks of these bronzes all had two grasping finger-like digits, whereas the trunk of *E. maximus* only ever has one, thus suggesting that the bronzes may depict *Palaeoloxodon*, not *Elephas*.

Not everyone agrees that this partial resurrection of *Palaeoloxodon* in China is valid, however, with fossil elephant experts Drs Adrian Lister and Victoria Herridge claiming that the supposed differentiating dental features are merely contrast artefacts created by the low resolution of the photographs as published in the Chinese team's paper and that these features do not appear in better-quality photographic reproductions. They also note that cultural and iconographical aspects appertaining to Chinese art at the time of the bronzes' creation might reconcile their double-digited trunks with *Elephas* after all, not requiring the need to resurrect *Palaeoloxodon*.

In short, it is likely that this intriguing subject will attract further palaeontological scrutiny and contention for some time to come.

In the Wake of the Water Elephant

What may be the most sensational example of a proboscidean prehistoric survivor—inasmuch as this one could still exist even today, yet still be eluding scientific discovery—made its Western debut in 1912, courtesy of an article by R.J. Cuninghame that appeared in the *Journal of the East Africa and Uganda Natural History Society*. In this, he referred to a Mr Le Petit, lately returned to Nairobi, Kenya, following five years of travelling within the French Congo (now the Republic of the Congo)—during which period he claimed to have encountered on two separate occasions an extraordinary animal known to the Babuma natives as the ndgoko na maiji, or water elephant.

His first sighting, which occurred around June 1907 while journeying down the River Congo near the River Kassai's junction with it, was brief and featured only a single animal. It was seen swimming with its head and neck above the water surface but at a considerable distance away.

In contrast, his second encounter featured five specimens seen close by, on land. This took place in an area nowadays situated within the borders of DR Congo, i.e. the swampy terrain present between Lake Mai-Ndombe (known as Lake Leopold II in Le Petit's time) and Lake Tumba, near to where the M'fini River finds its exit from the first of these lakes. After viewing the animals through binoculars while they stood about 400 yards away amid some tall grass, he shot one of them in the shoulder, but his native companions were unable to recover its body.

Le Petit described the water elephants as 6-8 ft tall at the shoulder, with relatively short legs whose feet had four toes apiece, a curved back, a smooth shiny skin like that of a hippo's and hairless too but darker, an elongate neck about twice the length of the African elephant's, plus ears that were similar in shape to those of the latter but smaller in size. Most distinctive of all was its head, conspicuously long and ovoid in shape, and which together with its short, 2-ft-long trunk and lack of tusks resembled that of a giant tapir.

According to the natives, the water elephant spends the daytime in deep water (where it is greatly feared by them, as it will sometimes rise upwards unexpectedly and capsize their canoes with its able if abbreviated trunk). Only at night does it emerge onto land, where it grazes upon rank grass. It is also very destructive to their nets and reed fish-traps, but is not a common species, and its distribution range is very restricted.

Confirming the natives' testimony, the five specimens under observation by Le Petit finally disappeared into deep water, and were not seen by him again.

Model of a deinothere (© Markus Bühler)

If Le Petit's detailed description is accurate, the water elephant does not belong to any of today's three known species of elephant (two African *Loxodonta*, one Asian *Elephas*). It has been likened by some to the deinotheres, an extinct proboscidean lineage whose members' diagnostic feature was a downward-curving lower jaw bearing a pair of long recurved tusks. The last known species survived until the late Pleistocene in Africa—but the water elephant

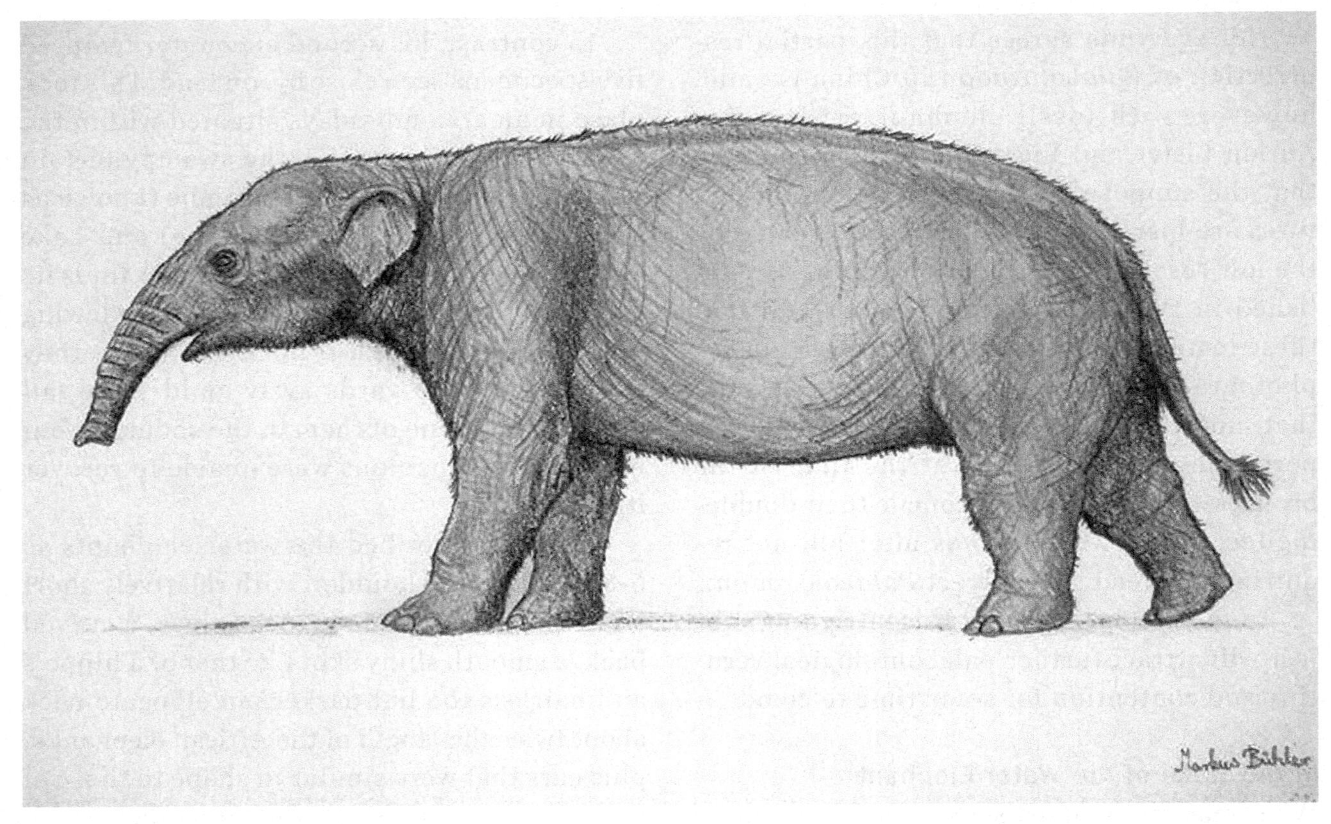

Artistic depiction of the water elephant, based upon eyewitness descriptions (© Markus Bühler)

bears little resemblance to these long-limbed forms with their curious lower jaw and tusks.

To my mind, it is much more similar to some of the most primitive proboscideans, such as *Phiomia* from Egypt's Oligocene, or even *Moeritherium* itself—the small tapir-like 'dawn elephant' from the late Eocene, 37-35 million years ago, whose fossils are known from various northern and western African countries, including Egypt, Algeria, Mali, and Senegal. Believed to have been a partially-aquatic swamp-dweller on account of its eyes' high, hippo-like position, it is at the base of the proboscidean evolutionary tree, but is thought to have died out without giving rise to any modern-day descendants. Yet if this beast *had* given rise to descendants, ones that had become much larger but had retained their ancestor's lifestyle and its attendant morphological attributes, the result would most likely be an animal greatly resembling the Congolese water elephant.

The author alongside a life-sized restoration of *Moeritherium* (© Dr Karl Shuker)

The concept of such a beast persisting unknown to science in the 1990s may not find favour among many scientists, but the Congo region of tropical Africa has already unveiled more than enough major zoological surprises during the past century or so for anyone with a knowledge of these matters to hesitate before discounting such a possibility entirely out of hand.

In April 2008, a study of the enamel of *Moeritherium* teeth published in the *Proceedings of the National Academy of Sciences* revealed that its diet corresponded with the diet of mammals known to be aquatic, thereby confirming that it was indeed a water-dweller. Consequently, if it did give rise to a reclusive, modern-day lineage of morphologically and behaviourally conservative representatives, these could constitute a very plausible water elephant.

On 25 July 2002, I received a fascinating email from Canada-based field cryptozoologist Bill Gibbons concerning what may be the mysterious water elephant, which I included in one of my Alien Zoo columns for *Fortean Times* and later in my book *Karl Shuker's Alien Zoo: From the Pages of Fortean Times* (2010). Here is what I wrote:

> In mid-2003, Bill Gibbons, a veteran seeker of cryptozoological curiosities, plans to visit the Democratic Republic of Congo (formerly Zaire) with a Belgian helicopter company operating there, in order to pursue claims by the company's president and CEO that a military helicopter flying over Lake Tumba spied a herd of very strange-looking elephants that the helicopter's pilots thought may be the legendary water elephants. According to Bill, the producer of a French TV documentary company is keen to

> film the expedition, so we wish everyone associated with this project the best of luck, and await further developments with interest.

Sadly, however, the planned expedition never took place. So the precise nature of those strange-looking elephants of Lake Tumba remains undetermined.

Most recently, the water elephant saga was revisited by British cryptozoological investigator Matt Salusbury in his extremely comprehensive book *Pygmy Elephants* (2013). After reviewing the Le Petit sightings, he pondered whether, confronted by environmental crises within the past century or so, isolated elephant populations in Africa could have undergone dramatic and highly accelerated bouts of evolution and behavioural changes, yielding in the Congo region a much-modified nocturnal, aquatic form—the water elephant.

A fascinating concept, but if this cryptid has been described accurately in those sightings, its morphological differences from Africa's typical, predominantly terrestrial elephants are, I feel, much too profound and wide-ranging to have plausibly arisen via evolution in such a short space of time. Consequently, and always assuming of course that the water elephant really does exist, I still consider it much more likely that this distinctive creature constitutes a wholly discrete species in its own right, one that may well have diverged long ago from the lineage leading to Africa's modern-day *Loxodonta* species.

A Giant Hyrax and the Beasts of Bronze

They superficially resemble small furry rodents, and were referred to as conies ('rabbits') in the Bible. However, hyraxes (or hyraces, as given in some zoological works) are actually ungulates—hoofed mammals—and, ironically, their closest living relatives are the mighty elephants. Also known as dassies or cavies, only four species of hyrax exist today, and they are all found only in Africa or in Africa and the Middle East, where they are often kept as pets, especially in rural areas. Joy Adamson of *Born Free* fame, for instance, kept a female common rock hyrax *Procavia capensis* called Pati-Pati as a much-loved pet during the early Elsa years. The other three modern-day species are the southern tree hyrax *Dendrohyrax arboreus*, the western tree hyrax *D. dorsalis*, and the yellow-spotted rock hyrax *Heterohyrax brucei*. As their names suggest, tree hyraxes predominantly live in trees, whereas rock hyraxes live among large rocks, boulders, and rocky hills or kopjes.

In prehistoric times, conversely, there were many other species, including some much larger ones and a number that existed far beyond the confines of Africa and Asia Minor. As will now be revealed, there is some remarkable iconographical evidence suggesting that at least one of these giant hyraxes, officially deemed extinct for over 2 million years, may in fact have survived in China into at least as recently as a famous period in this country's early civilised history.

As noted earlier in this chapter, Prof. Christine Janis is an ungulate expert based at Brown University in Providence, Rhode Island, U.S.A., and has a longstanding interest in whether some early iconographical objects, such as certain statuettes, figurines, and suchlike dating back many centuries, depict ungulate forms that are 'officially' believed to have died out long before such artworks were ever created, with the existence of these works thereby serving as potential evidence for believing that perhaps those ungulate forms actually survived into more recent times after all.

One day during the early 1980s, Prof. Janis

Rock hyrax (Amada44)

had a doctor's appointment, and was idly flicking through a copy of the magazine *Smithsonian* while sitting in her doctor's waiting room when her attention was drawn to an article featuring the photograph of a small but remarkable-looking bronze statuette from early China. Depicting some form of hoofed animal, it was one of 51 items present in an exhibition of early Chinese art from the Warring States Period (480-222 BC) that was being held at the Smithsonian Institution's Freer Gallery of Art in Washington DC. Several of these exhibits, including the animal statuette that so intrigued Janis, had been found in 1923 near the village of Li-Yu, in northern China's Shanxi province. So too had several additional animal statuettes of the same type as the Freer Gallery's interesting specimen.

A photograph of its own particular statuette is currently accessible on the Freer Gallery's official website (at: http://www.asia.si.edu/collections/ and search the digital collection for item F1940.23). The statuette is described there as being from the Middle Eastern Zhou dynasty (which existed during the Warring States Period), and, more specifically, to date from the early 5th Century BC. Its dimensions are given as: height, 11.2 cm (4.4 in); length, 18.3 cm (7.2 in); diameter, 6.3 cm (2.5 in). The following details are also included:

> This quadruped was crafted with considerable detail, especially in the modeling of the animal's face, its short, hoofed legs, and its scaled rotund body. Despite the clear features, the identity

of the animal remains obscure, as does the precise function of the bronze figure. The collar of cowrie shells, twisted rope bands, striations, and scale patterns suggest that this animal was also produced in the Houma workshops in southern Shanxi Province.

The Freer Gallery has made available for purchasing purposes official replicas of this statuette, produced by New York's Alva Studios. While visiting Sydney in Australia a while ago, Janis saw one for sale there, which she duly purchased.

In contrast to the gallery's circumspect attitude regarding the statuette creature's zoological identity, in the *Smithsonian* article seen by Janis it was referred to as a "fanciful tapir-like" animal.

Represented today by five species, of which only one, the Malayan tapir *Tapirus indicus*, exists in the Old World (but not China), tapirs are perissodactyls or odd-toed ungulates, as each of their feet possesses an odd (as opposed

Replica of the Chinese mystery beast statuette at the Freer Gallery (© Prof. Christine Janis)

to an even) number of toes. Hence they are most closely related to horses and rhinoceroses. Tapirs are noted for their short, vaguely elephantine trunk, their long slender claw-like hoofs, lengthy limbs, not overly large ears, and, in the case of the Malayan tapir, a very eye-catching 'saddle' of white across its back, flanks, and stomach, contrasting sharply with this species' black colouration elsewhere. Juvenile tapirs of all species can be readily distinguished from adults not only in terms of size but also inasmuch as they bear a striking pattern of white stripes all over their body, which gradually fade and vanish as they mature.

Adult Malayan tapirs in a 1903 chromolithograph

As an ungulate expert, however, Janis could see immediately that the exceedingly short but very sturdy-limbed, circular-hoofed, barrel-bodied, extremely big-eared, trunk-lacking, unstriped animal portrayed by the Chinese bronze statuette bore scant (if indeed any) resemblance to a tapir, whether adult or juvenile.

More recently, German cryptozoologist Markus Bühler has suggested that perhaps the statuette was meant to represent a young hornless specimen of the Javan rhinoceros *Rhinoceros sondaicus*—a species that did still exist in China back in those far-off times, does have

Female Javan rhinoceros and calf in illustration from 1839

scaly skin, and whose females are indeed sometimes hornless. Again, however, there is very little overall resemblance between the two, with the statuette's remarkably short limbs, lack of the well-demarcated skin-pleating exhibited by all modern-day rhinoceroses, fundamental differences in hoof shape and number, and the overall shape of its head all providing notable contrasts with the Javan rhinoceros.

If the Freer Gallery's statuette had been unique, one could have dismissed its creature as being an entirely imaginary beast, a one-off creation with no zoological significance whatsoever, but as noted earlier, several others had been found with it. By a remarkable coincidence, moreover, at much the same time that in the U.S.A. Janis had spotted the latter artefact's photograph within the *Smithsonian* magazine article, here in the U.K. I had made a fascinating, highly corroborative discovery while browsing through an early issue of a British magazine entitled *The Field*.

This country-themed magazine has a long tradition of including articles and letters documenting unusual, unidentified, or aberrant animal specimens, some of which are directly relevant to cryptozoology. Consequently, I have spent many long sessions in libraries perusing back issues in search of potential cryptozoological treasures. While doing so one day

Oppenheim's erstwhile Chinese mystery beast statuette

during the early 1980s, I came upon an article in the 30 November 1935 issue of *The Field* that had been written by F. Martin Duncan and was entitled 'A Chinese Noah's Ark'.

It surveyed a selection of Chinese works of art from an exhibition that was being staged at that time by London's Royal Academy. Duncan's article included photographs of several exhibits, one of which, loaned to the exhibition by its then-owner, a Mr H.J. Oppenheim, was a small bronze statuette of a short-legged, rotund-bodied hoofed animal—one that was almost identical in every way to the Freer Gallery's mystery beast statuette! It dated from the same time period too.

In 1987, Janis published a paper in *Cryptozoology*, the scientific, peer-reviewed journal of the now-defunct International Society of Cryptozoology, in which she documented a number of enigmatic iconographical artefacts that may depict ungulates officially deemed to have become extinct long before the dawn of human civilisation. One of those artefacts was the Freer Gallery's Chinese bronze statuette. Once I became aware of the latter statuette's existence from reading Janis's account of it in her paper, I sent details concerning Oppenheim's specimen to her. I have since discovered that in 1947, Oppenheim bequeathed his statuette to the British Museum in London, where it is still housed today, and where, despite all of its major morphological discrepancies, it is nonetheless labelled as a "bronze tapir"!

So too was an exquisite bronze statuette closely resembling the Freer and British Museum specimens in outward form and size but ornately decorated with inlaid gold and turquoise, and approximately 2,500 years old (thereby dating it towards the end of the Warring States Period), which was auctioned in March 2007 at the European Fine Art Fair in the Dutch city of Maastricht. Originally used as a wine cooler with a lid on its back to pour the wine in plus a spout on its snout, it was priced at a staggering £6.1 million, but it still sold—to a Chinese collector. Only one other statuette of this nature is known, and that is housed in a Taiwan museum.

Of course, and as could also be done in relation to the possibility of a young Javan rhino identity, one might argue that perhaps these statuette were simply very inaccurate portrayals of a tapir. However, other Chinese animal statuettes and figurines produced during the same time period and exhibited alongside the statuettes reveal that animals were portrayed very realistically, in an extremely natural, lifelike manner during that period (thus enabling their species to be readily identified), not inaccurately or even in a stylised, non-realistic manner. Therefore we have to assume that these statuettes (plus the others found with the Freer specimen) also depict in a natural, lifelife manner whatever the animal is that they represent—so what could it be?

As the tapir identity clearly does not correspond with the statuettes' morphology (nor does the young Javan rhino), and is certainly not favoured by Janis, what alternative did she offer in her paper? The answer is a hyrax, but no ordinary one . . .

The taxonomic order containing the greatest number of ungulate species alive today is Artiodactyla, the even-toed ungulates. Millions of years ago, however, before the diversification of artiodactyls took place that ultimately led to the great variety of present-day forms (which include pigs, hippopotamuses, peccaries, deer, camels, llamas, giraffes, chevrotains, antelopes, oxen, sheep, goats, and pronghorns, as well as a substantial number of now-extinct types), hyraxes were among the predominant forms of ungulate in existence, and occupied many ecological niches subsequently taken over by the artiodactyls. As a result, some hyraxes were different in form from and substantially bigger in stature than the rabbit-sized species existing today. Largest of all were those belonging to the aptly-named genus *Titanohyrax*, whose members were as big as modern-day rhinoceroses.

Also impressive were the hyraxes belonging to the genus *Pliohyrax*. Three species are currently known from fossils, which have been variously found in China, Afghanistan, Spain, Turkey, and France, and date back (as their generic name suggests) to the Pliocene epoch, which ended around 2.58 million years ago. No post-cranial remains belonging to these hyraxes have so far been uncovered, but based upon the dimensions of cranial fossils that have been disinterred, palaeontologists believe that *Pliohyrax* was as big as a pig. As someone who has kept hyraxes as pets, Janis was naturally very familiar with their appearance, and recognised at once that the head of the Freer Chinese bronze statuette was very hyrax-like. She also knew that the head of *Pliohyrax* was too—much greater in size than that of living hyraxes (one *Pliohyrax* skull in the American Museum of Natural History is 10 in long), yet very similar in basic outline.

But what would its body have looked like? In her paper, Janis noted that just like those of a hippopotamus, the orbits (eye sockets) of *Pliohyrax* were positioned high in its skull. In the hippo, this feature is evidence that the animal is semi-aquatic, because the eyes' elevated position enables this animal to see above the water surface while remaining almost entirely submerged. Consequently, the presence of this same feature in *Pliohyrax* encouraged Janis to speculate that it too may have been semi-aquatic, and that it might therefore have also sported a similar body shape to the hippo. This is because other fossil ungulates with elevated eyes do tend to have hippo-shaped bodies. And sure enough, the rotund, barrel-shaped body and sturdy pillar-like legs of the Chinese bronze statuettes' mystery animal is certainly very reminiscent of a hippo's body.

Even so, one notable problem swiftly presents itself when seeking to reconcile the bronze beasts with any type of hyrax. For whereas the latter have several hooves on each foot, the feet of the bronze beasts have only one hoof apiece. Janis, however, proposed that this discrepancy suggests the animal was one that was known from folk memory rather than one that had been personally seen by the sculptor(s). Presumably, therefore, as Janis concluded, *Pliohyrax* may have survived long beyond the Pliocene extinction date supplied for it solely by currently-known fossil evidence, and had once been well-known to humankind in China, even becoming incorporated into their mythology, but had died out prior to the Warring States Period.

Having said that, there is also an equally plausible alternative to this scenario, as proffered once again by Janis. A hyrax the size of *Pliohyrax* would be much heavier than the far smaller species alive today, and may therefore have evolved a more compact type of foot structure with a pad underlying the foot—resembling the feet of pigs and hippos. Perhaps this is what the sculptor(s) had attempted to convey when creating the bronze statuettes. If so, then *Pliohyrax* may have lingered into the Warring States Period after all.

As there is no evidence of its survival today in China, however, we must assume that even if this giant hyrax did significantly post-date the Pliocene, it still subsequently became extinct—or perhaps, like many other large, sluggish, inoffensive beasts with succulent flesh, it was exterminated by hunters—with evidence of its onetime existence confined entirely to the survival of the enigmatic little beasts of bronze exhibited in the Freer Gallery, the British Museum, and also in various public and private collections elsewhere. If only these silent statuettes could talk—what fascinating secrets concerning China's long-lost early fauna might they reveal to an astonished modern-day world?

Ethiopia's Dib—A Still-Living Giant Hyrax?

Although China's giant hyraxes appear to be long-vanished at least in the living state, even if preserved in iconography, Ethiopia may conceivably harbour a still-living species, as I documented in my book *Extraordinary Animals Revisited* (2007).

While carrying out Marine Corps training at a Californian bootcamp during the first half of 1999, American cryptozoological investigator Nick Sucik (see also Chapter 1) put to good use the opportunity to question his fellow recruits, many of whom were hunters and from a variety of different U.S. states, concerning mystery animals. On 3 July 1999, he posted an extremely interesting account to the cz@onelist.com cryptozoological chat group, presenting some of his findings. These included the following details concerning a sizeable but very mysterious form of herbivorous mammal:

> . . . one that may very well be a familiar animal but sounded unique I heard by [i.e. from] one recruit we had from Ethiopia. I asked him if they had anything unusual or mysterious where he was from. He said no, but told me about

> an animal called in their language a 'deep' [as pronounced, but 'dib' as written—see later], described as being "like a bear" except herbivore, they're about 2 ft high and 4 ft long found in the deserts of Ethiopia, light furred and very rare. He claimed it was considered dangerous even though it was herbivore. It's said to be incredibly strong and known to flip over vehicles by ramming them with its head! Unfortunately, he was unfamiliar with the English name with [i.e. for] this creature but brought it up because "there is nothing else like it in the world".

This elicited a reply from British cryptozoological researcher Allan Edward Munro, who voiced my own thoughts when he noted that Ethiopia's mystifying dib sounded like a very large hyrax. Even so, no known species alive today is anywhere near as big or as powerful as the dib; the largest, Johnston's rock hyrax *Procavia capensis johnstoni* from Central Africa, is no more than 2 ft in total length. Conversely, certain early northern African geniohyids (*Pliohyrax* relatives) were as big as tapirs or small horses, but these vanished millions of years ago.

Eager to learn more about the dib, I contacted Nick Sucik to request any additional information that he could supply to me, and on 5 July 1999 I received the following detailed reply:

> I'm not sure that's a cryptid at all. According to Deems (the recruit), it was a known animal. I wondered though, was it known to his people or regionally known or is it an actual scientific animal so to speak. I believe it may be the latter though he was unsure [of] the English name given to it. . . Demesa 21, migrated from Ethiopia when he was 14 and had been living in Las Vegas. Since he was originally from Africa I asked if he was familiar with Ethiopia having any mysterious animals. He didn't grasp what I meant by there being 'mysterious' or 'hidden animals'. I used Mokele-Mbembe as an example. He then understood but their philosophy of nature and the animals within tends to vary against ours. According to Demesa, whether an animal was unknown to science was irrelevant. Nature needed to be respected and exist unharmed by man. He told me that there was once a time in Africa when the animals lived harmoniously amongst men and that you could actually walk right past a lion without harm. After he was done with the Green Peace speech he did bring up a unique animal called in their language a "Deep". He described it as resembling a bear but smaller and herbivorous found in the desert part of the country. About 2.5 ft tall and 4 ft long and it didn't have [as] much hair as a bear he claimed. The strength of this animal was incredible, allegedly able to flip over a jeep ramming it with its head. Nowadays they're rare due to hunting. Demesa admired it because it was "one of a kind" like the lion, he said. That was about it. I think he did say he'd seen one once but at a distance. Allan Munro suggested this could be a "hyrax". I have no idea what that is.

If Demesa's description of the dib, as relayed by him to Sucik, is accurate, it does not appear to be a species presently known to science (at least in the living state). In any case, it would certainly warrant investigation by any future zoological visitors to Ethiopia—especially in view of Demesa's claim that this creature is nowadays rare, due to hunting. How wonderful it would be if a chance comment by Demesa ultimately led to the scientific unveiling and accompanying protection of a significant new mammal in his native Ethiopia.

Indeed, it may not even be confined to Ethiopia. On 25 August 1999, another American cryptozoologist, Chad Arment, noted on the cz@onelist.com group that he had just received an e-mail from a correspondent whose wife is Somalian. He stated that his wife apparently knows of this same creature (or at least of one resembling it). She claims that in Somalia it is referred to as the dewacco (in Somali, the 'c' in its name is pronounced like a deep 'h'). Moreover, Arment subsequently discovered that the written name for Ethiopia's version is in fact 'dib'—'deep' is how this word is pronounced.

Dib, deep, or dewacco, however, there is clearly a notable mammalian mystery awaiting a satisfactory resolution in parts of eastern Africa's more remote terrain.

Tapirs in Borneo?

The Malayan tapir *Tapirus indicus* is the only Old World tapir species alive today, and as noted earlier it is further differentiated from its four New World relatives by virtue of its 'saddle'—an area of striking white fur encompassing much of its torso and haunches. Already known to exist in mainland Malaysia, Burma (Myanmar), Thailand, and the large Indonesian island of Sumatra, there is a good chance that this species' current distribution extends even further afield—onto the island of Borneo, where it supposedly died out only a few millennia ago during the Holocene, fossil remains having been found during cave excavations in the Borneo-situated Malaysian states of Sarawak and Sabah. Moreover, as revealed by the Earl of Cranbrook and University of York palaeoecologist Dr P.J. Piper in a 2009 *International Journal of Osteoarchaeology* paper dealing with Bornean records of the Malayan tapir, its fossil representatives on Borneo are taxonomically indistinguishable from those elsewhere within this species' zoogeographical range, rather than constituting a separate species.

Over the years, a number of unconfirmed reports of tapirs existing in this extremely large yet still little-explored tropical island have been briefly documented in the literature (and summarised in the earlier-noted *Int. J. Osteoarchaeology* paper), but these have incited conflicting opinions. Whereas, for instance, Dutch biologist Prof. Lieven F. de Beaufort accordingly included Borneo within the accepted distribution range for *T. indicus*, Swedish zoologist Eric Mjöberg adopted a more circumspect stance, stating in his book *Forest Life and Adventures in the Malay Archipelago* (1930): "It is not yet certain that the tapir has been met with in Borneo, although there are persistent reports that an animal of its size and appearance exists in the interior of the country".

In 1949, however, Dr Tom Harrisson, then Curator of the Sarawak Museum, reported in two articles that tapirs were spied in Brunei (a small independent sultanate in Borneo) on two separate occasions some years earlier by Brunei resident E.E.F. Pretty, and he believed that the existence of such creatures in Borneo might yet be officially confirmed. Moreover, back in 1826, while touring Borneo in his capacity as an inspector of agriculture, French naturalist/explorer Pierre-Médard Diard allegedly obtained a tapir specimen in the interior of Pontianak, but its fate is apparently unrecorded.

Curiously, the Malayan tapir was featured on a 1909 postage stamp in a series issued by what was then North Borneo (now the Malaysian state of Sabah) that depicted species of animals supposedly *native* to this island. Wishful thinking perhaps—or an affirmation of reliable local knowledge that deserves formal acceptance by science?

Malayan tapir depicted on a North Borneo postage stamp issued in 1909

Another enigmatic item of putative evidence for the tapir's existence in Borneo can be found in the second volume of Norwegian explorer Carl Lumholtz's travelogue *Through Central Borneo: An Account of Two Years' Travel in the Land of Head-Hunters Between the Years 1913 and 1917* (1920). The item of evidence concerns a still-unidentified beast referred to as:

> . . . the giant pig, known to exist in Southern Borneo from a single skull which at present is in the Agricultural High School Museum of Berlin. During my Bornean travels I constantly made inquiries in regard to this enormous pig, which is supposed to be as large as a Jersey cow. From information gathered, Pa-au appears to be the most likely place where a hunt for this animal, very desirable from a scientific point of view, might be started with prospect of success. An otherwise reliable old Malay once told me about a pig of extraordinary size which had been killed by the Dayaks many years ago, above Potosibau, in the Western Division. The Dayaks of Pa-au, judging from the one I saw and the information he gave, are Mohammedans, speak Malay, and have no weapons but spears.

To non-zoological observers, a tapir could certainly be likened to a giant pig (indeed, this physical similarity actually makes tapir meat taboo to eat in much of Sumatra, where the population is predominantly Muslim). So, I wonder if that single skull is still preserved somewhere in Berlin? (Investigating this possibility, however, I learnt from German cryptozoologist Markus Bühler that the Agricultural High School Museum had closed down by the end of World War II, with its collections distributed among a number of other museums, these latter institutions probably including Berlin's Historical Museum, its Museum für Naturkunde, and the Botanical Museum, thereby making any attempt to trace this potentially significant specimen far more difficult.) For even if we assume that local comparisons of its unidentified species' total size with that of a Jersey cow are somewhat exaggerated, the fact that it was specifically termed a giant pig clearly suggests that this mystifying creature was no ordinary hog, of either the domestic or the wild variety, as will now be revealed.

When investigating the pedigree of any cryptozoological creature, we must always consider the possibility that the eyewitness reports ostensibly substantiating its existence are in truth nothing more than misidentifications of one or more species already known to science. In the case of Borneo's tapir, however, there is only one such species with which it might be confused—the bearded pig *Sus barbatus*. This Borneo-inhabiting beast possesses not only a long snout, but sometimes also a fairly pale 'saddle' over its back and haunches.

Even so, with a total head-and-body length of 3.5-5.5 ft, a body height not exceeding 3 ft, and a weight of 330 lb, it falls far short of the Malayan tapir's stature—as the head-and-body length of this latter species (the largest of all modern-day tapirs) can exceed 8 ft, its shoulder height can reach 3.5 ft, and its weight can typically be as much as 800 lb (but with certain exceptional specimens having weighed up to 1,190 lb). Nor could the bearded pig be mistaken for immature tapirs that have not attained their full size—because just like those of the four American species, juvenile Malayan tapirs are striped, only acquiring their characteristic white-backed, unstriped appearance when adult. In addition, the profuse—and very eyecatching—bunches of hair borne upon its snout's basal region and cheeks that give the bearded pig its common name are not sported by Malayan tapirs of any age, adult or juvenile.

In short, the bearded pig cannot be contemplated as a likely identity of supposed tapirs encountered in Borneo, but it does offer one item of interest with regard to this subject. Its distribution through Asia corresponds almost precisely with that of the Malayan tapir, except for one major difference—the bearded pig is *known* to exist on Borneo. Yet until near-Recent times during the present Holocene epoch, the Malayan tapir was known to occur here too (thereby greatly reducing the likelihood that any tapir living on this island today could belong to an *unknown* species). If the bearded pig could persist into the present day on Borneo, why not the tapir too? The island is more than

A bearded pig (© Markus Bühler)

sufficiently spacious to house a respectable number of tapirs, and their ecological requirements are adequately catered for. Indeed, the *absence* of the Malayan tapir from Borneo is really far more of a mystery than its disputed *existence* here!

Ironically, the Malayan tapir's modern-day occurrence in Borneo may well have been fully verified by now had it not been for an appalling instance of investigative apathy on the part of the scientific community, which led to the disappearance of what seems to have been conclusive evidence for this species' persistence there. In November 1975, as recorded soon afterwards by Jan-Ove Sundberg in *Pursuit*, Indonesia's Antara News Agency reported the capture in Kalimantan (Indonesian Borneo) of an extraordinary creature that appeared (at least from the description given of it) to be an impossible hybrid of several radically different types of animal.

According to the report (and always assuming that this entire episode was not a fabrication or a dramatic distortion of some much more commonplace event), the captured creature's body was similar to that of a tiger, its neck resembled that of a lion, it had an elephant-like trunk and the ears of a cow, its legs recalled those of a goat but its feet had chicken-like claws, and it reputedly sported a goatee beard like a billy goat's. Emphasising its composite appearance, the news agency referred to this amalgamated animal as a 'tigelboat'—a portmanteau word presumably derived from 'TIGEr', 'Lion', 'Bird', and 'gOAT'.

In spite of its bizarre appearance (or *because* of it?), the 'tigelboat' apparently failed to elicit any interest from scientists—which is a tragedy. If only someone had taken the trouble to analyse its description methodically, a major zoological discovery might well have been made. For as I have shown in various of my writings, when considered carefully its description can be readily translated to reveal a creature that can be identified after all.

Let us examine the features of the 'tigelboat' one at a time. If its body was similar to a tiger's, this must surely mean that it was striped. Equally, as the most noticeable aspect of a lion's neck is its mane, the tigelboat probably had a mane or a ridge of hair along its neck. An elephant-like trunk implies the presence of an elongated snout and upper lip, and cow-like ears would be relatively large and ovoid. If its legs recalled a goat's, then they must have been fairly long and sturdy (but not massively constructed), and chicken-like claws could be interpreted either as claws like those of certain mammalian carnivores or as distinctive, pointed hooves. As for its goatee, this might have been merely a tuft of hair on its chin.

The tigelboat is an impossible hybrid no longer. Combine most of the translated features noted here, and the result is a description that can now be seen to portray very accurately a particular creature already familiar to science—a juvenile tapir. As already noted, unlike the adult the juvenile tapir is striped. Moreover, its ears are certainly large and ovoid, its limbs fairly long and sturdy but not massive, its hooves distinctive and pointed, and its snout and upper lip drawn out into a conspicuous proboscis or trunk.

A juvenile tapir—the identity of the tigelboat?

It is true, of course, that whereas its four New World relatives are maned, the Malayan tapir normally lacks any notable extent of hair upon its neck (though juveniles are somewhat hairier than adult), as well as upon its chin. However, during its existence upon the island of Borneo for 11,700 years since the end of the Pleistocene, totally separated from all other populations of *T. indicus* elsewhere in Asia, it is probable that a Bornean contingent of Malayan tapirs would evolve one or two morphological idiosyncrasies (just as isolated populations of many other widely distributed animal species have done). Nothing very spectacular, but enough to permit differentiation from all other *T. indicus* specimens; such features could readily include a mane or a goatee or both.

Certainly, in view of the otherwise impressive correspondence between tigelboat and juvenile tapir, the mere presence of a mane and goatee is too insignificant morphologically to challenge the identification of the former beast as the latter one.

The capture in Borneo of what was assuredly a young Malayan tapir should have attracted attention from zoologists, especially as the animal was maintained alive for a time at a prison in Tengarong. Tragically, however, it received no attention at all, and eventually it 'disappeared'—the fate of so many mystery beasts. No further news has emerged regarding this monumentally missed opportunity, and the whereabouts of the tigelboat's remains are unknown—so there is no skeleton or skull available for identification, let alone the living animal itself. Not even a photograph of it has turned up. Consequently, the Bornean tapir is still a non-existent member of this island's fauna—at least as far as official records are concerned.

Even so, there is still hope that its existence will be confirmed one day. Borneo is a huge island, with extensive, little-penetrated rainforests and swamplands—ideal territory for secretive tapirs. And for absolute proof that large beasts can remain undetected here, look no further than the Sumatran rhinoceros *Dicerorhinus sumatrensis*, deemed extinct in Sarawak from the early 1940s—until a herd was found in a remote valley there in 1984.

If rhinoceroses can exist unknown to science in parts of Borneo, how much greater is the likelihood that the smaller, well-camouflaged Malayan tapir can elude discovery there too?

Making an Ass of the Zebro?

As I revealed in my book *Mirabilis* (2013), during the Middle Ages and the Renaissance several Spanish hunting treatises mentioned a mysterious, now-vanished equine creature called the zebro (or encebro, in Aragon), living wild in the Iberian Peninsula. One such work described it as "an animal resembling a mare, of grey colour with a black band running along the spine and a dark muzzle". Others likened it to a donkey but louder, stronger, and much faster, with a notable temper, and whose hair was streaked with grey and white on its back and legs. What could it have been?

Although largely forgotten nowadays, the zebro experienced a revival of scientific interest in 1992 when archaeologists Carlos Nores Quesada and Corina Liesau Vonlettow-Vorbeck published a very thought-provoking article in the Spanish periodical *Archaeofauna*. For they boldly proposed that the zebro may have been one and the same as an equally enigmatic fossil species—*Equus hydruntinus*, the European wild ass.

The precise taxonomic affinities of this latter equid have yet to be satisfactorily resolved. Although genetic and morphological analyses suggest that it was very closely related to the onager *E. hemionus* (one of several species of Asiatic wild ass), it can apparently be differentiated from these and also from African wild asses by way of its distinctive molars and its relatively short nasal passages. Arising during the mid-Pleistocene, approximately 300,000 years ago, the European wild ass is known from fossil evidence to have persisted into the early Holocene before finally becoming extinct. During the late Pleistocene, its zoogeographical distribution in western Eurasia stretched from Iran in the Middle East into much of Europe, reaching as far north as Germany, and it was particularly abundant along the Mediterranean, with fossil remains having been recovered from Turkey, Sicily, Spain, Portugal, and France.

According to Quesada and Vonlettow-Vorbeck, however, this enigmatic species may have survived in southernmost Spain and certain remote parts of Portugal until as late as the 16th Century (they look upon its disappearance as the Iberian Peninsula's last megafaunal extinction), where, they suggest, it became known locally as the zebro. More recently, their theory gained support from the discovery of *E. hydruntinus* remains at Cerro de la Virgen, Granada, dating from as late as the 9th Century.

Some researchers have also suggested that before dying out, the zebro gave rise at least in part to a primitive, nowadays-endangered Iberian breed of donkey-like domestic horse called the sorraia (which was once itself referred to as the zebro). Furthermore, many believe that it was from the term 'zebro' that 'zebra' originated as the almost universally-used common name for Africa's familiar striped equids.

Even today, many Iberian place-names still exist in which the mysterious but now-obscure zebro's name is preserved. These include Ribeira do Zebro in Portugal; and Valdencebro (in Teruel), Cebreros (in Ávila), Encebras (Alicante), and Las Encebras (Murcia) in Spain.

Thunder Horses and Titanotheres

Throughout this book, mythology and iconography have offered some intriguing evidence for the persistence into recent times of many different animals traditionally believed to have become extinct long before the advent of modern humans. However, they can also lay traps for the unwary and ingenuous who indiscriminately seek cryptozoological solutions to mysteries, failing to give equal consideration to more conservative, alternative explanations.

One of the most spectacular beasts from Amerindian mythology must surely be a huge, terrifying creature known as the thunder horse. According to Sioux legends emanating from Nebraska and South Dakota, thunder is the sound produced by the impact of its hooves when it leaps down from the skies to the ground during violent storms, and while on Earth it also uses its hooves to slay bison. It would be tempting to postulate that this extraordinary story was inspired by eyewitness accounts of some mysterious living creature—accounts perhaps passed down verbally from one generation to another for many centuries and subjected to much elaboration and exaggeration, but nonetheless derived ultimately from original encounters with a modern-day animal. In fact, the true solution is very different.

Titanothere model (© Jeff Johnson)

The thunder horse legend actually arose from occasional discoveries by the Sioux Nation of huge fossilised bones—which usually came to light by being washed up out of the ground during heavy rainstorms. Unable to explain their origin, but aware of their appearances' coincidence with rain, the Sioux assumed that they were the earthbound remains of some immense, storm-engendered sky beast—and thus was the legend born.

In reality, the bones were from an enormous rhinoceros-like ungulate, standing 8 ft tall at the shoulder and belonging to an extinct family of horse-related perissodactyls aptly called titanotheres. The type responsible for the thunder horse legend had existed roughly 35 million years ago, during the late Eocene epoch, and it had borne upon its nasal bones a massive V-shaped projection most closely resembling the horn-like structures (ossicones) of giraffes. When this creature's remains were examined during the 1870s by the celebrated American palaeontologist Prof. Othniel Charles Marsh, to whom Sioux hunters had brought various thunder horse relics, including in 1877 an amulet consisting of a huge molar teeth from one such beast, its intimately-associated myth inspired him to christen it *Brontotherium*—'thunder beast'.

In recent years, the genus *Brontotherium* has been synonymised with *Megacerops* by some researchers (as have certain other brontothere genera of titanothere, including *Brontops*). Yet as far as its legendary past is concerned, this spectacular beast will always be the thunder horse—a monumental physical embodiment of the thunderstorm's awesome power and terror.

Giant Black Unicorns and the Enigma of *Elasmotherium*

Down through the ages, a very considerable number of unicorn varieties have been differentiated in legends and folklore from around the world—everything from shape-shifting were-unicorns, carnivorous rabbit unicorns, polar bear unicorns with glowing horns, web-footed unicorns, swivel-horned unicorns, and man-eating unicorns with musical horns, to unicorn birds, unicorn snakes, unicorn snails, unicorn pigs, artificially-induced unicorns, and even two-horned unicorns (surely a contradiction in terms!).

One of the least-known members of this diverse array, however, is also among of the most fascinating—the giant black unicorn of Siberia. For this spectacular beast is not only of interest to students of animal mythology but may have significant cryptozoological relevance too as a prehistoric survivor.

The traditional lore of Siberia's Evenk people tells of a huge black bull-like creature bearing a single round, thick, tapering horn of immense size upon the middle of its head. This notoriously belligerent beast would charge at any Evenk rider that it spied, tossing the unfortunate man into the air if it could reach him, and spearing him when he fell back down to earth until he died. Moreover, its horn was so large and heavy that if one of these mega-unicorns were killed, the horn alone needed an entire sledge to transport it. Could this awesome but ostensibly fictitious animal have been inspired at least in part by a living species?

Pursuing a suggestion originally put forward by Alexander Brandt and Norman Lockyer in a *Nature* paper of 8 August 1878, in his book *The Lungfish, the Dodo, and the Unicorn* (1948) veteran cryptozoological writer Willy Ley presented a fascinating line of speculation as to the possible origin of the Evenk people's lore concerning giant black unicorns of ferocious demeanour:

> The paleontologists of around the year 1900 [expanding upon the idea of Brandt and Lockyer in 1878] began to believe that they had successfully discovered the original unicorn. In Russia and in Siberia bones and skulls of an extinct distant relative of the rhinoceros were discovered. This animal, *Elasmotherium sibiricum*, seemed to resemble the ancient reports [of unicorns] even more than does the living rhinoceros. It was noticeably larger than the largest Indian rhinoceros, its horn was much longer and was actually situated in the middle of the forehead of the animal. But after the first excitement had subsided, most scientists abandoned the idea that *Elasmotherium* had anything to do with

Artistic representation of a very belligerent *Elasmotherium sibiricum* (© Hodari Nundu)

the unicorn legend. However, as Melchior Neumayr wrote in his *History of the Earth* [published in 1887]: it is possible that in Siberia man and *Elasmotherium* actually lived together and that *Elasmotherium* was exterminated by man; at least one may explain in this way the ancient songs of the Tunguses [aka the Evenks], which tell that formerly there lived in their country a kind of terrible black ox of gigantic size and with only one horn in the middle of the forehead, so large that an entire sled was needed for the transportation of this horn alone.

Up to 15 ft long and standing over 6.5 ft at the shoulder, *Elasmotherium sibiricum* was truly a massive creature and sported proportionately long legs (at least for a rhinoceros). The last member of its genus, it survived in Russia's Siberian territory until around 350,000 years ago (and quite possibly even as late as 50,000 years ago in the opinion of some researchers), during the late Pleistocene epoch. However, it is now known to have definitely persisted until at least as recently as 27,000 years ago further south, in Kazakhstan's Pavlodar Priirtysh Region, part of the southern West Siberian Plain, thanks to the unexpected discovery there of *Elasmotherium* skull fragments dating from that time scale, as determined by a team of Russian and Kazakh researchers who made their findings public via an *American Journal of Applied Sciences* paper in March 2016.

Suddenly, the hypothesis of this spectacular rhino's survival for a time alongside modern humans as outlined by Ley seems neither impossible nor even overly implausible. An earlier species, *E. caucasicum*, had flourished in the Black Sea region of Europe during the early Pleistocene and was even bigger, up to 16 ft long and weighing an estimated 4-5 tons.

Certainly, a huge gracile rhinoceros bearing a gargantuan horn not upon its snout's nasal bones like all known living rhino species today but upon its brow's frontal bones instead might well make a very convincing giant unicorn. And indeed, *Elasmotherium* nowadays is commonly referred to informally by this exact name—the giant unicorn.

Nor is the huge black unicorn of Evenk lore the only evidence that has been put forward in support of the postulated if currently-unproven late survival of *Elasmotherium* and its contemporaneous, pre-extermination existence alongside humanity in Asia. In 1958, two additional *Elasmotherium* species were formally named—*E. inexpectatum* and *E. peii*, both of which had lived in northern China during the early Pleistocene, becoming extinct around 1.6 million years ago. However, they were subsequently reclassified as Chinese representatives of *E. caucasicum* rather than separate species in their own respective right (meanwhile, fossil remains of a still-valid third species, *E. chaprovicum*, have been obtained from Europe and Asia, and date from 2.6 million to 2.2 million years ago).

Nevertheless, the onetime occurrence of *any* species of *Elasmotherium* in China is particularly intriguing, due to the existence of a certain very unusual and greatly perplexing example of Chinese iconography.

While browsing online on 8 August 2012, I came upon a photograph of a small but very distinctive and highly unusual animal figurine, made of bronze and of Chinese origin. On first sight, it somewhat resembled a rhinoceros, with a short pointed horn upon its snout—but a closer look revealed that this horn curved forwards, whereas those of rhinos more often curve backwards. Moreover, its body did not bear any 'armour pleats' like those exhibited to varying degrees by all five rhino species alive today (unless the vertical lines that continue up onto its body the lines of its limbs are meant to represent the pleats present on the body of the great Indian rhinoceros *Rhinoceros unicornis*, a species present in ancient China before being hunted into extinction there; but if so, they are not portrayed at all accurately and entirely lack the transverse creases running across the limbs' proximal region). Even more striking, however, was something that it did

Photograph of the enigmatic Chinese bronze figurine of a mysterious horned beast (USC Title 17 § 107)

bear—a second, larger horn . . . but not upon its nasal bones. Instead, projecting slightly forwards, it arose directly from the centre of the creature's brow—just like the horn of a unicorn!

The photograph depicting this intriguing figurine appeared (but with no details of its origin) on the s8int.com website, which offers a biblical perspective on science (including cryptozoology), and is written by Chris Parker. Accompanying the photo were details of the figurine's origin and recent history, followed by Parker's thoughts as to what type of creature it may represent. (NB—in late April 2016, I attempted to revisit the specific page on the s8int.com website that contained Parker's analysis of this figurine, but I could no longer find it, so the page may have been deleted, but I retain on file a copy of it that I downloaded on a previous visit.)

Those details revealed that the figurine was in excellent condition and dated from China's Tang Dynasty (618-907 AD), an extremely formidable combination as far as its value as a collectible objet d'art was concerned, because in 2007 it was sold at auction by Christies in London for the staggering sum of US$ 216,000 !!

The Lot description for it that was provided by Christies read as follows:

A Rare and Small Bronze
Figure of a Rhinoceros
Tang Dynasty (618-907 Ad)

Shown standing four-square with tail flicked to the left, the head well cast with two horns of different length, ears pricked back, small eyes and downward curved, overlapping muzzle sensitively cast along the upper edges of the mouth with folds in the skin, which can also be seen in the skin of the neck and chest, the thick hide indicated by overlapping wave pattern diminishing in size on the head and legs, with a rectangular aperture in the belly, the dark brownish surface with some patches of dark red patina and green encrustation.

Lot Notes

The depiction of the rhinoceros in bronze is very rare, especially during the Tang period. Earlier depictions do exist, however, as evidenced by the late Shang rhinoceros zun [i.e. an ancient Chinese wine vessel made of bronze or ceramic] in the Avery Brundage Collection, illustrated by d'Argencè, The Ancient Chinese Bronzes, San Francisco, 1966, pl. XIX and another large zun (22 7/8in. long), ornately decorated, but quite realistic in its depiction of a rhinoceros, of late Eastern Zhou/Western Han dynasty date, found in Xingping Xian, Shaanxi province, included in the exhibition, The Great Bronze Age of China, Metropolitan Museum of Art, 1980, New York, Catalogue, no. 93

So, in spite of the fundamental differences in morphology exhibited by this figurine, it was nonetheless deemed to be a rhinoceros by whoever had prepared its Lot description at Christies—even though they had actually included references within that very same Lot description to much more accurate yet even earlier Chinese depictions of rhinoceroses. One can only assume, therefore, that the writer in question was determined to identify it, however tenuously, with some known animal type, and/or was not a zoologist!

Swiftly dismissing the rhinoceros from serious consideration as the figurine's basis, Parker offered up a very thought-provoking if dramatic alternative—a ceratopsian dinosaur. Known colloquially as the horned dinosaurs, this group included such prehistoric stalwarts as *Triceratops*, *Styracosaurus*, and *Monoclonius* (aka *Centrosaurus*). Officially, of course, like all non-avian dinosaurs, the last ceratopsians became extinct around 66 million years ago, at the end of the Cretaceous Period. But could a ceratopsian lineage have somehow survived beyond the Cretaceous and persisted right up and into the present day, undiscovered by science but represented iconographically by this perplexing work of ancient Chinese art?

Sadly for such an exciting prospect, the figurine lacks the bony neck frill characterizing quadrupedal ceratopsians, and its snout horn curves forward rather than backwards. Also, its tail does not appear to be as long and hefty as that of the latter dinosaurs (this characteristic is deduced from the fact that in the photograph it is hidden on the creature's left side, whereas if it were of ceratopsian dimensions it would have been too big to have been concealed in this manner). Conversely, the figurine does possess what Parker interpreted as a beak, which if so would certainly help to ally it morphologically with the ceratopsians, because these reptiles famously sported large beaks. However, it simply looks to me like a large pointed upper lip, very similar in shape, in fact, to that of Africa's black rhinoceros *Diceros bicornis* and also, albeit to a lesser degree, to that of the great Indian rhino.

Of particular interest is that Parker then compared the figurine with a certain cryptid that has already been popularly linked to the idea of a putative modern-day ceratopsian. Namely, tropical Africa's emela-ntouka ('killer of elephants')—see Chapter 1 of this present book for a detailed coverage. As documented there, it is an aquatic snout-horned mystery beast said by natives to disembowel with its long pointed horn any elephants recklessly trespassing within its swamp or lake domain. Unlike the Chinese figurine, the emela-ntouka does possess a very long, heavy ceratopsian-like tail, though it too is frill-less, but it also sports a pair of small, elephant-like ears—an incongruous feature indeed for any reptile to exhibit, thus indicating that the emela-ntouka is a mammal, not a reptile.

Parker concluded his account by briefly considering the mythological Chinese unicorn, noting that one Chinese researcher has proposed that depictions of this supposedly non-existent beast may in reality have been based upon sightings of a surviving species of *Elasmotherium*. Could the same identity thus explain the Chinese figurine too, he wondered?

Personally, I don't think so. To begin with, the figurine's snout-horn would be an anomaly for any *Elasmotherium* species currently known from the fossil record. In addition, even its brow horn, though correctly located, seems far too short and slender to be comparable to that of *Elasmotherium*. Conversely, its legs are indeed relatively long, like those of *Elasmotherium*, but its body is proportionately much too short and squat to be compatible with the huge 15-ft-plus length attributed by palaeontologists to the proportionately very long body of *Elasmotherium*. So I am certainly not persuaded personally by such an identity for it.

As far as I am concerned, unless it constitutes an inordinately stylised (or just simply inaccurate!) representation of the great Indian rhinoceros the mystery of this Chinese quasi-rhino figurine remains unresolved.

In any case, there is an issue of especial relevance concerning any attempt to link horned cryptids to *Elasmotherium* that has not been mentioned here so far, but which definitely needs to be—so here it is. All of the varied lines of speculation discussed in this section stem directly from one single, fundamental

Elasmotherium sibiricum (CC Stanton Fink)

assumption—namely, that *Elasmotherium* really did possess a horn, and a monstrously large one at that. Consequently, it will undoubtedly come as something of a surprise to discover now that despite many fossil remains of *Elasmotherium* having been found, and from many different sites too, not a single *Elasmotherium* horn, of any size, has ever been uncovered. So why do palaeontologists assume not only that it *was* horned but also that its presumed horn was so enormous?

The answer to this pertinent question is the presence of a very large hemispherical protuberance, 3 ft in circumference and 5 in deep, with a furrowed (and therefore vascularised?) surface, upon the frontal bones, which is typically interpreted as the base for a horn. Coupling this with the occurrence of irregular bony deposition normally indicating that an exceptionally firm attachment to something was required there, plus the extraordinarily large hump of muscle present to manage the head, the overriding conclusion is that an extremely large, heavy horn was indeed borne upon the frontal bones of *Elasmotherium*.

Moreover, there is a Palaeolithic cave painting in Rouffignac, France, of a mighty rhinoceros bearing an enormous horn, which is believed by some to represent *Elasmotherium* (although the horn of this cave painting rhinoceros looks to me more like a typical rhino snout-horn than a brow-borne horn).

But who knows—perhaps one day a fossil *Elasmotherium* skeleton will be unearthed that comes complete with its ostensibly real but currently invisible mega-horn. And if that happens, this huge enigmatic rhino's link to the Siberian giant black unicorn will suddenly take a tantalising step closer to veracity.

On 9 July 2014, German cryptozoologist Markus Bühler reminded me of an extremely bizarre Chinese illustration that had appeared in a wonderful book by Herbert Wendt, *Out of Noah's Ark* (1956), in which Wendt chronicled the history of discovery by humankind of our planet's major animal species. In his documentation of the rhinoceros's discovery, Wendt included two remarkable illustrations, both of them by the Japanese artist Katsushika Hokusai (1760-1849). The first was of a bizarre Asian unicorn, very different from the traditional Chinese deer-like ki-lin, because it sported a tortoise-shell upon its back and a single forward-curving horn upon the back of its head.

And the second was of a later version of this same mythical creature, but now as portrayed after the Chinese had carried out their large-scale transportations of southern Asian rhinoceroses. It still retained its tortoise-shell but now sported no less than three horns. According to Wendt, the two additional ones were derived from the rhinoceros. Yet whereas these were small, but were indeed nasal in location, the third (originally the only horn present and sited on the back of its head) was much longer and was now brow-borne (unlike any modern-day rhino horn). Moreover, all three were forward-pointing (just like those of the Chinese quasi-rhino bronze figurine documented by me earlier here), not rearward-pointing as is usually true of rhino horns.

Consequently, if Wendt was correct in his attribution of the tortoise-shelled unicorn's subsequently-gained trio of horns to the influence of Asian rhinoceros transportation by the Chinese, here is a precedent for rhinos being depicted inaccurately in Oriental art with forward-pointing and brow-borne horns directly comparable to the bronze figurine's. Having said that, however, the tortoise-shelled, triple-horned unicorn had also acquired a second, equally bizarre feature that again is not typical of rhinoceroses if accurately depicted. As can be seen from the two Hokusai illustrations reproduced here, in its original form the

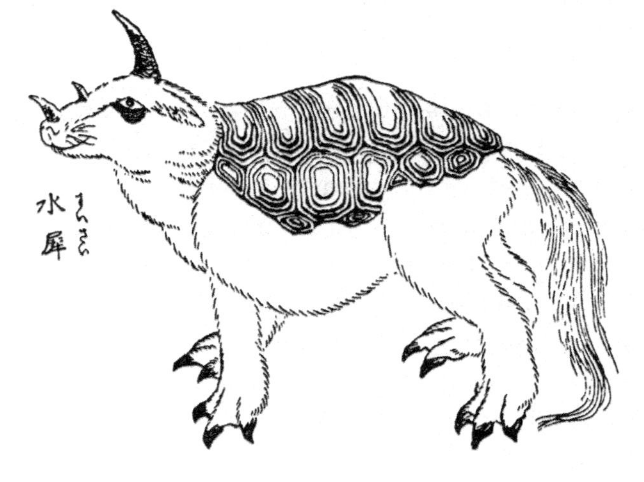

The earlier depiction of the tortoise-shelled Asian unicorn (top); and the later triple-horned, claw-footed version (bottom), the latter believed by Wendt to be the result of having been assigned certain morphological attributes from the rhinoceros—both illustrations produced by Katsushika Hokusai

tortoise-shelled unicorn had possessed ungulate hooves, whereas in its later, triple-horned incarnation these have been replaced by three black claws on each foot. True, rhinos are indeed triple-toed, but they possess clearly-formed hooves, not claws. In contrast, even the Chinese quasi-rhino bronze figurine was hoofed, not claw-footed.

So although intriguing, the tortoise-shelled unicorn's pertinence to the Chinese quasi-rhino figurine's identification as a rhinoceros is far from evident. Consequently, it is probably safest merely to say that it can be looked upon as yet another example of just how extremely stylised (or just plain inaccurate) artistic depictions of animals can be.

Finally: for the romantic zoologist in all of us. During the early Pleistocene's Villafranchian Period (ending about 1 million years ago), Europe harboured an extraordinary antelope called *Procamptoceras brivatense*. What made this creature so remarkable was the extreme nearness to one another of its two upward-pointing, in-line horns. Indeed, quoting from *The Ice Age* (1972) by vertebrate palaeontologist Prof. Björn Kurtén:

> The horns are very close together. In life the horn-cores were covered with horny sheaths that must have been so closely appressed that the animal looked like a unicorn.

Officially, *Procamptoceras* died out before humans reached Europe, but what if this species persisted into more recent times than currently indicated by available fossil evidence? If such a creature had been contemporary with humans, it would be tempting indeed to speculate that Western unicorn legends derive from a distant, hazily-recalled memory of *Procamptoceras*.

How wonderful it would be if palaeontology, and *Procamptoceras*, ultimately confirm that the unicorn—surely the quintessential fantasy beast—was once a real animal after all.

The Perplexing Pukau

The pukau is a mysterious pig-like beast that features in stories recounted by members of a tribe from Sabah in northern Borneo called the Saiap Dusuns. According to a concise coverage of it by Owen Rutter in his book *The Pagans of North Borneo* (1929), this creature is said to resemble a cross between a pig and a deer, to possess a sharp tongue, to occur in large numbers on Mount Madalong, and to give fierce chase if disturbed.

Adult male Sulawesi babirusa (CC Hirscheber)

The description of the pukau as a mixture of pig and deer irresistibly recalls the babirusa, the extraordinary wild pig of Sulawesi (Celebes) and certain nearby islets; in 2002, it was split into four species, with the Sulawesi species (the most famous one) becoming *Babyrousa celebensis*, and the original babirusa name *B. babyrussa* becoming that of the Buru babirusa, which is hairier than the others. Babirusas are very distinctive suids, noted for their very long and slender, deer-like limbs ('babirusa' translates as 'deer-pig'), and especially for their remarkable upper tusks, which in males grow upwards rather than downwards until they eventually pass through the upper jaw and curve backwards towards the brow.

Could tusks like these be the identity of the pukau's 'sharp tongue'? In short, is it possible that in pre-Holocene times babirusas entered Borneo? During the Pleistocene, there were transitory land-bridge connections between Borneo and the faunal region known as Wallacea (which includes Sulawesi), thereby facilitating such migration.

Katch Me a Kilopilopitsofy!

As fully documented by me in my book *Mirabilis* (2013), during late July and early August 1995 Fordham University biologist Dr David A. Burney and Madagascan archaeologist Ramilisonina conducted ethnographical research at three remote southwestern Madagascan coastal villages, in particular Belo-sur-mer. Here, interviewing the local people, they collected testimony from eyewitnesses describing three different mysterious beasts, two of which may well be species still unknown to science. One of these creatures apparently resembles a giant sifaka lemur, and was referred to locally as the kidoky. The second mystery beast was termed the kilopilopitsofy, is also called the tsomgomby or railalomena, and appears to be some form of ungulate.

As with the kidoky, Burney and Ramilisonina received consistent reports of the kilopilopitsofy from several different eyewitnesses, including Jean Noelson Pascou, an educated villager who was able to describe this cryptid in considerable detail, following his own sightings of it, one of which was made as recently as 1976. According to Pascou, it is cow-sized but hornless, has very dark skin, pink colouration around its eyes and mouth, fairly large floppy ears, big teeth, large flat feet, is nocturnal, and escapes from danger by running into the water.

When shown photos of various animals, another eyewitness selected a hippopotamus as bearing the closest resemblance to the kilopilopitsofy. Moreover, when asked to describe the sounds that this mystery beast makes, Pascou, who was well known in the village as a skilled imitator of local animal noises, gave voice to a series of deep, drawn-out grunts, which the startled scientists realised were very similar to the sounds made by the common hippopotamus *Hippopotamus amphibius* of the African mainland. Yet no species of hippopotamus is supposed to exist on Madagascar—or not any longer, that is.

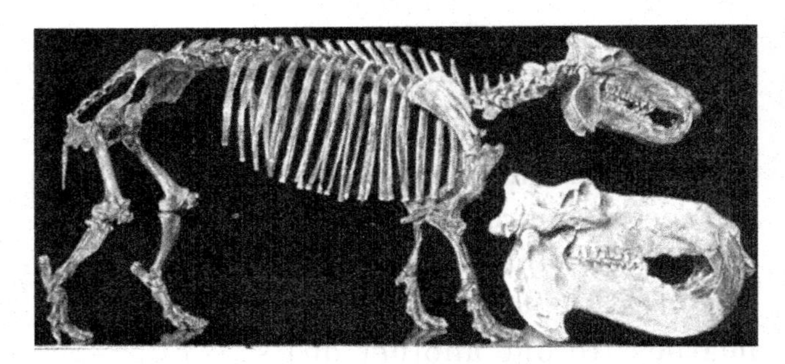

Skeleton of the Malagasy pygmy hippopotamus alongside the skull of the common hippopotamus

In fact, living alongside the giant lemurs that formerly existed on this island were once no less than three different species of endemic Malagasy hippopotamus. These were: the lesser Malagasy hippo *Hippopotamus laloumena* (a small relative of the common African mainland species); the Malagasy dwarf hippo *H. lemerlei* (an even smaller, dwarf species); and the Malagasy pygmy hippo *Choeropsis* [=*Hexaprotodon*] *madagascariensis* (a distinctive pint-sized species more closely related to the African mainland pygmy hippo *Choeropsis* [*H.*] *liberiensis*).

Just like the giant lemurs finally became extinct, however, after being hunted and eaten by humans (who first reached this island in around 2,000 BP), so too did all three Madagascan hippo species, once again having been killed by humans. Nevertheless, subfossil evidence confirms that at least one species, the Malagasy dwarf hippo *H. lemerlei*, was still alive as recently as 1,000 BP.

Furthermore, judging from the testimony of Pascou and others with regard to the mystifying kilopilopitsofy, mainstream zoologists as well as cryptozoologists now consider it plausible that at least one of Madagascar's diminutive trio of hippos actually persisted into much more recent times, and perhaps even into the present day. As for this mystery beast's floppy ears, they deem it possible that these were actually loose jowls and cheeks, misidentified as ears when seen fleetingly at night. Alternatively, it may be that they really are ears, larger than those of typical hippos, which have evolved to help dissipate heat if, as seems to be true, the Malagasy dwarf hippo in particular was (or is?) more terrestrial than typical hippos.

In 1876, the skin of an alleged kilopilopitsofy was shown to Westerner Josef-Peter Audebert, who likened it to that of an antelope (no species of which exists on Madagascar), and stated that it had supposedly come from the south of the island. If only this skin had been preserved—DNA analyses could have unmasked its owner's identity and thereby resolved the longstanding mystery of this very perplexing Madagascan cryptid.

An Anachronistic Camel from Utah

In the 1800s, dromedaries and Bactrian camels were introduced to North America for various purposes, but the last of this continent's native camelids was a very large, llama-like version called yesterday's camel *Camelops hesternus*, which occurred in western North America. Customarily assumed to have died out during the early Pleistocene around 2 million years ago, a notable challenge to this assumption was raised in 1928, when celebrated American palaeontologist Prof. Alfred Sherwood Romer published his conclusions concerning a very special camel skull sent to him by Utah University colleague Prof. A.L. Matthews.

While exploring the igneous buttes some 20 miles south and west of Fillmore, Utah, two high school boys had chanced upon the skull, about 200 ft back in a cave, buried under about 3-4 ft of fine dry eolian deposit. It consisted of a practically complete braincase and most of the palate, and was unfossilised—indeed, a strip of dried muscle was still attached to its basioccipital portion. Hence Romer initially assumed that it would simply prove to be from one of the dromedaries imported into the southwestern United States during the 1870s.

To his astonishment, however, after studying its structure in minute detail he realised that while differing markedly from the skulls of both species of Old World, modern-day camel, this extraordinary specimen corresponded very closely in many different features with North America's indigenous but supposedly long-extinct *Camelops*.

Only two explanations for this anachronistic anomaly seemed possible. The skull may have survived from the early Pleistocene in an unfossilised state due to some freak, hermetic process (like the Argentinian *Mylodon* ground sloth skin documented earlier in this chapter). Alternatively, the skull may truly have been of more recent origin, which would prove that *Camelops* had persisted much later than previously assumed.

As Prof. Christine Janis has noted to me, given that half a million years is considered excellent resolution in the fossil record, why should the discovery of relatively recent remains of any species with its last fossil record somewhere in the early Pleistocene be deemed so very astonishing? Especially when, as with *Camelops*, environmental conditions favouring its continued survival have persisted in some parts of the U.S.A. right up to the present day.

Nowadays, this case could be readily dealt with by applying radiocarbon-dating techniques to the skull, but for many years that couldn't be done, because the skull appeared to have been lost following Romer's donation of it to the University of Utah Geology Museum, where it was displayed for a time but never formally catalogued. During spring 1978, however, it was unexpectedly rediscovered by Drs Michael E. Nelson and James H. Madsen Jr in some unmarked collection drawers at the museum, being readily identifiable thanks to some excellent photographs of it taken by Romer. They duly submitted bone samples from it to Geochron Laboratories for a radiocarbon age determination, and the results obtained indicated that the skull dated from approximately 11,000 ago, i.e. late Pleistocene/early Holocene, thus confirming Romer's opinion that *Camelops* did indeed survive into much more recent times than had previously been supposed.

Llamas or Litopterns?—A Polydactylous Puzzle

Today's four recognised species of llama are the vicuna *Vicugna vicugna*, the alpaca *V. pacos* (a domesticated descendant of the vicuna), the guanaco *Lama guanicoe*, and the llama *L. glama* (a domesticated descendant of the guanaco). They are each distinguished from the other three by virtue of various readily-visible external characteristics, but, regardless of these differences, one significant morphological feature that the vicuna, alpaca, llama, and guanaco do all share is that they each possess just two toes per foot—which is crucial to the following mystery llama cases.

The great Tiahuanacan empire, spanning 300 to 1,000 AD, extended westwards from the Bolivian border of Lake Titicaca as far as northern Chile and the southern coast of Peru, and included Pisco. It was near Pisco, while excavating amidst the ruins of two coastal Tiahuanacan cities during the early 1920s, that Peruvian archaeologist Prof. Julio C. Tello uncovered some jugs depicting llama-like beasts with *five* toes on each foot.

In his book *In Quest of the White God* (1963), French archaeologist Pierre Honoré wondered if these fragments implied that the ancestral five-toed llamas had not become extinct many thousands of years ago, but had survived until much more recent times. In fact, this line of thought is founded upon a basic error. There has *never* been *any* species of five-toed llama.

The mammalian foot is fundamentally pentadactyl (equipped with five digits), but during the evolution of some mammal groups (including the ungulates) digit reduction has occurred. In the camelids, this began at a very early stage in their emergence from the ancestral ungulate stock. Indeed, even *Poebrotherium*, which existed in North America as far back in time as the late Eocene to the early Oligocene epochs (about 38-31 million years ago), had only two toes on each foot, each toe bearing a hoof.

Also: the camelids originated in North America, migrating into the Old World via land-bridges during the Pliocene, which began 5.3 million years ago. However, the modern-day South American camelids' ancestor did not reach South America until the Pliocene's close, about 2.7 million years later, because until then this continent had been an island for many millions of years, only rejoining North America when the isthmus of Panama rose up above sea-level at that time.

In short, whatever the Tiahuanacan five-toed beasts were, they were *not* ancestral llamas. However, Tello's finds were not limited to mysterious portrayals on pottery. Later, he also discovered some skeletons of llamas, with five toes, buried amidst the remains of temples.

According to American zoologist and explorer A. Hyatt Verrill, writing in his book *Strange Prehistoric Animals and Their Stories* (1948), the mummified bodies of five-toed llamas have also been found in the graves or tombs of the Paracas culture, an Andean society that existed between approximately 800 and 100 BC in what is now Peru's Ica Province. These enigmatic animals are depicted upon Paracas pottery and textiles too.

Peru's most contentious series of anomalous animal depictions also hail from Ica Province, and can be found engraved upon the immense collection of decorated stones obtained here by anthropologist/surgeon Dr Javier Cabrera Darquea—see Chapter 1 for a detailed account of these Ica stones. Some of them are engraved with portrayals of dinosaur-like, pterodactyl-like, and even mosasaur-like beasts, as well as creatures resembling ancient mammals, including what appears to be North America's long-necked (and long-extinct) giraffe camel *Aepycamelus*, plus some llamas with five toes. Not surprisingly, many archaeologists have deemed these depictions to be modern-day fakes, reputedly based upon comic-book art in some instances.

In 2000, Cabrera documented his controversial collection in a book entitled *The Message of the Engraved Stones of Ica*. This contains a photograph of a stone that is indeed engraved with five-toed llama-like beasts.

Not widely realised, however, is that many Ica stone discoveries are wholly unrelated to those made by Cabrera anyway, and one of these non-Cabrera finds is of significance here. During the 1960s, Peruvian archaeologist Alejandro Pezzia Assereto from the National Archaeology Department of Peru, who was also a trustee of the Ica Museum and in charge of archaeological investigations in Ica Province, conducted official excavations in some ancient Paracas and Ica cemeteries. On two different occasions, engraved Ica stones were disinterred from pre-Hispanic Indian tombs dating from 400 BC to 700 AD, the stones having been discovered embedded in the side of the tombs' mortuary chamber and also alongside various mummies. In 1968, he published his findings in Vol. 1, *Arqueología de la Provincia de Ica*, of his book *Ica y el Perú Precolombino*, containing descriptions and drawings of the Ica stones uncovered by him, one of which depicted a five-toed llama. These stones were deposited in the Ica Museum, where they remain today.

So, do all of these varied depictions of five-toed llamas by different cultures in different regions of Peru and its neighbouring South American countries mean, therefore, that despite all the dictates of palaeontology, pentadactyl camelids had not only existed, but had actually persisted in South America beyond the arrival here of humans?

No—because there is a much less radical, far more straightforward solution to this mystery. Polydactyly, a genetically-induced teratological condition, is the possession of more than the normal number of fingers or toes (or both). Polydactylous individuals occur quite frequently in many mammalian species (including humans), and some confirmed examples have been recorded from at least one species in the modern-day quartet of South American camelids.

At Tilcara, in the southern Argentinian province of Jujuy, for instance, the remains of two polydactylous guanacos were found inside Indian tombs dating back to the pre-Columbian era, as documented in 1917 by Dr G. Martinoli. In 1930, Argentinian naturalist and archaeologist Dr Carlos Rusconi recorded another such specimen, found in a tumulus (Indian burial mound) at Santiago del Estero, in northern Argentina. In both of these cases, the animals had three toes rather than five, but more extreme examples have been recorded from other mammalian species. Thus it is not unreasonable to suppose that the Tiahuanacan depictions and other similar ones reported here were inspired by polydactylous llamas (or guanacos?).

Freak specimens of animals have frequently attracted special attention from humankind—looking upon them variously as the harbingers of doom or the heralds of good fortune, but always as important (often magical) beasts. Hence it is not surprising that the remains of polydactylous camelids should be found in association with Indian relics; as Pierre Honoré suggested, because of their unique appearance they may well have been regarded as sacred, and deliberately maintained within the temples. Certainly, the presence of such creatures' remains inside Indian tumuli and tombs implies that they were believed to embody some special significance—capable, perhaps, of granting safe passage for the Indians into the next world?

Before concluding categorically that the five-toed llamas of Tiahuanaco and elsewhere were truly nothing more than polydactylous specimens of modern-day llama species, however, one further possibility (albeit rather more radical, but also very exciting) is well worth a consideration.

During its existence as an island continent—lasting over 60 million years—South America hosted the evolution of several unique taxonomic orders of ungulate, all of which regrettably died out prior or subsequent to its reconnection with North America. One such order constituted the litopterns.

Some of these hoofed mammals were remarkably faithful morphological imitators of horses, and hence are often termed pseudo-horses, whereas others were more similar to camelids, and were known as macraucheniids. Certain of the latter beasts, such as *Theosodon* from the early Miocene, were extremely llama-like—except that they had three toes on each foot instead of two.

Macrauchenia patachonica—was it really the last of the litopterns?

Macrauchenia patachonica of Argentina, which resembled a long-legged, long-necked humpless camel except for its short tapir-like nasal trunk, is the youngest litoptern species currently known from the fossil record, its most recent remains dating from the very end of the Pleistocene, just under 12,000 years ago. However, because the fossil record is by no means complete, this is not absolute proof that no litopterns persisted beyond that date, perhaps surviving for quite some time alongside humans (the earliest evidence for human existence here dates back to around 9,000 BC) before becoming extinct.

Indeed, Prof. Janis considers that a stylised representation of one of the three-toed llama-like litopterns is at least a possible explanation for the creatures depicted on the Tiahuanacan pottery; if so, this indicates that these survived into much more recent times than hitherto confirmed by fossils. Certainly, the visible discrepancy in toe count between the depicted beasts and their postulated litoptern models may be due to stylisation—perhaps even a deliberate attempt on the part of the artist to emphasise and exaggerate the llama-like litopterns' most noticeable external difference from true, camelid llamas?

A classic Charles Knight illustration, depicting *Toxodon* (left), three specimens of *Glyptodon* (right), and three *Macrauchenia* (background)

Frustratingly, however, it is here, possibly even on the very brink of proving that litopterns and humans were indeed contemporary for a time, that this saga currently draws to a close, lacking conclusive evidence for this fascinating theory—until (if ever), that is, a tumulus or temple is found that contains the conclusively identified remains of a llama-like litoptern and not a polydactylous llama or guanaco.

The Notoungulate and the Novel

Another Tiahuanacan riddle features some pieces of ceramic pottery and also some stonework uncovered there that apparently portray a very large hippopotamus-like beast unrecognisable among today's South American fauna. Until the Pleistocene's close, however, such a creature did exist here. Known as a toxodont, it belonged to another principally (though this time not exclusively) South American taxonomic order of hoofed mammals—the notoungulates.

Indeed, while acting as a zoological advisor to Sir Arthur Conan Doyle during the preparation of his novel *The Lost World* at the end of 1911, British Museum (Natural History) director Sir Edwin Ray Lankester suggested that a modern-day species of *Toxodon* may await formal scientific discovery. Bearing in mind that this is the same expert who correctly identified the remarkable okapi as a short-necked giraffid, his opinions can hardly be dismissed as ill-informed speculation. So perhaps palaeontologists should look out for post-Pleistocene remains of toxodonts too!

The Last of the Irish Elks?

One of the most spectacular members of the Eurasian Pleistocene megafauna was the Irish elk *Megaloceros giganteus*. Formally described in 1799, it is also aptly known as the giant deer, because its largest known representatives were only marginally under 7 ft tall at the shoulder and bore massive antlers spanning up to 12 ft, but did this magnificent species linger on into historic times? Below is an account of mine devoted to this tantalising subject and dating back to 1995, when it appeared in this present book's original edition, *In Search of Prehistoric Survivors*. It is followed by various fascinating updates, including some significant palaeontological discoveries made since that edition's publication that are of great pertinence to the question of post-Pleistocene survival for this species.

But first—here is the relevant excerpt from my 1995 book:

The Irish elk *Megaloceros giganteus* was one of the largest species of deer that ever lived. It was also one of the most famous—on account of the male's enormous antlers, attaining a stupendous span of 12 ft and a weight of over 100 lb in some specimens. Sadly, its common name is misleading, as this impressive species is only very distantly related to the true elk (moose), and, far from being an Irish speciality, was prevalent throughout the Palaearctic Region, from Great Britain to Siberia and China.

Nevertheless, it is to Ireland that we must turn for the majority of clues regarding *Megaloceros*—because in contradiction to the accepted view that it had died out here by the end of the Pleistocene, certain accounts and discoveries from the Emerald Isle have tempted researchers to speculate that this giant deer may still have been alive here a mere millennium ago.

According to accounts documented by H.D. Richardson in 1846, and reiterated by Edward Newman in the pages of *The Zoologist*, the ancient Irish used to hunt an extremely large form of black deer, utilising its skin for clothing, its flesh for food, and its milk for the same purposes that cow milk is used today. Supporting that remarkable claim is a series of bronze tablets discovered by Sir William Betham; inscribed upon them are details of how the ancient Irish fed upon the flesh and drank the milk of a great black deer.

These accounts resurfaced two decades later within an examination of the Irish elk's possible survival here into historic times by naturalist Philip Henry Gosse, in which he also

Benjamin Waterhouse Hawkins's magnificent Victorian-age statue of an adult male Irish elk *Megaloceros giganteus* at London's Crystal Palace Park (© Dr Karl Shuker)

documented an intriguing letter written by the Countess of Moira. Published in the *Archaeologia Britannica*, this letter recorded the finding of a centuries-old human body in a peat bog; the well-preserved body was completely clothed in garments composed of deer hair, which was conjectured to be that of the Irish elk.

Most interesting of all, however, was the discovery in 1846 by Dublin researchers Glennon and Nolan of a huge collection of animal bones surrounding an island in the middle of Lough Gûr—a small lake near Limerick. Among the species represented in it was the Irish elk, but of particular note was the condition of this species' skulls. Those lacking antlers each bore a gaping hole in the forehead, which seemed to have been made by some heavy, blunt instrument—recalling the manner of slaughtering cattle and other meat-yielding domestic animals with pole-axes, still practised by butchers in the mid-to-late 1800s. Conversely, those skulls that were antlered (one equipped with immensely large antlers) were undamaged.

Did this mean that the antler-less (i.e. female) Irish elks had actually been maintained in a domestic state by man in Ireland, as an important addition to his retinue of meat-producing species? Prof. Richard Owen sought to discount such speculation by stating that the mutilated skulls were in reality those of males, not females, and that the holes had resulted from their human killers wrenching the antlers from the skulls.

However, this was swiftly refuted by Richardson, whose experiments with intact skulls of male Irish elks showed that when the antlers were wrenched off they either snapped at their bases, thereby leaving the skulls undamaged, or (if gripped at their bases when wrenched) ripped the skulls in half. On no occasion could he obtain the curious medially sited holes exhibited by the Lough Gûr specimens. Clearly, therefore, these latter skulls were from female deer after all, explaining their lack of antlers—but what of the holes?

As Gosse noted in his coverage of Richardson's researches, it is significant that the skulls of certain of the known meat-yielding mammals present alongside the *Megaloceros* skulls at Lough Gûr had corresponding holes—and as Gosse very reasonably argued: "As it is evident that *their* demolition was produced by the butcher's pole-ax, why not that of the elk skulls?"

After presenting these and other accounts, Gosse offered the following conclusion:

> From all these testimonies combined, can we hesitate a moment in believing that the Giant Deer was an inhabitant of Ireland since its colonisation by man? It seems to me that its extinction cannot have taken place more than a thousand years ago. Perhaps at the very time that Caesar invaded Britain, the Celts in the sister isle were milking and slaughtering their female elks, domesticated in their cattle-pens of granite, and hunting the proud-antlered male with their flint arrows and lances. It would appear that the mode of hunting him was to chase and terrify him into pools and swamps, such as the marl-pits then were; that, having thus disabled him in the yielding bogs, and slain him, the head was cut off, as of too little value to be worth the trouble of dragging home . . . and that frequently the entire carcase was disjointed on the spot, the best parts only being removed. This would account for the so frequent occurrence of separate portions of the skeleton, and especially of skulls, in the bog-earth.

Although undeniably thought-provoking, the case of *Megaloceros*'s persistence into historic times in Ireland as presented by the above-noted 19th-Century writers has never succeeded in convincing me—for a variety of different reasons.

For instance, there is no conclusive proof that the large black deer allegedly hunted by the ancient Irish people really were surviving *Megaloceros*. Coat colour in the red deer *Cervus elaphus* is far more variable than its common name suggests; and, as is true with many other present-day species of sizeable European mammal, specimens of red deer dating from a few centuries ago or earlier tend to be noticeably larger than their 20th Century counterparts.

Similarly, the Lough Gûr skulls' ostensibly significant contribution to this case rests upon one major, fundamental assumption—that they are truly the skulls of *Megaloceros* specimens. But *are* they? Precise identification of fossil remains is by no means the straightforward task that many people commonly believe it to be.

Perhaps the greatest of all mysteries associated with this case, however, is that subsequent investigations of *Megaloceros* survival in Holocene Ireland as inspired by the researches of Gosse and company, and formally documented in the scientific literature, are conspicuous only by their absence. (In September 1938, A.W. Stelfox of Ireland's National Museum, in Dublin, did consider this subject, but without reference to any of the above accounts.) Yet if the case for such survival is really so compelling and conclusive, how can this investigative hiatus be accounted for?

Seeking an explanation for these assorted anomalies, I consulted mammalian palaeontologist Dr Adrian Lister [then at Cambridge University, England, now at London's Natural History Museum]—who has a particular interest in *Megaloceros*. Confirming my own suspicions, Dr Lister informed me that it is not unequivocally established that the female Lough Gûr skulls were from *Megaloceros* specimens, and he suggested that they might be those of female *Alces alces*, the true elk or moose, which did exist in Ireland for a time during the Holocene (though it is now extinct there). Certainly in general form and size, female *Alces* skulls seem similar to the Lough Gûr versions.

In contrast, Lister agreed that the enormous size of the antlers borne by the male Lough Gûr skulls indicated that these were bona fide *Megaloceros* skulls; but as he also pointed out, although their presence in the same deposits as the remains of *known* domesticated species is interesting, without careful stratigraphical evidence this presence cannot be accepted as conclusive proof of association between *Megaloceros* and man.

In his *Megaloceros* account, Gosse included some reports describing discoveries in Ireland of huge limb bones assumed to be from *Megaloceros*, which were so well preserved (and hence recent?) that the marrow within them could be set alight, and thereby utilised as fuel by the peasantry, or even boiled to yield soup! Yet once again, as I learnt from Dr Lister, these were not necessarily *Megaloceros* bones—especially as the limb bones of red deer, moose, and even cattle are all of comparable shape and form, and can only be readily distinguished from one another by osteological specialists (who do not appear to have been granted the opportunity to examine the bones in those particular 19th Century instances, and the bones were not preserved afterwards). Furthermore, on those occasions when exhumed bones used for fuel purposes *have* been professionally examined, none has been found to be from *Megaloceros*.

In conclusion: far from being proven, the case for post-Pleistocene survival of *Megaloceros* in Ireland is doubtful to say the least. Nevertheless, this is not quite the end of the trail. As noted by zoologist Dr Richard Lydekker, and more recently by palaeontologist Professor Björn Kurtén, the word '*Schelk*', which occurred in the famous *Nibelungenlied* ('*Ring of the Nibelungs*') of the 13th Century, has been considered by some authorities to refer to specimens of *Megaloceros* alive in Austria during historic times; other authorities, conversely, have suggested that a moose or wild stallion is a more plausible candidate.

Whatever the answer to the above proves to be, far more compelling evidence for such survival was presented in 1937 by A. Bachofen-Echt of Vienna. He described a series of gold and bronze engravings on plates from Scythian burial sites on the northern coast of the Black Sea. Dating from 600-500 BC and now housed at the Berlin Museum, the engravings are representations of giant deer-like creatures, whose antlers are accurate depictions of *Megaloceros* antlers! Undeniable evidence at last for Holocene survival?

The enigma of these engravings has perplexed palaeontologists for decades, but now a notable challenge to their potential significance has been put forward by Dr Lister, who has provided a convincing alternative explanation—postulating that the engravings were not based upon living *Megaloceros* specimens, but rather upon fossil *Megaloceros* antlers, exhumed by the Bronze Age people. This interpretation is substantiated by the stark reality that out of the hundreds of Holocene sites across Europe from which fossil remains have been disinterred, not a single one has yielded any evidence of *Megaloceros*.

True, absence of uncovered Holocene remains of *Megaloceros* does not deny absolutely the possibility of Holocene persistence (after all, there are undoubtedly many European fossil sites of the appropriate period still awaiting detection and study). But unless some such finds *are* excavated, it now seems much more likely that, despite the optimism of Gosse and other Victorian writers, this magnificent member of the Pleistocene megafauna failed to survive that epoch's close after all, like many of its extra-large mammalian contemporaries elsewhere.

That was where the matter stood back in 1995, when this present book's original edition was published—but not any longer!

On 15 June 2000, a paper published in the scientific journal *Nature* and co-authored by Lister revealed that a near-complete *Megaloceros* skeleton uncovered in the Isle of Man (IOM) and a fragmentary antler from southwest Scotland had recently been shown via radiocarbon dating to be only a little over 9,000 years old, i.e. dating from just inside the Holocene epoch—the first unequivocal proof that this mighty deer did indeed survive beyond the Pleistocene.

Intriguingly, however, as also disclosed in this paper, the IOM Holocene specimen's skeleton was statistically smaller (by over two standard deviations from its mean) than all Irish Pleistocene counterparts also measured in this study, indicating a diminution in body size for *Megaloceros* as it entered the Holocene, at least on the IOM. Conversely, the antlers for this specimen and also the Scottish antler were well within the Irish size range for adult males.

Adult male *Megaloceros* skeleton (© Jay Cooney)

The IOM separated from the British mainland around 10,000 years ago. Consequently, it may be that the decrease in body size recorded for the IOM specimen measured in this study (if typical and not merely a freak specimen) is a result of inhabiting a relatively small island, rather than a strictly chronological effect. There are, after all, many examples in the fossil record and also among present-day fauna

Megaloceros statue in Berlin's Tierpark, Germany (© Markus Bühler)

whereby large species inhabiting small islands have gradually decreased in size, becoming notably smaller than their mainland counterparts—a good example documented a little earlier in this present chapter being the dwarf woolly mammoths of Wrangel Island.

But that is not all. On 7 October 2004, once again via a *Nature* paper, a team of researchers that included Lister revealed via radiocarbon dating of uncovered skeletons that *Megaloceros* survived in western Siberia until at least circa 5,000 BC, i.e. some 3,000 years after the ice-sheets receded. Age-wise, these are currently the most recent *Megaloceros* specimens on record, and demonstrate that the Irish elk existed during the Holocene in two widely separate localities.

So who knows? Following these exciting finds, perhaps other Holocene specimens, and possibly some of even younger dates than those presently documented, still await scientific unfurling?

Also of note is that on 8 June 2015, the journal *Science Reports* published a paper from a research team including Drs Alexander Immel and Johannes Krause which revealed that *Megaloceros* remains recovered from cave sites in the Swabian Jura (Baden-Württemberg, southern Germany) dated to 12,000 years ago. Until now, it had been believed that this giant deer species had become extinct in Central Europe prior to, rather than after, the Ice Age. Moreover, the DNA techniques used in identifying the remains as *Megaloceros* showed that this species is more closely related to the fallow deer *Dama*

dama (as long believed in the past) than to the red deer (as more recently assumed).

One final *Megaloceros* mystery: On 4 July 2015, Hungarian cryptozoological blogger Orosz István posted a short but very interesting item about a supposed mythological beast that I had never heard of before—the hippocerf (a name combining the Greek for 'horse' with a derivation from the Latin for 'deer'). He stated that it was said to be half horse, half deer (hence its name) and, of particular interest, that some (unnamed) researchers believed that it was based upon a *Megaloceros* population surviving into historic times. István had obtained his information from a brief entry on this creature that appears on the Cryptids Wikia website (at: Cryptidz.Wikia.com).

Needless to say, I soon conducted some online research myself concerning this intriguing creature, but I was not exactly cheered by my findings. With the exception of the Cryptidz Wikia site and a few others giving only the barest information repeated one to another ad nauseam, plus some imaginative illustrations of it created by various artists on the deviantart.com site, the hippocerf seemed to be endemic to fantasy fighting and other fantasy-style game sites. On these sites, some of the fabulous creatures featured are bona fide mythological beasts but others are complete inventions, dreamed up exclusively for the games, with no basis whatsoever in world mythology. Hence I began to suspect that the hippocerf might be from the latter category, i.e. conceived entirely for fantasy fighting games.

Indeed, apart from its very frequent appearances in Final Fantasy and other fantasy game sites and its popularity as a subject for drawing/painting on deviantart, all that I have been able to trace about the hippocerf online is that it supposedly has the hindquarters of a horse and the forequarters, neck, and antlered head of a deer, and that because of its dual nature, in heraldry it represents indecision or confusion. However, I have yet to find any confirmation of this latter claim from standard sources on heraldry online or elsewhere. (I own several major works on this subject, and none of them contains any mention of the hippocerf.) Nor have I uncovered the names of any of the researchers who have purportedly suggested that this distinctive creature may have derived from *Megaloceros* sightings in historic times.

As for a claim repeated on several websites that the last known hippocerf sighting was in around 600 AD by an early archaeologist called Gregor Ishlecoff, I traced this to a book entitled *The Destineers' Journal of Fantasy Nations*, authored by N.A. Sharpe and Bobby Sharpe, and self-published in 2009—it proved to be a fantasy novel aimed at teenagers! I also own a considerable number of bestiary-type books on mythological beasts, and again none contains any information regarding the hippocerf.

In short, not very promising at all for the supposed reality of the hippocerf as a genuine (rather than a made-up) mythological beast. The only hope for its credibility is if a mention can be traced in an authentic bestiary pre-dating the coming of the internet and fantasy gaming (preferably one of the classic works from medieval or Renaissance times), or in some authoritative work on heraldry. If either or both of these possibilities result in positive info emerging, then it may be that the hippocerf was inspired by the imposing and somewhat equine form of the moose (which inhabited much of Central Europe until hunted into extinction in many parts there by the onset of the Middle Ages). To my mind, this seems a more plausible option than the survival of *Megaloceros* into historic times in Europe (i.e. into much more recent times than even the circa 7,000 BC date currently known for it there).

Having said that: I can't help but recall a certain noteworthy line from the Immel *et al.* paper of 8 June 2015 regarding the finding of post-Ice Age *Megaloceros* remains in Germany: "The unexpected presence of *Megaloceros giganteus* in Southern Germany after the Ice Age suggests a later survival in Central Europe than previously proposed". Interesting . . .

Incidentally, the hippocerf should not be confused (but sometimes is) with the hippelaphos (whose name also translates as 'horse-deer', but from the Greek for 'horse' and the Greek for 'deer'), which is a genuine creature of classical mythology.

Attempts to identify the hippelaphos with known animal species have been made down through the ages by many scholars, including Aristotle (whose account of it recalls a gnu), Cuvier (proposing the Asian sambar deer *Rusa unicolor*), and 19th-Century German zoologist Prof. Arend F.A. Wiegmann (the Indian nilghai *Boselaphus tragocamelus*). Another option is Africa's roan antelope *Hippotragus equinus*, a decidedly horse-like species, as emphasised by its taxonomic binomial, and also by its French name, antilope chevaline ('horse-like antelope').

In the original Latin version of Aristotle's work, the hippelaphos is termed the hippocervus (renamed the hippelaphos in the English translation version), a name that is sometimes applied to the hippocerf on various internet sites. Indeed, I wonder if the hippocerf may be nothing more than the hippelaphos (aka hippocervus) distorted and exaggerated by online invention, such as the unsourced claim noted earlier here—namely, that some researchers believe the hippocerf may be based upon a *Megaloceros* population surviving into historic times.

A roan antelope (© Dr Karl Shuker)

Modern-day reconstruction of *Sivatherium* on left, alongside a chalicothere model on right (© Markus Bühler)

Antlered Giraffes and Out-Of-Place Okapis

Sivatheres constitute a distinct taxonomic subfamily of giraffids very different in appearance from the familiar modern-day giraffe and okapi. Their bodies were massive and ox-like in shape, and their heads bore palmate, antler-like horns (modified ossicones—the bony, skin-covered, non-deciduous horns peculiar to giraffids), which conferred upon these animals a superficial resemblance to gigantic, long-limbed deer (especially the moose). Traditionally believed to have become extinct by the Pleistocene's close, there is some tantalising iconographical evidence to suggest that they may have survived into more recent times.

Early *Sivatherium* reconstruction (1896)

For instance, in the central Sahara's Tibesti Mountains is a rock-shelter containing a petroglyph of a possible sivathere—the petroglyph is no more than 8,000 years old. Additionally, for many years an even more compelling work of ancient art relative to putative sivathere portrayal was an enigmatic chariot-ring of Sumerian origin dating back to around 2,500 BC and which was unearthed at Kish, in Iraq, during the early 1930s by a joint archaeological expedition from Chicago's Field Museum and England's Oxford University. What makes this ring so intriguing is that it is ornamented with a sculpted figurine of a large antlered beast that bears a striking similarity to *Sivatherium*—a very large Asian and African sivathere.

The ring's figurine was initially thought to represent a robust-bodied deer, but when eminent vertebrate palaeontologist Dr Edwin Colbert examined it a few years after its discovery he perceived that it exhibited several morphological features not easily associated with any known form of deer, yet which were all characteristic of *Sivatherium*. These included a pair of small, conical, knob-like horns on the figurine's brow, in front of its large palmate 'antlers'; the posterior siting of the 'antlers' (those of deer are positioned further forward on the skull); the general shape and orientation of the cores of these 'antlers'; and the swollen shape of the figurine's nostrils.

What makes this figurine even more remarkable is that its sculpture incorporates what appears to be a representation of a rope or tether, attached to its muzzle. If that interpretation is correct, this implies that, irrespective of its taxonomic identity, the creature was a captured or domesticated beast. The concept of sivatheres not only lingering into the age of the Sumerian civilisation but also existing within it as domestic beasts is undoubtedly a radical one, as Colbert readily accepted, but the morphological match between the figurine and the likely appearance of *Sivatherium* (as based upon reconstructions from fossil evidence) seemed persuasive enough for this notion to warrant serious consideration.

In 1977, however, Heidelberg University graduate student Michael Müller-Karpe announced that the antlers of this famous Sumerian chariot-ring's figurine of an alleged sivathere were actually incomplete. In a small box of dried mud housed in a storeroom at Chicago's Field Museum, he had discovered some corroded copper artefacts resembling carved, branched antlers, which upon examination were found to be the hitherto-unknown, distal portions of this figurine's antlers!

They must have broken off at some stage prior to the figurine's discovery, so that until now they had not been recognised as part of it. As a result of this unexpected find, zooarchaeologist Dr David S. Reese of the Field Museum stated that the figurine conclusively depicted a Persian fallow deer *Dama dama mesopotamica*, not a sivathere. However, this was swiftly challenged by Prof. Christine Janis, who presented persuasive reasons for continuing to endorse a sivathere identity for it, such as the anomalous knob-like horns on its brow. Clearly, the dispute is far from over after all.

Moreover, Janis has also provided some additional evidence for possible sivathere survival into the Holocene—in the form of various Russian and early Middle Eastern carvings of deer-like beasts with antlers resembling the branched horns of another Pleistocene sivathere, *Bramatherium*. She wonders whether *Bramatherium* may have actually persisted long enough to leave behind a folk memory of its former existence, which could account for the unusual shape of the sculptures' antlers.

Traditional depiction of the branched- (and bi-) horned ki-lin

One sivathere-related aspect that had not been commented upon prior to the publication of this present book's original edition in 1995 is the resemblance between these unusual giraffids (especially *Sivatherium* itself) and certain depictions of the legendary Chinese unicorn or ki-lin. Ignoring readily identifiable flourishes of ornate stylisation, one example is the version reproduced here, portraying the male ki-lin.

The Persepolis beast as depicted upon the southern wall of the frieze along the eastern staircase of the Apadana Palace at Persepolis

The case of the Sumerian chariot-ring centred upon the possibility that a representation of a presumed non-giraffid (in this instance a deer) was in reality of a bona fide surviving prehistoric giraffid. In contrast, the next case under consideration here is the exact converse—revealing how a representation of an alleged surviving prehistoric giraffid may not have been of one at all.

Flanking the monumental eastern staircase of the Apadana Palace in the ancient Iranian city of Persepolis (aka Takht-e-Jamshid), ceremonial capital of the Achaemenid Empire (c.550–330 BC), is an enormous Achaemenien frieze in three parts or walls. Among the many depictions on the southern wall is one (slightly damaged) of a delegation (usually identified as either Ethiopian or Nubian) leading by a tether a superficially giraffid-like beast. (A second such frieze, associated with the northern staircase, is half-destroyed.) Erected at the time of Darius the Great (reigned 522-486 BC), the palace contains a number of friezes portraying the offering of tributes to the king from visiting foreign representatives—hence it is logical to assume that this animal was likewise a gift, transported here from elsewhere. Certainly, neither of today's surviving species of giraffid is known to have been native to this region at that time.

As for its taxonomic identity: in 1953, palaeontologist Dr Bryan Patterson was cited by archaeologist Erich Schmidt in his book *Persepolis* as suggesting that this bemusing animal might actually have been an okapi *Okapia johnstoni*. Self-proclaimed 'romantic zoologist' Willy Ley also favoured an okapi identity, as did German mammalogist Dr Bernhard Grzimek during the 1960s, and it is one that is still commonly attributed to the Persepolis beast today. Yet this hypothesis is beset by some very difficult problems. For example, modern-day experience in maintaining okapis in captivity must surely discredit any such identity for the depicted animal, because they can be notoriously difficult to sustain in zoos, thereby making the possibility of transporting a living specimen from its native Congolese forests all the way to Iran an exceedingly remote one.

An okapi (© Dr Karl Shuker)

In addition, the Persepolis beast and the okapi are far from exact morphological counterparts. The beast's neck is proportionately longer than the okapi's; its profile is more giraffine than okapine; it has a mane (like giraffes, but unlike okapis); and appears smaller in overall size (when compared with the human figures depicted alongside it). Yet, paradoxically, it does not wholly resemble a giraffe either—its ears are too short and small, and are neither broadly expanded nor fringed with hairs as in genuine giraffes; its neck and limbs are too short; and its body seems too robust in build, yet far too small in size.

When Prof. Janis showed a photo of the Persepolis beast to giraffid expert Dr Nikos Solunias at Baltimore's Johns Hopkins University, he suggested that it might be an extinct Turkish giraffid called *Palaeotragus rouenii*. This species is known to have existed around 7 million years ago, but could it have lingered in the Middle East until the time of Persepolis?

A very different solution was offered in 1978, when Robert G. Tuck (writing as Reza Gholi Takestani) identified the Persepolis beast as neither giraffe nor okapi—indeed, not a giraffid of any type—but rather as a male specimen of a long-necked Indian antelope called the nilghai or blue bull *Boselaphus tragocamelus*. Vaguely reminiscent of a bulky okapi, the nilghai is closer in overall morphology to the Persepolis beast than is either the giraffe or okapi. It even shares additional features that supposedly posed severe problems for the giraffid school of supporters—such as the beast's lateral digits (these vestigial toes are absent in giraffids); its sharp, hairless horns; and its long, extensively tufted tail. (Having said that, I have seen a fair number of photographs online depicting giraffes that possess tails sporting very bushy tufts, so I am not persuaded by any means that this particular feature poses a problem for a giraffid identity relative to the Persepolis beast.)

As the nilghai is much more hardy than the okapi and readily maintained in captivity, it is less difficult to believe that an adult male specimen, purchased by the visiting delegation during its journey to Iran, could (and did) survive the long journey from its native land to the Middle East. Having said that, how could such an explanation be reconciled with either an Ethiopian or a Nubian identity for the delegation—unless this identification is also in doubt?

To be honest, I'm not convinced by any of these identities—not an out-of-place okapi, a persisting *Palaeotragus*, nor even a northwest-bound nilghai. My discomfort with all of them is due in no small way to the inordinately-pronounced backwards slope of the Persepolis beast's back, far more so than occurs in either the okapi or the nilghai, or occurred in *Palaeotragus*. This in turn is a direct result of how remarkably short its hind limbs appear to be, yielding an outline unlike any ungulate species that one might even only remotely expect to see portrayed here. True, its depiction may be stylised, but in view of how realistic the representations of other species are in this same frieze, it is surely unreasonable, therefore, to assume that just this one particular animal has been inaccurately portrayed. Why would that be? There seems no logical reason for such a blatant artistic inconsistency. Consequently, in my view we have yet to discover the true identity of this most enigmatic ungulate brought from some still-unconfirmed location as an

Palaeotragus primeavus and *P. germaini* in background, with *Climacoceras africanus* and *C. gentryi* in foreground (CC Stanton Fink)

Adult male nilghai depicted upon a Moldovan postage stamp issued in 2001

exotic tribute to Darius at Persepolis so very long ago.

Incidentally, it is worth recalling here that an okapi is one of the identities proposed for the still-contentious species that served as the model for depictions of the ancient Egyptian deity Set. If such depictions were truly based upon the okapi, however, the issue of how Middle Easterners were familiar with this tropical African species once again arises to confound the cryptozoologist and antagonise the anthropologist.

Speaking of anthropologists: in 1967, Prof. Louis S.B. Leakey informed Dr Bernard Heuvelmans that an unidentified deer-like animal had been reported several times from fairly open habitat in southern Ethiopia, and he tentatively likened this crypto-ungulate to a primitive prehistoric antlered form called *Climacoceras*. Based upon Kenyan fossils dating from the Miocene, when *Climacoceras* was originally described in 1936 it was classified as a true deer. Conversely, later studies have revealed a giraffe kinship for it—like those of *Sivatherium*, its 'antlers' are ossicones, lacking the burrs of true, deciduous antlers; and its dentition exhibits giraffe characteristics too (e.g. its lower canines are bicuspid). However, *Climacoceras* (represented by two species) is deemed sufficiently distinct from the true giraffids to warrant its own separate taxonomic family within the giraffe superfamily.

As for the prospect of a modern-day continuation of *Climacoceras*, Janis favours the survival in Ethiopia of a North African subspecies of the fallow deer *Dama dama* (traditionally thought to have died out here by the end of the Pleistocene) as a more plausible explanation for Leakey's cryptid. Nevertheless, if this were confirmed it would be no less significant a discovery as the finding of a present-day population of *Climacoceras*, because the presence of deer on the African mainland is always very worthy of attention.

After all, the only known form of modern-day African deer is the highly endangered Barbary stag *Cervus elaphus barbarus*, nowadays numbering no more than 100 individuals as a result of severe hunting by man, and completely confined to the coniferous and cork oak forests on the border between Tunisia and Algeria, plus some specimens reintroduced during the 1990s from Tunisia into Morocco where it was once native. According to British zoologist Dr Richard Lydekker in Volume 4 of his *Catalogue of the Ungulata Mammals in the British Museum (Natural History)* (1915), however, this singular subspecies had formerly existed far beyond its present northern confines—alleging that it had also been reported from Senegambia, in tropical West Africa! Nevertheless, as no further statements substantiating this highly unexpected addition to this deer's distribution range in Africa have emerged, Lydekker's claim is nowadays discounted.

Resurrecting a Prehistoric Peccary

Anyone fearing that the undetected persistence of officially long-vanished ungulates is too unlikely a prospect to yield future cryptozoological successes should draw hope from this final example—which exposed the long-hidden secret of the exceedingly mysterious 'donkey peccary'.

In 1974, American zoologist Dr Ralph Wetzel from Connecticut University was in South America's Gran Chaco region (overlapping southeastern Bolivia, western Paraguay, and northern Argentina), studying those pig-like New World specialities known as peccaries. Both of the two modern-day species known to science at that time—the white-lipped peccary *Tayassu pecari* and the collared peccary *Pecari tajacu*—occur here. However, the Chaco's inhabitants claimed that a third, totally separate type was also present, burlier than the other two, and with longer ears—earning it its local name, curé-buro ('donkey peccary').

Wetzel was initially sceptical concerning this, until hunters showed him some skulls of this 'donkey peccary'—which were visibly different from those of the two known species. Excited by this highly unexpected but significant find, Wetzel returned to Connecticut with these precious skulls, and when he began studying them in detail he made a second, even more remarkable discovery. The donkey peccary was indeed new to science *in the living state*, but *not* to palaeontologists—who were already familiar with it, but only as a long-extinct species called *Platygonus wagneri*, which had supposedly died out at the end of the Ice Ages, around 10,000 years ago. Wetzel's discovery, however, showed that it had not died out at all.

The Chacoan peccary—a bona fide prehistoric survivor that became a major modern-day cryptozoological success story

Renaming it *Catagonus wagneri* to pinpoint more accurately its precise taxonomic affinities, Wetzel pointed out that the donkey peccary—nowadays termed the Chacoan peccary—not only represented a sensational modern-day resurrection of a prehistoric species, but at over 3 ft high and 100 lb in weight it was also the largest of the three modern-day peccaries now known to exist. (In 2007, a fourth, even larger species was formally described—the giant peccary *Pecari maximus* of Brazil—but not everyone accepts its status as a valid species.) This makes its success at eluding scientific detection until as late as the mid-1970s all the more remarkable.

Even so, not all Westerners had been so tardy in acknowledging this very sizeable ungulate's existence. With supreme irony, researchers subsequently learnt that for years before its *scientific* debut as a still-living species, the Chacoan peccary had been well known to New York furriers, who used its pelt to trim ladies' fur coats, hats, and handbags!

MENTIONING THE MAN-BEASTS

Finally: for the reasons given in this book's original introduction, an in-depth treatment of those man-beasts relevant to prehistoric survival is outside the intended scope of the present work, but the current views regarding this subject can be succinctly outlined as follows.

Unlike those reported from elsewhere, traditions of man-beasts in Madagascar are most likely based upon encounters with certain forms of giant lemur—traditionally believed to have died out several millennia ago, but now known to have survived into historic times before finally becoming extinct. Notable among these are *Palaeopropithecus ingens*—a slothful, ape-like lemur that may have inspired native reports of the calf-sized, man-faced tratratratra, documented in 1658 by a French traveller called Admiral Étienne de Flacourt;

Artistic rendition of *Palaeopithecus ingens* (© Markus Bühler)

and the giant aye-aye *Daubentonia robustus*, which conceivably survived until as recently as the 1930s (see my book *Mirabilis*, 2013, for an extensive documentation and consideration of these and other giant mystery lemurs reported in historic times).

The most famous man-beasts are unquestionably the yeti or abominable snowman of the Himalayas, and the bigfoot or sasquatch of North America. There would appear to be three quite distinct types of yeti. Smallest is the teh-lma, a 3-4-ft-tall, red-furred entity inhabiting the warmer valleys and probably constituting a race of proto-pygmy. The 'true' yeti, responsible for the well-known tracks frequently seen and photographed in the snow, is the meh-teh—5 ft tall, often quadrupedal, with reddish-brown fur and conical head, spending most of its time in the dense rhododendron forests. It is likely to be either a modified form of orang-utan, adapted for a more terrestrial mode of existence than those species currently known to science (this is also a popular identity for the yeren or Chinese wildman), or it might be a surviving representative of *Sivapithecus* (aka *Ramapithecus*). This was a prehistoric ape known from the Indian subcontinent, probably quadrupedal, with a face resembling a modern-day orang-utan's and a body size to match, which officially died out around 8.5 million years ago during the Miocene epoch.

The third type of yeti is the rimi or dzu-teh—a hulking, hairy giant, 7-9 ft tall, with dark hair and human-like feet. A Tibetan lama called Chemed Rigdzin Dorje Lopu claimed in 1953 that he had examined two mummified rimis—

one was housed in the monastery at Sakya, the other in the monastery at Riwoche in Kham Province. This was before the annexing of Tibet by China—followed by the razing of thousands of Tibetan monasteries.

The rimi is reminiscent of *Gigantopithecus blacki*—the largest known species of anthropoid ape. Estimated to have stood up to 10 ft tall if bipedal (as believed by some researchers), and distributed from Pakistan to southern China, it seemingly died out around 100,000 years ago—but for no good reason that palaeontologists have been able to uncover. The rimi's description also corresponds closely to that of North America's mysterious man-beast, the bigfoot (sasquatch)—and at the time of *Gigantopithecus*'s existence, a land-bridge connected Asia with North America.

Bigfoot—the elusive watcher in the woods? (© William M. Rebsamen)

Could some of these giant primates have survived in the Old World and others have migrated into the New World, with elusive modern-day descendants respectively explaining the rimi and the bigfoot? The late Prof. Grover S. Krantz, a noted anthropologist from Washington State University, was so convinced by this latter prospect that he classified the bigfoot as a living (if still-unrecognised) representative of *Gigantopithecus blacki*.

The author holding the cast of a bigfoot footprint discovered in Washington State during the early 1980s (© Dr Karl Shuker)

Having said that, in more recent times various other cryptozoological researchers have proffered certain anatomically-based objections to the *Gigantopithecus* option. In particular, they have claimed that the latter ape's huge weight (estimated to have been up to half a ton) would place immense stress upon its legs, ankles, and feet if it had walked bipedally on a habitual basis, as the bigfoot allegedly does.

In tropical Africa, reports have emerged from natives and European settlers alike concerning furry, bipedal man-beasts. These may either be undiscovered tribes of bushmen, or relict ancestral hominins called australopithecines—all of the latter species having supposedly died out at least 1 million years ago. The shorter, slender man-beasts, such as Tanzania's agogwe, Senegal's fating'ho, and DR Congo's kakundakari, do recall the gracile australopithecines, such as southern Africa's *Australopithecus africanus*, eastern Africa's *A. afarensis*, or even the earlier *A.* (*Ardipithecus*) *ramidus* from Ethiopia. The taller, burlier kikomba of DR Congo and Kenya's ngoloko, meanwhile, are closer to the robust australopithecines, like southern Africa's *A.* (*Paranthropus*) *robustus* or eastern Africa's *A.* (*Zinjanthropus*) *boisei*.

They may even be a surviving representative of our own species' ancestor—*Homo* (*Pithecanthropus*) *erectus*. This was a powerful, 5-6 ft-tall, chinless human with a thickened skull, heavy brow ridges, flat face, powerful projecting jaws, and large teeth.

A dwarf or miniature version of this latter identity has also been proffered in relation to southeast Asia's best-known man-beast—the orang pendek or gugu of Sumatra (though some reports assigned to this entity clearly refer to very large gibbons). Its Bornean equivalent is the batutut, and apparently there was once a similar form called the nittaewo on the island of Sri Lanka too. However, the notable discovery in 2004 of the diminutive Flores man or 'hobbit' *Homo floresiensis*, as represented by 3-ft-tall Pleistocene fossils uncovered on the small Indonesian island of Flores in the Lesser Sundas group, not far from Sumatra, offers a much more plausible identity for the orang pendek and batutut—especially as there are reports of small humanoid entities still living on Flores in modern times too, where they are known locally as the ebu gogu.

Unexpectedly, a man-beast recalling *Homo erectus* has even been reported from Australia, where it is termed the yowie. This identity seemed to have been strengthened by the excavation of hominin fossils at Victoria's Kow Swamp by Australian anthropologist Alan G. Thorne between 1968 and 1972 that shared certain cranial features with *Homo erectus*, such as protruding jaws, ridged brows, and thick skulls. According to radiocarbon dating, these enigmatic remains are between around 13,000 and 6,500 years old.

The author alongside a life-sized model of the orang pendek at the CFZ Weird Weekend 2013 (© Dr Karl Shuker)

But how could such a primitive form have reached this island continent? Illinois University anthropologist Dr Charles Reed proposed that *H. erectus* and modern man *H. sapiens* (which were contemporaries for a time) might have actually lived together—with *H. erectus* thus accompanying modern man when he made the first journeys by raft to Australia around 40,000 years ago, but then reverting to a wild state after becoming established here. Nowadays, however, the consensus is that the Kow Swamp hominins are modern man after all, rather than *H. erectus*.

Wooden yowie statue at Kilcoy, in Queensland, Australia (CC Seo75)

Perhaps the most exciting man-beasts on file include such beings as the Mongolian almas, Pakistan bar-manu, Siberian chuchunaa, and the hairy ape-men of Vietnam. This is because according to some native descriptions of these shy, reclusive mountain-dwellers, they bear more than a passing resemblance to Neanderthal man *Homo neanderthalensis*.

One recently-deceased man-beast of this type may even have been on public display—encased in a block of ice, in a Minnesota sideshow! This enigmatic hirsute exhibit, nicknamed the Minnesota iceman or Bozo, was closely examined in December 1968 by Dr Bernard Heuvelmans and Ivan T. Sanderson. When, however, the FBI also began to express an interest (the iceman had apparently been shot—killed by a bullet blasting away the back of its skull), the show reputedly substituted a plastic replica for it, and the original iceman has never been seen since—unless the iceman specimen auctioned on eBay in February 2013, allegedly purchased by Steve Busti, owner of Museum of the Weird in Austin, Texas, and currently doing the rounds of exhibitions across the U.S.A. is the original one, as some have suggested.

Many sceptics allege that the original iceman was itself a model, but Heuvelmans always disputed this. He believed that it had probably been shot in Vietnam, and was afterwards smuggled into the U.S.A. in one of the 'body bags' used at that time for bringing back home soldiers killed during the Vietnam War. Nearly 50 years have passed since the iceman was first displayed, but science is no closer to solving the mystery of its status. It could well have been nothing more than a cleverly-constructed fake—but it just may have been conclusive proof that *Homo sapiens* is not the only surviving hominin in the modern world.

And Finally. . .

There is a very interesting passage in the Bible that seems to allude to the existence in historic times of two separate types of human. One is modern man *Homo sapiens*, but the other recalls a hairy man-beast—possibly of the *Homo erectus* or *Homo neanderthalensis* type? The passage in question (*Genesis*, 25:21-7) refers to the brothers Esau and Jacob:

> And Isaac intreated the Lord for his wife . . . and Rebekah his wife conceived.
>
> And the children struggled together within her; and she said, If it be so, why

> am I thus? And she went to inquire of the Lord.
>
> And the Lord said unto her, Two nations are in thy womb, and two manner of people shall be separated from thy bowels; and the one people shall be stronger than the other people; and the elder shall serve the younger.
>
> And when her days to be delivered were fulfilled, behold, there were twins in her womb.
>
> And the first came out red, all over like an hairy garment; and they called his name Esau.
>
> And after that came his brother out, and his hand took hold on Esau's heel; and his name was called Jacob . . .
>
> And the boys grew: and Esau was a cunning hunter, a man of the field; and Jacob was a plain man, dwelling in tents.

Consider the countless wars and the tyrannical persecution that humans have inflicted upon humans since the dawn of history—despite the fact that we all belong to the same species. What would happen, therefore, if it were eventually proved beyond a shadow of doubt that we are sharing this planet with a totally separate, *second* species of human? He would be an Esau to our Jacob, cunning and wise in the ways of the forest, the field, and the mountain, but wholly incapable of standing against the mechanised, weaponised might of modern humans. What would be the outcome of such a revolutionary discovery?

When I contemplate such matters, I think once again about a pair of Biblical brothers—but not Esau and Jacob. The names that enter my mind this time are Abel . . . and Cain.

May God, and *Homo sapiens*, prove me wrong.

Esau depicted as a hairy man of the field by Johann Scheuchzer in 1731

CHAPTER 5

In Conclusion—Back to the Future?

We ought to make up our minds to dismiss as idle prejudices, or, at least, suspend as premature, any preconceived notion of what MIGHT, *or what* OUGHT TO, BE *the order of nature, and content ourselves with observing, as a plain matter of fact, what* IS.

SIR JOHN HERSCHEL—*PRELIMINARY DISCUSSION*

Herschel's advice should be taken to heart by anyone still sceptical about prehistoric survival. The countless examples of independently-submitted but extensively-corroborating testimony presented and assessed in this book yield a corpus of data so compelling and convincing that there really cannot be room for doubt that a number of visually spectacular and palaeontologically significant mystery beasts do frequent assorted expanses of land, freshwater, and ocean around the world. Such a concept may well disturb the more traditionally-minded members of the zoological community, but when the facts are examined in an objective, scientific manner, no satisfactory alternative explanation can be found.

Not even the faithful stalking horse of the sceptic and the cynic—discounting such reports as an unholy alliance of native folklore, imagination, inebriation, and wicked mendacity—can muster much enthusiasm or verisimilitude when its own pedigree is investigated. What will be found is (for its protagonists) an embarrassing exuberance of instances in which, even when confronted by the most virulent scientific scorn, such testimony has been fully vindicated.

Who in the West took seriously the Wambutti pygmies' claims that a donkey-like beast with the stripes of a zebra but the cloven feet of an antelope inhabited the humid depths of their Congolese rainforest homeland? Everyone—but only after Sir Harry Johnston in 1901 had obtained the skins and skulls of a hitherto-unknown and incongruously short-necked giraffe of prehistoric parentage that we nowadays refer to as the okapi.

Longstanding tribal claims that a gigantic hairy ogre inhabited the Virunga Volcano range bordering what is now DR Congo, Rwanda, and Uganda were patronisingly swept aside by Western scientists as the overblown product of too much myth and magic—until the mountain gorilla *Gorilla beringei beringei* was discovered there in 1902.

A mountain gorilla (USFWS/Richard Ruggiero)

Much more recently, who could possibly trust native stories of a bizarre beast resembling a whistling panda that flashes its chest at anyone passing by? Fortunately for cryptozoology, Australian zoologist Dr Tim Flannery did, and in 1994 he was rewarded with the discovery of the dingiso or bondegezou *Dendrolagus mbaiso*—a very primitive, dramatically new species of black-and-white tree kangaroo inhabiting the remote mountain jungles of Irian Jaya (Indonesian New Guinea).

The list goes on and on—as revealed in my book *The Encyclopaedia of New and Rediscovered Animals* (2012) and its two predecessors *The New Zoo* (2002) and *The Lost Ark* (1993)—proving something that should already have been self-evident to scientific scholars. Local people inhabiting a given area for countless generations are likely to know more about the types of creature sharing their region than visiting researchers who rarely penetrate far beyond its relatively tame fringes and borderlands.

So if there are plenty of precedents for the discovery even today of prehistoric survivors and other major new species, why are those documented in this book still awaiting formal scientific detection? Inaccessibility and inhospitability of their native habitat is one significant factor (especially in relation to mystery beasts likely to be living fossils), and was discussed in this book's Introduction, but it is not the only obstacle.

For instance, the nature of the creatures themselves can shed some light upon their famed ability to avoid discovery. A nocturnal species is clearly going to be more difficult to espy and ensnare than a diurnal counterpart, and an aquatic or aerial species will obviously prove more of a problem to track down than a terrestrial creature.

As noted in Chapter 3, monitoring a briefly-captive specimen revealed why the megamouth shark, up to 18 ft long and found in many parts of the world, succeeded in eluding scientific discovery until as late as 1976. It performs vertical migration—staying in the ocean depths during the day and rising closer to the surface only during the cover of darkness at night.

Even during the height of the Victorian era's passion for lepidoptery, Queen Alexandra's birdwing *Ornithoptera alexandrae*, the world's largest species of butterfly, remained unknown to science. Only in 1906 was it finally brought down to earth, quite literally—because on account of its size (as large as a pigeon) and its preference for lingering high in the jungle canopy, far beyond the reach of even the biggest butterfly net, earnest entomologists finally resorted to shooting down specimens with a rifle!

For very sound ecological reasons—elegantly elucidated in ecologist Prof. Paul Colvinaux's classic book *Why Big Fierce Animals are Rare* (1980)—large carnivorous species will necessarily be less common than their prey. Existing at the apex of their ecosystem's food pyramid, they cannot be more abundant than their prey, otherwise they would exhaust their own food supply and die of starvation. It follows, therefore, that these species will be less easy to observe—especially as in so many cases the success of their entire lifestyle depends upon their ability to remain unseen. It should come as no surprise, therefore, to discover that a disproportionate number of mystery beasts documented in this book are of the large carnivorous variety—spied so briefly, unexpectedly, and rarely that their observers have little chance of capturing them.

Of course, in some such cases we might actually be witnessing the natural extinction of a species—sightings of it are rare simply because it is itself rare. Heuvelmans suggested that the

The okapi (left) and the Vu Quang ox (right) (© William M. Rebsamen)

category of elongate sea monster dubbed by him the super-otter may well be extinct—modern-day reports of this creature virtually ceased after the mid-1800s. It would not be the first time that a contemporary species has vanished before science had even recognised its existence. By definition, many such animals will never be known to us—but a few have reached the zoological chronicles by virtue of preserved specimens or reliable anecdotal evidence coming to light some years after the last first-hand sightings.

The sea mink *Neovison* (*Mustela*) *macrodon* of eastern North America apparently died out during the 1880s. Yet not until some teeth and other remains were found in coastal Indian shell heaps at the turn of the 20th Century was its status as an undescribed giant species of mink finally recognised and rectified.

Much more recently, an agonising two years passed between the procurement of the first horns, skulls, and skins in 1992 and the capture in 1994 of the first living specimen ever recorded by science belonging to Vietnam's extraordinary Vu Quang ox or saola *Pseudoryx nghetinhensis*. In the meantime, some zoologists had begun to fear that it may have died out, perhaps just months before its scientific unveiling in 1992. The largest new mammal to have been discovered for over 50 years, this fascinating animal is believed by some to be related to a primitive group of ungulates called the hemibovids, which became extinct over 4 million years ago.

One aspect totally neglected by cryptozoological sceptics, but which is of profound importance when seeking reasons for the continuing evanescence of certain mystery animals, is the prevailing political situation within the country reputedly harbouring them. Many cryptozoological expeditions have found to their cost that obtaining the necessary visas and other rigmarole lavishly tied up in the correct length and thickness of bureaucratic red tape can prove more of a problem than the beasts that they hope to uncover!

In addition, there are areas of the world to which all outsiders and often even their own scientists are denied access for long periods of time. Consequently, there is no way in which traditional reports of mystery creatures in these localities can be pursued at all. The dinosaurian lake monsters in parts of Tibet off-limits to foreigners, the striped mega-fanged mystery cat of Colombia whose habitat is frequented by drug traffickers and terrorist groups, those unidentified beasts inhabiting whichever portions of Africa and the Middle East are, tragically, involved in warfare at the time of this book's publication—all are destined to remain a mystery for as long as the frontiers to their world remain closed to scientific exploration.

Since 1992, Vu Quang has hosted the discovery of not only *Pseudoryx* but also several distinctive new species of deer (including the very sizeable, aptly-named giant muntjac *Muntiacus vuquangensis*); and, from southern Vietnam has come the enigmatic, controversial 'holy goat' *Pseudonovibos spiralis* whose widely-splayed horns (if some are genuine and not all artificially-manipulated) resemble a pair of motorbike handlebars! It is surely no coincidence that Vu Quang (now a nature reserve) was previously out of bounds to the outside world and Vietnamese scientists alike for many years, due to the Vietnam War and accompanying political upheavals.

Whenever I am interviewed by the media concerning cryptozoology, there is one question that is voiced time and time again. Do cryptozoologists receive financial backing from scientific establishments, and, if not, why not? The answer to the first part of that question is

Rare early photograph of a giant forest hog

simple—no! As for the answer to the second part, scientific establishments and research granting bodies tend to view the funding of cryptozoological expeditions and research as a high-risk, low-returns venture.

Taking into account all of the reasons already outlined here concerning why mystery animals tend to stay undiscovered, there is an extremely sizeable chance that success will not come on the very first outing. With any difficult undertaking, practice makes perfect, but in the real world there is usually only one chance on offer—especially when large amounts of money are being sunk into it. In short, it is much easier (and much more profitable from a financial standpoint) for science to denounce mystery beasts as the product of native fable and fantasy than to provide the necessary backing to uncover the truth. Consequently, cryptozoologists tend to be very much on their own—which inevitably places great restrictions upon the extent of any project that they may wish to undertake.

Also worth bearing in mind is that many of those cryptozoological investigations which have actually achieved their goals tended to involve on-site researchers rather than those visiting from elsewhere. This clearly underlines the value of knowledge and experience concerning a particular terrain—thereby harking back to what I said earlier regarding the importance of paying serious attention to local testimony.

Thus, the Vu Quang panorama of new species was brought about by Dr John MacKinnon, who has been working in southeast Asia for many years. Sir Harry Johnston was governor of Uganda (which at that time encompassed part of what is now DR Congo) when he successfully investigated reports of the okapi. Captain Richard Meinertzhagen had been stationed in Kenya when he pursued and, in 1904, exposed the giant forest hog *Hylochoerus meinertzhageni*. Zoologist Dr Peter Hocking, who is currently investigating with great promise the apparent existence in Peru of several different mystery cats and a large baboon-like monkey called the isnachi, was born there and has been working professionally there for years. And so on.

Even so, it must be said that sheer luck also plays a noticeable part in the unveiling of mystery beasts. Indeed, there really ought to be a cryptozoological equivalent of the familiar old adage "A watched pot never boils"—something along the lines of "A sought cryptid is never found"—because there is no doubt that many remarkable animals have been discovered more by accident than by design. As on so many previous occasions in this book, the coelacanth *Latimeria chalumnae* is a case in point.

Were it not for the fact that a lone specimen wandering far from its normal Comoro seas was captured off South Africa at the very time that the local museum had taken on a young curator who was particularly interested in fishes, and who just so happened to spot its fin sticking out from beneath a pile of sharks that had otherwise completely concealed it from view, science may still be unaware of the existence of a living species of coelacanth there—and the Comoro natives may still be using its scales to mend their bicycle tyres' punctures in blissful ignorance of this species' immeasurable zoological significance. If this storyline had been used in a work of fiction, it would have been severely criticised for being hopelessly contrived—yet it is of course all perfectly true!

A pygmy chimpanzee or bonobo (© Dr Karl Shuker)

Even more bizarre is the fact that some important new animals had actually been represented by specimens in museums for many years before their identity as undescribed species was finally realised. These include such notable examples as the distinctly primeval Congo peacock *Afropavo congensis*, the 'living fossil' crustacean *Neoglyphea inopinata*, the giant pied-billed grebe *Podilymbus gigas*, the gastric-brooding frog *Rheobatrachus silus*, Madagascar's giant striped mongoose *Galidictis grandidiensis*, the greater yellow-headed vulture *Cathartes melambrotus*, Delcourt's giant gecko *Hoplodactylus delcourti*, and the olinguito *Bassaricyon neblita*.

Even more remarkable is that during the early 1920s, a living pygmy chimpanzee or bonobo *Pan paniscus* was masquerading in public as a common chimp *P. troglodytes* at Amsterdam Zoo several years *before* its species was 'officially' discovered by science!

Who knows what other unrecognised zoological treasure trove may be lurking, unlisted and unlooked for, ignored and incognito, within the vaults of the world's museums? Such proof of prehistoric persistence as a dusty but genuine chipekwe horn, a bona fide makalala skull, or the pelt of a waheela? Perhaps even—wonder of wonders—a copy of the elusive thunderbird photo! Stranger things have happened in cryptozoology. After all, as recently as 2010, German cryptozoologist Markus Hemmler and I jointly uncovered a series of photographs published more than 80 years earlier yet remaining wholly unknown to cryptozoologists and mainstream zoologists alike that depicted the beached carcase of what had hitherto been one of the most mystifying marine entities of all—Trunko—but which, after I'd studied these long-overlooked images, enabled me to determine beyond any doubt that it had merely been a globster, i.e. the greatly decomposed remains of a sperm whale.

Indeed, one of the most dispiriting facets of this emerging discipline's complex history is the surprising frequency with which science has been within a fingertip's reach of obtaining incontrovertible physical evidence for the reality of various prehistoric survivors—only for that selfsame evidence to lose no time in slipping back into irretrievable obscurity.

What could zoology have learnt from the Gambian sea serpent's decapitated head, or

An example of what may yet rise forth out of the past and into the present some day in the future? (© William M. Rebsamen)

from the baby Caddy that spent a day swimming in a bucket aboard its captor's ship? What would the jailed 'tigelboat' have proven to be? Would the Tampa souvenir seller's barrel of scales, or the artist's necklace, have marked the debut of a New World species of living coelacanth? If the missing thunderbird photo really did exist, what happened to the creature that it depicted? And where do the butchered mokele-mbembe's bones from 1959 lie concealed?

If any of the prehistoric survivors surveyed here are somehow discovered one day, the palaeontological texts will require a major rewrite—and for two quite different reasons.

No amount of fossils, however well-preserved, can compete with a living specimen for providing biological information. At the same time, however, any prehistoric survivor discovered alive and well will be an evolved species—as opposed to one plucked directly from the Cretaceous or wherever else in geological time its most recent known fossil antecedents existed. In many ways, therefore, its correspondence to its fossilised antecedents will be analogous to the image we receive of ourselves when gazing into any of the trick, distorting mirrors at a funfair.

All in all, there promises to be some exciting and unpredictable times ahead for zoology, should any of this book's beasts decide to abandon its anonymity and seek fame (if not fortune) on the scientific stage.

And Finally . . .

It is strange, even somewhat paradoxical, to consider that many major zoological discoveries of the future may prove to have been drawn forth from the long-distant past—but perhaps we ought not to be greatly surprised. After all, as Pinero astutely recognised:

> The future is only the past again,
> entered through another gate.

STOP PRESS

A 15-FT-LONG LIZARD IN AUSTRALIA'S INTERIOR?

Reports of giant Australian lizards have been documented in the cryptozoological literature since the late 1970s, but have appeared in print elsewhere for some considerable time before that. Here is one such report, hitherto obscure and dating back to 1931, which was kindly brought to my attention very recently by long-standing cryptozoological correspondent Jason McAllister:

> LIZARD 15 FEET LONG
>
> Some prospectors from Central Australia who are now in Sydney, have reported having seen a gigantic lizard in the interior. Mr F. Blakeley, brother of the Federal Minister, who was with the party, said it grew there to a length of 15 feet and had been seen by white men, though natives shunned the country where it lived. He believed that it was a survival of an ancient form of life.
>
> Big Jim, an old prospector, had been attacked by one of these giant lizards, which uttered a noise described as a mixture of the bark of a dog and the roar of a lion. Big Jim hurled a stone at the animal and fled. Mr Blakeley said he next day measured the distance of the claw marks in the sand and found they were 15 feet apart.

Interestingly, varanids—the most popular identity for Australia's giant lizards—rarely vocalise, except for hissing, and even that is infrequent. So if the above report is genuine and not journalistic hyperbole or a total invention (although in view of the Federal Minister's brother being among the party, this would appear unlikely), then it would seem that something other than a varanid (over-sized or otherwise) was encountered by those prospectors 85 years ago.

ANON., 'Lizard 15 Feet Long', *Northern Miner* (Charters Towers), 14 August (1931).

A LIVING PTERODACTYL AT MALIBU, CALIFORNIA?

On 26 August 2016, British cryptozoological researcher Richard Muirhead kindly sent me a copy of a short but remarkable *San Francisco Call* newspaper report from 17 May 1908 concerning the alleged sighting at close range in California of a supposed living pterodactyl by two eyewitnesses linked to the British Museum. The report reads as follows:

> Los Angeles, May 16.—Two attachés of the British national museum, C. J. F. Browne and Alex MacDonald, both of London, are here on a still hunt for a live pterodactyl, a flying reptile supposed to have been extinct for centuries [sic—in reality, at least 66 million years!].
>
> In a cave on the Malibu ranch, west of the city, a few days ago, while searching for prehistoric fossils, the two men were driven out by a strange animal, described as a combination of bat and lizard, which escaped into the mountains, half flying, half running.
>
> They say it resembled the pterodactyl of history, and they have gone back to the cave with big steel traps and heavy chains, determined to capture it. Only recently the fossil remains of a neosaurus, said to have been contemporaneous with the pterodactyl, were found on the Malibu.

As the pterodactyls are still currently represented only by fossils dating back at least as far as the Cretaceous Period (and earlier), we can rest assured that no living specimen was ever captured by those steel traps and heavy chains. Moreover, as the American newspaper media were notoriously rife with outlandish journalistic spoofs and hoaxes featuring monsters and mystery beasts during the 19th and early 20th Century, it is highly likely that the alleged sighting of the pterodactyl-like beast documented in the above report never took place anyway. Having said that, it would be interesting nonetheless to research this report, if only to determine whether the two named persons really were linked to the British Museum, but in view of how recently I received it there was not enough time for me to do so and include any findings within the present book. But if anyone reading the report here has information concerning it, I would welcome any details that they could send to me.

> ANON., 'Hunt Flying Reptile Long Thought Extinct—Attaches of British Museum Observe Combination of Bat and Lizard—Special Dispatch to the Call', *San Francisco Call* (San Francisco), 17 May (1908).

A SEA MONSTER'S ROCKY END

On 29 April 2016, a YouTube member with the user name Wowforreeel uploaded a short video and photograph of a very large 'something' at the centre of a sea disturbance near Antarctica, which had first been seen by Scott C. Waring via Google Earth on 9 April, at the co-ordinates 63° 2′56.73″ S, 60° 57′32.38″ W. Wowforreeel had also viewed it via Google Earth and suggested that it may be a rock formation but also likened it to a plesiosaur's fin. Moreover, after the video went viral in mid-June, other claims made online included an underwater UFO, and (a very popular identity) the tail fin of a truly ginormous giant squid, claimed on one website to be 60-120 m (approximately 200-400 ft) long—a veritable kraken!

Following an investigation of this maritime mystery by deep-sea biologist/conservationist Andrew David Thaler (among others), however, in contrast to the subject of a certain John Wyndham sci-fi novel it proved not to have been the kraken rising after all. Instead, it was merely a rock, albeit a large one—as suspected by Wowforreeel. One of the South Shetland Islands of Antarctica is Deception Island—

which was very close indeed to the coordinates of the mystery object, and also was aptly named, in view of how deceiving one of its minor neighbours had been.

For as Thaler revealed on 17 June, after homing in on the object's coordinates using a nautical chart he had discovered that arising up above the ocean surface at their exact point just southwest of Deception Island was a tall rocky outcrop, known as Sail Rock due to its sail-like shape, and which also provided an exact morphological match with the mystery object seen in the video. There was no plesiosaur, no gargantuan squid, merely a far-from-monstrous, well-mapped tower of rock (and one, moreover, which had previously been measured incorrectly in the 'monster' image, because Sail Rock only stands 100 ft high).

LIBERATORE, Stacy, 'Has a KRAKEN Been Spotted on Google Earth? Monster Hunters Claim to Have Found 120m Long Giant Squid-Like Creature', *Daily Mail* (London), http://www.dailymail.co.uk/sciencetech/article-3643828/Has-KRAKEN-spotted-Google-Earth-Giant-squid-like-create-120m-long.html 16 June (2016).

THALER, Andrew D., 'Did Monster Hunters Find a 120 Meter Long Giant Squid on Google Maps?', *Southern Fried Science*, http://www.southernfriedscience.com/did-monster-hunters-find-a-120-meter-long-giant-squid-on-google-maps/ 17 June (2016).

WOWFORREEEL, 'Massive Unknown Creature Near Antarctica?', *YouTube*, https://www.youtube.com/watch?v=RsijTye9LRA 29 April (2016).

LOCH NESS—IN AT THE (NOT SO) DEEP END

According to Nessie researcher Dick Raynor, the record depth of 813 ft measured some years ago at Edwards' Deep in Loch Ness by George Edwards, the skipper of a local tourist boat, was most probably due to a sonar anomaly known as a side-wall echo rather than being an accurate depth measurement, as it has never been recorded again, not even by the comprehensive joint Loch Ness Project/BBC Science Unit survey in 2002. As for the even greater depth of 889 ft claimed in January 2016 by Keith Stewart at a hitherto-unrecorded trench in the loch bottom: a state-of-the-art underwater robot named Munin, operated by Norwegian company Konigsberg Maritime and conducting a very extensive two-week survey of the loch during April 2016, found no trace of this trench. Consequently, the greatest confirmed depth recorded at Loch Ness currently remains at 755 ft.

ANON., 'A Real Monster', *Loch Ness & Morar Project*, http://www.lochnessproject.org/explore_loch_ness/MUNIN%20SURVEY/Loch%20Ness%20Project%20Munin.html April (2016).

RAYNOR, Dick, 'An Examination of the Claims and Pictures Taken by George Edwards', *Loch Ness Investigation*, http://www.lochnessinvestigation.com/georgeedwardsclaims.html 25 August (2012); most recent update 18 June (2013).

A LIVING AMMONITE?

Among the most familiar of invertebrate fossils today due to their ornate coiled shells, ammonites constitute a prehistoric subclass,

Ammonoidea, of marine cephalopod molluscs, whose closest living relatives are the octopuses and squids. The earliest known ammonites occurred during the Devonian, with the last known ones becoming extinct at the end of the Cretaceous—or did they? The indefatigable English cryptozoological researcher Richard Muirhead kindly brought to my attention recently the following very intriguing newspaper report that had appeared in the *Niagara Falls Gazette* on 1 May 1878:

> The Elmira Advertiser is responsible for the following statement: "An extinct species of Mollusk, an 'ammonite,' was recently picked up in Watkins Glen near the 'Glen Hole.' Another proof of the wide range of Flora and Fauna attributed to this glen by the celebrated Agassiz. NO other living ammonite is known according to the books."

What are we to make of this short but—if correct—spectacular statement? Firstly, the Agassiz referred to in it is none other than the highly celebrated Swiss-born American zoologist Prof. Louis Agassiz, whose many publications include a significant study on fossil molluscs published during the 1840s, from which one might assume that the above newspaper report must surely be trustworthy. However, Watkins Glen is a state park in New York, famed for its Devonian rocks (including fossiliferous limestone) in the vicinity of a deep gorge cut through the rock by a stream—i.e. it is an inland area where one might expect to find fossils, not a marine environment where a living ammonite could exist. Even the expression "was recently picked up" as used in the report implies a fossil—a living ammonite would have been dredged up or netted from an expanse of sea.

Statue of ammonite portrayed in the living state (CC0 N. Steffens)

Most crucial of all, however, is that any mention of what would be a major zoological discovery if genuine—a living ammonite—is conspicuous only by its absence from modern-day zoological works, in which ammonites are resolutely confined to the realm of prehistory. So, clearly, no such find was made after all, with the newspaper's claim presumably being the product of confused reporting or even a journalistic hoax—by no means an uncommon occurrence back in the late 19th and early 20th Centuries.

ANON., [Untitled], *Niagara Falls Gazette* (Niagara Falls), 1 May (1878).

HUNGARIAN FISHER PIG—ENTELODONTS EMERGENT?

Whereas Arthurian legend had its Fisher King, rural Hungarian lore apparently once included an ostensibly real but presently-unidentified mystery beast known as the fisher pig. Also termed the swamp pig, this hitherto-obscure creature, seemingly undocumented in mainstream cryptozoological literature until now, was kindly brought to my attention by Facebook colleague and Hungarian crypto-investigator Orosz István via a series of FB communications during early July 2016, and which can be summarised as follows. The old shepherding folk of his country still speak of this mysterious animal, which they claim to be extinct now (allegedly dying out during the 1880-1890s, according to a mention of it by famous Hungarian agricultural writer Imre Somogyu in his celebrated 1942 book *Kertmagyarország Felé*), but which once lived in marshes around the rivers Tisza and Körös. It did not graze like normal wild boars, and its diet consisted of crabs and fishes. When I asked Orosz if any illustrations of fisher/swamp pigs existed, he replied that he was not aware of any, but added that it was said to be very big, with a curved back, and lived in large herds.

Restoration of *Daeodon shoshonensis* (Jay Matternes/Smithsonian)

This interesting account attracted a wide range of speculation on FB, including whether it may actually have constituted a late-surviving species of entelodont. These omnivorous pig-like ungulates (but constituting a separate taxonomic family from true pigs) existed in Eurasia and North America from the middle Eocene to the early Miocene (37 million to 16 million years ago), culminating in their last but largest representative *Daeodon shoshonensis* (aka *Dinohyus hollandi*). Distributed widely across the U.S.A., this monstrous so-called 'hell pig' or 'terminator pig' stood around 6 ft tall at the shoulder and sported a massive 3-ft-long skull. However, it seems highly unlikely that such conspicuous creatures as entelodonts could have survived into modern times in Europe without having attracted very appreciable, sustained attention from the sporting fraternity, for whom they would have made extremely noteworthy targets and thence trophies (i.e. mounted heads, preserved pelts, etc), to be displayed proudly in hunting lodges and country estates across the continent. And yet no such specimens seemingly exist; none, at least, has been brought to public notice so far.

Much more plausibly, Orosz István felt that the fisher pig was probably nothing more than a local variety of the familiar European wild boar *Sus scrofa*, whereas fellow Hungarian crypto-enthusiast Tötös Miklós considered that it may have been a feral (run-wild) variety of domestic pig. Both wild boars and feral domestic pigs will indeed inhabit swamps and marshes, are famously omnivorous, and are known to enter shallow water to devour fishes and invertebrates. Yet as wild boars and feral domestic pigs are such well known creatures in this region of Europe, why would any that lived in the Tisza and Körös marshes be delineated with their own name by the local shepherds, unless they had evolved a distinctive morphology and lifestyle that separated them from more typical wild boar and ferals at least in the eyes of the shepherds (if not in those of zoologists)? For now, therefore, the Hungarian fisher pig remains a thought-provoking cryptozoological conundrum.

SELECT BIBLIOGRAPHY

To restrict this bibliography to a manageable length, I have confined its listing wherever possible to cryptozoological references. Those strictly palaeontological and other non-cryptozoological sources included here are ones whose content has been specifically discussed or incorporated within this book's main text.

General References

ALEXANDER, R. McNeill, *Dynamics of Dinosaurs and Other Extinct Giants* (Columbia University Press: New York, 1989).

BORD, Janet & BORD, Colin, *Alien Animals* (rev. edit., Panther: London, 1985).

BROOKESMITH, Peter (Ed.), *Creatures From Elsewhere* (Orbis: London, 1984).

BURTON, Maurice, *Living Fossils* (Thames & Hudson: London, 1954).

COHEN, Daniel, *The Encyclopedia of Monsters* (Dodd, Mead: New York, 1982).

COLVINAUX, Paul A., *Why Big Fierce Animals Are Rare: An Ecologist's Perspective* (Princeton University Press: Princeton, 1978).

COX, Barry, *et al.* (Consultants), *Macmillan Illustrated Encyclopedia of Dinosaurs and Prehistoric Animals* (Macmillan: London, 1988).

EBERHART, George M., *Mysterious Creatures: A Guide to Cryptozoology* (2 vols) (ABC-Clio: Santa Barbara, 2002; 2nd edit. (2 vols), CFZ Press: Bideford, 2013, 2015).

GILROY, Rex & GILROY, Heather, *Out of the Dreamtime: The Search For Australasia's Unknown Animals* (URU Publications: Katoomba, 2006).

HALL, Mark A., *Natural Mysteries* (rev. edit., Mark A. Hall Publications and Research: Bloomington, 1991).

HEUVELMANS, Bernard, *On the Track of Unknown Animals* (Rupert Hart-Davis: London, 1958; rev. edit., Kegan Paul International: London, 1995); *Les Derniers Dragons d'Afrique* (Plon: Paris, 1978); 'Annotated Checklist of Apparently Unknown Animals With Which Cryptozoology is Concerned', *Cryptozoology*, 5: 1-26 (1986).

LAMB, Simon, *et al.* (Consultants), *Prehistoric Life: The Definitive Visual History of Life on Earth* (Dorling Kindersley: London, 2009).

LEY, Willy, *The Lungfish, the Dodo and the Unicorn: An Excursion Into Romantic Zoology* (Viking Press: New York, 1948); *Exotic Zoology* (Viking Press: New York, 1959).

McGOWAN, Christopher, *Dinosaurs, Spitfires, and Sea Dragons* (Harvard University Press: Oxford, 1991).

MACKAL, Roy P., *Searching For Hidden Animals* (Doubleday: Garden City, 1980).

MAREŠ, Jaroslav, *Svet Tajemných Zvírat [The World of Mysterious Animals]*, (Littera Bohemica: Prague, 1997).

MICHELL, John & RICKARD, Robert J.M., *Living Wonders: Mysteries and Curiosities of the Animal World* (Thames & Hudson: London, 1982).

NEWTON, Michael, *Encyclopedia of Cryptozoology: A Global Guide* (McFarland: Jefferson, 2005).

NORRELL, Mark & MENG, Jin (Consultants), *Dorling Kindersley Encyclopedia of Dinosaurs and Prehistoric Life* (Dorling Kindersley: London, 2001).

OAKES, Ted, *Monsters We Met: Man's Battle With Prehistoric Animals* (BBC: London, 2003).

RICH, Pat V., *et al.*, *Kadimakara: Extinct Vertebrates of Australia* (Pioneer Design Studio: Lilydale, 1985).

SCHOUTEN, Peter, *The Antipodean Ark* (Angus & Robertson: London, 1987).

SHUKER, Karl P.N., 'Living Fossils: A *Fortean Times* Guide to Dinosaurs in the 20th Century', *Fortean Times*, No. 67 (February-March) [a 4-page colour supplement] (1993); *In Search of Prehistoric Survivors: Do Giant 'Extinct' Creatures Still Exist?* (Blandford Press: London, 1995); *From Flying Toads to Snakes With Wings: From the Pages of FATE Magazine* (Llewellyn Publications: St Paul, 1997); *The Beasts That Hide From Man: Seeking the World's Last Undiscovered Animals* (Paraview Press: New York, 2003); *Extraordinary Animals Revisited: From Singing Dogs to Serpent Kings* (CFZ Press: Bideford, 2007); *Karl Shuker's Alien Zoo: From the Pages of Fortean Times* (CFZ Press: Bideford, 2010); *The Encyclopaedia of New and Rediscovered Animals: From The Lost Ark to The New Zoo—and Beyond* (Coachwhip Publications: Landisville, 2012); *Mirabilis: A Carnival of Cryptozoology and Unnatural History* (Anomalist Books: San Antonio, 2013); *The Menagerie of Marvels: A Third Compendium of Extraordinary Animals* (CFZ Press: Bideford, 2014); *A Manifestation of Monsters: Examining the (Un)Usual Suspects* (Anomalist Books: San Antonio, 2015).

SHUKER, Karl P.N. (Consultant), *Man and Beast* (Reader's Digest: Pleasantville, 1993); *Secrets of the Natural World* (Reader's Digest: Pleasantville, 1993).

SHUKER, Karl P.N. *et al.* (Consultants), *Almanac of the Uncanny* (Reader's Digest: Surry Hills, 1995).

VERRILL, A. Hyatt, *Strange Prehistoric Animals and Their Stories* (L.C. Page: Boston, 1948).

VICKERS-RICH, Patricia & RICH, Thomas H., *Wildlife of Gondwana* (Reed: Chatswood, 1993).

WARD, Peter D., *On Methuselah's Trail: Living Fossils and the Great Extinctions* (W.H. Freeman: Oxford, 1991).

WILKINS, Harold T., *Secret Cities of Old South America* (Library Publishers: New York, 1952).

WOOD, Gerald L., *The Guinness Book of Animal Facts and Feats* (3rd edit., Guinness Superlatives: London, 1982).

Chapter 1: How Dead are the Dinosaurs?

ANON., 'A Bolivian Saurian,' *Scientific American*, 49: 3 (1883).

ANON. [SCHMIDT, Franz H.], 'Prehistoric Monsters of the Amazon', *Globe* (Sydney), 19 April: 1-2 (1913).

ANON., 'To Bring Home a Dinosaur Dead or Alive', *Literary Digest*, 64 (28 February): 76-77, 80 (1920).

ANON., 'An Iguanodon From Dahomey', *Pursuit*, 3 (January): 15-16 (1970).

ANON., 'Dinosaur Hunt', *Fortean Times*, No. 42 (autumn): 27 (1984).

ANON., 'Teenager Joins Trek to Search For Mokele', *Leicester Mercury* (Leicester), 23 November (1992).

ANON., 'Dinosaur Kangaroos Spotted in Chile', *Ananova*, http://www.ananova.com/news/

story/sm_1080697.html 27 August (2004) [no longer online].

ANON., "Tianchi Monster' Caught On Film', *China.Org.Cn*, http://www.china.org.cn/english/China/223790.htm 10 September (2007).

AGNAGNA, Marcellin, 'Results of the First Congolese Mokele-Mbembe Expedition', *Cryptozoology*, 2: 103-112 (1983).

'AN AMERICAN IN PARIS', 'English Nobleman on a Secret Expedition to Alaska [re Partridge Creek monster]', *Evening Star* (Washington DC), 9 May (1908).

ANSELME, Jean-Pierre, 'Les Créatures de l'Ombre', *VSD Nature*, No. 5 (December): 62-63 (1995).

ARMENT, Chad, 'Dinos in the U.S.A.: A Summary of North American Bipedal "Lizard" Reports', *North American BioFortean Review*, 2(2): 8 pp (2000).

AVERBUCK, Philip, 'The Congo Water-Dragon', *Pursuit*, 14 (autumn): 104-106 (1981).

BAKKER, Robert, *The Dinosaur Heresies* (Longman: London, 1986).

BALLOT, Michel, 'La Sculpture de l'Emela N'Touka', *Mokele-Mbembe Expeditions*, http://mokelembembeexpeditions.blogspot.co.uk/2009/10/la-sculpture-de-lemela-ntouka-et-les.html 21 October (2009).

BAYLESS, Mark K., 'Giant Lizards, Salamanders, Snakes, and Turtles', *Fate*, 53 (November): 25-27 (2000).

BEER, Endymion (Ed.), 'River Monsters [re South American Living Dinosaurs]', *Athene*, No. 6 (spring): 28 (1995).

BLANCOU, Lucien, 'Notes sur les Mammifères de l'Equateur Africain Fran¢ais. Un Rhinocéros de Forêt?', *Mammalia*, 18 (December): 358-363 (1954).

BLANTON, John, 'The Acambaro Dinosaurs', *The North Texas Skeptic*, 13 (October) http://www.ntskeptics.org/1999/1999october/october1999.htm (1999).

BLOMBERG, Rolf, *Rio Amazonas* (Gebers: Stockholm, 1966).

BRIGHT, Michael, 'Meet Mokele-Mbembe', *BBC Wildlife*, 2 (December): 596-601 (1984).

CABRERA DARQUEA, Javier, *The Message of the Engraved Stones of Ica* (Plaza De Armas: Bolivar, Peru, 2000).

CARRIVEAU, Gary W. & HAN, Mark C., 'Thermoluminescent Dating and the Monsters of Acambaro', *American Antiquity*, 41(4): 497-500 (1976).

CHEESMAN, Evelyn, *Six-Legged Snakes in New Guinea* (George Harrap: London, 1949).

CLARK, Jerome, 'Monstrous Monitors', *Omni*, 6 (July): 98 (1984).

COPENS, Filip, 'Jurassic Library [re the Ica stones]', *Fortean Times*, No. 151 (October): 28-32 (2001).

CZERKAS, Stephen, 'Discovery of Dermal Spines Reveals a New Look for Sauropod Dinosaurs', *Geology*, 21 (December): 1068-1070 (1992).

DAVIES, Adam, 'I Thought I Saw a Sauropod', *Fortean Times*, No. 145 (April): 30-32 (2001).

DESMOND, Adrian J., *The Hot-Blooded Dinosaurs: A Revolution in Palaeontology* (Blond & Briggs: London, 1975).

DUPUY, Georges, 'Le Monstre de "Partridge Creek"', *Je Sais Tout*, 39 (15 April): 403-409 (1908).

FASSETT, James E., 'New Geochronologic and Stratigraphic Evidence Confirms the Paleocene Age of the Dinosaur-Bearing Ojo Alamo Sandstone and Animas Formation in the San Juan Basin, New Mexico and Colorado', *Palaeontologia Electronica*, http://palaeo-electronica.org/2009_1/149/149.pdf (April) 146 pp (2009).

FAWCETT, Percy H, 'Brontosaurus Hunt', *Daily Mail* (London), 17 December (1919); *Exploration Fawcett* (Hutchinson: London, 1953).

'FULAHN' [HICHENS, William], 'On the Trail of the Brontosaurus and Co.', *Chambers's Journal* (Series 7), 17 (1 October): 692-695 (1927).

GETTINGER, Dan, 'Cryptoletter', *ISC Newsletter*, 3 (winter): 11 (1984).

GIBBONS, William J., *Mokele-Mbembe: Mystery Beast of the Congo Basin* (Coachwhip: Landisville, 2010).

GIBBONS, Bill [=William J.] & RICKARD, Robert J.M., 'Operation Congo Returns', *Fortean Times*, No. 47 (autumn): 22-25 (1986).

GILROY, Rex, 'Australia's Lizard Monsters', *Fortean Times*, No. 37 (spring): 32-33 (1982); *Burrunjor! The Search for Australia's Living Tyrannosaurus* (URU Publications: Katoomba, 2013).

GREENWELL, J. Richard (Ed.), 'Congo Expeditions Inconclusive', *ISC Newsletter*, 1 (spring): 3-5 (1982); 'Special Interview' [with Marcellin Agnagna], *ISC Newsletter*, 3 (summer): 7-9 (1984); 'Mokele-Mbembe: New Searches, New Claims', *ISC Newsletter*, 5 (autumn): 1-7 (1986).

GUERRASIO, John, 'Dinosaur Graffiti—Hava Supai Style', *Pursuit*, 10 (April): 62-63 (1977).

GUILFORD, Gwynn, 'We're Not Even Close to Discovering All the Dinosaur Types That Ever Existed', *Quartz*, http://qz.com/430625/were-not-even-close-to-discovering-all-the-dinosaur-types-that-ever-existed/ 30 June (2015).

HAGENBECK, Carl, *Beasts and Men* (Longmans, Green, and Co: London, 1909).

HALL, Mark A., 'Pinky, the Forgotten Dinosaur', *Wonders*, 1 (December): 51-59 (1992); 'Sobering Sights of Pink Unknowns', *Wonders*, 1 (December): 60-64 (1992).

HICHENS, William, 'Africa's Mystery Beasts', *Wide World*, 62: 171-6 (1928); 'African Mystery Beasts', *Discovery*, 18 (December): 369-373 (1937).

HOCKNULL, Scott A., *et al.*, 'Dragon's Paradise Lost: Palaeobiogeography, Evolution and Extinction of the Largest-Ever Terrestrial Lizards (Varanidae)', *PLoS ONE*, http://journals.plos.org/plosone/article?id=10.1371/journal.pone.0007241 30 September (2009).

HUGHES, J.E., *Eighteen Years on Lake Bangweulu* (The Field: London, 1933).

IRWIN, J. O'Malley, 'Is the Chinese Dragon Based on Fact, Not Mythology?', *Scientific American*, 114 (15 April): 399, 410 (1916).

JACOBS, Louis, *Quest For the African Dinosaurs* (Villard Books: New York, 1993).

JAMES, C.G., 'Congo Swamps Mystery', *Daily Mail* (London), 26 December (1919).

JOLLY, Ben, 'Enter a Dragon That's Queerer Than Nessie', *Daily Telegraph* (London), 28 May (1994).

KNOLL, Fabien, *et al.*, 'Paleoneurological Evidence Against a Proboscis in the Sauropod Dinosaur *Diplodocus*', *Geobios*, 39: 215-221 (2006).

KOLDEWEY, Robert, *Das Wieder Erstehende Babylon. . .* (J.C. Hinrichs: Leipzig, 1913); *Das Ischtar-Tor in Babylon. . .* (J.C. Hinrichs: Leipzig, 1918).

LORENZONI, Silvano, 'Extant Dinosaurs: A Distinct Possibility', *Pursuit*, 10 (spring): 60-61 (1977); 'More on Extant Dinosaurs', *Pursuit*, 12 (July): 105-109 (1979).

MACKAL, Roy P., *A Living Dinosaur? In Search of Mokele-Mbembe* (E.J. Brill: Leiden, 1987).

MACKAL, Roy P. & POWELL, James H., 'In Search of Dinosaurs' [Mackal-Powell Likouala Expedition Report (Original Text) 26 February 1980 Impfondo, Africa], *Fortean Times*, No. 34 (winter): 8-9 (1980).

MACKAL, Roy P., *et al.*, 'The Search for Evidence of Mokele-Mbembe in the People's Republic of the Congo', *Cryptozoology*, 1: 62-72 (1982).

MAGIN, Ulrich, 'Living Dinosaurs in Africa: Early German Reports', *Strange Magazine*, No. 6 (November): 11 (1990).

MAY, John, 'Lost Dinosaurs of the Congo', *Curious Facts Monthly*, (August): [unpaginated] (1981).

MAYOR, Adrienne, 'Griffin Bones: Ancient Folklore and Paleontology', *Cryptozoology*, 10: 16-41 (1991).

MILLAIS, John G., *Far Away Up the Nile* (Longmans, Green, & Co: London, 1924).

MILLER, Charles, *Cannibal Caravan* (Lee Furman: New York, 1939).

MILLER, Leona, *Cannibals and Orchids* (Sheridan House: New York, 1941).

MOLNAR, Ralph E., 'New Cranial Elements of a Giant Varanid From Queensland', *Memoirs of the Queensland Museum*, 29: 437-444 (1990); *Dragons in the Dust: The Paleobiology of the Giant Monitor Lizard Megalania* (Indiana University Press: Bloomington, 2004).

MONKS, John, 'Riddle of Dinosaur in Cave Pictures', *Sunday Express* (London), 7 December (1969).

MORELL, Virginia, 'Roy Mackal, Dinosaur Hunter' *Reader's Digest*, 123 (July): 108-113 (1983).

MORGAN, Branwen, 'Dinosaurs Didn't Hibernate, Says Study', *Australian Geographic*, http://cms.ausgeo.bauer-media.net.au/news/2011/09/dinosaurs-didnt-hibernate,-says-study/ 26 September (2011).

MURAY, Leo, 'Cheer Up Nessie, You're Not Alone', *Daily Post* (Liverpool), 3 January (1976).

MYERS, George S., 'Asiatic Giant Salamander Caught in the Sacramento River. . .', *Copeia*, (8 June): 179-180 (1951).

NUGENT, Rory, *Drums Along the Congo: On the Trail of Mokele-Mbembe, the Last Living Dinosaur* (Houghton Mifflin Co.: New York, 1993).

PRICE, Gilbert J., *et al.*, 'Temporal Overlap of Humans and Giant Lizards (Varanidae; Squamata) in Pleistocene Australia', *Quaternary Science Reviews*, 125 (1 October): 98-105 (2015).

REGUSTERS, Herman A., 'Mokele-Mbembe. An Investigation Into Rumors Concerning a Strange Animal in the Rupublic [sic] of the Congo, 1981', *Munger Africana Library Notes*, 12 (July): 1-27 (1982).

RICKARD, Robert J.M., 'Monster Man [interview with Bill Gibbons]', *Fortean Times*, No. 67 (February-March): 28-31 (1993).

RIZZI, Jorge A. Livraga, 'Existen aun Animales Prehistoricos?', *Nueva Acrópolis Revista*, No. 158 (March): 23-30 (1988).

RODGERS, Thomas L., 'Report of Giant Salamander in California', *Copeia*: 646-647 (1962).

ROTHERMEL, Mark; GIBBONS, William J.; WALLS, Jonathan G.F.; & DELLA-PORTA, Joe B., 'Operation Congo—The First British/Congolese Mokele-Mbembe Expedition of 1985/6', Expedition report (unpublished) for Operation Congo (1986).

RUSSELL, William N., 'Report On Acambaro', *Fate*, 6 (June): 31-35 (1953).

SAHNI, Ashok, 'Cretaceous-Palaeocene Terrestrial Faunas of India: Lack of Endemism During Drifting of the Indian Plate', *Science*, 226 (26 October): 441-443 (1984).

SANDERSON, Ivan T., 'There Could Be Dinosaurs', *Saturday Evening Post*, 220 (3 January): 17, 53, 56 (1948).

SASS, Herbert R., quoted in: ANON., 'The Pink What-Is-It?', *Saturday Evening Post*, 221 (4 December): 10 (1948).

SCANLON, John D. & YEO, Michael S.Y., 'The Pleistocene Serpent *Wonambi* and the Early Evolution of Snakes', *Nature*, 403 (27 January): 416-420 (2000).

SCHMIDT, Franz H., 'Prehistoric Monsters in Jungles of the Amazon', *New York Herald* (New York), p. 5 of magazine section, 29 January (1911).

SHUKER, Karl P.N., 'Living Dinosaurs', *Me*, (28 July): 56-57 (1993); *Dr Shuker's Casebook: In Pursuit of Marvels and Mysteries* (CFZ Press: Bideford, 2008); *ShukerNature*'s Top Ten Living Dinosaurs of Cryptozoology', *ShukerNature*, http://karlshuker.blogspot.co.uk/2013/01/shukernatures-top-ten-living-dinosaurs.html 22 January (2013); *Dragons in Zoology, Cryptozoology, and Culture* (Coachwhip Publications: Greenville, 2013); 'The Emela-Ntouka—New Corroborative Evidence For the Congo's Cryptic 'Killer of Elephants'?', *ShukerNature*, http://karlshuker.blogspot.co.uk/2014/07/the-emela-ntouka-new-corroborative.html 5 July (2014); 'From Mini-Rex to Moon Cow—Unravelling the Riddle of America's Modern-Day 'River Dinosaurs'', *ShukerNature*, http://karlshuker.blogspot.co.uk/2015/12/from-mini-rex-to-moon-cow-unravelling.html 9 December (2015); 'In Search of a Stegosaur', *Practical Reptile Keeping*, (February): 54-59 (2016).

SIEVEKING, Paul, 'Monsters of New Britain—Sightings of Dinosaurs and Other Strange Reptiles in Papua New Guinea', *Fortean Times*, No. 184 (June): 13 (2004).

SLOAN, Robert E., *et al.*, 'Gradual Dinosaur Extinction . . . in the Hell Creek Formation', *Science*, 232 (2 May): 629-633 (1986).

SMITH, Alfred A., *Trader Horn* (Readers Library Publishing Co.: London, 1931).

SMULLEN, Ivor, 'Do Dinosaurs Still Exist?', *Weekend* (London), 7 October (1981).

SUCIK, Nick, '"Dinosaur" Sightings in the United States', *In*: ARMENT, Chad (Ed.), *Cryptozoology and the Investigation of Lesser-Known Mystery Animals* (Coachwhip Publications: Landisville, 2006): 137-168.

SWIFT, Dennis, 'Are the Ica Stones Fake? Skeptics Under Fire', *Cryptozoology Research Team*, http://livingdinos.com/2011/07/are-the-ica-stones-fake-skeptics-under-fire/ July (2011).

THESIGER, Wilfred, *The Marsh Arabs* (Longmans: London, 1964).

VAN VALEN, Leigh & SLOAN, Robert E. 'Contemporaneity of Late Cretaceous Extinctions', *Nature*, 270 (10 November): 193 (1977).

VAŠÍCEK, Arnošt, *Tajemná Minulost* [*Mysterious Past*] (Baronet: Prague: 1998).

VINE, Brian, 'I Aim to Bring Back a Live Dinosaur. Dr. Roy Mackal Talking to Brian Vine: New York', *Daily Express* (London), 26 June (1981).

WAVELL, Stewart, *The Lost World of the East: An Adventurous Quest in the Malayan Hinterland* (Souvenir Press: London, 1958).

WEDEL, Mathew J., 'Vertebral Pneumaticity, Airsacs, and the Physiology of Sauropod Dinosaurs', *Paleobiology*, 29(2): 243-255 (2003).

WILKINS, Harold T., *Monsters and Mysteries* (James Pike: London, 1973).

WIRTH, Diane E., 'Dinosaurs in Pre-Columbian Art', *Pursuit*, 17 (spring): 13-16 (1984).

Chapter 2: Things with Wings

ANON., 'A Singular Bird', *Columbus Daily Enquirer* (Columbus, Georgia), 2 September (1868).

ANON., 'Terror Bird in S.R. Swamp', *Rhodesia Herald* (Salisbury), 26 March (1957); 'Ape-Man in City Office More Likely Than Pterodactyl in S.R. Swamp—Museum Director', *Rhodesia Herald*, 29 March (1957); 'Pterodactyls Seen Near Northern Rhodesian River', *Rhodesia Herald*, 2 April (1957); 'Museum Director Says There are No Flying Reptiles', *Rhodesia Herald*, 5 April (1957).

ANON., 'Last of Condors Captured in Bid to Save Species', *Express and Star* (Wolverhampton), 1 June (1987).

ANON., 'Andean Condor Flaps Wings and Takes Off', *Plain Dealer* (Cleveland, Ohio), 20 December (1988).

ADMIN, 'Centre NAD Reassures Montrealers: No Danger of Being Snatched by a Royal Eagle', *NAD*, http://nad.ca/centre-nad-reassures-montrealers-no-danger-of-being-snatched-by-a-royal-eagle/ 19 December (2012).

ARMSTRONG, Perry A., *The Piasa, Or the Devil Among the Indians* (E.B. Fletcher: Morris, 1887).

BAKER, Craig S., '9 Unsolved Mysteries of the Wild West', *Mental_Floss*, http://mentalfloss.com/article/56759/9-unsolved-mysteries-wild-west 14 October (2015).

BALOUET, Jean-Christophe, 'Les Étranges Fossiles de Nouvelle-Calédonie', *La Recherche*, 15 (March): 390-392 (1984).

BENEDICT, W. Ritchie, 'Benedict Hot on Thunderbird Photo Trail', *Strange Magazine*, No. 6: 44

(1990); 'Thunderbird Photograph Investigation', *Strange Magazine*, No. 11 (spring-summer): 39 (1993); 'The Search For the Thunderbird Photo', *Strange Magazine*, No. 12 (fall-winter): 39 (1993).

BERGER, L.R. & CLARKE, R.J., 'Eagle Involvement in the Accumulation of the Taung Child Fauna', *Journal of Human Evolution*, 29: 275-299 (1995).

'BONOMI', 'On a Gigantic Bird Sculptured on the Tomb of an Officer of the Household of Pharaoh', *American Journal of Science*, 49 (October): 403-405 (1845).

BURTON, Maurice, 'Gone For Ever?', *Animals*, 3 (18 February): 308 (1964).

BURTON, Maurice & BENSON, C.W., 'The Whale-Headed Stork or Shoe-Bill: Legend and Fact. Parts I & II', *Northern Rhodesia Journal*, 4: 411-426 (1961).

CAMPBELL, Kenneth E. & STENGER, Allison T., 'A New Teratorn (Aves: Teratornithidae) From the Upper Pleistocene of Oregon, USA', In: ZHOU, Z. & F. ZHANG (Eds), *Proceedings of the 5th Symposium of the Society of Avian Paleontology and Evolution Beijing, 1–4 June 2000* (China Science Press: Beijing, 2000): 1-11.

CAMPBELL, Kenneth E. & TONNI, Eduardo P., 'Size and Locomotion in Teratorns (Aves: Teratornithidae)', *Auk*, 100: 390-403 (1983).

CLARK, Jerome & COLEMAN, Loren, *Creatures of the Outer Edge* (Warner: New York, 1978).

COLEMAN, Loren, *Curious Encounters* (Faber & Faber: London, 1985).

FEDUCCIA, Alan & VOORHIES, Michael R., 'Miocene Hawk Converges on Secretarybird', *Ibis*, 131(3): 349–354 (1989).

FISCHER, G.A. & A. REICHENOW, 'Briefliche Reiseberichte aus Ost-Afrika III', *Journal für Ornithologie*, 26(3): 268-297 (1878).

GEGGEL, Laura, "Winged Monster' Rock Art Finally Deciphered', *LiveScience*, http://www.livescience.com/51886-winged-monster-rock-art-deciphered.html 18 August (2015).

GERHARD, Ken, *Big Bird! Modern Sightings of Flying Monsters* (CFZ Press: Bideford, 2007).

GOSS, Michael, 'Pterosaurs Over Texas', *The Unknown*, No. 5 (November): 46-50 (1985).

GREENWELL, J. Richard, 'Flights of Speculation [re living pterosaurs]', *BBC Wildlife*, 13 (March): 25 (1995).

GREENWELL, J. Richard (Ed.), 'Big Bird is Back', *ISC Newsletter*, 2 (winter): 8-9 (1983); 'Highlights of Galveston Meeting [including Mackal's talk re Namibian pterosaurs]', *ISC Newsletter*, 10 (summer): 4-5 (1991).

GRISCELLI, Paul, 'Deux Oiseaux Fossiles de Nouvelle-Calédonie', *Bulletin de la Société d'Etudes Historiques de Nouvelle-Calédonie*, 29: 3-6 (1976).

HALL, Mark A., *Thunderbirds! The Living Legend of Giant Birds* (Mark A. Hall Publications and Research: Bloomington, 1988); 'Thunderbirds!', *Wonders*, 1 (April): 1-8 (1992); 'Thunderbirds are Go', *Fortean Times*, No. 105 (December): 34-38 (1997); *Thunderbirds: America's Living Legends of Giant Birds* (Paraview Press: New York, 2004).

HEUVELMANS, Bernard, 'Of Lingering Pterodactyls', *Strange Magazine*, No. 6 (November): 8-11, 58-60 (1990).

HILDEBRAND, Norbert, 'The Monster on the Rock', *Fate*, 7 (March): 13-19 (1954).

HOCH, Ella, 'Reflections on Prehistoric Life at Umm an-Nar (Trucial Oman) Based on Faunal Remains From the Third Millennium BC', *Geological Museum of the University of Copenhagen, Contributions to Palaeontology*, No. 283: 1-50 [589-638] (1979).

HSU, Jeremy, 'Creators of 'Golden Eagle Snatches Kid' Video Admit Hoax', *LiveScience*, http://www.livescience.com/25697-golden-eagle-video-hoax.html 19 December (2012).

JEFFREYS, Mervyn D.W., 'African Pterodactyls', *Journal of the Royal African Society*, 43: 72-74 (1943).

LARSEN, Carl, 'The Tuscarora Mountain T-Birds', *Pursuit*, 15 (fall): 106-107 (1982).

LAWSON, Douglas A., 'Pterosaur From the Latest Cretaceous of West Texas: Discovery of the Largest Flying Creature', *Science*, 187 (14 March): 947-948 (1975).

LE QUELLAC, J-L., *et al.*, 'The Death of a Pterodactyl', *Antiquity*, 89 (August): 872-884 (2015).

LYMAN, Robert R., *Amazing Indeed: Strange Events in the Black Forest, Vol. 2* (Potter Enterprise: Coudersport, 1973).

MARSCHALL, Count, 'Oiseau Problématique [makalala]', *Bulletin de la Société Philomatique* (7th Series), 3: 176 (1878-1879).

MARSHALL. Larry G., 'The Terror Bird', *Field Museum of Natural History Bulletin*, 49: 6-15 (1978).

MARUNA, Scott, 'Substantiating Audubon's Washington Eagle', *Biofort*, http://biofort.blogspot.co.uk/2006/10/substantiating-audubons-washington.html 14 October (2006); 'Of Washington Eagles, Ivory-bills and 'Thunderbird' Sightings', *Biofort*, http://biofort.blogspot.co.uk/2006/10/of-washington-eagles-ivory-bills-and.html 18 October (2006); 'Witness Claims a Washington Eagle Sighting', *Biofort*, http://biofort.blogspot.co.uk/2006/10/witness-claims-washington-eagle.html 23 October (2006).

MELLAND, Frank H., *In Witchbound Africa* (Seeley Service: London, 1923).

MOURER-CHAUVIRÉ, Cécile & BALOUET, J.C., 'Description of the Skull of the Genus *Sylviornis* Poplin, 1980 (Aves, Galliformes, Sylviornithidae New Family), a Giant Extinct Bird From the Holocene of New Caledonia', *In:* ALCOVER, J.A. & BOVER, P. (Eds), *Proceedings of the International Symposium "Insular Vertebrate Evolution: the Palaeontological Approach, Monografies de la Societat d'Història Natural de les Balears*, 12: 205–218 (2005).

MOURER-CHAUVIRÉ, Cécile & POPLIN, François, 'Le Mystère des Tumulus de Nouvelle-Calédonie', *La Recherche*, 16 (September): 1094 (1985).

NORMAN, Scott [Ed. ARMENT, Chad], 'The Scott Norman "Pterosaur" Sighting: His Own Words', *BioFortean Review*, http://www.strangeark.com/bfr/articles/scott-norman-sightings.html No. 16 (March): [unpaginated] (2008).

PADIAN, Kevin & BRAGINETZ, Donna, 'The Flight of Pterosaurs', *Natural History*, 97 (December): 58-65 (1988).

PEARL, Jack, 'Monster Bird That Carries Off Human Beings!', *Saga*, (May): 29-31, 83-85 (1963).

PERCIVAL, A. Blayney, *A Game Ranger On Safari* (Nisbet & Co: London, 1928).

PITMAN, Charles R.S., *A Game Warden Takes Stock* (James Nisbet: London, 1942).

POPLIN, François & MOURER-CHAUVIRÉ, Cécile, '*Sylviornis neocaledoniae* . . . Oiseau Géant éteint de l'Ile des Pins (Nouvelle-Calédonie)', *Geobios*, No. 18 (February): 73-97 (1985).

POUCHET, Felix A., *The Universe* (Blackie & Son: London, 1873).

PRICE, G. Ward, *Extra-Special Correspondent* (George Harrap: London, 1957).

RASMUSSEN, D.T., *et al.*, ' Hindlimb of a Giant Terrestrial Bird from the Upper Eocene, Fayum, Egypt', *Palaeontology*, 44 (March): 325-337 (2001).

RICHARDS [=RICHBURG], Mike, 'Giant Bird', *Fortean Times*, no. 276 (June): 74 (2011).

SANDERSON, Ivan T., *Investigating the Unexplained* (Prentice Hall: Eaglewood Cliffs, 1972).

SCHAEFFER, Claude E., 'Was the California Condor Known to the Blackfoot Indians?', *Journal of the Washington Academy of Sciences*, 41 (June): 181-191 (1951).

SENTER, Phil & KLEIN, Darius M., 'Investigation of Claims of Late-Surviving Pterosaurs: The Cases of Belon's, Aldrovandi's, and Cardinal Barberini's Winged Dragons', *Palaeontologia Electronica*, 17(3; 41A): 1-19 (2014).

SHUKER, Karl P.N., 'The Search For the Thunderbird Photo', *Strange Magazine*, No. 12 (fall-winter): 38 (1993); 'Flying Graverobbers [re ropen]', *Fortean Times*, No. 154 (January): 48-49 (2002); 'Washington's Eagle and Other Giant Mystery Eagles of North America', *ShukerNature*, http://www.karlshuker.blogspot.co.uk/2012/08/washingtons-eagle-and-other-giant.html 2 August (2012); 'I Thought I Saw a Terror Saur!—Do Prehistoric Flying Reptiles Still Exist?', *ShukerNature*, http://karlshuker.blogspot.co.uk/2013/10/i-thought-i-saw-terror-saur-do.html 7 October (2013); 'Seeking the Missing Thunderbird Photograph—One of Cryptozoology's Most Tantalising Unsolved Cases', *ShukerNature*, http://karlshuker.blogspot.co.uk/2014/11/seeking-missing-thunderbird-photograph.html 25 November (2014); 'Meet the Monstrous Makalala—a Tanzanian Terror Bird?', *Shuker Nature*, http://karlshuker.blogspot.co.uk/2015/10/meet-monstrous-makalala-tanzanian.html 31 October (2015).

SMITH, J.L.B., *Old Fourlegs: The Story of the Coelacanth* (Longmans, Green, & Co: London, 1956).

SPITZ-BOMBONNEL, Zo,, 'Animaux Perdus et Non Retrouvés', *Le Chasseur Français*, (June): 375 (1959).

'STANY', *Loin des Sentiers Battus, Vol. 4—Douze Femmes* (Table Ronde: Paris, 1953).

STEYN, Peter, *Birds of Prey of Southern Africa: Their Identification and Life Histories* (David Philip: Cape Town, 1982).

STIFFY, Jonathan D., 'The Thunderbirds of Western Pennsylvania—Mistaken Identity or Migratory Cryptids?', *Journal of Cryptozoology*, 3 (December): 9-20 (2014).

SUTHERLY, Curt, 'Great Birds of the Allegheny Plateau', *Caveat Emptor*, No. 20 (winter): 21-24 (1989).

TRAJANO, E. & de VIVO, M., '*Desmodus draculae* . . . Reported for Southeastern Brasil [sic], With Paleoecological Comments . . .', *Mammalia*, 55: 456-459 (1991).

TURNER, Mark, 'Monsters of the Skies! Thunderbird—Legends, Sightings & Evidence', *Mark Turner's Mysterious World*, http://markturnersmysteriousworld.blogspot.co.uk/2011/06/thunderbird-legends-sightings-evidence.html 23 June (2011).

VEMBOS, Thanassis, 'A Prehistoric Flying Reptile?', *Strange Magazine*, No. 2: 29 (1988).

WELLNHOFER, Peter, *The Illustrated Encyclopedia of Pterosaurs* (Salamander: London, 1991).

WHITCOMB, Jonathan, *Searching For Ropens: Living Pterosaurs in Papua New Guinea* (BookShelf Press: Livermore, 2006); 'Living Pterodactyls: Ropens of the Southwest Pacific: Giant Flying Creature, Lizard-Like, Seen Over Perth, Australia', *Ropens*, http://www.ropens.com/main/ (2008); *Live Pterosaurs in America: Sightings of Apparent Pterosaurs in the United States* (CreateSpace: Charleston, 2009); 'Gitmo Pterosaur: Living Pterosaurs at Guantanamo Bay', *Live Pterosaur*, http://www.live-pterosaur.com/Prodigy/cuba_pterosaur/ (2011); 'Ropen Sighting in Florida', *Live Pterosaur*, http://www.livepterosaur.com/LP_Blog/archives/4119 19 November (2012).

WIMAN, Carl, 'Ein Gerücht von einem Lebenden Flugsaurier', *Natur und Museum*, 58: 431-432 (1928).

WORTHY, Trevor H., *et al.*, 'Osteology Supports a Stem-Galliform Affinity For the Giant Extinct Flightless Bird *Sylviornis neocaledoniae* (Sylviornithidae, Galloanseres)', *PLoS ONE*, http://journals.plos.org/plosone/article?id=10.1371/journal.pone.0150871 30 March (2016).

Chapter 3: Monsters from the Ancient Waters

ANON., 'The Great Sea-Serpent—Evidences of the Former Appearance of the Sea-Serpent—The Great American Sea-Serpent', *Illustrated London News*, 13 (28 October): 264-266 (1848).

ANON., 'A Large Turtle', *Scientific American*, 48: 292 (1883).

ANON., 'The Squids From Onondaga Lake, N.Y.', *Science*, 16 (12 December): 947 (1902).

ANON., 'Head of Sea Serpent Recently Captured Displayed in Honolulu', *Jonesboro Evening Sun* (Jonesboro), 11 May (1905).

ANON., 'Snake That Roars Like a Lion [re huillia]', *Kingston Gleaner* (Kingston), 23 June (1934).

ANON., 'Giant Shark Photographed', *Times* (San Mateo), 15 August (1966).

ANON., 'Scientists to Search For Swedish Monster', *The Trentonian* (Trenton, New Jersey), 5 April (1987).

ANON., 'Nahuelito: Creature Story Makes Waves', *Times* (Picayune), 28 March (1989).

ANON., 'Jeg var Søslangen i Loch Ness', *Hjemmet*, No. 47 (1992).

ANON., 'Des Coelacanthes dans le Golfe du Mexique?', *Science et Vie*, No. 911 (August): 17-18 (1993).

ALDERTON, David, *Breakfast With a Bigfoot* (Beaver: London, 1986).

BAUER, Aaron M. & RUSSELL, Anthony P., 'A Living Plesiosaur? A Critical Assessment of the Description of *Cadborosaurus willsi*', *Cryptozoology*, 12: 1-18 (1996).

BAUER, Henry H., *The Enigma of Loch Ness: Making Sense of a Mystery* (University of Illinois Press: Urbana, 1986, updated 1988); 'The Case For the Loch Ness "Monster": The Scientific Evidence', *Journal of Scientific Exploration*, 16(2): 225-246 (2002); 'Genuine Facts About "Nessie", the Loch Ness "Monster"', *Henry H. Bauer—Homestead*, http://henryhbauer.homestead.com/LochNessFacts.html 30 November (2014).

BERNARD, Aurélien, *et al.*, 'Regulation of Body Temperature by Some Mesozoic Marine Reptiles', *Science*, 328 (11 June): 1379-1382 (2010).

BIRD, Sheila, 'Morgawr Joke is a Monstrous Lie', *Packet* (Falmouth), 30 August (1985).

BORGES, Jorge Luis, *The Book of Imaginary Beings* (rev. edit., Penguin: Middlesex, 1974).

BOUSFIELD, Edward L. & LeBlond, Paul H, 'An Account of *Cadborosaurus willsi*, New Genus, New Species, a Large Aquatic Reptile From the Pacific Coast of North America,' *Amphipacifica*, 1 (Supplement 1; 20 April): 1-25 (1995).

BOWEN, Henry, 'A Monstrous Sea Serpent', Privately printed information sheet (1817).

BRENTJES, B., 'Eine Vor-Entdeckung des Quastenflossers in Indien?', *Naturwissenschaftliche Rundschau*, 25: 312-313 (1972).

BRIGHT, Michael, 'Salty Tales II', *BBC Wildlife*, 5 (January): 13-14 (1987); *There are Giants in the Sea* (Robson: London, 1989).

BUFFETAUT, Eric, 'Vertical Flexure in Jurassic and Cretaceous Marine Crocodilians and Its Relevance to Modern "Sea-Serpent" Reports', *Cryptozoology*, 2: 85-89 (1983).

BURTON, Maurice, *The Elusive Monster: An Analysis of the Evidence From Loch Ness* (Rupert Hart-Davis: London, 1961).

CAHILL, Tim, 'A la Recherche du Continent Perdu', *GEO*, No. 35 (January): 78-101 (1982).

CHONO, Kenji, 'Issie of Japan's Lake Ikeda', *Elsewhen*, 2(4): 9 (1991).

CHORVINSKY, Mark, 'Tony "Doc" Shiels: Magic and Monsters—Photographs Investigation', *Strange Magazine*, No. 8 (fall): 8-11, 46-48 (1991); 'Our Strange World [re Sir Arthur Conan Doyle's sea serpent sighting]', *Fate*, 49 (March): 12-14 (1996).

COLEMAN, Loren, 'Screen Shots of Caddy Image', *Cryptomundo*, http://cryptomundo.com/cryptozoo-news/caddy-screen/ 19 July (2011).

COLEMAN, Loren & HUYGHE, Patrick, *The Field Guide to Lake Monsters, Sea Serpents, and Other Mystery Denizens of the Deep* (Tarcher/Penguin: New York, 2003).

COONEY, Jay, 'A Most Compelling 'Sea Serpent' Case: The Alvin Submersible Encounter', *Bizarre Zoology*, http://bizarrezoology.

blogspot.co.uk/2013/04/a-submarine-pilot-plesiosaur-sighting.html 21 April (2013).

COSTELLO, Peter, *In Search of Lake Monsters* (Garnstone Press: London, 1974).

CRUICKSHANK, Arthur, *et al.*, 'Dorsal Nostrils and Hydrodynamically Driven Underwater Olfaction in Plesiosaurs', *Nature*, 352 (4 July): 62-64 (1991).

DASH, Mike, 'The Dragons of Vancouver', *Fortean Times*, No. 70 (August-September): 46-48 (1993).

DAY, Charles W., *Five Years' Residence in the West Indies* (Colburn: London, 1852).

DINSDALE, Tim, *The Leviathans* (Routledge & Kegan Paul: London, 1966; rev. edit., Futura: London, 1976); 'The Rines/Edgerton Picture', *Photographic Journal*, (April): 162-165 (1973); *Loch Ness Monster* (4th edit., Routledge & Kegan Paul: London, 1982).

DONOVAN, Stephen, '*Cruziana* and *Rusophycus*: Trace Fossils Produced by Trilobites . . . In Some Cases?', *Lethaia*, 43 (no. 2; June): 283-284 (2010).

DOYLE, Alister, '"Arctic Ocean Survey May Reveal Lost World"—Experts', *Yahoo! News*, http://story.news.yahoo.com/news?tmpl=story&cid=570&ncid=753&e=1&u=/nm/20040624/sc_nm/environment_arctic_dc 24 June (2004) [no longer online].

DOYLE, Arthur Conan, *Memories and Adventures* (Hodder & Stoughton: London, 1924).

DRUCKENMILLER, P.S. & RUSSELL, A.P., 'Skeletal Anatomy of an Exceptionally Complete Specimen of a New Genus of Plesiosaur From the Early Cretaceous (Early Albian) of Northeastern Alberta, Canada', *Palaeontographica Abteilung A*, 283: 1–33 (2008).

DUVAL, Mathieu, 'Dating Fossil Teeth by Electron Paramagnetic Resonance: How is That Possible?', *Spectroscopy Europe*, http://www.spectroscopyeurope.com/articles/55-articles/3328-dating-fossil-teeth-by-electron-paramagnetic-resonance-how-is-that-possible 14 February (2014).

ECKERT, S.A., 'Swim Speed and Movement Patterns of Gravid Leatherback Sea Turtles (*Dermochelys coriacea*) at St Croix, US Virgin Islands', *Journal of Experimental Biology*, 205: 3689–3697 (2002).

EIGHTS, James, 'Description of a New Crustaceous Animal Found on the Shores of the South Shetland Islands, With Remarks on Their Natural History', *Transactions of the Albany Institute*, 2(1): 53-69 (1833).

ELLIS, Richard, *Monsters of the Sea* (Alfred A. Knopf: New York, 1994).

EVANS, Mark, 'An Investigation Into the Neck Flexibility of Plesiosauroid Plesiosaurs: *Cryptoclidus eurymerus* and *Muraenosaurus leedsii*', M.Sc. Research Project in Vertebrate Palaeontology, University College London (May 1993; minor corrections September 2014).

EVERHART, Mike, *Sea Monsters: Prehistoric Creatures of the Deep* (National Geographic Society: Washington, D.C., 2007).

'F, Mary', 'Sir: Kindly Find Enclosed One Sea Serpent', *Packet* (Falmouth), 5 March (1976).

FANTI, Federico, *et al.*, 'The Largest Thalattosuchian (Crocodylomorpha) Supports Teleosaurid Survival Across the Jurassic-Cretaceous Boundary', *Cretaceous Research*, 61 (June): 263-274 (2016).

FELL, H. Barry, *Bronze Age America* (Little, Brown: London, 1982).

FORSTNER, Freiherr von, 'Das Schottische Seeungeheuer schon von U 28 Gesichtet', *Deutsche Allgemeine Zeitung* (Tübingen), 19 December (1933).

FRAIR, W., *et al*,. 'Body Temperature of *Dermochelys coriacea*: Warm Turtle From Cold Water', *Science*, 177 (1 September): 791-793 (1972).

FRICKE, Hans, 'Quastie im Baskenland?', *Tauchen*, No. 10 (October): 64-67 (1989).

FRICKE, Hans & PLANTE, Raphael, 'Silver Coelacanths From Spain are not Proofs of a Pre-Scientific Discovery', *Environmental Biology of Fishes*, 61: 461-463 (2001).

GAAL, Arlene, *Ogopogo: The True Story of the Okanagan Lake Million Dollar Monster* (Hancock House: Surrey, British Columbia, 1986).

GEORGE, Uwe, 'Venezuela's Islands in Time', *National Geographic*, 175 (May): 526-561 (1989).

GIBSON, J.A. & HEPPELL, David (Eds.), *The Search For Nessie in the 1980's* [sic] (The Scottish Natural History Library: Kilbarchan, 1988).

GILES, Jacqueline, 'The Underwater Acoustic Repertoire of the Long-Necked, Freshwater Turtle *Chelodina oblonga*', Thesis presented for the degree of Doctor of Philosophy in the School of Environmental Science, Murdoch University, Perth, Western Australia, http://research repository.murdoch.edu.au/39/2/02Whole.pdf September: 241 pp. (2005).

GILES, Jacqueline C., *et al.*, 'Voice of the Turtle—The Underwater Acoustic Repertoire of the Long-Necked Freshwater Turtle, *Chelodina oblonga*', *Journal of the Acoustical Society of America*, 126 (no. 1; July): 434-443 (2009).

GINGERICH, Philip D., *et al.*, 'Hind Limbs of Eocene Basilosaurus: Evidence of Feet in Whales', *Science*, 249 (13 July): 154-157 (1990).

GOSS, Michael, 'Do Giant Prehistoric Sharks Survive?', *Fate*, 40 (November): 32-41 (1987).

GOTTFRIED, Michael D., COMPAGNO, Leonard J.V., & BOWMAN, S. Curtis, 'Size and Skeletal Anatomy of the Giant "Megatooth" Shark *Carcharodon megalodon*', *In:* KLIMLEY, A. Peter & AINLEY, David G. (Eds), *Great White Sharks: The Biology of* Carcharodon carcharias (Academic Press: San Diego/London, 1996): 55-66.

GOULD, Rupert T., *The Case For the Sea Serpent* (Philip Allan: London, 1930).

GRACHEV, Guerman, 'Mysterious Water Mammoths Inhabit Siberian Lakes', *Komsomolskaya Pravda* (Moscow), 6 August: 1-4 (2007).

GREENWELL, J. Richard, 'An Endothermic "Nessie"?', *Science*, 200 (19 May): 722-723 (1978);'Prehistoric Fishing', *BBC Wildlife*, 12 (March): 33 (1994).

GREENWELL, J. Richard (Ed.), 'Lake Champlain Monster Draws Worldwide Attention', *ISC Newsletter*, 1 (summer): 1-4 (1982); 'Champ Photo Analysis Supports Animal Hypothesis', *ISC Newsletter*, 1 (autumn): 5-6 (1982); '"Sea Serpents" Seen Off California Coast', *ISC Newsletter*, 2 (winter): 9-10 (1983); 'Close Encounter in Lake Okanagan Revealed', *ISC Newsletter*, 6 (spring): 1-3 (1987).

GREY, Zane, *Tales of Tahitian Waters* (Harper Brothers: New York, 1931).

HELM, Thomas, *Shark! Unpredictable Killer of the Sea* (Dodd, Mead: New York, 1961).

HEMINGWAY, Sam, 'Lake's First 'Champ-Hearing' Recorded', *Burlington (VT) Free Press* (Burlington), 20 July (2003).

HEPPLE, Rip (Ed.), *Nessletter*, Nos. 1-current issue (1974-present day).

HEUVELMANS, Bernard, *Le Grand Serpent-de-Mer* (Plon: Paris, 1965); *In the Wake of the Sea-Serpents* (Rupert Hart-Davis: London, 1968).

HLIDBERG, Jón B. & ÆGISSON, Sigurdur, *Meeting With Monsters: An Illustrated Guide to the Beasts of Iceland* (JPV Útgáfa: Reykjavik, 2008).

HOLIDAY, F.W. & WILSON, Colin, *Goblin Universe* (Llewellyn Publishers: St Paul, 1986).

JOSEPH, Edward L., *History of Trinidad* (A.K. Newman: London, 1838).

KARPA, Robert, 'Ken Chaplin's Monstrous Obsession', *West* [magazine with *Globe and Mail*, Toronto], (November): 73-80 (1989).

KEAR, Benjamin P., et al., 'An Archaic Crested Plesiosaur in Opal From the Lower Cretaceous High-Latitude Deposits of Australia', *Biology Letters*, 2: 615–619 (2006).

KELLOCK, Andrew, 'More Sightings of Loch Ness Monster in 2015 Than Anytime in Last Ten

Years', *Press & Journal* (Aberdeen), https://www.pressandjournal.co.uk/fp/news/inverness/773587/more-sightings-of-loch-ness-monster-in-2015-than-anytime-in-last-ten-years/ 8 December (2015).

KINGSLEY, Charles, *At Last: A Christmas in the West Indies* (Bernhard Tauchnitz: Leipzig, 1871).

KIRK, John, '15 Cadborosauruses? Maybe a Few Less', *Cryptomundo*, http://cryptomundo.com/sea-serpents/15-caddy-2/ 16 August (2010).

KUBAN, Glen J., 'Lake Erie Sea Monster?', *Paluxy*, http://paleo.cc/paluxy/eriebaby.htm (2006, updated 2007).

LANGTON, James, 'Revealed: The Loch Ness Picture Hoax', *Sunday Telegraph* (London), 13 March (1994).

LARRABEE, John, 'Monster Mania: Champ No Chump in Japan', *USA Today*, 8 September (1993).

LASCOW, Sarah, 'Found: An Ancient Whale Fossil With a Smaller Whale Inside', *Atlas Obscura*, http://www.atlasobscura.com/articles/found-an-ancient-whale-fossil-with-a-smaller-whale-inside 3 June (2015).

LeBLOND, Paul H. & BOUSFIELD, Edward L., *Cadborosaurus: Survivor From the Deep* (Horsdal & Schubart: Victoria, 1995).

LeBLOND, Paul H. & COLLINS, Michael J., 'The Wilson Nessie Photo: A Size Determination Based on Physical Principles', *Cryptozoology*, 6: 55-64 (1987).

LIU, S., *et al.*, 'Computer Simulations Imply Forelimb-Dominated Underwater Flight in Plesiosaurs', *PLoS Computational Biology*, 11 (18 December): 1-18 (2015).

LOVELL, Selina, 'Letter [re moha-moha]', *Land and Water*, (3 January): 145-147 (1891).

LUCAS, Spencer B. & REYNOLDS, Robert E., 'Putative Paleocene Plesiosaurs From Cajon Pass, California, U.S.A.', *Cretaceous Research*, 14: 107-111 (1993).

McCORMICK, Cameron, 'The Flexibility of Plesiosaur Necks', *The Lord Geekington*, http://cameronmccormick.blogspot.co.uk/2008/09/flexibility-of-plesiosaur-necks.html 28 September (2008).

McEWAN, Graham J., *Sea Serpents, Sailors and Sceptics* (Routledge & Kegan Paul: London, 1978).

MACKAL, Roy P., *The Monsters of Loch Ness* (Futura/Macdonald and Janes: London, 1976).

MAGIN, Ulrich, 'Forstner Sea Serpent Sighting: A Possible Hoax?', *Strange Magazine*, No. 2: 4 (1988); 'New Material on the Moha-Moha', *Journal of Cryptozoology*, 3 (December): 21-31 (2014).

MAKÁDI, L., *et al.*, 'The First Freshwater Mosasauroid (Upper Cretaceous, Hungary) and a New Clade of Basal Mosasauroids', *PLoS ONE*, 7 (no. 12; December), e51781; doi: 10.1371/journal.pone.0051781 (2012).

'MANDRAKE' [PURSER, Philip], 'Making of a Monster', *Sunday Telegraph* (London), 7 December (1975).

MANGIACOPRA, Gary S., *Theoretical Population Estimates of the Large Aquatic Animals in Selected Freshwater Lakes of North America* [Thesis submitted to the School of Graduate Studies in partial fulfilment of the requirements for the degree of Master of Arts] (Southern Connecticut State University: New Haven, 1992 [unpublished]).

MARDIS, Scott, 'Plesiosaurs on Ice: Perspectives on the "Living Plesiosaur" Controversy', *Bizarre Zoology*, http://bizarrezoology.blogspot.co.uk/2014/05/plesiosaurs-on-ice-perspectives-on.html 9 May (2014); 'The Prehistoric Survivor Paradigm and More on Purported Post-Cretaceous Plesiosaurs', *Bizarre Zoology*, http://bizarrezoology.blogspot.co.uk/2014/07/the-prehistoric-survivor-paradigm-and.html 15 July (2014).

MARTIN, David & BOYD, Alistair, *Nessie: The Surgeon's Photograph Exposed* (Martin and Boyd: East Burnet, 1999).

MATTERS, Leonard, 'An Antediluvian Monster', *Scientific American*, 127 (July): 21 (1922).

MATTISON, David, 'An 1897 Sea Serpent Sighting in the Queen Charlotte Islands', *British Columbia Historical News*, 17(2): 15 (1964).

MAWNAN-PELLER, A., *Morgawr—The Monster of Falmouth Bay* (Morgawr Prod.: Falmouth, 1976).

MEURGER, Michel & GAGNON, Claude, *Lake Monster Traditions: A Cross-Cultural Analysis* (Fortean Tomes: London, 1988).

MORPHY, Rob, 'Kodiak Dinosaur [sic] (Alaska, USA)', *Cryptopia*, http://www.cryptopia.us/site/2010/03/kodiak-dinosaur-alaska-usa/ 6 March (2010).

MOTANI, Ryosuke, 'Warm-Blooded "Sea Dragons"?', *Science*, 328 (11 June): 1361-1362 (2010).

MURPHY, Patrick, 'Breeding Caddys Disturbed by Plane', *Times-Colonist* (Victoria, B.C.), 28 July (1993); 'More Sightings Add to Caddy's Stack', *Times-Colonist* (Victoria, B.C.), 31 July (1993); 'Saanich Inlet Attracts Caddy For Breeding, Says Cryptozoologist', *Times-Colonist* (Victoria, B.C.), 16 August (1993).

'N., J.', 'Pas de Coelacanthes dans le Golfe du Mexique!', *Science et Vie*, No. 913 (October): 6 (1993).

NAISH, Darren, 'Humping Archaeocetes (Snigger)', *Dinosaur Mailing List*, http://dml.cmnh.org/1995Sep/msg00849.html 20 September (1995); 'Sea Serpents, Seals, and Coelacanths', *Fortean Studies*, 7: 75-94 (2001); 'Swan-Necked Seals', *Tetrapod Zoology*, http://darrennaish.blogspot.co.uk/2006/02/swan-necked-seals.html 4 February (2006); 'The Cadborosaurus Wars', *Tetrapod Zoology*, http://blogs.scientific american.com/tetrapod-zoology/the-cadborosaurus-wars/ 16 April (2012); 'As a 'Mysterious Skeleton is Washed Up on a British Beach. . . Do Sea Monsters REALLY Exist?', *Daily Mail Online* (London), http://www.daily mail.co.uk/sciencetech/article-2017101/As-mysterious-skeleton-washed-British-beach—Do-sea-monsters-REALLY-exist.html #comments 21 July (2011); Plesiosaurs and the Repeated Invasion of Freshwater Habitats: Late-Surviving Relicts or Evolutionary Novelties? http://blogs.scientific american. com/tetrapod-zoology/plesiosaurs-invaded-freshwater-habitats/ 9 January (2013); 'Plesiosaur Peril—The Lifestyles and Behaviours of Ancient Marine Reptiles', *Tetrapod Zoology*, http://blogs.scientificamerican.com/tetrapod-zoology/plesiosaur-peril-the-lifestyles-and-behaviours-of-ancient-marine-reptiles/ 3 March (2014).

NICHOLLS, Elizabeth & RUSSELL, Anthony, 'The Plesiosaur Pectoral Girdle: The Case For a Sternum', *Neues Jahrbuch für Geologie und Paläontologie Abhandlungen*, 182: 161-185 (1991).

O'KEEFE, F. Robin & CHIAPPE, Luis M., 'Viviparity and K-Selected Life History in a Mesozoic Marine Plesiosaur (Reptilia, Sauropterygia)', *Science*, 333: 870-873 (2011).

PARK, Penny, 'Beast From the Deep Puzzles Zoologists [Naden Harbour carcase]', *New Scientist*, 137 (23 January): 16 (1993).

PAXTON, Charles G.M., 'A Cumulative Species Description Curve For Large Open Water Marine Animals', *Journal of the Marine Biological Association U.K.*, 78: 1389-1391 (1998).

PAXTON, Charles G.M., *et al.*, 'Cetaceans, Sex and Sea Serpents: An Analysis of the Egede Accounts of a "Most Dreadful Monster" Seen Off the Coast of Greenland in 1734', *Archives of Natural History*, 32(1): 1-9 (2005).

PIMIENTO, Catalina & CLEMENTS, Christopher F., 'When Did *Carcharocles megalodon* Become Extinct? A New Analysis of the Fossil Record', *PLoS One*, http://journals.plos.org/plosone/article?id=10.1371/journal.pone.0111086 22 October (2014).

PIMIENTO, Catalina, *et al.*, 'Geographical Distribution Patterns of *Carcharocles megalodon* Over Time Reveals Clues About Extinction Mechanisms', *Journal of Biogeography*, doi:10.1111/jbi.12754 (2016).

POGAN, Charles, [='POGSQUATCH'], 'Enter the PlesiOturtle', *Enter: The PlesiOturtle*, http://aquaticandaerialanomolyassociation.blogspot.co.uk/2014_04_01_archive.html 8 April 2014; 'What is Echolocating in Lake Champlain?', *Enter: The PlesiOturtle*, http://aquaticandaerialanomoly association.blogspot.co.uk/2014_07_01_archive.html 8 July 2014; 'Are There Beluga Whales in Lake Champlain?', *Enter: The PlesiOturtle*, http://aquaticandaerialanomoly association.blogspot.co.uk/2015/02/are-there-beluga-whales-in-lake.html 24 February 2015; 'Stupendemys, the World's Largest Turtle', *Enter the PlesiOturtle*, http://aquaticandaerial anomolyassociation.blogspot.co.uk/2016/05/stupendemys-worlds-largest-turtle.html 16 May (2016).

RADFORD, Benjamin & NICKELL, Joe, *Lake Monster Mysteries: Investigating the World's Most Elusive Creatures* (University Press of Kentucky: Lexington, 2006).

RAYNOR, Dick, 'The Flipper Pictures Re-examined', *Loch Ness Investigation*, http://www.lochness investigation.com/flipper.html 6 August (2002) (revised May 2009).

RINES, Robert H., *et al.*, *Underwater Search At Loch Ness* (Academy of Applied Science: New York, 1972).

RIORDAN, James, 'Hell's Teeth [re megalodon]', *New Scientist*, 162 (12 June): 32-35 (1999).

SANDERSON, Ivan T., *Follow the Whale* (Bramhall House: New York, 1956).

SCOTT, Peter, 'Why I Believe in the Loch Ness Monster', *Wildlife*, 18 (March): 110-111, 120-121 (1976).

SCOTT, Peter & RINES, Robert, 'Naming the Loch Ness Monster', *Nature*, 258 (11 December): 466-468 (1975).

SERREC, Robert le, 'The Barrier Reef Monster', *Everybody's Magazine*, (31 March): 8-10 (1965).

SHUKER, Karl P.N., 'The Gambian Sea-Serpent', Parts 1 and 2, *The Unknown*, No. 15 (September): 49-53 (1986), and No. 16 (October): 31-36 (1986); 'Gambo—The Beaked Beast of Bungalow Beach', *Fortean Times*, 67 (February-March): 35-37 (1993); '"Bring Me the Head of the Sea Serpent!"', *Strange Magazine*, No. 15 (spring): 12-17 (1995); 'The Ark Down Under [re Australian mystery beasts]', *Fortean Times*, No. 199 (August): 54-55 (2005); 'The Truth Behind the Monster [re Lake Khaiyr monster]', *Fortean Times*, No. 232 (February): 58-59 (2008); 'Contemplating the Con Rit', *ShukerNature*, http://karlshuker.blogspot.co.uk/2014/02/contemplating-con-rit.html 16 February (2014); 'Deeply Dippy Over *Diplocaulus*', *ShukerNature*, http://karlshuker.blogspot.co.uk/2015/10/deeply-dippy-over-diplocaulus.html 12 October (2015); *Here's Nessie! A Monstrous Compendium From Loch Ness* (CFZ Press: Bideford, 2016).

SMITH, Adam S., 'Why Did Elasmosaurids Have Such a Long Neck?', *Plesiosaur Bites*, http://plesiosauria.com/news/index.php/why-did-elasmosaurids-have-such-a-long-neck/ 23 November (2014).

SMITH, Andrew B., *et al.*, 'Sea-Level Change and Rock-Record Bias in the Cretaceous: A Problem For Extinction and Biodiversity Studies', *Paleobiology*, 27 (spring): 241-253 (2001).

STEAD, David G., *Sharks and Rays of Australian Seas* (Angus & Robertson: London, 1964).

SWEENEY, James B., *A Pictorial History of Sea Monsters and Other Dangerous Marine Life* (Nelson-Crown: New York, 1972).

SYLVA, Donald P. de, 'Mystery of the Silver Coelacanth', *Sea Frontiers*, 12: 172-175 (1966).

TAYLOR, Michael A., 'Plesiosaurs—Rigging and Ballasting', *Nature*, 290 (23 April): 628-629 (1981); 'Lifestyle of Plesiosaurs', *Nature*, 319 (16 January): 179 (1986); 'Stomach Stones For Feeding or Buoyancy? The Occurrence and Function of Gastroliths in Marine Tetrapods', *Philosophical Transactions of the Royal Society B*, 341 (29 July): 163-175 (1993).

THAN, Ker, 'Newfound Reptile Swam in Dinosaur Era [re *Umoonasaurus* plesiosaur]', *Live Science*, http://www.livescience.com/4116-newfound-reptile-swam-dinosaur-era.html 7 July (2006).

THEWISSEN, J.G.M., *et al.*, 'Fossil Evidence For the Origin of Aquatic Locomotion in Archaeocete Whales', *Science*, 263 (14 January): 210-212 (1994).

THOMAS, Lars, *Mysteriet om Havuhyrerne* (Gyldendal Boghandel: Copenhagen, 1992).

TSCHERNEZKY, Wladimir, 'Age of *Carcharodon megalodon*?', *Nature*, 184 (24 October): 1331-1332 (1959).

VAVREK, Matthew J., *et al.*, 'Arctic Plesiosaurs From the Lower Cretaceous of Melville Island, Nunavut, Canada', *Cretaceous Research*, 50: 273-281 (2014).

WATSON, Roland, *The Water Horses of Loch Ness* (CreateSpace: London, 2011); 'Is There Enough Food For Nessie?', *Loch Ness Monster*, http://lochnessmystery.blogspot.co.uk/2012/02/is-there-enough-food-for-nessie_12.html 12 February (2012).

WESTRUM, Ron, 'Knowledge About Sea-Serpents', *Sociological Review Monograph*, No. 27: 293-314 (1979).

WHITLEY, Gilbert P., *The Fishes of Australia Part 1: The Sharks, Rays, Devil-Fish, and Other Primitive Fishes of Australia and New Zealand* (Royal Zoological Society of New South Wales: Sydney, 1940).

WHYTE, Constance, *More Than a Legend: The Story of the Loch Ness Monster* (Hamish Hamilton: London, 1957).

WILLIAMS, Gareth, *A Monstrous Commotion: The Mysteries of Loch Ness* (Orion Books: London, 2015).

WITCHELL, Nicholas, *The Loch Ness Story* (3rd edit., Corgi: London, 1989).

WITZKE, Brian J., 'The Age of Dinosaurs in Iowa', *Iowa Geology*, No. 26: 2–7 (2001).

WOODLEY, Michael A., *In the Wake of Bernard Heuvelmans: An Introduction to the History and Future of Sea Serpent Classification* (CFZ Press: Bideford, 2008).

WOODLEY, Michael A., *et al.*, 'How Many Extant Species of Pinniped Remain to be Described?', *Historical Biology*, 20(4): 225-235 (2009).

WOODLEY, Michael A., *et al.*, 'A Baby Sea Serpent No More: Reinterpreting Hagelund's Juvenile "Cadborosaur" Report', *Journal of Scientific Exploration*, 25(3): 497-514 (2011) [also available online at: https://lordgeekington.files.wordpress.com/2012/02/jse-253-woodley.pdf].

YOUNG, Noel, 'Loch Ness Monster: The Unsolved Mysteries of Collisions With Nessie on the Loch', *Scotland Now* [in *Daily Record*], http://www.scotlandnow.dailyrecord.co.uk/lifestyle/loch-ness-monster-unsolved-mysteries-5008548 25 January (2015).

ZAMMIT, Maria, *et al.*, 'Elasmosaur (Reptilia: Sauropterygia) Neck Flexibility: Implications For Feeding Strategies', *Comparative Biochemistry and Physiology, Part A*, 150: 124-130 (2008).

ZARZYNSKI, Joseph W., *Champ—Beyond the Legend* (2nd edit., M-Z Information: Wilton, 1988).

ZIEGLER, Jon, 'Megalodon Sighting?', *Strange Magazine*, no. 23 (2005).

Chapter 4: Lions with Pouches and Horses with Claws

ANON., 'A New Underground Monster', *Nature*, 17 (21 February): 325-326 (1878); 'Underground Monsters', *Nature*, 18 (8 August): 389 (1878).

ANON., 'Mastodons Still Living', *Winnipeg Daily Free Press* (Winnipeg), 28 March (1893).

ANON., 'A Live Mammoth', *Evening Express* (California?), 10 September (1903).

ANON., 'Hunting a Live "Prehistoric Monster" [gazeka]', *Stevens Point* (Wisconsin), 31 August (1910).

ANON., 'American Kills Mystery Beast in E.Africa', *San Diego Union* (San Diego), 9 July (1923).

ANON., 'Een Merkwaardige Vondst ['A Strange Find'—re alleged Bornean mastodont]', *Indische Courant* (Batavia, Java), 31 March (1926).

ANON., 'This Way, Folks, It's Here—A Real Mexican Whatizit', *Morning Oregonian* (Portland), 2 August (1928).

ANON., 'Shot Near Tully—Marsupial Tiger', *Daily Mail* (Brisbane), 12 December (1932).

ANON., 'On Mt. Bellenden-Ker—Hazardous Climb—"Marsupial Tiger" and Mysterious Waterfall', *Richmond River-Herald and Northern Districts Advertiser* (Richmond, New South Wales), 24 November (1936).

ANON., 'One Ton Armadillo Caught', *Science News*, 92 (7 October): 347 (1967).

ANON., 'Digging For Man Finds Megaroos', *New Scientist*, 79 (28 September): 920 (1978).

ANON., 'Hunt For the Monster of the Amazon [mapinguary]', *New Scientist*, 141 (22 January): 9 (1994).

ANON., 'Sabeltigeren Spøger Endnu', *Illustreret Videnskab*, no. 8: 64 (1998); 'Le Félin aux Dents de Sabre', *Science Illustrée*, no. 9 (September): 62 (1998).

ANON., 'Smallest Mammoths Found On Crete', *BBC News*, http://www.bbc.co.uk/news/science-environment-18003093 9 May (2012).

AFSHAR, Ahmed, *et al.*, 'Giraffes at Persepolis', *Archaeology*, 27 (April): 114-117 (1974).

AGUSTI, Jordi & ANTÓN, Mauricio, *Mammoths, Sabertooths, and Hominids: 65 Million Years of Mammalian Evolution in Europe* (Columbia University Press: New York, 2002).

ANDREWS, Charles W., 'An African Chalicothere', *Nature*, 112: 696 (1923).

ARMAN, Samuel D. & PRIDEAUX, Gavin J., 'Behaviour of the Pleistocene Marsupial Lion Deduced From Claw Marks in a Southwestern Australian Cave', *Science Reports*, http://www.nature.com/articles/srep21372 doi:10.1038/srep21372 (2016).

BACHOFEN-ECHT, A., 'Bildliche Darstellung des Riesenhirsches aus Vorgeschichtlicher und Geschichtlicher Zeit', *Zeitschrift für Säugetierkunde*, 12: 81-88 (1937).

BARRETT, Katherine & BARRETT, Robert, *A Yankee in Patagonia: Edward Chace, His Thirty Years There, 1898-1928* (W. Heffer & Sons: Cambridge, 1931).

BECK, Jane C., 'The Giant Beaver: A Prehistoric Memory?', *Ethnohistory*, 19 (spring): 109-122 (1972).

BERGMAN, Sten, 'Observations on the Kamchatkan Bear', *Journal of Mammalogy*, 17: 115-120 (1936).

BOAZ, Noel T. & CIOCHON, Russell L., 'The Scavenging of 'Peking Man'', *Natural History*, 110: 46–52 (2001).

BRACE, Matthew, 'The Stone Menagerie [re Australian fossil mammals unearthed at Riversleigh]', *Geographical*, 75 (July): 17-21 (2003).

BRAIT, Ellen, 'Extinct 'Siberian Unicorn' May Have Lived Alongside Humans, Fossil Suggests', *Guardian*, https://www.theguardian.com/science/2016/mar/29/siberian-unicorn-extinct-humans-fossil-kazakhstan 29 March (2016).

BRANDT, Alexander & LOCKYER, Norman, 'The Elasmotherium', *Nature*, 18 (No. 458): 387-389 (1878).

BROWNE, Malcolm W., 'Legendary Giant Sloth Sought by Scientists in Amazon Rain Forest', *New York Times* (New York), 8 February (1994).

BURNEY, David A. & RAMILISONINA, 'The *Kilopilopitsofy*, *Kidoky*, and *Bokyboky*: Accounts of Strange Animals From Belo-Sur-Mer, Madagascar, and the Megafaunal "Extinction Window"', *American Anthropologist*, 100 (No. 4; December): 957-966 (1998).

BURTON, Adrian, 'The Echidna Enigma', *Frontiers of Ecology and the Environment*, 14 (No. 3; 4 April): 172 (2016).

BURTON, Maurice & BURTON, Robert (Eds), *Purnell's Encyclopaedia of Animal Life*, 6 vols (Purnell: London, 1968-1970).

BUXTON, Cara, 'The 'Gadett' or Brain-Eater', *Journal of the East Africa and Uganda Natural History Society*, 15: 498 (1919).

CASE, E.C., 'The Mastodons of Båråboedaer', *Proceedings of the American Philosophical Society*, 81 (September): 569-572 (1939).

CHATWIN, Bruce, *In Patagonia* (Jonathan Cape: London, 1977).

COLBERT, Edwin H., 'Was the Extinct Giraffe (*Sivatherium*) Known to the Early Sumerians?', *American Anthropologist*, 38: 605-608 (1936); 'The Enigma of *Sivatherium*', *Plateau*, 51(1): 32-33 (1978).

COLEMAN, Loren, 'An Answer From the Pleistocene [re possible *Panthera atrox* survival]', *Fortean Times*, No. 32 (summer): 21-22 (1980); 'Red Elephants of New Guinea', *Cryptomundo*, http://cryptomundo.com/cryptozoo-news/red-elephants/ 4 September (2007).

COURTNEY, Roger, *Africa Calling* (George Harrap: London, 1936).

CUNINGHAME, R.J., 'The Water-Elephant', *Journal of the East Africa and Uganda Natural History Society*, 21: 97-99 (1912).

DARIUS, Jon, 'Last Gape of the Tasmanian Tiger', *Nature*, 307 (2 February): 411 (1984).

DAVENPORT, Elaine, 'Hunters Lured By Texas Ranch Safaris', *Sunday Times* (London), 12 June (1988).

DAVIES, Ella, 'Ancient Bear Had the Strongest Bite', *BBC*, http://www.bbc.co.uk/nature/15559929 4 November (2011).

DAY, David, *The Doomsday Book of Animals* (Ebury: London, 1981).

DELABORDE, Jean, 'The Mylodon of Patagonia', *Animal Life*, No. 20 (April): 32-34 (1964).

DEUSEN, Hobart M. van, 'First New Guinea Record of *Thylacinus*', *Journal of Mammalogy*, 44: 279-280 (1963).

DEVANEY, James, 'The Queensland "Tiger"', *Daily Mercury* (Mackay), 17 June (1929).

DONOVAN, Paul M., 'Buckley's Bunyip', *Journal of Cryptozoology*, 4 (in press) (2016).

DOUGLAS, Athol M., 'The Thylacine: A Case For Current Existence on Mainland Australia', *Cryptozoology*, 9: 13-25 (1990).

DRUELLE, F. & BERILLON, G., 'Bipedal Behaviour in Olive Baboons: Infants Versus Adults in a Captive Environment', *Folia Primatologica*, 84(6): 347-361 (2013).

DUNCAN, F. Martin, 'A Chinese Noah's Ark', *The Field*, 166 (30 November): 1286-1287 (1935).

EARL OF CRANBROOK & PIPER, P.J., 'Borneo Records of Malay Tapir, *Tapirus indicus* Desmarest: A Zooarchaeological and Historical Review', *International Journal of Osteoarchaeology*, 19: 491-507 (2009).

'FABIAN', 'The "Marsupial Tiger"', *Brisbane Courier* (Brisbane), 22 December (1928).

FARIÑA, Richard A. & BLANCO, R. Ernesto, '*Megatherium*, the Stabber', *Proceedings of the Royal Society B*, 263 (22 December): 1725-1729 (1996).

FARIÑA, Richard A., *et al.*, *Megafauna: Giant Beasts of Pleistocene South America* (Indiana University Press: Bloomington, 2013).

FARQUHARSON, R.J., 'The Elephant Pipe', *American Antiquarian*, 2: 67-69 (1879).

FELL, H. Barry, *America B.C.: Ancient Settlers in the New World* (Wildwood House: London, 1982).

FLANNERY, Tim F. & PLANE, Michael D., 'A New Late Pleistocene Diprotodontid (Marsupialia) From Pureni, Southern Highlands Province, Papua New Guinea [*Hulitherium*]', *BMR Journal of Geology and Geophysics, Australasia*, 10: 65-76 (1986).

FORGE, Laurent, 'Un Marsupial Géant Survit-Il en Nouvelle Guinée', *Amazone*, No. 2 (January): 9-11 (1983).

FORTELNY, Joseph G., 'Gibt es einen Australischen Tiger?', *Kosmos*, 63: 292-294 (1967).

GANN, Thomas W.F., *Discoveries and Adventures in Central America* (Duckworth: London, 1928).

GILLETTE, David D. & MADSEN, David B., 'The Short-Faced Bear *Arctodus simus* From the Late Quaternary in the Wasatch Mountains of Central Utah', *Journal of Vertebrate Paleontology*, 12 (March): 107-112 (1992).

GILROY, Rex, 'The Queensland Tiger', *Fortean Times*, No. 62 (April-May): 55-56 (1992); 'Mystery Lions in the Blue Mountains', *Nexus*, (June-July): 25-27, 64 (1992).

GOLOVANOV, Yaroslav, 'Is the Mammoth Really Extinct?', *Sputnik*, (January): 124-127 (1975).

GONZALEZ, Silvia, *et al.*, 'Survival of the Irish Elk Into the Holocene', *Nature*, 405 (15 June): 753-754 (2000).

GOSS, Michael, 'Do Mammoths Survive?', *Fate*, 38 (October): 58-67 (1985).

GOSSE, Philip H., *The Romance of Natural History, Second Series* (James Nisbet: London, 1867).

'GOUGER' [IDRIESS, Ion L.], '[Article re Queensland tiger]', *Bulletin* (Australia), 8 June (1922).

GOUGH, Myles, 'Fearsome Australian 'Lion' Could Climb Trees, Study Says', *BBC News*, http://www.bbc.co.uk/news/world-australia-35557269 15 February (2016).

GRAVES, Frank, 'The Valley Without a Head', [originally published in undetermined periodical during the early 1960s, reprinted in:] *North American BioFortean Review*, 6 (No. 2; October): 45-50 (2004).

GRAYSON, Mike, 'Pleistocene Panthers', *Fortean Times*, No. 36 (winter): 58-59 (1982).

GREENWELL, J. Richard, 'The Muskox Mysteries', *BBC Wildlife*, 11 (June): 37 (1993); 'The Whatsit of Oz', *BBC Wildlife*, 12 (February): 53 (1994).

GREENWELL, J. Richard (Ed.), 'Giant Bear Sought by Soviets', *ISC Newsletter*, 6 (winter): 6-7 (1987).

HARRISSON, Tom, 'The Large Mammals of Borneo', *Sarawak Gazette*, 75: 62–64 (1949); 'The Large Mammals of Borneo', *Malayan Nature Journal*, 4: 70–76 (1949).

HAY, Oliver P., 'An Extinct Camel From Utah', *Science*, 68 (28 September): 299-300 (1928).

HEALY, Tony & CROPPER, Paul, *Out of the Shadows: Mystery Animals of Australia* (Pan Macmillan Australia: Chippendale, 1994).

HELGEN, Kristofer M., *et al.*, 'Twentieth Century Occurrence of the Long-Beaked Echidna *Zaglossus bruijnii* in the Kimberley Region of Australia', *ZooKeys*, 255 (28 December): 103-132 (2012).

HEMMER, Helmut, 'Socialisation by Intelligence: Social Behaviour in Carnivores as a Function of Relative Brain Size and Environment', *Carnivore*, 1: 102–105 (1978).

HEPTNER, W.G., 'Über das Java-Nashorn auf Neu-Guinea', *Zeitschrift für Säugetierkunde*, 25: 128-129 (1960).

HERRIDGE, Victoria L. & LISTER, Adrian M., 'Extreme Insular Dwarfism Evolved in a Mammoth', *Proceedings of the Royal Society B*, http://rspb.royalsocietypublishing.org/content/early/2012/05/04/rspb.2012.0671 9 May (2012).

HESKETH-PRICHARD, Hesketh V., '*Through the Heart of Patagonia*' (D. Appleton: New York, 1902).

HEUVELMANS, Bernard, *Les Félins Encore Inconnus d'Afrique* (L'Oeil du Sphinx: Paris, 2007); *Les Ours Insolites d'Afrique* (L'Oeil du Sphinx: Paris, 2015).

HINSON, David H., 'Cryptoletter', *ISC Newsletter*, 10 (summer): 11 (1991).

HOBLEY, C.W., 'On Some Unidentified Beasts', *Journal of the East Africa and Uganda Natural History Society*, 3(6): 48-52 (1912); 'Unidentified Beasts in East Africa', *Journal of the East Africa and Uganda Natural History Society*, No. 7: 85-86 (1913).

HOCKING, Peter J., 'Large Peruvian Mammals Unknown to Zoology', *Cryptozoology*, 11: 38-50 (1992).

HOLLOWAY, Marguerite, 'Beasts in the Mist [mapinguary]', *Discover*, 20 (September), http://discovermagazine.com/1999/sep/beastsinthemist1682/ (1999).

HONORÉ, Pierre, *In Quest of the White God* (Hutchinson: London, 1963).

HRALA, Josh, 'A Fossilised Skull Has Revealed When the Last 'Siberian Unicorn' Lived on Earth', *Science Alert*, http://www.sciencealert.com/a-fossilised-skull-has-revealed-when-the-last-siberian-unicorn-lived-on-earth 27 March (2016).

IMMEL, A., *et al.*, 'Mitochondrial Genomes of Giant Deers [sic] Suggest Their Late Survival in Central Europe'. *Science Reports*, 5 (8 June): doi: 10.1038/srep10853 (2015).

JACOBY, Charlie, 'A Century On and the Adventure Continues in the Hunt For the Giant Sloth', *Daily Express* (London), 20 November (2000); 'Giant Sloth—A Century On and the Hunt is Still Continuing For This Mystical Creature', *Express* (London), 8 February (2001); 'Ground Sloth', *Charlie Jacoby*, http://www.charliejacoby.com/giantsloth.htm (2006).

JANIS, Christine, 'Fossil Ungulate Mammals Depicted on Archaeological Artifacts', *Cryptozoology*, 6: 8-23 (1987); 'A Reevaluation of Some Cryptozoological Animals', *Cryptozoology*, 6: 115-118 (1987); 'Hurrah For Hyraces!', *Cryptozoology*, 7: 104-106 (1988); '*Sivatherium* Defended', *Cryptozoology*, 9: 111-115 (1990).

JANIS, Christine M., *et al.*, 'Locomotion in Extinct Giant Kangaroos: Were Sthenurines Hop-Less Monsters?', *PLoS One*, 9(10), http://journals.plos.org/plosone/article?id=10.1371/journal.pone.0109888 15 October (2014).

JENKIN, Robyn, *New Zealand Mysteries* (A.H. & A.W. Reed: Wellington, 1970).

JENSEN, A.S., 'The Sacred Animal of the God Set', *Biologiske Meddelelser*, 11(5) 1-19 (1934).

JOHNSON, Chris, *Australia's Mammal Extinctions: A 50 000 Year History* (Cambridge University Press: Port Melbourne, 2006).

JOHNSON, Ludwell H., 'Men and Elephants in America', *Scientific Monthly*, 75: 215-221 (1952).

JORDAN, John A., 'Unknown Animals of the African Wilds', *Wide World Magazine*, 39 (November): 187-197 (1917); 'The Brontosaurus—Hunter's Story of Tusked and Scaly Beast [re dingonek]', *Daily Mail* (London), 16 December (1919); *Elephants and Ivory* (Rinehart: London, 1956); *The Elephant Stone* (Nicholas Kaye: London, 1959).

JOYE, Eric, 'Le Mourou-ngou se Porte Bien!!!', *Cryptozoologia*, No. 6 (September): 1-5 (1994).

KEELING, Clinton H., 'The British Nandi Bear?', *Animals and Men*, No. 6 (July): 32-33 (1995).

KNIPPENBERG, Jim, 'Tristater Searches For Giant Sloth', *Cincinnati Enquirer* (Cincinnati), 31 May (2001).

KURTÉN, Björn, *Pleistocene Mammals of Europe* (Weidenfeld and Nicolson: London, 1968); *The Ice Age* (Rupert Hart-Davis; London), 1972).

KURTÉN, Björn & ANDERSON, Elaine, *Pleistocene Mammals of North America* (Columbia University Press: New York, 1980).

LANKFORD, George E., 'Pleistocene Animals in Folk Memory', *Journal of American Folklore*, 93: 293-304 (1980).

LEAKEY, Louis S.B., 'Does the Chalicothere—Contemporary of the Okapi—Still Survive?', *Illustrated London News*, 97 (2 November): 730-733, 750 (1935).

LEY, Willy, 'Is There a Nandi Bear?', *Fate*, 16 (July): 42-50 (1963).

LI, Ji, *et al.*, 'The Latest Straight-Tusked Elephants (*Palaeoloxodon*)? "Wild Elephants" Lived 3000 Years Ago in North China', *Quaternary International*, 281 (19 December) 84-88 (2012).

LISTER, Adrian M., 'The Evolution of the Giant Deer, *Megaloceros giganteus* (Blumenbach)', *Zoological Journal of the Linnean Society*, 112: 65-100 (1994).

LIU, Alexander G.S.C., *et al.*, 'Stable Isotope Evidence For an Amphibious Phase in Early

Proboscidean Evolution', *Proceedings of the National Academy of Sciences*, 105 (No. 15; 14 April): 5786-5791 (2008).

LONG, John, *et al.*, *Prehistoric Mammals of Australia and New Guinea: One Hundred Million Years of Evolution* (University of New South Wales Press Ltd: Sydney, 2002).

LUMHOLTZ, Carl, *Among Cannibals. . .* (John Murray: London, 1889); *Through Central Borneo: An Account of Two Years' Travel in the Land of Head-Hunters Between the Years 1913 and 1917* (2 vols) (Charles Scribner's Sons: New York, 1920).

LYDEKKER, Richard, *Catalogue of the Ungulata Mammals in the British Museum (Natural History), Vol. IV* (BMNH: London, 1915).

McGEEHAN, John, 'Striped Marsupial Cat—Description of Wild Animal Seen on Atherton Tableland', *North Queensland Naturalist*, 6: 3-4 (1938).

McISAAC, Gerald, *Bird From Hell and Other Mega Fauna* (Trafford Publ.: Bloomington, 2010).

MAHAUDEN, Charles, *Kisongokimo: Chasse et Magie Chez les Balubas* (Flammarion: Paris, 1965).

MAKEIG, Peter, 'Is There a Queensland Marsupial Tiger?', *North Queensland Naturalist*, 37: 6-8 (1970).

MANGIACOPRA, Gary A. & SMITH, Dwight G., 'Canada's Headless Valley Revisited: Troglodytes, Bigfoot, Mystery Bears, and Dire Wolves', *North American BioFortean Review*, 8 (January): 8-14 (2006).

MARTIN, Larry D. & GILBERT, B. Miles, 'An American Lion, *Panthera atrox*, From Natural Trap Cave, North Central Wyoming', *Contributions to Geology, University of Wyoming*, 16(2): 95-101 (1978).

MARTIN, Paul S., *Twilight of the Mammoths: Ice Age Extinctions and the Rewilding of America* (University of California Press: Berkeley/Los Angeles, 2005).

MARTIN, Paul S. & KLEIN, Richard G. (Eds.), *Quaternary Extinctions: A Prehistoric Revolution* (University of Arizona Press: Tucson, 1984).

MARTINOLLI, G., 'Huesos Anormales de Llama y de Cóndor Exhumados en el Pucará de Tilcara', *Revista "Physis"*, 3: 69-74 (1917).

MATTHIESSEN, Peter, *The Cloud Forest* (Pyramid Books: New York, 1966).

MENZIES, Gavin, *1421: The Year China Discovered the World* (Bantam Press: London, 2002).

MENZIES, James I., 'Reflections on the Ambun Stones', *Science in New Guinea*, 13: 170-173 (1987).

MEYER, Adolf B., 'The Rhinoceros in New Guinea', *Nature*, 11 (4 February): 268 (1875).

MJÖBERG, Eric, *Forest Life and Adventures in the Malay Archipelago* (G. Allen & Unwin: London, 1930).

MONCKTON, Charles A.W., *Some Experiences of a New Guinea Resident Magistrate* (John Lane/The Bodley Head: London, 1920); *Last Days in New Guinea* (John Lane/The Bodley Head: London, 1922); *New Guinea Recollections* (John Lane/The Bodley Head: London, 1934).

MONTGOMERY, G. Gene (Ed.), *The Evolution and Ecology of Armadillos, Sloths, and Vermilinguas* (Smithsonian Institution Press: Washington, 1985).

MORENO, Francisco P. & WOODWARD, A. Smith, 'On a Portion of Mammalian Skin, Named *Neomylodon listai*, From a Cavern Near Consuelo Cove, Last Hope Inlet, Patagonia, With a Description of the Specimen', *Proceedings of the Zoological Society of London*, (21 February): 144-156 (1899).

MORESBY, John, *Discoveries & Surveys in New Guinea and the D'Entrecasteaux Islands: A Cruise in Polynesia & Visits to the Pearl-Shelling Stations in Torres Straits of H.M.S. Basilisk* (John Murray: London, 1876).

MORGAN, Greg, 'Gone? Maybe Not [re thylacines in New Guinea]', *Garuda*, 30-34 (1993?).

MÜLLER, Fritz, 'Der Minhocao', *Zoologische Garten*, 18: 298-302 (1877).

MURRAY, Peter & CHALOUPKA, George, 'The Dreamtime Animals: Extinct Megafauna in Arnhem Land Rock Art', *Archaeology in Oceania*, 19: 105-116 (1984).

MUSEUM OF THE WEIRD, 'The Long Lost "Minnesota Iceman" Resurfaces. . .in Austin, Texas!', *Museum of the Weird*, http://www.museumoftheweird.com/news/2013/06/26/the-long-lost-minnesota-iceman-resurfaces-in-austin-texas/ 26 June (2013).

NAISH, Darren, 'Identifying 'Jaws', the Margaret River Mammal Carcase', *Journal of Cryptozoology*, 1: 45-55 (2012).

NAISH, Darren, *et al.*, "Mystery Big Cats' in the Peruvian Amazon: Morphometrics Solve a Cryptozoological Mystery', *PeerJ*, http://www.ncbi.nlm.nih.gov/pmc/articles/PMC3961146/ 6 March (2014).

NELSON, Michael E. & MADSEN, James H., 'The Hay-Romer Camel Debate: Fifty Years Later', *Contributions to Geology, University of Wyoming*, 18 (No. 1; December): 47-50 (1979).

NEWMAN, Edward, 'The Mammoth Still in the Land of the Living', *The Zoologist* (Series 2), 8: 3731-3733 (1873).

OPIT, Gary, 'The Bunyip Mystery', *Nexus*, 9 (No. 1; December-February): 55-62 (2001-2002).

OREN, David C., 'Did Ground Sloths Survive in Recent Times in the Amazon Region?', *Goeldiana Zoologia*, No. 19 (20 August): 1-11 (1993).

PEIGNÉ, Stéphane, *et al.*, 'A New Machairodontine (Carnivora, Felidae) From the Late Miocene Hominid Locality of TM 266, Toros-Menalla, Chad', *Comptes Rendus Palevol*, 4 (No. 3; April): 243-253 (2005).

PEZZIA ASSERETO, Alejandro, *Ica y el Perú Precolombino: Vol. 1—Arqueología de la Provincia de Ica* (Editora Ojeda: Ica, 1968).

PFEIFFER, Pierre, *Bivouacs à Borneo* (Flammarion: Paris, 1963).

PICKFORD, Martin, 'Another African Chalicothere', *Nature*, 253 (10 January): 85 (1975).

PITMAN, Charles R.S., *A Game Warden Among His Charges* (James Nisbet: London, 1931); *A Game Warden Takes Stock* (James Nisbet: London, 1942).

POCOCK, Reginald I., 'The Story of the Nandi Bear', *Natural History*, 2 (no. 13; January): 162-169 (1930).

PRINGLE, Heather, 'Bulldog Bears: Lords of Ice-Age Canada', *Equinox*, 5 (March-April): 11 (1988).

QUESADA, Carlos Nores & VONLETTOW-VORBECK, Corina Liesau, 'La Zoologia Historica Como Complemento de la Arqueozoologia. El Caso del Zebro', *Archaeofauna*, 1: 61-71 (1992).

REESE, David S., 'Paleocryptozoology and Archaeology: A Sivathere No Longer', *Cryptozoology*, 9: 100-107 (1990).

REVENKO, Igor A., 'Brown Bear (*Ursus arctos piscator*) Reaction to Humans on Kamchatka', *International Conference on Bear Research and Management*, 9(1): 107-108 (1994).

RICHARDSON, H.D., *Facts Concerning the Natural History . . . of the Gigantic Irish Deer* (Orr: Dublin, 1846).

ROHTER, Larry, 'A Huge Amazon Monster is Only a Myth. Or is It?', *New York Times* (New York), http://www.nytimes.com/2007/07/08/world/americas/08amazon.html?_r=0 8 July (2007).

ROMER, Alfred S., 'A "Fossil" Camel Recently Living in Utah', *Science*, 68 (6 July): 19-20 (1928); 'A Fresh Skull of an Extinct American Camel', *Journal of Geology*, 37: 261-267 (1929).

ROSE, M.D., 'Bipedal Behavior of Olive Baboons (*Papio anubis*) and Its Relevance to an Understanding of the Evolution of Human Bipedalism', *American Journal of Physical Anthropology*, 44 (No. 2; March): 247-261 (1976).

ROSEN, Baruch, 'Mammoths in Ancient Egypt?', *Nature*, 369 (2 June): 364 (1994).

RUSCONI, Carlos, 'Un Nuevo Caso de Polidactilia en un Guanaco Hallado en un Túmulo Indígena de Santiago del Estero (Argentina)', *Revista Chilena de Historia Natural*, 34 (1 August): 224-227 (1930).

RUTTER, Owen, *The Pagans of North Borneo* (Hutchinson: London, 1929).

SAINT HILAIRE, Auguste de, 'On the Minhocao of the Goyanes', *American Journal of Science* (Series 2), 4: 130-131 (1847).

SALUSBURY, Matt, *Pygmy Elephants: On the Track of the World's Largest Dwarfs* (CFZ Press: Bideford, 2013).

SANDERSON, Ivan T., 'The Riddle of the Mammoths', *Saturday Evening Post*, 219 (No. 23): 26-27, 142, 144, 147 (1946); 'Giant "Armadillo"?', *Pursuit*, 1 (June): 10-11 (1968); 'Ends Glyptodont', *Pursuit*, 1 (September): 3-4 (1968); 'The Dire Wolf', *Pursuit*, 7 (October): 91-94 (1974).

SAVAGE, Robert J.G. & LONG, Michael R., *Mammal Evolution* (British Museum—Natural History: London, 1986).

SCOTT, Walter J., ' Letter Addressed to the Secretary, Respecting the Supposed "Native Tiger" of Queensland', *Proceedings of the Zoological Society of London*, (5 March): 355 (1872); 'Second Letter on the Existence of a "Native Tiger" in Queensland', *Proceedings of the Zoological Society of London*, (5 November): 796 (1872).

SELVEY, L.A., *et al.*, 'Infection of Humans and Horses by a Newly Described Morbillivirus', *Medical Journal of Australia*, 162 (no. 12; 19 June): 642-645 (1995).

SHERIDAN, Brinsley G., 'Notice of the Existence in Queensland of an Undescribed Species of Mammal', *Proceedings of the Zoological Society of London*, (7 November): 629-630 (1871).

SHIELDS, Michael J., 'Hope For the Glyptodont?', *Fate*, 21 (May): 123 (1968).

SHPANSKY, Andrei V., *et al.*, 'The Quaternary Mammals From Kozhamzhar Locality (Pavlodar Region, Kazakhstan)', *American Journal of Applied Sciences*, 13 (No. 2; March): 189-199 (2016).

SHUKER, Karl P.N., *Mystery Cats of the World: From Blue Tigers to Exmoor Beasts* (Robert Hale: London, 1989); 'Hoofed Mystery Animals: A Collection of Crypto-Ungulates', Parts 1-III, *Strange Magazine*, No. 9 (spring-summer): 19-21, 56-57 (1992), No. 10 (fall-winter): 25-27, 48-51 (1992), and No. 11 (spring-summer): 25-27, 48-50 (1993); 'Zebro—An Equine Mystery From Iberia', *Flying Snake*, 1 (no. 1; April): 46-47 (2011); *Cats of Magic, Mythology, and Mystery: A Feline Phantasmagoria* (CFZ Press: Bideford, 2012); 'The New Guinea Thylacine—Crying Wolf in Irian Jaya?', *ShukerNature*, http://www.karlshuker.blogspot.co.uk/2013/05/the-new-guinea-thylacine-crying-wolf-in.html 8 May (2013); 'Giant Black Unicorns, A Chinese Quasi-Rhino Figurine', and the Enigma of *Elasmotherium*', *ShukerNature*, http://karlshuker.blogspot.co.uk/2014/07/giant-black-unicorns-chinese-quasi.html 9 July (2014); 'Dung-Heaps, Devil-Pigs, and Monckton's Gazeka', *ShukerNature*, http://karlshuker.blogspot.co.uk/2014/08/dung-heaps-devil-pigs-and-moncktons.html 22 August (2014); 'Caterpillar Bears, Bulldog Bears, and God Bears—Ursine Cryptids of Kamchatka', *ShukerNature*, 7 September (2014);

'A Giant Hyrax and China's Mystery Beasts of Bronze', *ShukerNature*, http://karlshuker.blogspot.co.uk/2015/05/a-giant-hyrax-and-chinas-mystery-beasts.html 8 May (2015); 'Marsupial Sabre-Tooths, Queensland Tigers', Blue Mountain Lions, and a Most Elusive Crypto-Cutting', *ShukerNature*, http://karlshuker.blogspot.co.uk/2015/06/marsupial-sabre-tooths-queensland.html 6 June (2015); 'The Last of the Irish Elks?—Investigating Some *Megaloceros* Mysteries', *ShukerNature*, http://karlshuker.blogspot.co.uk/2015/07/the-last-of-irish-elks-investigating.html 6 July (2015); 'The Yukon Beaver Eater, and Ground Sloths in New Zealand?', *ShukerNature*, http://

www.karlshuker.blogspot.co.uk/2016/03/the-yukon-beaver-eater-and-ground.html 8 March (2016); 'Fire Beasts, Lightning Beasts, and the Elephant of Tiahuanaco: Investigating a Crypto-Archaeological Anomaly', In: DOWNES, Jonathan (Ed.), *CFZ Yearbook 2016* (CFZ Press: Bideford, 2016): 29-35.

SISINYAK, Mancy, 'The Biggest Bear . . . Ever', *Alaska Fish & Wildlife News*, http://www.adfg.alaska.gov/index.cfm?adfg=wildlifenews.view_article&articles_id=232 August (2006).

SMITH, G. Elliot, 'Pre-Columbian Representations of the Elephant in America', *Nature*, 96 (25 November): 340-341 (1915), (16 December): 425 (1915), and 96 (27 January): 593-595 (1916).

SMITH, Malcolm, 'Thylacines in Indonesian New Guinea?', *Malcolm's Musings*, http://malcolmscryptids.blogspot.co.uk/2011/10/thylacines-in-indonesian-new-guinea.html 18 October (2011); 'The Queensland Tiger: Further Evidence on the 1871 Footprint', *Journal of Cryptozoology*, 1 (November): 19-24 (2012); 'Thylacines in Indonesian New Guinea—Further Evidence', *Malcolm's Musings*, http://malcolmscryptids.blogspot.co.uk/2013/10/thylacines-in-indonesian-new-guinea.html 3 October (2013); 'The Great North Queensland Tiger Hunt of 1923', *Malcolm's Musings*, http://malcolmscryptids.blogspot.co.uk/2013/11/the-great-north-queensland-tiger-hunt.html 8 November (2013).

SOMEREN, G.R. Cunningham van, 'The Nandi Bear', *East Africa Natural History Society Bulletin*, (September-October): 91-93 (1981).

SOUEF, Albert S. Le & BURRELL, Harry, *The Wild Animals of Australasia. . .* (George Harrap: London, 1926).

SPASSOV, Nikolai, 'The Musk-Ox in Eurasia: Extinct at the Pleistocene-Holocene Boundary or Survivor to Historical Times?', *Cryptozoology*, 10: 4-15 (1991).

SPILLMANN, F., 'Das Letzte Mastodon von Südamerika', *Natur und Museum*, 2: 119-123 (1929).

STELFOX, A.W., 'The Problem of the Irish Elk', *Irish Naturalists' Journal*, 5 (No. 4; July): 74-76 (1934); 'Did the Giant Deer Survive the Last Glaciation in Ireland?', *Irish Naturalists' Journal*, 7 (No. 3; September): 70-71 (1938).

STOW, George W. & BLEEK, Dorothea, *Rock-Paintings in South Africa* (Methuen: London, 1930).

STRUM, S.C., *Almost Human: A Journey Into the World of Baboons* (Elm Tree: London, 1987).

STUART, A.J., et al., 'Pleistocene to Holocene Extinction Dynamics in Giant Deer and Woolly Mammoth', *Nature*, 431 (7 October): 684-689 (2004).

SUTCLIFFE, Anthony, *On the Track of Ice Age Mammals* (British Museum—Natural History: London, 1985).

SWITEK, Brian, 'Bronze Art Sparks Debate Over the Extinction of the Straight-Tusked Elephant', *Laelaps*, http://phenomena.nationalgeographic.com/2012/12/27/bronze-art-sparks-debate-over-the-extinction-of-the-straight-tusked-elephant/ 27 December (2012).

TAKESTANI, Reza G. [=TUCK, Robert G.], 'The Mystery Beast of Persepolis', *Tehran Journal*, 9 (23 May): 9 (1978).

TERRY, Michael, *War of the Warramullas* (Rigby: Adelaide, 1974).

THOME, Wolfgang H., 'Wildlife and Art—Guy Combes Combines the Two Par Excellence', *ATC News*, https://wolfganghthome.wordpress.com/2015/08/11/wildlife-and-art-guy-combes-combines-the-two-par-excellence/ 11 August (2015).

TUCK, Robert G. & VALDEZ, Raul, 'Persepolis: Nilghai—Not Okapi', *Cryptozoology*, 8: 146-149 (1989).

TURNER, Alan & ANTÓN, Mauricio, *The Big Cats and Their Fossil Relatives* (Columbia University Press: New York, 1997).

VALDEZ, Raul, 'La Véritable Identité de l'Okapi, Représenté à Persépolis (Iran)', *Zoo*, 44 (No. 1; July): 31 (1978).

VALDEZ, Raul & TUCK, Robert G., 'On the Identification of the Animals Accompanying the "Ethiopian" Delegation in the Bas-Reliefs of the Apadana at Persepolis', *Iran*, 17: 156-157, 170 (1980).

VARTANYAN, Sergey L., *et al.*, 'Holocene Dwarf Mammoths From Wrangel Island in the Siberian Arctic', *Nature*, 362 (25 March): 337-340 (1993).

WALKER, Alfred O., 'The Rhinoceros in New Guinea', *Nature*, 11 (28 January): 248 (1875), and (4 February): 268 (1875).

WALLACE, David R., *Beasts of Eden: Walking Whales, Dawn Horses, and Other Enigmas of Mammal Evolution* (University of California Press: Berkeley/Los Angeles, 2004).

WANG, Sebastian H.F., 'Revisiting the Mystery of the "Beaver Eater"—The Possible Survival of North American Ground Sloth in Yukon', *British Columbia Scientific Cryptozoology Club Newsletter*, No. 60 (winter): 7-9 (2006).

WENDT, Herbert, *Out of Noah's Ark: The Story of Man's Discovery of the Animal Kingdom* (Weidenfeld & Nicolson: London, 1956).

WEST, Peyton, 'The Lion's Mane', *American Scientist*, http://www.americanscientist.org/issues/pub/2005/3/the-lions-mane/1 May-June (2005).

WESTFALL, Scottie, 'Why Bigfoot Cannot Be Gigantopithecus', *Natural History*, https://retrieverman.net/2012/01/18/why-bigfoot-cannot-be-gigantopithecus/ 18 January (2012).

WETZEL, Ralph M., 'The Chacoan Peccary *Catagonus wagneri* (Rusconi)', *Bulletin of Carnegie Museum of Natural History*, no. 3: 1-36 (1977); 'The Hidden Chacoan Peccary', *Carnegie Magazine*, 55: 24-32 (1981).

WETZEL, Ralph M., *et al.*, '*Catagonus*, an "Extinct" Peccary, Alive in Paraguay', *Science*, 189 (1 August): 379-81 (1975).

WIGNELL, Edel (Ed.), *A Boggle of Bunyips* (Hodder & Stoughton: Sydney, 1981).

WROE, Stephen, *et al.*, 'Bite Club: Comparative Bite Force in Big Biting Mammals and the Prediction of Predatory Behaviour in Fossil Taxa', *Proceedings of the Royal Society B*, http://rspb.royalsocietypublishing.org/content/272/1563/619 DOI: 10.1098/rspb.2004.2986 22 March (2005).

ZIMMER, Carl, *At the Water's Edge: Fish With Fingers, Whales With Legs, and How Life Came Ashore but Then Went Back to Sea* (Touchstone: New York, 1998).

INDEX OF ANIMAL NAMES

AUTHOR BIOGRAPHY

Born and still living in the West Midlands, England, Dr Karl P.N. Shuker graduated from the University of Leeds with a Bachelor of Science (Honours) degree in pure zoology, and from the University of Birmingham with a Doctor of Philosophy degree in zoology and comparative physiology. He now works full-time as a freelance zoological consultant to the media, and as a prolific published writer.

Dr Shuker is currently the author of 25 books and hundreds of articles, principally on animal-related subjects, with an especial interest in cryptozoology and animal mythology, on which he is an internationally-recognised authority, but also including two volumes of poetry. In addition, he has acted as consultant for several major multi-contributor volumes as well as for the world-renowned *Guinness Book of Records/Guinness World Records* (he is currently its Senior Consultant for its Life Sciences section), and he has compiled questions for the BBC's long-running cerebral quiz *Mastermind*. He is also the editor of the *Journal of Cryptozoology*, the world's only peer-reviewed scientific journal dedicated to mystery animals.

Dr Shuker has travelled the world in the course of his researches and writings, and has appeared regularly on television and radio. Aside from work, his diverse range of interests include motorbikes, travel, world mythology, quizzes, poetry, philately, the life and career of James Dean, collecting masquerade and carnival masks, musical theatre, and the history of animation.

He is a Scientific Fellow of the prestigious Zoological Society of London, a Fellow of the Royal Entomological Society, and a member of several other wildlife-related organisations, he is Cryptozoology Consultant to the Centre for Fortean Zoology, and is also a Member of the Society of Authors.

Dr Shuker's personal website can be accessed at http://www.karlshuker.com and he has an extremely popular anomalous wildlife/animal mythology blog, *ShukerNature* (at: http://www.karlshuker.blogspot.com), a separate blog, *Star Steeds*, devoted to his poetry (at: http://www.starsteeds.blogspot.com), and a third blog, *The Eclectarium of Doctor Shuker*, devoted to his diverse interests outside wildlife and poetry (at: http://www.eclectariumshuker.blogspot.com).

There is also an entry for Dr Shuker in the online encyclopedia Wikipedia (at: https://en.wikipedia.org/wiki/Karl_Shuker), and a fan page on Facebook.

Author Bibliography

Mystery Cats of the World: From Blue Tigers To Exmoor Beasts (Robert Hale: London, 1989)

Extraordinary Animals Worldwide (Robert Hale: London, 1991)

The Lost Ark: New and Rediscovered Animals of the 20th Century (HarperCollins: London, 1993)

Dragons: A Natural History (Aurum: London/Simon & Schuster: New York, 1995; republished Taschen: Cologne, 2006)

In Search of Prehistoric Survivors: Do Giant 'Extinct' Creatures Still Exist? (Blandford: London, 1995)

The Unexplained: An Illustrated Guide to the World's Natural and Paranormal Mysteries (Carlton: London/JG Press: North Dighton, 1996; republished Carlton: London, 2002)

From Flying Toads To Snakes With Wings: From the Pages of FATE Magazine (Llewellyn: St Paul, 1997; republished Bounty: London, 2005)

Mysteries of Planet Earth: An Encyclopedia of the Inexplicable (Carlton: London, 1999)

The Hidden Powers of Animals: Uncovering the Secrets of Nature (Reader's Digest: Pleasantville/Marshall Editions: London, 2001)

The New Zoo: New and Rediscovered Animals of the Twentieth Century [fully-updated, greatly-expanded, new edition of *The Lost Ark*] (House of Stratus Ltd: Thirsk, UK/House of Stratus Inc: Poughkeepsie, USA, 2002)

The Beasts That Hide From Man: Seeking the World's Last Undiscovered Animals (Paraview: New York, 2003)

Extraordinary Animals Revisited: From Singing Dogs To Serpent Kings (CFZ Press: Bideford, 2007)

Dr Shuker's Casebook: In Pursuit of Marvels and Mysteries (CFZ Press: Bideford, 2008)

Dinosaurs and Other Prehistoric Animals on Stamps: A Worldwide Catalogue (CFZ Press: Bideford, 2008).

Star Steeds and Other Dreams: The Collected Poems (CFZ Press: Bideford, 2009).

Karl Shuker's Alien Zoo: From the Pages of Fortean Times (CFZ Press: Bideford, 2010).

The Encyclopaedia of New and Rediscovered Animals: From The Lost Ark to The New Zoo—and Beyond [fully-updated, greatly-expanded, third edition of *The Lost Ark*] (Coachwhip Publications: Landisville, 2012).

Cats of Magic, Mythology, and Mystery: A Feline Phantasmagoria (CFZ Press: Bideford, 2012).

Mirabilis: A Carnival of Cryptozoology and Unnatural History (Anomalist Books: San Antonio, 2013).

Dragons in Zoology, Cryptozoology, and Culture (Coachwhip Publications: Greenville, 2013).

The Menagerie of Marvels: A Third Compendium of Extraordinary Animals (CFZ Press: Bideford, 2014).

A Manifestation of Monsters: Examining the (Un)Usual Suspects (Anomalist Books: San Antonio, 2015).

More Star Steeds and Other Dreams: The Collected Poems—New, Expanded Edition (Fortean Words: Bideford, 2015).

Here's Nessie! A Monstrous Compendium From Loch Ness (CFZ Press: Bideford, 2016).

Still In Search Of Prehistoric Survivors—The Beasts That Time Forgot? (Coachwhip Publications: Greenville, 2016).

Consultant and also Contributor

Man and Beast (Reader's Digest: Pleasantville, New York, 1993)

Secrets of the Natural World (Reader's Digest: Pleasantville, New York, 1993)

Almanac of the Uncanny (Reader's Digest: Surry Hills, Australia, 1995)

The Guinness Book of Records/Guinness World Records 1998-present day (Guinness: London, 1997-present day)

Consultant

Monsters (Lorenz: London, 2001)

Contributor

Of Monsters and Miracles CD-ROM (Croydon Museum/Interactive Designs: Oxton, 1995)

Fortean Times Weird Year 1996 (John Brown Publishing: London, 1996)

Mysteries of the Deep (Llewellyn: St Paul, 1998)

Guinness Amazing Future (Guinness: London, 1999)

The Earth (Channel 4 Books: London, 2000)

Mysteries and Monsters of the Sea (Gramercy: New York, 2001)

Chambers Dictionary of the Unexplained (Chambers: Edinburgh, 2007)

Chambers Myths and Mysteries (Chambers: Edinburgh, 2008)

The Fortean Times Paranormal Handbook (Dennis Publishing: London, 2009).

Folk Horror Revival: Field Studies (Wyrd Harvest Press/Lulu, 2015).

Numerous contributions to the annual *CFZ Yearbook* series of volumes.

Editor

Journal of Cryptozoology (the world's only peer-reviewed scientific journal dedicated to mystery animals, published annually by CFZ Press).

COACHWHIP PUBLICATIONS
COACHWHIPBOOKS.COM

The Encyclopaedia of New and Rediscovered Animals
Dr. Karl P. N. Shuker

ISBN 1616461306

COACHWHIP PUBLICATIONS
CoachwhipBooks.com

Dragons in Zoology, Cryptozoology, and Culture
Dr. Karl P. N. Shuker

ISBN 1616462159

STRANGE CREATURES
SELDOM SEEN
GIANT BEAVERS, SASQUATCH, MANIPOGOS,
AND OTHER MYSTERY ANIMALS
IN MANITOBA AND BEYOND
JOHN WARMS

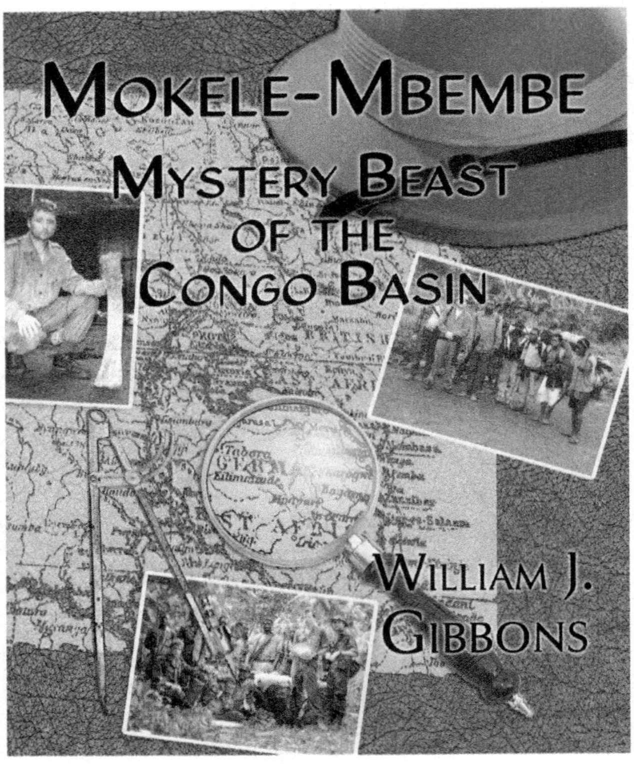
MOKELE-MBEMBE
MYSTERY BEAST
OF THE
CONGO BASIN
WILLIAM J.
GIBBONS

The
Historical
Bigfoot
RANCHERS BEGIN
'WILD MAN' HUNT
ALONG ROCKIES
Prospectors Say
Wild Man Rules
Arctic Kingdom
TERRIBLE SASQUATCH
ABROAD IN THE LAND
Chad Arment
Cavemen Roam the Rockies?

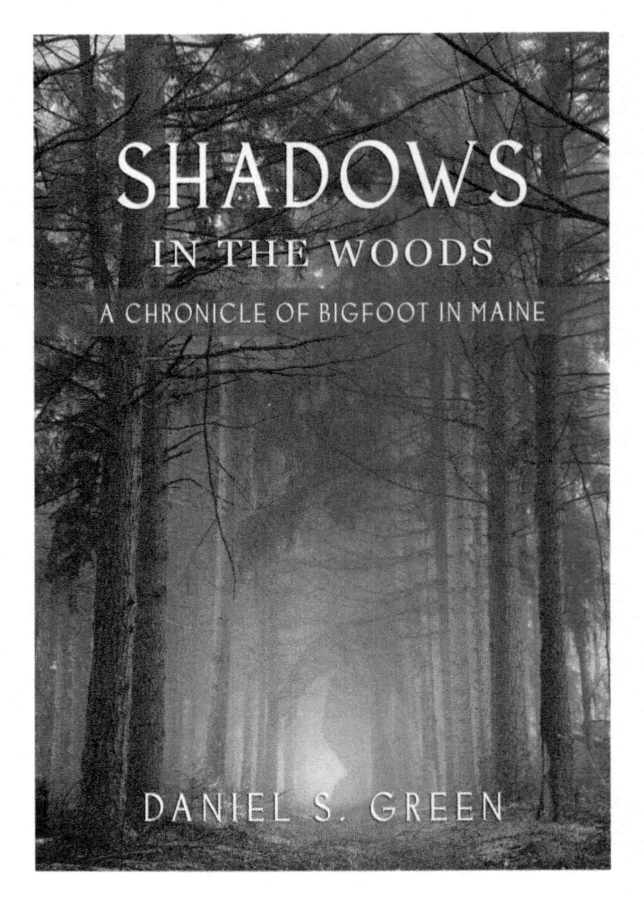
SHADOWS
IN THE WOODS
A CHRONICLE OF BIGFOOT IN MAINE
DANIEL S. GREEN

VARMINTS
CHAD ARMENT
MYSTERY CARNIVORES
OF NORTH AMERICA

FLESH FALLS
& BLOOD RAINS
JOHN HAIRR

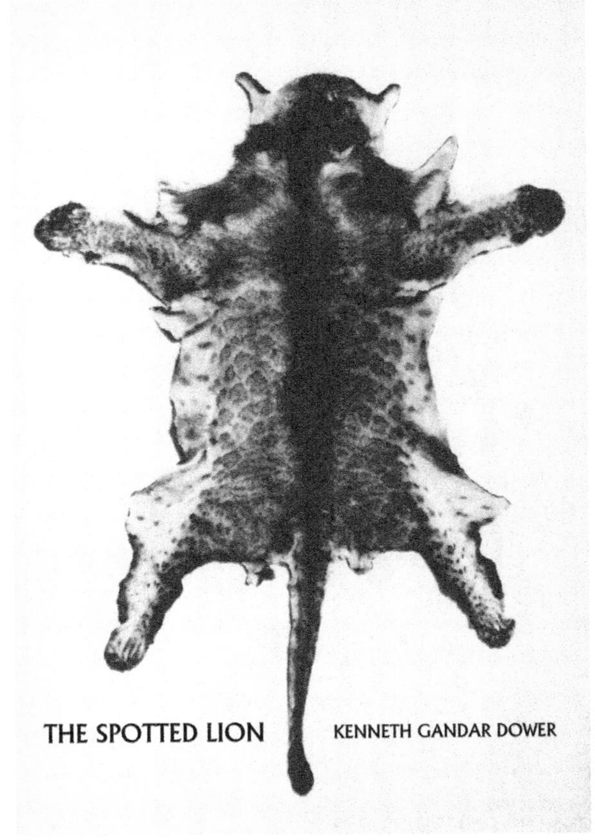
THE SPOTTED LION
KENNETH GANDAR DOWER

CPSIA information can be obtained
at www.ICGtesting.com
Printed in the USA
BVOW07*1507180118
504951BV00023B/22/P